"The *Routledge International Handbook of Play, Therapeutic Play and Play Therapy* is a delightful collection of stories, research, experiences from play practitioners across multicultural contexts. What is most fascinating about this play potpourri is the range it is able to cover, from children as play experts in co-research to use of play in classroom, work place, health space to play as core medium for healing and therapy. This innovative handbook by Sue and Clive is an ultimate encyclopedia on play that will treasured by therapists, teachers, paediatricians, parents and play campaigners across the world for many generations."

– **Shelja Sen**, *Family Therapist*

"Bringing perspectives from around the world to consider play across time and cultures and in a broad range of disciplines, artistic forms, settings and creative spaces, this handbook offers an indispensable point of entry and a timely journey through this multifaceted field."

– **Dr Maggie O'Neill**, *NUI Galway, Ireland*

"I celebrate Clive Holmwood and Sue Jennings in showing how 'play and the arts are intrinsically linked.' The silos of professions, and of the arts themselves, too often block access to this natural interdependence of imagination advancing the whole of life – even the workplace. Children model a life-affirming creative intelligence involving all of the senses and ways of playing needed to create, heal and be in the world in the most complete and organic ways. The *Routledge International Handbook of Play, Therapeutic Play and Play Therapy*, and its global community of authors, reinforce the vital role of play in our common humanity."

– **Shaun McNiff**, *Professor, Lesley University, USA*

"The importance of play for human development is paramount. Neurosciences showed that play is ingrained in our brain, generating the basic emotion of joy, and, at least since Froebel, we know that a large amount of what children learn is learnt through play. Notwithstanding that, we can perceive everywhere what Roger Caillois called the 'debasement of play,' the lacking of spontaneity, creativity and freedom. How can these virtues of play be recovered? This precious book offers a whole range of reflections on the matter in a multicultural perspective, and suggests many powerful ways to engage with play and restore its therapeutic potentiality."

– **Salvo Pitruzzella**, *Centro ArtiTerapie, Lecco, Italy*

ROUTLEDGE INTERNATIONAL HANDBOOK OF PLAY, THERAPEUTIC PLAY AND PLAY THERAPY

The *Routledge International Handbook of Play, Therapeutic Play and Play Therapy* is the first book of its kind to provide an overview of key aspects of play and play therapy, considering play on a continuum from generic aspects through to more specific applied and therapeutic techniques and as a stand-alone discipline.

Presented in four parts, the book provides a unique overview of, and ascribes equal value to, the fields of play, therapeutic play, play in therapy and play therapy. Chapters by academics, play practitioners, counsellors, arts therapists and play therapists from countries as diverse as Japan, Cameroon, India, the Czech Republic, Israel, USA, Canada, Ireland, Turkey, Greece and the UK explore areas of each topic, drawing links and alliances between each. The book includes complex case studies with children, adolescents and adults in therapy with arts and play therapists, research with children on play, work in schools, outdoor play and play therapy, animal-assisted play therapy, work with street children and play in therapeutic communities around the world.

The *Routledge International Handbook of Play, Therapeutic Play and Play Therapy* demonstrates the centrality of play in human development, reminds us of the creative power of play and offers new and innovative applications of research and practical technique. It will be of great interest to academics and students of play, play therapy, child development, education and the therapeutic arts. It will also be a key text for play and creative arts therapists, both in practice and in training, play practitioners, social workers, teachers and anyone working with children.

Sue Jennings is Professor of Play (European Federation); Distinguished Scholar at the University of Witwatersrand, South Africa; Senior Research Fellow, The Shakespeare Institute, University of Birmingham, UK; and a play therapist and dramatherapist. She is the originator of Neuro-Dramatic-Play (NDP) and is a pioneer of dramatherapy around the world.

Clive Holmwood is an associate professor, lecturer, researcher, author and doctoral supervisor in the Discipline of Therapeutic Arts at the University of Derby, UK. He is a consultant dramatherapist with 25 years post-qualifying experience working in the public and voluntary sectors and in private practice as a director of Creative Solutions Therapy Ltd. He is also an NDP practitioner and associate director of NDP.

ROUTLEDGE INTERNATIONAL HANDBOOKS

ROUTLEDGE INTERNATIONAL HANDBOOK OF DRAMATHERAPY
Edited by Sue Jennings and Clive Holmwood

ROUTLEDGE HANDBOOK OF COUNTER-NARRATIVES
Edited by Klarissa Lueg and Marianne Wolff Lundholt

ROUTLEDGE HANDBOOK OF ART, SCIENCE, AND TECHNOLOGY STUDIES
Edited by Hannah Star Rogers, Megan K. Halpern, Kathryn de Ridder-Vignone, and Dehlia Hannah

ROUTLEDGE HANDBOOK OF BOUNDED RATIONALITY
Edited by Riccardo Viale

ROUTLEDGE INTERNATIONAL HANDBOOK OF CHARISMA
Edited by José Pedro Zúquete

ROUTLEDGE INTERNATIONAL HANDBOOK OF WORKING-CLASS STUDIES
Edited by Michele Fazio, Christie Launius, and Tim Strangleman

ROUTLEDGE HANDBOOK OF DIGITAL MEDIA AND COMMUNICATION
Edited by Leah A. Lievrouw and Brian D. Loader

ROUTLEDGE INTERNATIONAL HANDBOOK OF RELIGION IN GLOBAL SOCIETY
Edited by Jayeel Cornelio, François Gauthier, Tuomas Martikainen and Linda Woodhead

THE ROUTLEDGE HANDBOOK ON THE INTERNATIONAL DIMENSION OF BREXIT
Edited by Juan Santos Vara and Ramses A. Wessel; Assistant Editor, and Polly R. Polak

ROUTLEDGE HANDBOOK OF CRITICAL FINANCE STUDIES
Edited by Christian Borch and Robert Wosnitzer

ROUTLEDGE HANDBOOK ON THE GOVERNANCE OF RELIGIOUS DIVERSITY
Edited by Anna Triandafyllidou and Tina Magazzini

THE ROUTLEDGE HANDBOOK OF CRITICAL EUROPEAN STUDIES
Edited by Didier Bigo, Thomas Diez, Evangelos Fanoulis, Ben Rosamond and Yannis A. Stivachtis

HANDBOOK ON ARCTIC INDIGENOUS PEOPLES IN THE ARCTIC
Edited by Timo Koivurova, Else Grete Broderstad, Dorothée Cambou, Dalee Dorough and Florian Stammler

For more information about this series, please visit:
www.routledge.com/Routledge-International-Handbooks/book-series/RIHAND

ROUTLEDGE INTERNATIONAL HANDBOOK OF PLAY, THERAPEUTIC PLAY AND PLAY THERAPY

Edited by Sue Jennings and Clive Holmwood

Routledge
Taylor & Francis Group
LONDON AND NEW YORK

First published 2021
by Routledge
2 Park Square, Milton Park, Abingdon, Oxon OX14 4RN

and by Routledge
52 Vanderbilt Avenue, New York, NY 10017

Routledge is an imprint of the Taylor & Francis Group, an informa business

British Library Cataloguing-in-Publication Data
A catalogue record for this book is available from the British Library

Library of Congress Cataloging-in-Publication Data
Names: Jennings, Sue, 1938– editor. | Holmwood, Clive, editor.
Title: Routledge international handbook of play, therapeutic play and play therapy / edited by Sue Jennings and Clive Holmwood.
Description: Milton Park, Abingdon, Oxon ; New York, NY : Routledge, 2021. | Series: Routledge international handbooks | Includes bibliographical references and index.
Identifiers: LCCN 2020026868 (print) | LCCN 2020026869 (ebook) | ISBN 9780367347017 (hardback) | ISBN 9780429327230 (ebook)
Subjects: LCSH: Play therapy—Handbooks, manuals, etc. | Child psychotherapy—Handbooks, manuals, etc.
Classification: LCC RJ505.P6 R68 2020 (print) | LCC RJ505.P6 (ebook) | DDC 618.92/891653—dc23
LC record available at https://lccn.loc.gov/2020026868
LC ebook record available at https://lccn.loc.gov/2020026869

ISBN: 978-0-367-34701-7 (hbk)
ISBN: 978-0-429-32723-0 (ebk)

Typeset in Bembo
by Apex CoVantage, LLC

To our mothers, who taught us how to play.

CONTENTS

FIGURES

TABLES

ABOUT THE EDITORS AND CONTRIBUTORS

Editors

Sue Jennings (UK) is Professor of Play, a life-time award from the European Federation of Dramatherapists, and Distinguished Scholar, University of the Witwatersrand and a Senior Research Fellow, The Shakespeare Institute, University of Birmingham. She has pioneered dramatherapy and Neuro-Dramatic-Play for many years, both in the UK and overseas. Her doctoral fieldwork was carried out with the Temiars, a tribal people who inhabit the rain forest of Malaysia. She spent 18 months living in the forest, together with her three children. She has kept in continuous contact with the tribal villages and with the children she helped to deliver as assistant village midwife. Recently her focus has been writing for young children with the stories of Moose and Mouse. These stories place an emphasis on empathy, attachment, diversity and curiosity. These books are also recommended for children on the spectrum. Sue has written over 50 books on group work, trauma and play and dramatherapy, and many have been translated into Korean, Chinese, Hebrew, Russian, Norwegian, Swedish, Italian and Greek. Her passion is Shakespeare, especially *A Midsummer Night's Dream*, which she explores constantly in her work.

Clive Holmwood (UK), PhD, is a dramatherapist with 25 years' experience, and an associate professor in the Discipline of Therapeutic Arts at the University of Derby, where he lectures and researches in dramatherapy and in the area of creative arts health and wellbeing. He gained his PhD in education from the University of Warwick, which was subsequently published by Routledge in 2014 as *Drama Education and Dramatherapy*. He is also the co-editor of The *Routledge International Handbook of Dramatherapy* with Sue Jennings, published by in 2016. His most recent edited publication is *Learning as a Creative and Developmental Process in Higher Education*, published by Routledge in 2019. He is an NDP (Neuro-Dramatic-Play) associate director, practitioner and trainer. He also runs his own private practice – Creative Solutions Therapy Ltd. He has known and has collaborated with Sue Jennings for 20 years.

Contributors

Emma Allen (UK) is the professional lead for Arts Psychotherapies at Rampton Hospital, one of three high secure hospitals in the UK. Emma has been working at Rampton since 2009 as an HCPC registered art psychotherapist, and she is the founder, pioneer and researcher of Forensic Sandplay Therapy, qualifying with BISS (British & Irish Sandplay Society) and ISST (International Society for Sandplay Therapy) in September 2019. Emma also works in a prison, (HMP/YOI Swinfen Hall) in an enabling environment, assessing and treating young men at risk of developing personality disorders and who struggle with interpersonal relationships related to their trauma histories. Emma has published several book chapters and journal papers; is a clinical supervisor and lecturer; and has presented at BISS, ISST and IAFP (International Association of Forensic Psychotherapy) conferences.

Paul AnimbomNgong (Cameroon) is a senior lecturer in theatre, television, film and communication studies in the University of Bamenda Cameroon. He is Chair of the Department of Performing and Visual Arts at the University of Bamenda. He holds a PhD in information and communication from the Université Libre de Bruxelles in Belgium. He has published widely on/around communication, film, television, theatre and dramatherapy.

Neal Brodsky (USA), LMFT, uses expressive therapies to help young people, adults and families create lives they love. He is one of ten therapists to contribute a chapter to the 2015 book edited by Dennis McCarthy, *Deep Play: Exploring Depth in Psychotherapy with Children*, edited by Dennis McCathy from Jessica Kingsley Publishers. Neal is licensed in marriage and family therapy and certified in both Core Energetics and Exceptional Marriage Mentoring. Awarded his professional master's degree by Fairfield University, he also holds a master's degree from what is now the Wagner School of Public Service at New York University. Past work includes writing grants to support family self-sufficiency in subsidized public housing and the marketing of national television series. His original training was in screenwriting.

Fraser Brown (UK), PhD, was the first Professor of Playwork in the UK. He was also the joint course leader of the BA (Hons) Playwork degree at Leeds Beckett University, and the specialist link tutor for the postgraduate play therapy courses run by the Academy of Play and Child Psychotherapy (APAC). He has presented at conferences across the UK and around the world and has produced several key texts in the field of play and playwork. He is the chair and co-founder of the Aid for Romanian Children charitable trust, and a member of the executive board of the Association for the Study of Play (TASP). He is well-known for his research into the therapeutic impact of playwork on a group of abandoned children in a Romanian paediatric hospital. Originally studying politics at the University of Leeds, he spent three years as a playworker on an adventure playground in Runcorn. He then managed a range of projects for the North West Play Association. For two years he was District Leisure Officer in Middlesbrough, and subsequently held posts with Playboard and the National Playing Fields Association. Before joining Leeds Beckett University, he was Director of the playwork training agency Children First for ten years. He has published many books in the field of playwork, including *Aspects of Playwork* (2018), *101 Stories of Children Playing* (2014), *Rethinking Children's Play* (2013) and *Foundations* of Playwork (2008). He was also a contributor to the *Handbook of the Study of Play* (2015).

Irina Calota (Romania) is a play therapy and dramatherapy counsellor, a member of the Romanian Association of Play Therapy and Dramatherapy's directory board, Adlerian psychotherapist and clinical psychologist. In the last eight years she organized many play and dramatherapy workshops and camps for children and adolescents. Prior to this she worked as a school counsellor and with children with autism, using play and drama resources. Since 2016 she is part of Magicamp association, a team of psychotherapists who offers group counselling support for children and teenagers who lost a significant member of their family. She is currently the co-organizer of play therapy and dramatherapy workshops for professionals in Bucharest for the Romanian Association of Play Therapy and Dramatherapy. She also works in private practice.

Leeanne Casey (Ireland), MIAPTP, BA ECCE, MA psychotherapy and play therapy, is a child and adolescent psychotherapist specialising in play therapy. She obtained her master's degree in humanistic and integrative psychotherapy and play therapy at the Children's Therapy Centre and is now running her own private practice. Before this training she obtained her undergraduate degree in early childhood care and education from Mary Immaculate College. Since the completion of this degree she has gained experience working as an early childhood educator in a variety of early year's settings including preschools and primary school settings. Working in these settings and training as a therapist stimulated Leeanne's passion for messy play as she identified its developmental and therapeutic benefits for children within the early years.

Filiz Çetin (Turkey), MA, CertCCPT-I, CertFT-I, is a Certified Child-Centred Play Therapist-Instructor and Certified Filial Therapist-Instructor in Istanbul, Turkey. She got her BA and MA degrees in guidance and psychological counselling from Bogazici University in 1996 and 1999 respectively. Her MA thesis was on "Social Skills Training Program for Children" which was then published as a book that was the first of its kind in Turkey. Her study also received the "Best Research and Application Award" from YORET (a leading NGO in the field). In addition, Filiz translated the book *A Parent's Handbook of Filial Therapy* (VanFleet, 2012) into Turkish. Filiz is the leading certified Filial Therapy Instructor in Turkey and has been running child-centred play therapy and Filial Therapy certification programs since 2009. She has trained hundreds of colleagues who are interested in play therapy and filial therapy. Since 2000, Filiz has been working with families and children who have emotional and behavioural problems, particularly attachment and trauma in Istanbul. Her work is dedicated primarily to healing the relationship between children and their families. She currently continues studies with children and families; and gives trainings about play therapy and filial therapy at her own newly established centre. Recently, she has also focused on healing touch, body-mind relations and somatic re-experiencing. This year, she is furthering her background with training in the certification programs for FirstPlay® Child Kinesthetic Storytelling and Neuroaffective Developmental Psychology and Therapy. She is a member of the Turkish Guidance and Psychological Association and the Association for Play Therapy.

Jaff Choong Gian Yong (Malaysia) is a registered counsellor under the Board of Counsellors, Malaysia, and a certified NDP trainer from UK. After obtaining a bachelor's degree in psychology from University Tunku Abdul Rahman, he has practised therapeutic play for the past ten years in educational, hospital and welfare settings (including juvenile schools, orphanage homes and prison). In 2016, he completed his master of counselling from Open University Malaysia, and since then he started his first company known as NDP Consultancy. In 2018, he and his business partners decided to establish BePsych-U Wellness Resources. Apart from running the companies, he also has a private practice at Silver Lining Psychology Specialist. Jaff has been appointed as specialist career consultant in the Education Fair and was a panel counsellor of EPF (M'sia) in the year 2017–2019; occasionally, he has been invited as a radio guest speaker.

Alison Chown (UK) is a play therapist, supervisor, trainer and qualified and experienced specialist teacher for children and young people with complex BESD. As a play therapist, Ali has worked with children and young people with complex behaviour, emotional and social needs (BESD), who have experienced early childhood trauma and loss and have attachment difficulties and pupils with autism, language and communication and other complex special needs. She is developing therapeutic work in outdoor environments and is the author of *Play Therapy in the Outdoors*, published by Jessica Kingsley Publishers.

Maura L. Collins, Psy.D. (USA) is a clinical child psychologist currently doing her postdoctoral fellowship at the Kennedy Krieger Institute's Center for Autism and Related Disorders. She has experience of conducting psychotherapy and psychological evaluations with children, adolescents and adults for a variety of mental health concerns across diverse settings. Maura's primary clinical interest is conducting psychological evaluations with children and adolescents to diagnose autism and other neurodevelopmental disorders, such as learning disorders, ADHD and intellectual disabilities. Maura also has clinical interests in providing psychotherapeutic treatment for childhood trauma through the use of evidence-based protocols, such as Trauma-Focused Cognitive Behavioral Therapy (TF-CBT) and Parent Child Interaction Therapy (PCIT). Her primary research interest is childhood alexithymia, although she has conducted and presented research on a diverse range of topics.

Katerina Couroucli-Robertson (Greece), PhD, is a registered DTA, dramatherapist, psychotherapist, teacher in special education, supervisor. She is a full member of the British Association of Dramatherapists, a founder member and secretary of the Panhellenic Professional Association of Dramatherapy and Playtherapy and a certified member of the European Association for Psychotherapy. She works as a dramatherapist and supervisor at the 'Herma' Institute of Dramatherapy and Playtherapy, where she is Head of Studies. She is a visiting lecturer at the University of Thessaly in the "Counselling in Special Education, Education and Health" programme. She also works as a volunteer, running a music and theatre group consisting of people with a range of disabilities under the auspices of V.S.A. HELLAS. She has published several articles in arts therapy magazines and has contributed chapters to several books. Her latest books, written in collaboration with her husband, are *Larger than life* (Karnac, 2017) and *Before You Let the Sun In* (Sphinx, 2018). She has also published Δραματοθεραπεία (*Dramatherapy, a Psychotherapeutic Intervention for Children, Teenagers and Adults*) (Herma, 2018) in collaboration with her daughter.

David A. Crenshaw (USA), PhD, ABPP, RPT-S, is clinical director of the Children's Home of Poughkeepsie, New York. He is a board-certified clinical psychologist, Fellow of the American Psychological Association (APA), Fellow of APA's Division of Child and Adolescent Psychology, and a registered-play therapist-supervisor (RPT-S). He has taught graduate courses in counselling and play therapy at Johns Hopkins, Columbia University and Marist College. He was President of the New York Association for Play Therapy (2004–2008). He is the author of 15 books, numerous journal articles and book chapters on child trauma, child abuse and youth violence.

Shyla Dauria, LMHC, NCC, CASAC is an Addiction Program Specialist at The New York State Office of Addiction Services and Supports (NYS OASAS), where she works on increasing the availability and quality of addiction services for youth, young adults and their families throughout the state. She operates a private practice, where she works with adolescent and young adults with mental health and co-occurring issues. Prior to her work at OASAS, she was the supervisor at Lexington Center for Recovery for all adolescent programs in Dutchess County, NY. She has been trained as an instructor for the NYS Impaired Driver Program, where she co-instructed classes for individuals with impaired driving convictions. She has been trained as a Certified Leader in the Seven Challenges, in TF-CBT, ASIST Suicide Prevention, Youth Mental Health First Aid, Narcan Administration and CRAFT.

Athena A. Drewes (USA), PsyD, MA, RPT-S is a licensed Psychologist and Registered Play Therapist-Supervisor. She is former Director of Clinical Training and Accredited Doctoral Internship at Astor Services for Children and Families, a large multi-service non-profit mental health agency in New York. She has over 45 years of clinical experience in supervision and clinical work with complex trauma, sexually abused, traumatized and attachment-disordered children and adolescents in school, outpatient, foster care and inpatient settings. She is a former Board Director of the Association for Play Therapy and Founder and President Emeritus of the New York Association for Play Therapy. She has written extensively on play therapy and is an invited guest lecturer in the US, and Taiwan, England, Ireland, Argentina, Mexico, Denmark, Canada, and Italy. She has edited/co-edited twelve books including *School-based Play Therapy*; *Cultural Issues in Play Therapy*; *Supervision Can be Playful: Techniques for Child and Play Therapy Supervisors*; *Blending Play Therapy with Cognitive Behavioral Therapy: Evidence-Based and Other Effective Treatments and Techniques*; *Integrative Play Therapy*; *The Therapeutic Powers of Play: 20 Core Agents of Change*; *Puppet Play Therapy, and Play Therapy in Middle Childhood.* This latter book, published by the American Psychological Association, has a companion video DVD of Dr. Drewes' integrative prescriptive play therapy work. Dr. Drewes now lives in Ocala, Florida, is semi-retired, and continues to write, supervise, consult and give training in play therapy. She volunteers as a Guardian ad Litem to the courts for children in foster care.

Marrianne Dubouchet (India) was born in 1962 in Switzerland. She has lived in Auroville since 1993. She studied psychology and is a teacher by profession. Before coming to Auroville, she worked for the ICRC (International Committee of the Red Cross) for five years mainly in Africa, in the Relief department (administration, coordination, distribution of goods and managing local workers). In Auroville, she has been teaching French for first and second language and later took over the section "Arts and Crafts". She has also been an Individual Tutor for children with special needs aged 8–12. In 2005 she created a project of recreational horse riding called "Persian sky". It was aimed at children with different physical or mental or emotional capacities at Red Earth Riding School where she was one of the project holders, in charge of the accounts, lessons and all other activities related to running a riding school comprising 25 horses for 50 clients.

Sandra El Gemayel (UK) holds a PhD in education from the UCL Institute of Education. Sandra's research concerns the importance of play for young refugee children and seeks to promote their rights and play opportunities in Lebanon. She draws on Froebelian Principles, Vygotskian sociocultural theory, child rights and Victor Turner's theory of liminality and adopts a 'Day in the Life' approach to provide unique insights into the constructions of young Iraqi and Syrian refugee children's play and childhoods in Lebanon.

Caroline Essame (Singapore) is a British-trained art therapist and occupational therapist with over 35 years' experience in the field of health, therapy, education, play and the arts. She runs her own company – CREATE, Creative Arts, Therapy and Training Pte Ltd – and is a consultant at Kaleidoscope Therapy Centre in Singapore where she works with children with social and emotional issues and special needs. She pioneered the first master's in art therapy in Asia for LASALLE College of the Arts, trains at the University of Social Sciences and National University of Singapore in creativity, play and human development, and her master's thesis in education is on the role of play in social development and theory of mind. She is a visiting creative arts and play consultant for Deepti Special School and Rehabilitation Centre in Kerala, India, and in 2012 she developed the first creative arts and play centre for children on the autistic spectrum in Malaysia (www.developmental-play.com).

Rosie Flewitt (UK) is Professor of Early Childhood Communication in the Education and Social Research Institute (ESRI), Manchester Metropolitan University, and former Founding Co-director of the Helen Hamlyn Centre for Pedagogy (0–11 years), University College London, Institute of Education. She is an international expert in young children's communication, language and literacy development in mainstream and inclusive education, with extensive experience in ethnographic, multimodal and participatory research methods. Recent publications include *Research Methods for Early Childhood Education* (Flewitt and Ang, 2020), *The Routledge Handbook of Digital Literacies in Early Childhood* (Erstad, Flewitt, Kümmerling-Meibauer and Pires Pereira, 2020), and the award-winning *Storytelling in Early Childhood: Enriching Language, Literacy, and Classroom Culture* (Cremin, Flewitt, Marsden and Swann, 2017).

Larry Galloway (UK) is an experienced play therapist who is based in Somerset in the UK where she has a dedicated play barn equipped for working with children and young people using art, sand, clay, music and drama. Larry has specialised over the last 20 years in working with young people within the residential care sector as well as in schools and those who have been adopted. Larry works in a non-directive way but combines this approach with Theraplay techniques where appropriate.

Ken Gardner (Canada), MSc, R Psych, CPT-S, is a registered clinical child psychologist with a background in clinical, school and community psychology. As a certified child psychotherapist and play therapist/supervisor, Ken specializes in the areas of play therapy for children with developmental, emotional and behavioural concerns. His practice also includes psychoeducational evaluations for students experiencing learning and developmental difficulties. As the co-director of the Rocky Mountain Play Therapy Institute (RMPTI) Ken has extensive experience as a consultant and as a play therapy trainer, working both nationally and internationally.

Sue Gascoyne (UK) is a qualified creative arts and play therapist working primarily with children with emotional and behavioural issues who have experienced trauma. Sue specialises in messy play, sensory processing, autism and sensory attachment, and particularly favours the use of creative media to explore and express emotions, build resilience and help children bring about positive change. She has ten years' experience in the education sector and 25 years as an award-winning trainer. She launched multi-award-winning sensory equipment company Play to Z Ltd in 2006, specialising in providing quality, environmental and ethically made resources to support wellbeing and learning. Maintaining an active role as creative director, this ethical and environmental focus remains a key driver. Convinced of the value of messy play and dissatisfied with the lack of resources and insight available, in 2018 Sue made this her research focus. As well as the publication of her latest book – *Messy Play in the Early Years, Learning through Material Engagements* – Sue was proud to launch in 2019 an enriching outdoor therapy space, within which she provides play therapy and joint adult/child sessions as well as practitioner training and workshops.

Audrey Gregan (Ireland) is an accredited play therapist and child psychotherapist and clinical supervisor with 40 years' experience of working with children on the autistic spectrum, developmental delay and eating difficulties. Audrey is a AutPlay Therapy provider a practicing DIR®/Floortime™ practitioner who integrates sensory integration in her work. She has worked with parents of children in the school and residential centre and training care staff and teachers using play therapy with children with autism. Her greatest interest is in attachment, child development and early intervention. She is accredited by the British Association of Play Therapists, Play Therapy International, Play Therapy Ireland, Irish Association for Play Therapy & Psychotherapy and abides by their ethical framework.

Sylwyn Guilbaud (UK) is an independent researcher, artist and home educating mother of three. She qualified as a Playworker in 1997 and returned to Leeds Beckett University to complete one of the first PhDs in the sector in 2011. She lives by the sea, where she loves to wander and wonder, drawing inspiration for part of her current play-advocacy—the creation of small stitched magical beings in support of the playing relationship between children and the elements.

Natalie Hayek Damouni (Palestine/Israel) is a social worker and dramatherapist, a supervisor at the Drama Therapy Program at Tel Hai and Haifa University. Born and raised in Nazareth, she has been working for about ten years with abused adolescent Palestinians in Israel and also with abused women. Natalie is a director of a multidisciplinary professional staff that works with children with autism and their parents, also with children with developmental diffculties/disabilities and unsupportive environmental backgrounds.

Priscilla Ho (Malaysia) is a certified play therapist after training with Play Therapy International (PTI UK) and has an advanced certificate in Theraplay. She has worked in the area of therapeutic play skills with chronically ill, abused and traumatised children since 2005. She runs her own creative arts centre and also works in children's homes with disadvantaged children and teenagers on remand in the juvenile home. She is the author and publisher of the book *I Am Special*, a therapeutic and creative workbook that helps children to express their thoughts and feelings through drawing and words. Priscilla leads an anti-bullying program and creative learning program for children in Penang and had published 13 children's storybooks written and illustrated by the children in the program. She also helped to organize Neuro Dramatic Play training by Dr Sue Jennings in Penang in early 2000.

Phil Jones (UK) is Professor of Children's Rights and Wellbeing and UCL IOE Head of Research Ethics and Governance at University College London, Institute of Education. He has held the roles of Director of Research, School of Education at the University of Leeds; Visiting Professor at Concordia University, Montreal, Canada; and was a Mellon Distinguished Scholar at the University of the Witwatersrand, Johannesburg, South Africa. Recent publications include *Rethinking Children's Rights* (with Welch, 2018); *The Arts Therapies: Theory, Research and Practice* (2nd edition, 2020); *Child Agency and Voice in Therapy: New Ways of Working in the Arts Therapies* (with Cedar, Coleman, Haythorne, Mercieca and Ramsden, 2020) and he is editor of *Bloomsbury's New Childhoods Series*.

Ankita Khanna (India) is a psychologist and arts-based therapist at Children First, a multidisciplinary child and adolescent mental health service in New Delhi and Gurgaon, India. As an arts-based therapist, Ankita uses the arts in an integrated way – music, drama, play and stories, in therapeutic work with children and young people, as well as in training with adults.

Amanda Krülle (USA), LMHC, is a licensed clinician currently working for Northern Rivers, Intensive Aftercare Program (IAPP) in Hudson, New York. Amanda helps to support families and their children, providing therapy and social work connections. She has experience working with children with speech and developmental disorders as well as experience providing therapy to trauma survivors. Amanda is trained in dialectical behavioural therapy and has provided individual and group therapy. Amanda graduated from Marist College and was a graduate intern at the Children's Home of Poughkeepsie; working with the foster care and residential populations of children ages birth to two. She worked under the supervision of clinical director, David A. Crenshaw, PhD, ABPP.

Chika Matsudaira (Japan). After finishing her bachelor's degree at Kyoto University of Foreign Studies, Chika started her professional research for social work at Bukkyo University master course of social welfare. Chika studied at Bukkyo University doctorial course after finishing her master's. In 2011, Chika became a qualified hospital play specialist due to her professional studies at Stevenson College Edinburgh, and found placement at Royal Children's Hospital, Leicester. Chika has always kept her strong interest in how to help other people who are in pain or in trouble and in the education of professionals to help them. She has additionally trained in the US with Dr Terry Kottman to become an Adlerian play therapist. She lives in Shizuoka, Japan, and she is a wife and a mother of two.

Dennis McCarthy (USA), LMHC, DMT-R, has worked in the field of play therapy for over 40 years, developing his own unique approach that is body centred and imagination driven. He trained in dance therapy, bioenergetic analysis and Jungian analysis. His understanding of the necessity of healthy aggression in ego development and relationship formation is the centrepiece of his work, described in numerous articles and books, including *If You Turned into a Monster . . . A Body-Centered Approach to Play Therapy*. He runs training programs and workshops in the US and abroad.

Penny McFarlane (UK) worked as a teacher in primary and secondary education for over 15 years before qualifying as a dramatherapist with the objective of helping children in schools. She founded a creative arts therapy project in inner city schools and has subsequently worked as a dramatherapist and supervisor on a multi-agency support team within education. She co-founded a children's bereavement charity and has worked extensively in the field of parental separation and its effect on children. Penny has published four books in the field of dramatherapy for emotional support. She lives in Devon where she is involved in running pilot projects as well as training, supervising and writing.

Leong Min See 梁茗思 (Malaysia) is a consultant counsellor and trainer, a play/creative arts therapist, special and inclusive educator, MA special and inclusive education specialist (UK, Netherlands, Czech Republic) with a postgraduate diploma in play therapy (Roehampton University, UK). Leong Min See is a full member of BAPT (British Association of Play Therapists) and secretary of the Centre for Creative Arts Therapy, Penang (CCAT).

Sonia Murray (UK) holds a diploma in social work and is a registered play therapist. She has over 30 years' experience as a practitioner in the fields of child abuse, social, emotional and behavioural difficulties, children's mental health, childhood trauma, parenting programmes and play therapy. Sonia trains nationally and internationally on topics related to play therapy, understanding behaviour, play, communicating with children, behaviour management and positive parenting. Her work history includes employment as a childcare social worker, parent worker and play therapist in education and social care and health settings. Sonia provides consultation, supervision and training in both statutory and voluntary fields of children's services. Sonia has also contributed to several publications, including *The Use of Therapeutic Stories*, *The British Journal of Play Therapy*, *Turning Points in Play Therapy* and the *Emergence of Self*. Sonia served on the Board of Directors for British Association of Play Therapists in a variety of capacities, including treasurer and chair of business and research sub-committees for nearly ten years.

Anna Oijevaar (India) has lived in Auroville since 1987. Born in Holland in 1951 she studied art and education and is a teacher and teacher trainer by profession. In Holland, she worked in experimental schools and schools for children with learning and behaviour difficulties. When she came to Auroville in 1985, she became one of the project holders of an indigenous coastal reforestation project. Within this setting, she raised her five children. They took her on a magical journey through the realms of nature, reconnecting her with her inner child. Through the years, she played in this forest with children, adults and families from all over the world. She developed stories, activities and games through which she reconnects with nature, listening and becoming aware of the sacred life cycle we share. Nine years ago, she joined the Lilaloka project and found there an atmosphere of joy and stimulation for her research and development.

Galila Oren (Israel) trained as a dramatherapist at New York University. Currently she is a clinical supervisor and Director of the dramatherapy training program at College, Israel.

Susana Pendzik (Israel), PhD, RDT, is a dramatherapist and supervisor, lecturing at the Department of Theatre Studies at Hebrew University of Jerusalem, the Drama Therapy Program at Tel Hai College (Israel), and the Swiss Institute of Dramatherapy. Susana has done extensive international work and teaching in Europe, Latina America, and the US. She is the author of *Action Techniques for Working with Battered Women* (published in German and Spanish), co-editor of a book *Assessment in Drama Therapy* (2001, Charles. C. Thomas) and the author of many papers on dramatherapy and of two poetry books.

Siobhán Prendiville (Ireland), MIAHIP, SIAPTP, BEd, MEd., MA, is a child and adolescent psychotherapist, play therapist, clinical supervisor, author, presenter and trainer. Siobhán is MA course leader on the Master of Arts in Creative Psychotherapy (humanistic and integrative modality) at the Children's Therapy Centre. She also maintains a private child and adolescent psychotherapy and play therapy practice.

Bridget Rees (UK) is an HCPC registered art psychotherapist working in independent practice in a rural area of Somerset in the UK. She trained in MA art psychotherapy at the University of South Wales and has a background in art teaching. She works with people of all ages, but mainly with children and families in adoption or kinship care.

Yaspia Salema (UK) researches and lectures in the area of children's free play in early years education. She successfully completed her PhD research at the UCL Institute of Education which examined the relationships over time between children's free play cultures and practices at home and in nurseries in London. She has previously researched the impacts of social protection programmes on children from families that were ethnic minorities and socio-economically marginalised in rural communities in Bangladesh.

Grace Schuchner (Argentina) is an actress, a dramatherapist and a mindfulness meditation practitioner located in Buenos Aires, Argentina. Grace gained her education at international institutes. She graduated from Tel Hai Regional College, Israel, at 2000. She studied mindfulness meditation at the Sounds True institute in Colorado, US. Grace is a pioneer in the field of dramatherapy in South America. She introduced the field of dramatherapy to the Universidad Nacional de Cuyo (UNCUYO) in the province of Mendoza, Argentina. She has published the first dramatherapy book in Argentina: *Dramaterapia. Teoría y práctica* (Schuchner and Ferrandis, 2017). Grace also regularly teaches dramatherapy at the Faculty of Art at the University of Chile. She is currently conducting research in mindfulness and dramatherapy for groups and individuals.

Jitta Skoles (India) was born in Holland in 1963 where she attended a teacher training school. She grew up in Auroville in an indigenous coastal reforestation project. She attended the Auroville schools as well as being home schooled. She was one of the initiators of Kailash a project for independent living for teenagers age 15–20. She has two children, who received one year of home schooling and are now attending Auroville high schools. One of them lives in Kalish for the last four years she has worked in Lilaloka doing group activities, individual sessions and she is the main person running financial affairs. She is one of the project holders.

Jonas Torrance (UK) is a registered private practitioner and supervisor with ADMP (Association of Dance Movement Psychotherapists UK). He works with children and young adults in schools and colleges across Oxfordshire. Jonas has worked with autistic children and adults for over 35 years. He lectures nationally and internationally about movement therapy, challenging behaviour and autism. He is a guest lecturer at various universities and national conferences and delivers training to therapists, parents and teaching staff. Jonas has co-written and advised on various policies for schools and behaviour within Oxfordshire County. In 2003 he published an article "Autism, Aggression and Creating a Therapeutic Contract" for *The American Journal of Dance Therapy*. In 2018 he published a book entitled *Therapeutic Adventures with Autistic Children, Connecting through Movement, Play and Creativity*, Jessica Kingsley Publishers.

Risë VanFleet (USA), PhD, RPT-S, CDBC, is a licensed psychologist, registered play therapist-supervisor and certified dog behaviour consultant based in Boiling Springs, Pennsylvania, USA. She is well-known for her 45 years of clinical service and training work with mental health professionals throughout the world. She first studied Filial Therapy with its co-creators for several years, Drs Bernard and Louise Guerney, 38 years ago, and she is the author of a number of Filial Therapy- and play therapy-related books, including *Child Centered Play Therapy* (2010 with Sywulak & Sniscak), *Filial Therapy: Strengthening Parent-Child Relationships through Play* (3rd edition, 2014), *Casebook of Filial Therapy* (with L. Guerney, 2003), *A Parent's Handbook of Filial Therapy* (2nd edition, 2012, translated into several languages) and others. She has also been instrumental in developing the field of Animal Assisted Play Therapy® and has numerous publications in that area, including the award-winning book, *Animal Assisted Play*

Therapy (with Faa-Thompson, 2017). She was also featured on the American Psychological Association's DVD, *Filial Play Therapy* (2008). She has received numerous national and international awards for her work in play therapy, Filial Therapy, and Animal Assisted Play Therapy® as well as for her training programs and writing.

Monique Van Loggerenberg (South Africa), PhD, worked in the corporate arena for 12 years before starting private practice in the Free State, Central South Africa. Currently she is conducting research on the efficacy of equine assisted therapy on bullied children. She has noticed that children's self-esteem is elevated with horse-riding activities that are combined with play therapeutic techniques and is in the process of writing-up these findings in a formal study.

Lorri Yasenik (Canada), PhD, RSW, RPT-S, CPT-S, is specialist in working with children and families in the areas of treatment of trauma, high conflict separation and divorce, and child development. In addition to being a certified supervisor of child psychotherapy and play therapy, Lorri is a registered family mediator and a registered parenting coordinator. Lorri has a keen interest in legal issues that affect children's lives and has completed research in the area of "Including the Voices of Children in the Legal System". As the co-director of the Rocky Mountain Play Therapy Institute in Calgary, Alberta, Canada, Lorri provides approved play therapy training nationally and internationally.

FOREWORD

How wonderful to have a new, comprehensive text devoted to play! I believe that play is at the heart of all that is good, joyful, connecting and enriching. Playing has always been a central component in my family relationships and professional life. Now, I am lucky enough to have grandchildren, Emilia and Andrew, to provide additional excuses and opportunities for playfulness at every family gathering, major occasion and mini-event. For children, no day is complete without play. I suggest that this is so for children of all ages. What day is not better for the experience of playing?

In this engaging text, Sue and Clive, both enthusiastic play professionals with vast experience and knowledge, bring forth the power of play in a multitude of settings, and highlight the role of play within a diverse range of professions, and with people of all ages. Far from being "just play", play is highlighted as being significant for recreational, developmental, educational and therapeutic purposes internationally and across cultural boundaries. We see how all creative expression has its roots in play. This has always been a key feature in Sue's work and has inspired my own practice for decades. Her developmental playtherapy model (Jennings, 1999), encompassing her Embodiment-Projection-Role paradigm, exemplifies the early roots of each of the expressive arts within the development of the dramatized self. We see the roots of movement and dance therapy within her Embodiment stage, art therapy within her Projective stage, and dramatherapy within the Role play stage. From these roots spring forth creative expression and the potential for both development and repair. It is lovely to see how Clive has been influenced by this model, and by Sue's subsequent work on Neuro-Dramatic-Play. As a dramatherapist, it is clear that Clive was open to the richness of this approach and fully embraced it as a way to expand the focus and context of his direct work with children. We see this come alive in Chapter 15.

The four sections of this book bring us on a journey through the world of play in its many forms. With its 35 chapters it is a big book! It is not a text to sit and read sequentially from beginning to end. Rather, let yourself play with it. See where your interest takes you on any particular day. Dip in and out. Explore chapters linked by a research focus, or groupwork interventions, or maybe go on a cross-cultural trip and select chapters authored by those from various continents. Perhaps you will be enticed by chapters that explore developmental models or specific conceptualising frameworks. Some chapters engage us in exploration of the content of actual play sessions, and/or the materials utilised. Others present solid theoretical knowledge to inform and educate. Many do both. Where to start? The choice is yours. You may also decide to choose any of its four sections as your initial focus. Go where your heart leads you. You will certainly read all chapters along the way.

Chapters in Part I invite us to consider play itself through a variety of lenses, in diverse settings, across international borders and with reference to developmental and cultural considerations. We are invited to consider play and playfulness from the perspective of the researcher, the child and professionals from many corners of the globe who make use of play to scaffold the child's development; to facilitate communication; to assist ill children; to enhance adults understanding of, and capacity to meet, the needs of children in their care; and to facilitate workplace relationships. The richness of play is truly evident in this section.

Part II brings us into the world of therapeutic play and demonstrates, through all six chapters, how playwork, including the provision of a playfully rich environment and a playful, attentive, attuned and responsive adult, can be utilised to promote acquisition of new skills; positive mental health; social and emotional development; and assist children in overcoming struggles arising from heightened anxiety, play deprivation, and adjusting to changed circumstances. Therapeutic play interventions include one-to-one and groupwork sessions. Both are powerful vehicles to foster personal development. This section highlights just how much fun this personal development, utilising, for example, sensory play, rhythmic play, story, performance and messy play, can be!

With its ten chapters, Part III, is a big read! It brings us into the world of play in therapy. This is a great concept to bridge the gap between therapeutic play and play therapy. It informs us of the need to develop clarity in regard to the services offered to children and vulnerable adults. Chapter 16 highlights the need for public awareness of the need for appropriate training and describes the difficulties that are emerging in Malaysia related to non-mental health professionals, with little or no training in play therapy, unethically working outside their competency levels. Chapters in the Play in Therapy section highlight the use of play in counselling, dramatherapy, dance movement therapy, forensic Sandplay, dyadic art psychotherapy, family therapy, filial therapy and approaches bridging dramatherapy and play therapy. Contributors describe clinical practice in various countries, including Malaysia, Israel, UK, USA and Greece; with clients of all ages, and with various referral issues including behavioural and emotional difficulties, panic, addiction, gender dysphoria and relationship difficulties. There is definitely something for all readers in this section.

The final section, Part IV, focuses directly on play therapy as practised by those with specialised training and registration in the field. The differences that sometimes exist between UK- and USA-trained play therapists, linked, perhaps, to the crossover between dramatherapy and play therapy that exists in the UK, are noted. We are brought on a riveting journey through case material and exploration of the external influences on play therapy practice in various settings and with diverse client groups. We are exposed to very structured interventions, more spontaneous practice which relies on the person of the play therapist and their capacity to make therapeutic use of self in their interactions with their clients, and all levels in between. We discover that there is a place for all. The richness of the play therapy experience and the environments in which it is practised is celebrated.

This book truly has something for everyone who uses, or aspires to use, play to enrich their work with children, adolescents or adults.

Eileen Prendiville

Reference

Jennings, S. (1999). *Introduction to Developmental Playtherapy: Playing and Health*. London: Jessica Kingsley Publishers.

ACKNOWLEDGEMENTS

Sue and Clive feel indebted to many people. Firstly the families they came from, including their mothers and grandmothers, and the families they have now for their patience and forbearance. Clive would particularly like to thank Clare for her continued forbearance.

They also thank all the contributors for the dedicated work and for their clients who generously gave permission for their histories to be included.

They appreciate the support they receive from Routledge Publishers for believing in their judgement, ideas and writing skills.

They would also like to thank each other for generosity of time and ideas. They are delighted that their working partnership continues to strengthen, book by book.

INTRODUCTION

When Clive Holmwood and I edited our first International Handbook on the subject of dramatherapy, we thought a subsequent one would be a doddle, or at least far easier second time around. However we were faced with a very different situation, like being handed a map where the key staging posts have been forgotten.

The broad definitions of dramatherapy and its practice are fairly straightforward. There are those orientated towards a more verbal, psychotherapeutic model contrasted with people who favour a ritualistic practice with its historic roots in shamanism (Jennings 1986). However play has a much more complex heritage, definitions and models. Play-based learning has been around for a century although play therapy has emerged much more recently. Understanding play is very challenging. No one would dispute that early play for children is important as a means of development, learning and socialising. We do not challenge when it is applied to very young children. But as Cozolino (2014) says, saying goodbye to the kindergarten is a big mistake. He says that:

> Perhaps every school should be like a start-up business, where those in control are not tied to rigid ideas and rules, but, like a tribe, re-create the learning environment in reaction to the day-to-day realities of life.
>
> Enriched, challenging environments result in more resilient brains, which contain more neurons, more complex connections, and greater resilience to stress.
>
> *(p. 160)*

We were also faced with an anomaly. Whereas plenty of people are very experienced in play in its many different aspects, authorities both local and national do not give it much credence. How often do we hear that "play is recreation" or "play is not real learning" or "when will you sit down and learn instead of messing about"? There seems to be a perversity at management level to avoid any proof or scientific credibility or neuroscience research that shows the importance of play-based learning and therapy.

Huizinga (1944), one of the earliest writers on play suggests the idea of the "magic circle" of play, an activity standing outside the sphere of ordinary life. It is not serious but absorbs the player intensely and utterly. Very relevant to our purpose is the writing of Stephen Nachmanovitch (1990) in his book *Free Play: Improvisation in Life and Art*. He says:

> Improvisation, composition, writing, painting, theatre, invention, all creative acts are forms of play, the starting place of creativity in the human growth cycle, and one of the great primal life functions. Without play, learning and evolution are impossible. Play is the taproot from which original art springs; it is the raw stuff that the artist channels and organizes with all his [*sic*] learning and technique.
>
> *(p. 42)*

We hope that this wonderfully diverse book of chapters will show this very diversity, that play and the arts are intrinsically linked. Our contributors include both a range of therapists who integrate play into their work and various play specialists who approach play from very different backgrounds. We are celebrating "difference in play" within the cross-cultural contexts of play having primacy.

Clive and I do not ascribe any value judgements to the separate headings for sections in this book: Play, Therapeutic Play, Play in Therapy and Play Therapy. All are important and perhaps for the first time show an alliance and a sequence between them all. We have deliberately not discussed training as this could well be a contentious area where there is disagreement about what constitutes "appropriate and adequate" training for practising play people and play therapists.

Finally we hope that readers will embark on a playful journey and be ready to take risks, find new notions and generally have fun on the way.

I would like to thank Clive my co-editor for being a superb partner to work with in the multitasking role of "editor".

Sue Jennings
Wells, UK and Gua Musang, Malaysia

References

Cozolino, L. 2014 *Attachment-Based Teaching: Creating a Tribal Classroom*. New York and London: W. W. Norton.

Huizinga, J. 1944 *Homo Ludens: A Study of the Play Element in Culture*. London: Routledge and Kegan Paul.

Jennings, S. 1986 *Creative Drama in Groupwork*. Abingdon: Routledge, Speechmark.

Nachmanovitch, S. 1990 *Free Play: Improvisation in Life and Art*. London: Penguin.

PART I

Play

Introduction

Primarily Section 1 is interested in play in its most generic form. We bring a range of perspectives from around the world that consider play in its most basic but essential form, its ability to engage with children in a wide range of diverse and culturally appropriate perspectives. From research in play through to a range of uses of play in such places as South America, Africa, India, Malaysia and Japan, you will notice that as we move through this section we move towards a more "applied" use of play within specific contexts, which could be defined as moving towards more of a therapeutic intent at the end of Section 1.

We begin with Jones et al. and their groundbreaking research considering how children play and what goes on within play from a child's perspective, and the importance for children and young people to have full involvement in "play-based" research. Chapter 2 considers the first of many uses of play within this book, in an educational context in Argentina, asking how play can help teachers teach, and children learn, in the classroom. Our third chapter considers the importance of play from Children First, a specialist mental health centre in Delhi, considering the use of play throughout the context of their organisation, staff down, working with children.

Chapter 4 switches continents and we consider participatory play from a theatrical and social context working with "street children" in the village of Mbalmayo, near Yaoundé, in Cameroon, acknowledging aspects of catharsis and theatrical conventions that relate to play. The following chapter, "The Ness of Being Playful", reflects from a more philosophical perspective what we mean by playful "ness". Chapter 6 takes us to Malaysia where we consider a particular perspective, "developmental play", from a specific cultural and social perspective, working with children who are "differently abled".

Chapter 7 offers us some practical applications and theoretical orientation towards aspects of play that might be considered to be leaning towards a "therapeutic" context using such notions as Sue Jennings' Neuro-Dramatic play (NDP) (2011). Whilst the following chapter returns us to India where a whole social and cultural movement has developed in the village of Lilaloka, where play is considered absolutely core to everything they do. Chapter 9, our final chapter in Section 1 considers the complex and culturally and socially sensitive area of working with sick children in Japan, and the experience and struggles of an individual hospital play specialist and the children and families she works with. Matsudaira pleasingly finishes her chapter by mentioning the famous Buddhist priest Ryoukan, who insists on the importance of finding time to simply "play" with children. Something which appears to sum up this first section of the book well.

Reference

Jennings, S. (2011) *Healthy Attachments & Neuro-Dramatic-Play*. London: Jessica Kingsley Publishers.

1

PLAY AND CHILDHOODS

How are the relationships between researching play and children changing?

Phil Jones, Sandra El Gemayel, Yaspia Salema and Rosie Flewitt

Introduction

Play does not happen in a sealed vacuum, and it is practised in myriad ways across time and cultures. One way of thinking about research in the field is how shifts in culture, politics, policy and environment change how children play and how adults relate to child play. Innovations in play research have been versatile and responsive to the emergent contexts of play: from enquiry that explores the impact of gender or poverty on play, to research that is constructed appropriately to conduct sensitive enquiry into play therapy. This chapter explores and problematises interdisciplinary connections between play, the new sociology of childhood and children's rights. It examines how this relationship creates questions and new opportunities concerning how children and adults engage in research together. Three examples from contemporary projects illustrate how recent developments are resulting in important changes and innovation in how research, children and play relate to each other. The first concerns a 'day in the life' methodology (Gillen et al., 2007; Gillen & Cameron, 2010), the second 'child conferencing' (Huser, 2015) and the third a children as researchers approach (Jones et al., 2018).

Research, play, the new sociology of childhood and child rights

Recent literature on research involving children has included the evaluation of a particular 'phase' of theory and related enquiry, often described as being informed by the 'new sociology of childhood' and by children's rights (Larkins et al., 2015). As Murray notes, children's participation in 'research in matters affecting them has become increasingly articulated. This development aligns closely with Articles 12 and 13 of the United Nations Convention on the Rights of the Child (UNCRC, 1989)' (Murray, 2011, 92). These articles concern state parties assuring children the 'right to express' their views 'freely in all matters affecting' them and the 'the right . . . to seek, receive and impart information and ideas of all kinds' (UNCRC, 1989). Kellett summarises this phase as a 'paradigm shift', where children began to be seen as 'participants' with rights rather than 'objects', and that this is manifested in changed practices, such as children having places on advisory groups to guide research or children being researchers themselves (Kellett, 2010). Kellett describes key aspects of this change:

> part of our responsibility in researching with and for children and young people entails developing their capacity for judgment, for communicating their views and agency for action. Good practice aspires to a partnership in which adults, children and young people generate a body of child research knowledge. Here, research with, for and by children and young people are complementarities that inform and interact with each other.
> *(2010, 4)*

This approach is often framed in the literature by concepts such as power, collaboration and control. For example, Fargas-Malet et al. note that the 'new approach has meant a methodological shift' which has engaged children as collaborating with adult researchers within the 'various stages of the research process, such as formulating the research questions, planning the methodology, collecting and/or analysing data, drafting recommendations and disseminating findings' (2010, 175). They position this as mediated power and control: 'differing levels of control-sharing and of participation in the research process' between children and adults (2010, 176). Recent play-related research has begun to reflect this shift in attention and approach, for example, by exploring what children *themselves* think about play. In Barnett's (2013) US study, children aged 8–11, identified as Caucasian, African American, Asian American and Hispanic, were invited to define what play meant to them. The research reveals that the children emphasised play as being fun and active, the importance of being able to play in the ways they wanted to, alone or with others, and of having time away from things they were obliged to do (such as school). Other studies have responded to the views of children concerning their play spaces, for example (Burke, 2005), or utilised play based interview methods in work with young children (Koller & San Juan, 2015).

Researchers have begun to problematise the nature of such child involvement in research and the ways in which adults and children participate and collaborate together. Invitations to re-evaluate participation have highlighted particular issues connected to children, adults and research (Flewitt & Ang, 2020; Larkins et al., 2015; McCarry, 2012; Powell et al., 2016). These concern a need to be especially aware of the relationships between the context and any act of participation in order to engage reflectively, rather than to essentialise the process. Larkins et al. (2015), for example, argue that there is a 'lack of critique' in much extant literature on participatory 'rights based' research. Authors such as Buhler-Niederberger, have warned against the danger of 'children's actorship being essentialised rather than analysed and therefore affecting the quality as well as the credibility of research' (2010, 160). Einsdottir notes the particular 'complexity' and diversity of power issues within a research context with children, as 'unequal power can exist in terms of age, status, competence and experience' (2007, 204). Gaps in the literature concerning the need to review and evaluate the process and outcomes of participation and collaboration, from both adult and child perspectives, have been identified. Powell et al. (2016, 197) call for 'a deeper engagement' with the ways in which children are constructed in and through research, with greater reflexivity and professional dialogue creating 'improved practice' through 'critical engagement'.

Three examples from research

This background illustrates how the interactions between play, rights and the new sociology of childhood offer new perspectives to approach the research process in relation to children's play.

These include how children are seen as participants rather than subjects, how the agendas for research emanate from children's perspectives, and how the data and findings can be interpreted or responded to by children to enrich any enquiry and to recognize their participation rights.

A cornerstone for these new perspectives concerns how adults and children construct their research relationships with each other. The following research examples illustrate three different ways of working with relationships between researcher and children that are informed by such changed agendas. Rather than essentialising the participation of children in research as a given 'good', our presentation of data and analysis responds to the critiques of Buhler-Niederberger (2010) and McCarry (2012) by examining the nature of research conducted in specific contexts, and the particular benefit to participating children. Our approach to each example responds to calls in the literature to address gaps in our understanding of children's participation by offering insights into the contextual details of the relationships between research, children and play.

Each example illuminates different facets of how researchers develop relationships with children and play:

- In the first example, the researcher creates a relationship with children who are in the complex situation of being 'temporarily displaced' (Government of Lebanon and United Nations, 2019, 4) in Lebanon as a result of armed conflict in their birth countries. In this study, a 'day in the life' approach is used to generate and share data with children, with a view to empower them to share and reflect on their play.
- The second example involves a researcher working alongside two young children, developing relationships over time to enable each child to work as a co-researcher of their experiences and perceptions of play. The child-researcher interactions explore the development of a 'child-conferencing' approach to meaning-making.
- The third example involves adult researchers training and mentoring young children as researchers. This extract illustrates children drawing on play as a data collection method in their research design. It shows the ways in which play can be an empowering method, how adult researchers draw on its potentials in their training of young researchers and how the young researchers reflect on its qualities in their enquiry.

All research was undertaken in concordance with BERA's *Ethical Guidelines for Educational Research* (2018) and was approved by University College London Institute of Education's Ethics Committee. All children and parents or guardians gave consent to for participation and for material to be published in anonymised form, using pseudonyms.

Research example 1: 'day in the life', play and children's perspectives

This research investigated the impact of armed conflict and displacement on the play and childhoods of young Iraqi and Syrian children who were living as 'temporarily displaced' persons in Lebanon. The study, funded by the Froebel Trust, explored how armed conflict and displacement shaped the childhoods, play opportunities and constructions of play of young Iraq and Syrian child refugees in Lebanon, and how their opportunities for play could be improved.

The research involved conducting case studies with two Iraqi and two Syrian young refugee children (4–8 years old) and their families in their homes in Beirut's northern suburbs. As a way of gaining children's own perspectives on their childhood experiences and on their play, the researcher adopted a 'day in the life' approach (Gillen & Cameron, 2010) to fit the study aims. This involved visiting each family on four occasions, helping to establish a trusting and comfortable relationship with both adults and children. During the visits, the researcher conducted semi-structured interviews with parents and semi-structured interviews using participatory methods with young children, observed children's play for a whole day (around six hours) using a video camera, and watched and discussed with the children and their families a 30-minute compilation video of footage taken during the day of filming. Additionally, interviews were conducted with professionals working with child refugees in Lebanon, an observation of four hours (the length of the school day) was made in a local school for refugee children, and the researcher kept a research diary throughout the period of data generation. This chapter focuses on one component of this work. The next section explores an integral part of the 'day in the life' methodology which includes sharing a selection of video clips (approximately 30 minutes in length) with participants to gain their perspectives on the recorded data. However, child participants are not always involved in this phase of participant consultation. In this study, the researcher involved parents and children in consultation as a way of broadening and deepening children's opportunities to work with the researcher to convey their meanings and perceptions.

Looking to gain the case study children's perspectives as 'experts in their own lives' (Langsted, 1994, 35), the researcher also engaged with the participating children in a playful manner throughout the study and devised diverse approaches to prompting their self-expression. Drawing on Pyle and Danniels' (2016) picture book idea, the researcher created information sheets for the children in the form of colouring books with cartoon image outlines, explaining the study to the children in a way that interested them and was accessible to them (see El Gemayel, 2019, for more on colouring book information sheets). Following parental consent, children's consent was gained by asking them, as depicted in the colouring books, to make a 'thumbs up' sign if they wanted to take part in the study and a 'thumbs down' sign if they did not. This method of consent was well received by the children who, at times, played with the process, by, for example, alternating very quickly between the thumbs up and down signs to toy with the researcher.

This approach helped build a playful bond and a degree of trust between researcher and child participants. Children, in turn, gave the researcher access to their play worlds as depicted in the following extract from a 'day in the life' of Kefa, an Iraqi boy, aged 5 years and 7 months. In August 2014, Kefa and his family were forced to flee their home in Northern Iraq's Nineveh Plains when ISIS was on the verge of invading their village. After three months of internal displacement, Kefa and his family moved to Lebanon with the intention of being resettled to a third country via the United Nations High Commissioner for Refugees (UNHCR). At the time of the study, Kefa had been in Lebanon for two and a half years, living in poverty with his family of mother, father and younger brother, in a one-bedroom apartment, while they all anxiously awaited resettlement overseas to Australia.

First extract: Kefa's play

The parents and children agreed that the researcher could set up a camera, mounted on a tripod, in the main living area in order to film a day in Kefa's life, and the researcher encouraged Kefa and the family to go about their daily life as usual. However, rather than being observed, Kefa preferred to stand behind the camera beside the researcher and take on the role of commentator, giving the researcher insight into aspects of his play, as illustrated in the following account which was compiled from the researcher's diary and video recordings:

> *The researcher follows Kefa as he leads her upstairs to his cousins' apartment, who live on the second floor in the same building. Kefa, his 3-year-old brother and 4- and 5-year-old cousins*

immediately run out onto the balcony where they start to play. Kefa joins the other children for a short while but then retakes his position behind the camera. As they both stand watching the other children play, the researcher asks:

Researcher: What do you play here?

Kefa: At night we, we come up here and play. [. . .] We bring these (blankets) and cover this (a pram lying on the balcony floor) and sleep in it.

Kefa: (*suddenly runs over to his cousins exclaiming*) Come on let's bring those! *He hurries into the bedroom, and carries cushions and blankets onto the balcony for his cousins to play with. After a quick conversation with his cousins, he returns to his position behind the camera and explains to the researcher* He is going to play sick person. [. . .] Now he is going to sleep here like he is sick.

Kefa's interest in manipulating and standing behind the camera decreased as the day wore on, particularly when his 7-, 5- and 4-year-old cousins visited him later. Kefa brought out play blocks and started building a road with his brother and cousins, instructing them how to build it, controlling what blocks they could use and voicing his frustration when they went against his wishes. While they played, the children mainly spoke in Chaldean, their mother tongue. Knowing that the researcher did not speak Chaldean, Kefa regularly looked up at her as she sat in the corner taking notes, voluntarily translating what they were saying into Arabic and explaining 'now we are building a very big road! [. . .] This is for the cars to drive on [. . .] this is my car [. . .] now we are driving the cars'.

Second extract: Kefa's reflections on play

On the researcher's final visit with Kefa, after re-watching the 'building of the road' play episode on video, Kefa explained that he loved cars and that his father was going to buy him a car when they moved to Australia. The following conversation ensued:

Kefa: We left Iraq so we could come to Lebanon and then go to Australia.

Researcher: Why did you leave? What happened?

Kefa: ISIS came.

Researcher: Do you know who ISIS are?

Kefa: Do you mean what do they do?

Researcher: Yes, what do they do?

Kefa: They explode the . . . they explode and kill the people.

Although adopting a 'day in the life' approach provided the researcher with a structured process to investigate the home lives of young refugee children in Lebanon, flexibility on behalf of the researcher was pivotal when conducting this study with Kefa. The relationship between Kefa and the researcher evolved throughout the study as power relations played out between them. When filming began, Kefa felt empowered to challenge the researcher's structured approach and he sought to gain control by varying his position from the 'observed' to the 'observer'. Recognizing Kefa as an expert in his own life, the researcher strived to gain his perspectives and knowledge by facilitating his understanding of, and participation, in the study. Therefore, the researcher, who initially intended to film Kefa as he played, instead followed his lead and listened to him as he provided her with his insider knowledge and unique perspectives about his play, his traumatic experiences of ISIS, and his concerns over challenges that hindered him from attaining his future aspirations.

Kefa balanced out power relations and exerted his agency by standing behind the camera beside the researcher, positioning himself as a co-researcher. Instead of only being filmed, Kefa chose to give the researcher insight into his own play by commenting on and directing his cousins' play of 'sick person', for example, providing them with the necessary play props, and explaining to the researcher that he usually played this game with them. Later on in the day while he built a road with his cousins, Kefa thoughtfully translated their play, yet in so doing he exercised power in choosing what information to divulge about the children's play, giving her his own perspective on what was unfolding and in so doing, exercising control over how the play was represented. Re-watching and discussing the compilation video gave Kefa further opportunities to elaborate on certain points and turn the researcher's focus to aspects of his play that he deemed important.

Kefa had been, and continued to be, exposed to life-changing events that were beyond his control. He was forced to flee his home, was separated from his extended family and was anxiously awaiting his resettlement to Australia. Through observations and conversations, the researcher found that although Kefa had very limited access to play resources (space for play, and toys and people to play with), play provided Kefa with an arena where he could be in control. By imagining play scenarios, transforming objects to meet his play needs when he had limited access to toys, and directing his cousins' play. Kefa had the opportunity to transport himself out of his liminal state, which was riddled with fear and insecurity, into a world where he could exercise his agency, express his perceptions and fulfil his desires through play.

Research example 2: child conferencing and play

This research examined the relationships over time between 3- and 5-year-old children's free play cultures and practices at home and in the nursery. The study addressed the following research questions: 'What are the relationships between children's free play cultures and practices at home and in the nursery?', 'How are these relationships perceived and responded to from multiple child and adult (parents, practitioners and researcher) perspectives?' and 'How do these relationships, perceptions and responses develop over time and impact children's free play experiences in the nursery?'

Eighteen children from two state-maintained nurseries in London were invited to take part in this qualitative research to share aspects of their free play at home and in the nursery. The researcher was introduced to the children by the lead-practitioner as a learner who attended a school for grown-ups, and who was interested in how children play at home and in the nursery. The researcher introduced the children to the study using a combination of talk and activities, where the children could try out showing and sharing their play with the researcher, before deciding to take part. The children's consent was understood as provisional, and each child was invited to take part at the beginning of each observation (Flewitt, 2005). The children were informed that they could take part in, or withdraw from, the study and ask questions, at any time. They indicated consent and dissent on paper through mark making of their choice such as drawing 'smileys' in columns labelled 'yes' and 'no'. During the study the researcher made it clear that at any time children could say that they did not want to research on a particular day, and at times children chose not to.

Six hour-long observations of children's free play were made during playtime in the nursery, over six months, in addition to photography and child-conferences (Huser, 2015) as participatory tools that best suited the individual children's skills and preferences. Photography, for example, meant that children could take photographs of areas they enjoyed playing in as part of their work with the researcher. Child-conferences included children's answers to a short interview consisting of structured and age-appropriate questioning at the beginning, middle and final stages of research. These questions pertained to aspects of their free play experiences at home and in the nursery, such as their likes and dislikes concerning play resources and spaces. All child participants were asked the same questions in the same order and could respond with talk or showing. The aim was to capture consistencies and developments in answers over the six months and to gain insight into their perceptions of their experiences through a comparatively structured medium of communication. Each interview ended with an open-ended question that provided an opportunity to add any additional information. Engagement in the child-conferences was optional and children were encouraged to choose if, and how, they wanted to respond to the questions.

The researcher refrained from replicating patterns of adult behaviour that were typical of the nursery cultures and practices in order to communicate to the children that her role as a researcher was different from the roles of other adults in the setting. The researcher responded sensitively to each child's unique engagement with the research process. This led to the development of diverse researcher-child relationships as both children and researcher practised co-reflexivity by examining each other's, as well as their own, responses in an ongoing manner. As the children engaged in play during the study, they communicated as co-researchers by sharing aspects of their free play at home and in the nursery as the following extracts will reveal.

Data generated by the children were supplemented by in-depth semi-structured interviews with their parents and practitioners, which provided insight into multiple adult and child perspectives from which the temporal, contextual and relational dimensions of each child's free play journey was perceived and responded to. The following three extracts are selections from this data and focus on the child-researcher/adult-researcher dynamic. They illustrate how two child participants, Clare (aged 3 years and 5 months) and Irene (aged 3 years and 7 months) made meaning of the researcher-child relationship, their own role in the study over time and how their understandings shaped how they shared particular aspects of their play during the research process. Three excerpts are presented in succession, and then discussed.

First extract: what do you like about. . .?

The researcher asked Clare the following questions about her play at home and in the nursery during their first child-conference:

Researcher: What do like about playtime in the nursery?

Clare: (*Clare's eyes widened as she exclaimed*): Everything! All the toys. Doll's house, water station, paint, doll's house!

Researcher: What do you not like about playtime in the nursery?

Clare looked around the room before answering and exclaimed when she saw the book corner.

Clare: Reading!

She walked up to the bookshelf.

Clare: Let's go to the book corner. I don't like Friday (*children read for half an hour with their parents every Friday morning in this nursery and were instructed by practitioners at playtime to select a book of their choice to take home to read over the weekend*). Because some of these books are. . . (*she pulled out a book called* Kipper and the Egg *from the shelf, flipped through the pages*) this book is rubbish (*she wrinkled her nose and turned one corner of her mouth up*).

Researcher: What do you like about playtime at home?

Clare: Doll's house! (*Clare grinned and her eyes widened again*)

Researcher: What do you not like about playtime at home?

Clare: Nothing. *She shook her head and shrugged her shoulders.*

Second extract: do you want me to do that again? Slowly? So you can write?

This also occurs at the time of the first child-conference. The researcher reminded Irene about what the research was looking at, and asked Irene what she wanted to share about her play at home and in the nursery. Irene led the researcher into the nursery play kitchen.

Irene placed one Teletubby on her lap and another on the researcher.

Irene:	These are babies. Let's hold them on our laps. This one is Shayla. Yours is Cici.
	Irene speaks to Shayla.
Irene:	You're too heavy to sit on my lap. I'll take you to baby school where you'll have friends, ok?
	Irene picked up the phone.
Irene:	*Calling policeman.* She's been punching people. Not eating properly. Baby is being naughty. *Calling Mum.* Mummy, I don't think my hairband is right for school. You have to come pick it up. Ok bye alligator. See ya later alligator.
	Irene hung up the phone and turned to the researcher.
Irene:	How was that? Was that alright?
Researcher:	Yes. That was lovely. Thank you.
	Irene looked into the researcher's notebook where observation of her play was documented.
Irene:	Do you want me to do that again? Slowly? So you can write?

Third extract: 'Redacted'

Four months into the study, the following interaction took place while the researcher observed Irene play under the shed in the reception playground, as it was raining.

Irene:	Let's go out shopping?
Researcher:	I have to stay in because I can't get my notebook wet in the rain.
Irene:	Why??
Researcher:	Because I need to show this to my teacher.
Irene:	Oh yeah what's his name again?
Researcher:	Phil.
	Irene instructed the researcher not to report on the next section.
	What Irene said next is redacted, as per her instructions.

In the first extract, among various resources within the nursery play provision that Clare preferred, her repetition of the doll's house while discussing playtime in the nursery and playtime at home revealed her fondness for it. It also showed parallels in her play experiences between home and nursery. The recurrence of doll's house in later child-conferences (not included in this chapter) showed its consistent presence in Clare's play journey over time. In addition to her overt verbal expression, Clare's dislike for activities related to reading at playtime was also visible in her facial expressions as she flipped through the book. Clare's act of looking around the room and stopping to mention reading as she saw the book corner shows the impact of context on her engagement in the research process. Clare showed agency, for example, in the way she used space within the researcher-child relationship, as she led the researcher to the book corner and spatially relocated the research activity.

In the second extract, Irene showed agency in relationship to space and resources as she led the researcher into the play kitchen. Using open-ended resources within the nursery provision, Irene developed a selected demonstration that included themes of mum, babies, hairband, policeman, being reprimanded for bad behaviour, taking care of babies. These all related to her play experiences at home and in the nursery in response to the child-conference questions. These themes recurred in Irene's talk and play, and were mentioned by practitioners and her mother over time. In extract two, Irene picked up and hung up the phone to move in and out of the pretend situation and reality, as she overtly attempted to co-reflect with the researcher on how they conducted the research; and ensured that the play that she displayed matched the research agenda and was documented. This showed her awareness of sharing her experiences of play with the researcher. In extract three, as Irene discussed the researcher's teacher, she showed awareness of what happened to the observation notes documented on her play. As she instructed the researcher to omit parts of their interactions from the notes, Irene exercised agency in the researcher-child relationship by determining who was able to access information relating to her play.

In extract one, by following Clare's suggestion to relocate the child-conference to the book corner, the researcher communicated to Clare her intention to share control over the research process. In extract two, the researcher's agenda to learn about Irene's free play at home and in the nursery served as a focus for the child-conference. However, the researcher deliberately did not give explicit instructions to Irene on how to interact with the provision in the nursery: the aim was to share control over the research agenda with Irene by enabling her to decide what aspects of her play she chose to share and how she wished to communicate them. The events that unfolded were a consequence of child-conference interactions that took place moments earlier when Irene was asked questions relating to her play at home and in the nursery. The researcher intended Irene's choices and actions to be rooted in her interpretation of the purpose of the research; and Irene's resultant intentions for it. This process of interpretation involved co-reflexivity between Irene and the researcher, as the researcher carefully considered how she embodied and communicated her intended researcher role; and Irene responded through reflexivity by thinking about the implications of her choices. Based on these interpretations, Irene communicated aspects of her play repertoires at home and in the nursery that she considered to be of value for the research. Similarly in extract three, by answering Irene's questions and following her instructions to omit part of their interaction from her observation notes, the researcher shared control over the research agenda with Irene.

Research example 3: training children as researchers and play

This final research example illustrates the potentials for play as a research language, in a study where an adult research team trained primary age children to design and conduct small-scale research. The examples illustrate how the researchers developed relationships as trainers and mentors to the child researchers and how play language and process was key to this.

The project, funded by the LankellyChase Foundation, aimed to develop ways of working that maximised the potential of participatory research to enable children and young people experiencing severe and multiple disadvantage to represent their experiences and views. UCL Institute of Education (UCL IOE) and the Digital Arts Research in Education Collaborative (DARE), which is part of the UCL Knowledge Lab, joined together with voluntary sector local organisations to develop the project. The research was conducted in four sites in different parts of England, and included children aged 7 to 18 years (Flewitt et al., 2018; Jones et al., 2018). Each site offered the opportunity for children to become involved as researchers and offered sample workshops and information sessions, with age-appropriate information sheets and consent forms to enable children to make an informed decision about taking part. The research questions included 'What are effective ways to *develop and sustain relationships* with young, hard to reach individuals, and establish effective ways to initiate and sustain participatory research for them to document and report on their own lives?' and 'How can participants be involved as active and empowered agents in every stage of the process?' The project involved training the children in research techniques, and supporting them to develop questions they wanted to explore with their peers, and then mentoring them in designing, recruiting, conducting, analysing and presenting the results of their enquiry. The project was supported by a 'reference group' of children who advised on the project's design and development (Jones et al., 2018). One of the areas of advice from the group concerned the activities and methods used within the project. The advice of the reference group was to 'make the process enjoyable' and 'fun', to use 'play', to create 'characters to make confidentiality easier' and to use 'age appropriate languages'. In designing the workshops to train the young researchers, the adult researchers reflected the child reference group's advice by creating a variety of research activities that were playful or based in play processes. These included games, imaginative play and creating stories and characters.

Workshops with the child researchers then supported them in designing research for them to conduct with their peers. The adult researchers worked with the young researchers to create a mind map of options of areas to focus on as a subject for their research and then, similarly, offered a 'menu' of diverse conventional (such as questionnaires) and participatory data collection methods (such as creating images or using role play). The child researchers were then supported in designing their project by choosing from the data collection 'menu' or developing their own ideas about new research methods. The following data is from one of the four sites, where we trained child researchers aged 7–11 years in collaboration with MD Productions, a voluntary sector organisation that works with the arts in disadvantaged inner city areas. The young researchers decided as the focus of their project to explore child participants' experiences of the streets they live in. The young researchers then developed a research project, researching with a peer group of eight children of the same age who were already involved in MD Productions arts workshops in the same city. The adults facilitated design workshops to support the children in planning their research with their peers and mentored them as they implemented their project. During these workshops, the young researchers developed data collection methods, building on those explored during the training phase of the work.

One activity they designed and used involved participants splitting into subgroups, with each group drawing round one of their peers on a large piece of paper in order to create the shape of a generalised character of a child living in their part of the city. The child participants within the project created one male (named 'Wilfie' by the group) and one female (named 'Majesty' by the group). Participants then added words or images to their character's outline, depicting some of the issues they thought the imaginary child might face. They then were invited by the young researchers to act out how the character might be interviewed about their life and experiences in an imaginary TV show, with a child participant playing the role of a TV interviewer and other child participants taking it in turns to play the role of the character that had been created. The following is a sample of the data generated when child participants used the method.

First extract: interview with Wilfie

Child Interviewer: Hello, and welcome back to the Ellen show. I'm with a young boy called Wilfie.

Child in role as Wilfie: Hi!

Interviewer:	So, Wilfie. What is it like where you live?
Wilfie:	It's very sad and, like, lonely. No one really plays out. And, like, you know, they all bully me cause, like, you know, how I look.
Interviewer:	How do you feel about the area where you live?
Wilfie:	I feel like that no one really cares about it and that no one really knows about the area so we can't get any help.
Interviewer:	Is your area clean?
Wilfie:	Kind of and kind of not. It's because, like, not that many people go in our street so there's not that much, like, litter on the floor. But there are bins everywhere what are like overflowing. We've got loads of bins there, but people just throw it on the floor and think, 'well, it's not my business'.

Figure 1.1 'Wilfie', created by a group of five child participants

The young researchers collated their results and represented their findings to their peers and also to Liverpool counsellors making the case for further involvement of children's views on areas such as safety, the streets and play in their city.

Second extract: feedback

Feedback from child participants in the research was obtained by a short anonymised questionnaire, designed by the young researchers. This included questions such as 'What did you enjoy?' 'What did you not enjoy?' The participants said that there was nothing that they did not enjoy. Table 1.1 shows a sample of the data about enjoyment, which confirmed the accuracy of the reference group's perceptions and advice on the use of play-based activities.

This anonymised feedback from the participants (Table 1.1) was shared with the young researchers and with the child reference group to enable them to be given evaluative information on what it was like for the children to participate in the activities they had facilitated or helped design.

The project included a review meeting involving the young researchers reflecting on their experiences of the research; part of the session involved a discussion of the data collection methods they had used. The child researcher reflections shown in Table 1.2 offer perceptions on the value of playful, participatory methods.

The evaluative comments of the participants and of the child-researcher perceptions of the research and their responses to it (Tables 1.1 and 1.2) illustrate their perceptions of play-based participatory methods, such as children creating imaginary characters, stories and 'age appropriate' forms of expression to engage with complex themes, whilst experiencing enjoyment and having 'fun'. The adult researchers created a relationship with children where the research training and mentoring empowered the children to design and implement their own research, drawing on play languages and processes to enable their peers to express and communicate their experiences and views. The research did not approach participation and play as an essentialised

Table 1.1 Feedback on the research workshops from child participants

• We had fun activities.
• We got to enjoy it whilst discussing serious things.
• It wasn't completely serious but we had fun.
• It wasn't completely serious and we all got to share our ideas.

Table 1.2 Feedback from the young researchers on the methods they developed and used

• The art and games and stuff they weren't boring.
• It wasn't boring.
• The activities kept people interested.
• It's not just charts and words.
• The games and drawings helped people do things that just talking wouldn't.
• It kept them moving, it gave us things to make and start things off.
• They said after how much it had been good, doing things like that.
• They could talk about things but not saying it's me.
• It was imagination, too, not just blah blah.

good – but attempted to facilitate evaluative feedback from child participants and from the young researchers on their specific experiences of the participatory methods used.

Conclusions: reflections on play, children and researchers

This chapter has illustrated new perspectives and opportunities for practice in approaching research in relation to children and play. The analysis addressed the tendency in research, identified in our review of literature, to position children's participation as a 'given good', rather than analysing the specific context and acts of participation of the research in order to engage reflectively, rather than to essentialise the process. The extracts and data illustrate the innovations created in each example, made possible by dialogue between the design and implementation of research with children and theories concerning power, collaboration and control, also identified in our review of literature. These include how play enables children and researchers to form relationships where children are valued as participants rather than subjects, how ways of working with research are created that foreground and empower children's perspectives and voice, and how views on the meanings of data can be responded to by children, rather than by adult researchers alone. The following summarises the nature of these innovative insights offered within the analysis of each extract from our research.

The first example illustrates how researchers and young children who have experienced armed conflict and forced displacement can work together to reveal children's invaluable insight into play and the unique knowledge and understandings of their own lives and childhood experiences. The research process and trusting child-researcher relationship offered Kefa age-appropriate methods to share his experiences, with the adult researcher adopting a flexible approach to data collection and actively listening to, and following, the child's lead. This example illustrates how, within the specific approach taken, Kefa was not the object of adult attention and interpretation: the analysis revealed how a child can work differently with the power relationships between participant and researcher. The combination of filming a 'day in the life' of Kefa alongside the opportunity for Kefa to comment on the data, both during and after filming, enabled his voice to feature in shared meaning making and empowered him to work alongside the researcher to communicate and express his ideas (Kellett, 2010). We argue that this approach can empower children who have lost control over many aspects of their lives to regain some control and exercise their agency by choosing to bring to light aspects of their childhood and play that they consider important, but which would otherwise remain silenced and invisible to the researcher.

The second example illustrates particular aspects of the relationships between play, research, power and control as children and the adult work together, with Claire and Irene choosing what to share with the researcher about their play experiences. This develops particular insights into how very young children can be facilitated though adaptation of child conferencing (Huser, 2015) to engage with Fargas-Malet et al.'s concept of 'differing levels of control-sharing and of participation in the research process' between children and adults (2010, 176). For example, Irene shares the research process by working alongside the adult researcher; both negotiate and share power and control over the research agenda. The extracts illustrate the complex role of the researcher – as an adult who differed from both practitioners and children in terms of power; and how her intentions to share control over the research agenda were communicated to Irene by practising co-reflexivity. Co-reflexivity between the researcher and the children involved the researcher carefully considering how she embodied and communicated her intended researcher role and how children responded reflexively by thinking about the implications of their choices. This extract also shows how children were involved in the decision-making process regarding particular aspects of their research engagements. In Irene's case, these included the selection of aspects of her play that she considered to be of importance to the research; and decisions regarding what aspects of her play were to be recorded as research data and made accessible to people beyond the nursery who were involved in the research. We argue that experiencing the process of collaborative enquiry and the development of the researcher-child relationship over time can enable young children such as Irene to engage as co-researchers of their experiences of play and to make decisions about what they value and want to share.

The third example portrays how play as a process enabled the young researchers to design and conduct their own research. The research illustrates how young researchers developed methods using play and were facilitated to reflect and give feedback on their perceptions of the ways of working they developed. The data also shows how the children were empowered by adults through training and mentoring, illuminating Kellett's concepts of changes in research where children develop 'their capacity for judgment, for communicating their views and agency for action' (2010, 4). The extract shows, for example, how the adult researchers adopted the role of research trainer and mentor to facilitate the children's choice of research topic, design, conduct and analysis of their own research. In this instance, play is integral to research as a conduit for children to draw on in designing their enquiry. Play enables the adult researcher to show the child researchers methods they feel they could inhabit and value. The reflective feedback from the child participants and young researchers on the research process are testimony to their perceptions of the potency of play as a method. It also responds to calls for the need to gain insight through specific reflection on the nature and benefit for the children and the research's specific context and use of play, rather than essentialising the participation of children in the research as a given 'good'. As the feedback extracts show (Tables 1.1 and 1.2), the children considered that play offers them particular values: to communicate their awareness of their own lives, to conduct research in innovative ways and to explore and communicate their experiences.

Our presentation of data and analysis has responded to the concerns of researchers such as Buhler-Niederberger (2010) and McCarry (2012) by addressing gaps in our understanding of children's participation in research by offering insights into the detail of our practices and the relationships between research, children and play. Each example has examined the nuanced ways innovative relationships between play, adult researchers and children can be formed. The analysis has illustrated how the researchers and children in each context are exploring the 'paradigm shift' identified by Kellett (2010). The chapter has shown, for example, how the right for children to express their views becomes realised in different configurations of the researcher-child relationship; and how play can feature in such innovative explorations of the creation of 'partnerships' that facilitate children 'communicating their views' and developing their 'capacity for judgment' and 'agency' (Kellett, 2010, 4).

The research by El Gemayel was made possible by doctoral funding from the Froebel Trust and the Overseas Research Scholarship funding from University College London.

References

Barnett, L.A. (2013) Children's Perceptions of Their Play: Scale Development and Validation, *Child Development Research*, Article ID 284741, https://doi.org/10.1155/2013/284741.

British Educational Research Association [BERA]. (2018) *Ethical Guidelines for Educational Research*, fourth edition, London, www.bera.ac.uk/researchers-resources/publications/ethicalguidelines-for-educational-research-2018.

Buhler-Niederberger, D. (2010) Defining the State of the Art and Ensuring Reflection, *Current Sociology*, 58 (2), 155–164.

Burke, C. (2005) Play in Focus: Children Researching Their Own Spaces and Places for Play, *Children, Youth and Environments*, 15 (1), 27–53.

Einarsdóttir, J. (2007) Research with Children: Methodological and Ethical Challenges, *European Early Childhood Education Research Journal*, 15 (2), 197–211.

El Gemayel, S. (2019, March 12) *The Colouring Book Information Sheet: Involving Young Children in the Research Process Through Informed Assent* [Blog Post], https://blogs.ucl.ac.uk/global-youth/.

Fargas-Malet, M., McSherry, D., Larkin, E., & Robinson, C. (2010) Research with Children: Methodological Issues and Innovative Techniques, *Journal of Early Childhood Research*, 8 (2), 175–192.

Flewitt, R. (2005) Conducting Research with Young Children: Some Ethical Considerations, *Early Child Development and Care*, 175 (6), 553–565.

Flewitt, R., & Ang, L. (2020) *Research Methods for Early Childhood Education*, London: Bloomsbury.

Flewitt, R., Jones, P., Potter, J., Domingo, M., Collins, P., Munday, E., & Stenning, K. (2018) 'I Enjoyed It Because . . . You Could Do Whatever You Wanted and Be Creative': Three Principles for Participatory Research and Pedagogy, *International Journal of Research and Method in Education*, 41 (4), 372–386.

Gillen, J., & Cameron, C. A. (Eds.) (2010) *International Perspectives on Early Childhood Research: A Day in the Life*, Basingstoke: Palgrave Macmillan.

Gillen, J., Cameron, C.A., Tapanya, S., Pinto, G., Hancock, R., Young, S., & Gamannossi, B.A. (2007) 'A Day in the Life': Advancing a Methodology for the Cultural Study of Development and Learning in Early Childhood, *Early Child Development and Care*, 177 (2), 207–218.

Government of Lebanon, & United Nations. (2019) *Lebanon Crisis Response Plan 2017–2020 (2019 Update)*, http://data.unhcr.org/syrianrefugees/regional.php.

Huser, C. (2015) Children's Voices on Play in a Mosaic Approach Study: Children as Conscious Participants in a Case Study, *Boğaziçi University Journal of Education*, 26 (1), 35–48.

Jones, P. Mercieca, D., & Munday, E. (2018) Research into the Views of Two Child Reference Groups on the Arts in Research Concerning Wellbeing, *Arts & Health*, doi:10.1080/17533015.2018.1534248.

Kellett, M. (2010) *Rethinking Children and Research*, London: Continuum.

Koller, D., & San Juan, V. (2015) Play-Based Interview Methods for Exploring Young Children's Perspectives on Inclusion, *International Journal of Qualitative Studies in Education*, 28 (5), 610–613.

Langsted, O. (1994) Looking at Quality from the Child's Perspective, in P. Moss & A. Pence (Eds.), *Valuing Quality in Early Childhood Services: New Approaches to Defining Quality*, London: Paul Chapman Publishing.

Larkins, C., Thomas, N., Carter, B., Farrelly, N., Judd, D., & Lloyd, J. (2015) Support for Children's Protagonism: Methodological Moves Towards Critical Children Rights, *The International Journal of Children's Rights Research*, 23 (2), 332–364.

McCarry, M. (2012) Who Benefits? A Critical Reflection of Children and Young People's Participation in Sensitive Research, *International Journal of Social Research Methodology*, 15 (1), 55–68.

Murray, J. (2011) Knock, Knock! Who's There? Gaining Access to Young Children as Researchers: A Critical Review, *Educate*, 11 (1), 91–109.

Powell, M., Graham, A., & Truscott, J. (2016) Ethical Research Involving Children: Facilitating Reflexive Engagement, *Qualitative Research Journal*, 16 (2), 197–208.

Pyle, A., & Danniels, E. (2016) Using a Picture Book to Gain Assent in Research with Young Children, *Early Child Development and Care*, 186 (9), 1438–1452.

United Nations Convention on the Rights of the Child (UNCRC). (1989) Convention on the rights of the child. http://www.ohchr.org/en/professionalinterest/pages/crc.aspx. Retrieved on August 09, 2020.

2

LILA

Teacher's play through the eyes of a child

Grace Schuchner

My intention to write this chapter came from my role as a mother. I would like to contribute to raising awareness of the importance of adult behaviour and attitude in education – the crucial role of the teacher in the classroom. Just as the planet is an ecosystem, the classroom is a system in which one part affects the other. We are relational beings, we grow and develop through our relationships with others and in the formative years with our main caregivers. Our nervous system depends on those delicate bonds with other people. From my professional experience as an actress and dramatherapist, play allows us to *connect and relate* more deeply and honestly with ourselves and with other people.

This chapter is about the playful nature of an English teacher in a school located in the city of Buenos Aires. It is a bilingual inclusive school with an intensive workload of ESL (English as a Second Language).

Although there is a lot of literature about play and its benefits for children and adults, there is not much literature about the importance and benefits for the teacher-pupil bond when it is the teacher who has a playful behaviour in the learning process, in this case also to learn a foreign language. Since teachers have an important role in the child's ability to learn, and based on the assumption that mothers (fathers or caregivers) "donate" their playfulness to their children; at school, teachers "donate" their playful ability to enable it in their pupils.

Research suggests that teachers' playfulness encourages divergent thinking[1] in the classroom and influences the behaviour of pupils. Lieberman (1977) carried out a small exploratory study in which sixteen teachers completed an open-ended questionnaire aimed at defining playfulness in teachers. A content analysis produced eight component traits: a) sense of humour; b) kindness, sensitivity; c) cheerfulness, laughter; d) enthusiasm, active participation; e) flexibility; f) imagination; g) at ease, or relaxation; and h) entertainment (Lieberman 1977, p. 52). Lieberman concluded that teacher's playful learning approach can create an environment that enables children to express greater joy and be more flexible and creative.

My desire to write about the teacher's playfulness came one night while my eleven-year-old daughter and I were having dinner and she told me "Mom, we have an English teacher who 'teaches in a different way'". I said, "It sounds interesting and what do you mean with 'she teaches in a different way'"? She replied, "For example, she does not shout at us, if a child does something that is not right, she does not shout at him, she starts singing".

She starts singing the name of a pupil imitating the teacher.

The conversation with my daughter sparked my curiosity to learn this different way of teaching. I asked the teacher and the school principal for permission to observe the classes and to write a chapter about her playful behaviour. I decided to call the teacher Lila, which in Sanskrit literally means "play".

The classroom and its layout

Children enter at 8:20 a.m. in the morning. They have the syllabus in Spanish, they have lunch at school and in the afternoon they learn in English. English classes are from 3 p.m. to 4.30 p.m. The classroom is located on the third floor, next to an open playground. There are eighteen pupils in the class between ten and eleven years old; nine boys and nine girls. There are two boards in the classroom, one is green and it is used with chalk and the other is a digital board. There are twelve double desks and twenty-two green chairs. In the corner next to the digital board is the teacher's desktop with a laptop with Internet connection and a fixed installation with a video projector.

The children's lockers are on one of the walls and at the top, there is a window with a transparent pink curtain. The top of the wall next to the entrance door to the class is also made of glass. The classroom has natural light. The desks are distributed in horizontal lines.

The wall on the green board and the one in front are decorated with different coloured cards, many of them written in English with the vocabulary they are learning.

Scene one: "the classroom as a place to get together" (Duarte 2003, p. 105)

The transforming role of the classroom is up to the teacher, encouraging an environment that enables a communicative interaction and the meeting between the teacher and the pupils as a group.

> Once the doors of the classroom are closed, there start interactions that only their actors can account for. It is here where teachers constitute and show themselves, here desires become true, it is no longer the world of what could be, but the space of what is.
>
> *(Duarte 2003, p. 105)*

The bell rings. The children enter the classroom after the break and sit in pairs.

The teacher walks to her desk and from the computer she plays the music tuned to 432hz,[2] she asks the pupils to close their eyes or leave them half closed and softly the heads of the children go down and they lean on the desk. On the green board, are the class rules written on a small card (see Figure 2.1). The children repeat after the teacher while their heads are still leaning on the desk.

Lila:	Repeat after me "listen carefully".
Pupils:	Listen carefully.
Lila:	Follow directions.
Pupils:	Follow directions.
Lila:	Work quietly.
Pupils:	Work quietly.
Lila:	Respect others.
Pupils:	Respect others.

Figure 2.1 Classroom rules

Lila:	Play safely.
Pupils:	Play safely.
Lila and the pupils: all together	We are going to have a beautiful day!

Once the affirmations are finished, all the pupils raise their heads and there is a peaceful atmosphere in the classroom.

Lila: (*Standing in front of the boards with a cheerful tone and waving her arms as if she were a dancer*) Now, listen to what we are going to do today. First, let's work with the workbook.
Pupil: (*Challenging*) No, I don't want.
Lila: (*Kindly*) We are going to work a little with the book and then we are going to do something else.
Pupil: (*Provokingly*) I'm not going to take out the book.
Lila: (*Playing with a serious tone of voice*) Please take out the book.
The pupil makes a gesture indicating that he does not care what the teacher says.
Lila: (*Firmly*) I will not tolerate disrespect.

The behaviour of children, particularly at school, still causes concern. It could be said that disruptive behaviour is a challenge for the teacher, a challenge to transform it into something playful to disarticulate the pupil's discomfort.

Lila: (*Patiently*) Please take out your book, otherwise you will have to leave the class.
Pupil: (*Mockingly*) Much better, I'm going home!

Lila: (*Less patiently, she approaches the pupil so as not to raise her voice and looks into his eyes, just as a mother would with her son*) It seems to me that you do not understand what I am saying, you are going to work with the book in the principal's office.

Pupil: (*He laughs ironically*) I'm going home, I'm going home.

The teacher takes her cell phone from her desk and writes a text message.

A few minutes later, the principal enters the class and asks the pupil to go to her office with his book.

However, it is not always easy to be warm and supportive, especially when provocative pupil behaviours thwart the teacher's efficacy to perform his or her task (Jennings & Greenberg 2009). "Teachers would need to be specially trained, and with an understanding of attachment work they could see that their 'acting out' pupils are trying to communicate distress rather than just being 'difficult'" (S. Jennings 2011, p. 235).

Once the pupil leaves, they begin to work on the book.

Meanwhile, one pupil is doing the exercises in the book.

Pupil: (*Out loud*) I'm dumber than a rat.

Lila: *Closing her fists, she asked them to be silent like an orchestra conductor and with a smile that seemed to cover her entire face*) Please, stop working and listen to me.

With a tone as if she were telling a story

Do you remember that a few weeks ago I told you that I had to go to live in my parents' house because I had no gas and I could not turn on the stove to cook and did not have hot water to take a bath?

Pupils: (*Different voices and volumes are heard with the word yes*) Yesssss.

Lila: (*Smiling, she begins to walk with steps to generate expectations in front of the boards*) Well, one afternoon I went back to my house because I had a private pupil and I bought some muffins for tea before he came. There are three muffins in the pack, have you seen?

Pupils: (*Excited, many of them have something to say about the muffins*)

Yes, I eat the three of them.

I don't like them.

I eat one and I die for thirst.

Do not mention that now because it makes me want to eat

Lila: (*Closing one fist, she asked them to keep silent*) I ate only one because my pupil just arrived. I finished the class and went to my parents' house. And I forgot to put the muffins in the cupboard. (*Speaking slower*) The next day I went back to my house, because I had a pupil, (*pauses*) went to the kitchen, where I had left my two muffins, (*she pauses again*) and I saw that they were bitten, a rat had eaten them.

Pupils: (*Shrieking*) Noooo, puaj!

Lila: (*Pretending that she is crying*) Have you seen how smart rats are? They ate my muffins!!!!

The children laugh.

Thinking about the children who are in school from 8:20 a.m. and learn a second language in the afternoon, when they are already tired, you can only do it creatively with humour, games and dramatization. A story, told with humour, also produces an important improvement in relation and when the kids laugh and soften their defences they are more open and pay more attention. Sense of humour and laughter help reduce anxiety and stress in unpleasant situations. The authors of the book *Hey Teacher, let's skip recess!* write that "different strategies and humour can coexist in the context of the classroom and be successful tools to motivate pupils" (Ottobre &

Temporelli 2015, p. 23). Research has evidenced that "the mixture of fantasy and real, jokes and lessons are a very effective combination" (Brown 2009, p. 103). Also a very important benefit of humour is that it integrates the two cerebral hemispheres.

The bell rings.
All the children grab their backpacks, the teacher says good-bye and they leave the classroom.
The teacher takes her stuff and leaves the classroom and goes to the principal's office.
At the principal's office.

Lila: (*Very distressed*) If I put myself in the shoes of the pupil to whom I asked to leave the class, this is not how I would like them to interact with me. *I think it is very important to build a bond and empathy with the children, as well as their wellbeing*, and I cannot find the right tools to motivate, especially the pupils who misbehave.

Principal: (*Trying to calm down the teacher*) First of all, I want to tell you that your classes are very good and in no way can I think that you are not a good teacher because of what happened today, actually this is the first time.
Tears fall down Lila's cheeks.

Principal: (*In a soft tone*) These pupils with behavioural problems are those who most need a supportive relationship from their teacher. Sometimes children lead us to stressful situations, they create challenges we have to overcome. It is important not to lose your temper, think that they are developing their subjectivity, and highlighting something that is not right, is correct, but we must encourage them to do it thoughtfully and not in anger. Neither him, nor the teacher.

A teacher's support and sensitive reactions to their challenging behaviours may have lasting positive effects on the pupils' social and emotional development, especially in the early grades (Lynch & Cicchetti, 1992).

Scene 2: there are no games without rules

The pupils repeat after the teacher the class rules with their heads resting on the desk, background music can be heard tuned to 432hz.

Lila: (*She walks to her desk and takes her cell phone*) Do you know. . .?

Pupil: (*Raising her head from the desk*) We haven't said that we are going to have a beautiful day!

Lila: (*With a smile and moved*) It is true! Thank you!

Lila and her: pupils together (*Pupils raise their heads from the desk*) We are going to have a beautiful day!

Lila: (*Enthusiastically*) First, you are going to write down the vocabulary that you have to study for the test and then we will do the theatre performance you wanted to do.
(*Playing with her tones of voice, she changes to a higher tone and says*) Please take out your folders and write down.
The teacher reads out the list of words from her cell phone, while the children write them down on their folders. In the classroom there is an atmosphere of calm, concentration and silence.

Lila: (*Rubbing her hands, raising their expectations*) Are you ready?

Pupils: (*Some of them answer in English and others in Spanish*) Yes, sí!

Lila: (*She asks one of the pupils who has more behavioural problems to go to the corner diagonally to where she is*) I'm going to stay in this corner and you are going to stand in the other corner and you are going to repeat after me.
The classroom environment should be thought of as a dynamic environment, with the possibility of recreating and changing it, depending on the projects that the teacher is dealing with. The idea is that there is not a single front. Children should be able to reach with their sight and body the different corners and each should be a learning area.

Pupil: (*He stands with a slight smile on his face and stands where the teacher told him*).

Positive relationships between teachers and pupils can be seen through a variety of psychological models, in this chapter the focus is through play. Mainly, in the playful behaviour of the teacher. The characteristics of teachers also play an important role in building close bonds with pupils.

Lila: (*Looking at all the pupils from the corner of the classroom*) Do you know what my eyes can see?
The pupils look at the teacher and look at their classmate who is standing diagonally to the teacher.

Lila: (*Putting her arms in first ballet position, arms in a rounded shape at the elbow joint and laughing*) I see all of you ready to prepare the play! Yessssss! But first, the rules of the game; and (*changing to a roguish tone of voice, points to the pupil standing diagonally*) you are going to repeat after me.

"It's not about playing to exercise, but to play so that learning circulates, so that the stagnant aspects can be put into play from another point of view" (Öfele 2017, p. 40). The commitment in games in the mother-child relationship, can be transferred to the teacher–pupil relationship; when there is commitment in play, inevitably, there is an understanding that humanizes the bond. The commitment in the game by both parties will equally facilitate getting involved in school learning (Öfele, 2013).

Lila: (*Reading from some small cards, making pauses for the pupil to repeat the phrases that later on they will have to complete and include in the scenes*)
The phrases are: The weather. . .
You are on. . .
You go to. . .
You eat. . .
You want. . .
The pupil repeats after the teacher with enthusiasm.

Once the teachers and pupils detach from the order of daily dynamics in the classroom and are ready to play, it is necessary to create a new order, the playful order. Graciela Scheines says that "order ensures the freedom of the player. The relationship between order and freedom is clear in all games" (Scheines 1998, p. 28). Once the order and the rules of the game are set, children can rehearse and unfold their imagination, creativity and inner freedom.

The teacher divides them into groups of five, hands out the cards with the phrases to be included in the scene.

The aspects of commitment, understanding and joy that are found in games allow pupils to relate and meet each other and with the teacher by means of new communication channels and at the same time raise the confidence of the pupils for meaningful learning.

You cannot compel others to learn, but you can offer the means and generate the conditions to arouse interest and motivation. When a "climate of optimism and hope" is created, pupils can acknowledge and value their own abilities (Ottobre y Temporelli 2015, p. 30).

Scene 3

> The school environment must be diverse, it must transcend the idea that all learning takes place between the four walls of the classroom.
>
> *(Duarte 2003, p. 12)*

The pupils are sitting on the floor in groups of five in the playground. There is wholeness in the spaces between the interior and exterior of the physical environment of the classroom. There are a few clouds in the sky and the winter sun that hints at the arrival of spring. It is 3:30 p.m. As children prepare their performances, they have to talk in English, no Spanish.

The only possibility of freedom and power that children have is through play. At last while they are playing, they invent a world according to their dimension, there they are powerful, and they are free (Scheines 1998). Children create a world in which they have control and can go out in search of the uncertain to conquer it. Enabling a space and time where they can create without a guide that indicates and establishes the plan to be developed, will favour a much more significant learning and production (Öfele 2007).

The teacher goes from one group to another offering her help.

The teacher's work is important, it will be the actor that will enable "that intrinsic motivation has a fundamental role and allows rescuing creativity" (Ottobre y Temporelli 2015, p. 53).

Lila sits on the floor next to a group that is working, and in her gestures you can see the joy that she feels from what the pupils are creating in a committed and interesting way. Her attitude is in tune with the game of the pupils.

According to Daniel Siegel one of the basic elements of the secure attachment process is *attunement*, which means harmony between the internal state of the parents or caregivers and the internal state of the child. "It is often accomplished by the contingent sharing of nonverbal signals" (Siegel y Hartzell 2012, p. 154), such as eye contact, tone of voice, facial expression, gestures, body posture and the intensity of response. Siegel writes that most probably "Non-verbal resonance likely involves a connection between the right hemispheres as they mediate non-verbal signals in both people" (Siegel y Hartzell 2012, p. 172).

A pupil of another class approaches the teacher and greets her with a hug.

It is essential *not to forget* bonds and not to be afraid to generate a bond of tenderness with the children. Affection can be shown in different ways, verbal or non-verbally, with a caress, with a word of encouragement or congratulations and it does not mean that the teacher loses her "area of power" or that she leaves aside the rules and academic curricular proposals. *Horizontality* in the teacher-pupil relationship it means "treating all the pupils warmly" (Ottobre y Temporelli 2015, p. 52).

Scene 4: playing is our first learning model

It is October in Buenos Aires and in the classroom is hot. The pupils seated at their desks fan themselves with their hands. Lila approaches the computer and plays music tuned to 432hz. It is loud. She walks at the side of the classroom and ties her hair into a high ponytail. She keeps walking towards the back of the classroom and raising her arms towards the ceiling, she walks towards the boards making a space between the desks. When she reaches the board, she lowers her arms, turns to her left and makes another space between the desks, walking again towards the back of the classroom with her arms raised. The classroom is divided into three horizontal lines of desks with two aisles that separate the rows. Once the desks are arranged in the classroom, Lila closes the windows and turns on the air conditioning.

It is important the transformation and flexibility of the physical area of the classroom, in addition to pedagogical knowledge, teachers adequately distribute the areas inside the classroom. A well-distributed classroom allows adequate communication between the teacher and pupils, pupil-pupil interaction, learning of personal privacy and improves teaching processes.

Lila kindly invites the children to close their eyes or to leave them half closed and to rest their heads on the desk to perform the ritual (Schuchner 2016), to repeat the rules, as they do at the beginning of every class.

Lila: (*Happily sings*) We are going to have a beautiful day! (*She finishes the sentence with a higher tone*) Now, everybody look at me.

She starts walking with quick steps making a circle around the classroom while snapping her fingers creating a rhythm

Lila: I have good news and bad news.

Rhythmic game influences the emotional brain. Rhythm regulates life, it creates a predictable and stable pattern (S. Jennings 2011). We are incubated in rhythm; our mother's heartbeat is the first concert in the womb. There is an innate connection to rhythm. "The rhythm and their repetition in games, songs and rhymes connect the child with their own inner rhythm and, thus, they can grow in harmony" (Öfele 2014, p. 79).

Lila: (*She stops at the front of the class, puts the fists under her chin*) The bad news is that next Thursday you will have a test.

Voices of : the children Nooooo!

Pupil: Bad luck day!

Lila: And the good news (*she starts moving her hips gently and dancing*) is that on that same Thursday is my birthday!

The pupils are very happy with Lila's birthday, they stand up and jump. Some girls talk to each other quietly sharing something in secret.

Lila: (*She walks to the computer, lowers the music and writes some words on the computer so that the kids can see them on the digital board*)

The words: gardening, chatting, eating, walking the dog, playing football, jogging, shopping.

Lila: (*With a hand on her hip*) Now, listen to me. You probably know how to play charades. (*She winks an eye and mimics as if she were walking a dog.*)

The pupils: (*shout*) Walking the dog, walking the dog!

Lila claps and invites a pupil to play. She secretly tells him the word he has to perform. Lila and the pupil pretend they are playing soccer.

The playful spirit of the teacher is important for the pupils to play. In order to accompany the play of the children, teachers have to set in motion their own playful ability; it means that both the teacher and the pupils must be fully involved in this activity. Highlighting the idea of playful ability as a framework and support while children learn (Öfele 2009).

As mother-child relationship is built by playing, exchanging looks, physical movements and swings, songs and vocal rhythms; and in that playful meeting babies discover themselves and mothers come to know and discover their babies, when teachers and their pupils "are enjoying themselves" together in the play, a bond of trust that enables personal, social and learning development is generated. Stuart Brown writes "play activity is actually helping sculpt the brain. We make new cognitive connections that find their way into our everyday lives" (Brown 2009, p. 34).

Scene 5: Lila's birthday: celebration and ritual play are one of the types of social play

Lila enters the classroom while the pupils are still in the break. She approaches the computer and plays music tuned to 432hz. She writes on the board: rules, reading, video, writing, party. She goes back to the classroom door to wait for the pupils.
Lila welcomes them and kindly invites them to enter the classroom as they arrive from the break. The pupils are glad to see her waiting for them at the door.

Children have an innate biological tendency to create an attachment bond with the adult who spends time with them and cares for them; Daniel Siegel says that "Attachment is an innate system of the brain designed to preserve the safety of children and that leads them to seek *proximity*" (Siegel 2012, p. 155). That is why they will also seek to build an attachment relationship with the teacher.

Once all the pupils are sitting at their desks, Lila, with a high tone of voice, pretends to make a "vibrato".

Lila: Please, close your eyes. Relax (*she raises her arms with an inspiration and puts them down while exhaling*), relax.

This time, Lila says the rules rapping. The pupils repeat the rules with the rhythm proposed by Lila.
When they finish saying the rules, Lila approaches the computer, lowers the volume of the music and leaves it as background music.
They start doing the activities in the syllabus written on the board. Once they are done, Lila takes a deep breath, puts a hand on her heart.

Lila: Now we can celebrate my birthday!

An important characteristic in a safe teacher-pupil relationship is confidence, it gives pupils peace of mind and a secure background that provides them with an internal feeling of well-being that also enables them to explore.

The children, happily, put their school stuff in their backpacks and ask Lila to leave to prepare the classroom for the birthday. Some girls place on the wall a garland in the shape of a pennant written with dedications from the children. Other girls count the number of plastic cups and put them next to the bottle of Coke. Some boys hang the garland that says Happy Birthday.

Lila: (*Standing in the doorway looking up pretending she is absent-minded, she asks*) Can I see?
Her joy was as pure as that of a child.

Pupils: Noooo, nooot yet!

Pupils: (*Whispering*) The cake, the cake, we have to light the candle.

Figure 2.2 Birthday cake

They carefully light the candle, start singing Happy Birthday and Lila comes closer.
Lila: (*She looks to all the children around her with affection*) I love you!

These social rituals are important for children because they store good memories and this helps them develop the taste of social ritual as adults (Brown 2009, p. 91).

After blowing out the birthday candle, the pupils stand in front of the green board with a sheet of paper in their hands and begin to sing a song they had written.
Believer
We love you
You are the best teacher in the world you are very kind happy birthday ouuh
We love you you're beautiful ouuh
You are an example, no mistakes we love you, you are very pretty There are no such good people and WE ARE GOING TO RAP LIKE THIS
This year shared with you is amazing
Amazing
Ouuh
You are kind, funny, very funny

You are wonderful
You are . . . the best teacher in the world
Worldwide!!
You sing very well too well
We love you
Sing us as if it were the last time
Come and teach us those lyrics again
Taki taki x2 Rumba!

Lila: (*When they finish singing, she puts her hands on her cheeks, opens her mouth with amazement and emotion*) I love you so much! (*All the children approach her and hug her. One of the girls gives her a present. It is a letter that they wrote in English. The children watch Lila as she reads it silently.*)

In a safe teacher–pupil relationship, children spontaneously hug the teacher and seek to read the face of the teacher to receive information, just like babies do with their mother. That is, it is more likely to develop a secure attachment relationship between teacher and pupil when the teacher engages sensitively and has frequent positive interactions with pupils.

Pupils: (*When Lila finishes reading the letter they start singing*) Lila, Lila, Lila!

The harmonious understanding of minds begins with the visual contact between the adult and the child, and what happens in the brain is even more incredible. When the eyes meet, as happens between a mother and her baby, they synchronize neuronal activity in the right cortex of the brain. "This is attunement. Their brain rhythms are getting in tune, performing a kind of meld that is a very pure form of intimacy" (Brown 2009, p. 82).

Lila (approaches the observer): I am so happy that someone has witnessed this.

Scene 6: interviewing Lila – teacher's attachment history

I was born in 1992 in a family where education has always been crucial; my mom and my dad, as well as my sister and brother are teachers and work in different levels of schooling.

My mom has always been a great help, mainly with her example of perseverance and patience; it is from her that I've learned, above all, loving and inclusive teaching. I love to play to imagine if I were the girl that is sitting in the class and if I were my own teacher, would I like to be in class? That's a "key" question for me to be able to create and put imagination at the service of education. The school is the place where I feel happiest and I come to contribute, as I can, to the training of people who value, respect themselves and, therefore, can respect others. I am in school, quoting the Brazilian pedagogue and educator Paulo Freire, with the intention of "educating for freedom". I think that 21st century children have the inner freedom to be curious and to question everything; I also think that you can learn while having fun, I always try to create a space for joy in my classes, I honestly get frustrated when I cannot get it. "Those things you learn without joy you will forget easily." – A Finnish saying, and something very important, nothing remains the same after playing

(*Teacher Interview*)

That's right, playing is not against learning, it goes with learning. It improves knowledge retention, playing is "like fertilizer for brain growth" (Brown 2009, p. 101). The authentic, pure play originates from an innate motivation from the deepest part of our being and arises from the strength of imagination that is inside. "Without play, Panksepp suggests, optimal learning, normal social functioning, self-control, and other executive functions may not mature properly" (Brown 2009, p. 100). The classroom is one of the spaces that replaces the play between the baby and the mother. That is why some researchers link the secure attachment histories of teachers with their primary caregivers to the quality of the teacher-pupil relationship. Play makes it possible to open the door to the relationship with another person and facilitates group relations and communication.

Last scene: interviewing the children – qualities of an attachment-focused teacher

"She helped me in English. I learned a new language".
"It helped me with my behaviour, to pay attention".
"The way she talks, she says: come on, pay attention that the class is great and if you are getting bored I can look for a game so you can learn having fun".
"Sometimes she sings to ask for silence".
"She asks us what we like and then relates it to a subject that we have to learn in English".
"She makes learning fun".
"If somebody cannot do something, she encourages him/her. For example, she tells personal thing to help solve it".
"Lila turns boring things into fun".
"When she is in a bad mood, she also plays".
The end

The following poem is a tribute to the teacher and the pupils that allowed me to write this chapter:

Play with Me

I tried to teach my child from books;
He gave me only puzzled looks.
I tried to teach my child with words;
They passed him by, oft unheard.
Despairingly I turned aside;
"How shall I teach this child?" I cried.
Into my hand he put the key;
"Come" he said, "play with me."
~ *Author Unknown*

Notes

1 **Divergent thinking:** The physician and psychologist Dr Edward de Bono is the creator of the term "divergent thinking" and it means the use of a creative thinking.

2 432 Hz Frequency of Nature's Music. The frequency with which our body naturally resonates. The music tuned to 432Hz automatically becomes more calm and soothing. Those positive vibrations and healing energy for 432Hz music helps in aligning our energy, our vibrations to that of earth and nature.

References

Brown, S. (2009) *Play: How It Shapes the Brain, Opens the Imagination, and Invigorates the Soul.* New York: Penguin Group.

Duarte D., Jakeline. (2003) Ambiente de Aprendizaje: una aproximación conceptual. *Estud. pedagóg.*, no. 29, pp. 97–113, citado 12 Febrero 2019, World Wide Web, [online], http://mingaonline.uach.cl/scielo.php?script=sci_arttext&pid=S0718-07052003000100007&lng=es&nrm=iso, ISSN 0718-0705

Jennings, P., & Greenberg, M. (2009) The Prosocial Classroom: Teacher social and emotional competence in relation. *Review of Educational Research*, 79, no. 1, PDF Available.

Jennings, S. (2011) *Healthy Attachments and Neuro – Dramatic – Play*. London: Jessica Kingsley Publishers.

Lieberman, J.N. (1977) *Playfulness: Its Relationship to Imagination and Creativity*. New York: Academic Press.

Lynch, M., & Cicchetti, D. (1992) Maltreated Children's Reports of Relatedness to Their Teachers. In R. C. Pianta (Ed.), *Beyond the Parent: The Role of Other Adults in Children'S Lives: New Directions for Child Development* (pp. 81–108). San Francisco: Jossey-Bass.

Öfele, M.R. (2007) Jugar, crear e imaginar. *Newsletter "En Juego"*, 5, no. 3, www.instituto.ws/iifj/esp/cap09.htm.

Öfele, M.R. (2009) Acerca de la capacidad lúdica. *Newsletter "En Juego"*, 7, no. 3, www.instituto.ws/iifj/esp/cap09.htm.

Öfele, M.R. (2013) Juego, humor y creatividad en el aula. *Revista Novedades Educativas*, 25, no. 272, Agosto.

Öfele, M.R. (2014) Juego, ternura y encuentro. Fundamentos en la primera infancia. In *Play, Tenderness and Encounter: Fundaments in Early Childhood (Espacios en Blanco: Revista de Educación, 24* (pp. 71–80). Buenos Aires: Universidad Nacional del Centro de la Provincia de Buenos Aires.

Öfele, M.R. (2017) *El valor del juego en el aprendizaje infantil*. Lima: Congreso Internacional "Cuando la Educación sí es un juego", www.perueduca.pe/recursosedu/ponencias/valor-del-juego-en-el-aprendizaje-infantil.pdf.

Ottobre, S., y Temporelli, W. (2015) *¡Profe, no tengamos recreo!* Buenos Aires: Ediciones La Crujia.

Scheines, G. (1998) *Juegos inocentes, juegos terribles*. Buenos Aires: Editorial Universitaria de Buenos Aires.

Schuchner, G. (2016) *Routledge International Handbook of Dramatherapy*. London: Routledge.

Siegel, D.J., y Hartzell, M. (2012) *Ser Padres Conscientes*. Barcelona: Ediciones La Llave.

3
CULTIVATING PLAYFULNESS IN THE WORKPLACE

Ankita Khanna

For the longest time, the concept of playfulness was associated predominantly with children, childhood, and child-centric spaces. In recent years, however, an emerging narrative of bringing 'playfulness' into the serious business of adult work and work places has been gaining prominence. As George Bernard Shaw said, "We don't stop playing because we grow old; we grow old because we stop playing".[1] But what do we mean, really, when we talk about 'play' and 'playfulness'? Is this simply about bringing 'play props' and opportunities for play into our adult lives, or is it something else, something deeper, and more intangible?

Just like concepts such as 'art' and 'happiness', play has an inherently experiential quality that makes it almost impossible to define. Play is and can be so many things – but we recognise it best when we see or experience it. Researcher Stuart Brown, MD, describes play as "time spent without purpose" (2009). In one interview, he says that

> Play is an ancient, voluntary, inherently pleasurable, apparently purposeless activity or process that is undertaken for its own sake and that strengthens our muscles and our social skills, fertilizes brain activity, tempers and deepens our emotions, takes us out of time, and enables a state of balance and poise.
>
> *(Brown, 2009, p. 412)*

Some of the other words and characteristics smattered across the play literature that I came across included curiosity, spontaneity, vulnerability, and silliness. Further described characteristics included a sense of timelessness, freedom, joy, and the sheer lightness of being. The aspect I resonated with most and that rang most true for me was that "Play is a state of mind, rather than an activity" (Brown, 2010).

In a recent article titled 'Forget IQ & EQ, It Is All about PQ' in *The Indian Express*, Dr. Shelja Sen, child and adolescent psychologist, stated, "When I talk about being playful, I mean getting in touch with the free, spontaneous child within each one of us, and just being uninhibited" (Sen, 2019). Since the predominant source of data for this chapter is meant to be anecdotal, I also asked several of my colleagues at Children First, a multidisciplinary child and adolescent mental health service in Delhi, India, what they meant by 'play' and 'playfulness'. Some of the responses that stood out were:

Y: Being myself; not taking myself or others too seriously! Not allowing myself to be defined by someone else's expectations. Pranks, humour. . .

K: It means an element of surprise and wonder, and a little something 'extra' that usually touches the happiness, or excitement, spot in me. Or just makes me discover a different dimension of my own emotion.

A: It's not about the activity but the joy you feel within, while doing it. It could be an exaggerated voice or a tickling game, which brings forth blissful contagious laughter. . .

The essence of play through the lifespan

So – is play only for children? Not at all – the literature on this seems to shout unanimously! Let us try to examine the conviction of this rationale a little more closely. Brown (2010) has spoken about the concept of 'neoteny' (Brown and Vaughan, 2010), which offers an explanation for this phenomenon from an evolutionary lens. The word 'neoteny' comes from the Greek word for 'stretch' or 'extend' and essentially means the 'stretching' of juvenile or childlike periods, and sometimes the retention of juvenile characteristics in adulthood. This essentially means that our biology has designed us to keep playing throughout our lives, which is not the case in many other mammals, whose adulthood is centred far more around mate selection and species dominance, after they attain sexual maturity. The ways in which neotenous humans are wired to play throughout their lives include ongoing curiosity and openness to change and newness, and the ability to keep receiving and incorporating new information to what we already know. In fact, the complexity of play actually increases along with the increasing complexity of the brain.

Conversely, several studies have shown that our brains actually start to undergo degeneration if we stop playing. Other documented effects include depression, rigidity, and a lack of adaptability. In this sense, then, lifelong play seems to be central to our continued health, wellbeing, and sanity as adults!

If play is so important, and we are both *wired* and *able* to keep on playing, why do we stop doing it as we grow older? One reason seems to be that we just forget to play as we grow older – this wasn't the case culturally in ancient times though. While contemporary representations of play mostly feature only children and seem to lack adults, this wasn't the case in ancient cultures. In one TED talk, Brown (2008) shows us a 15th-century courtyard picture that depicts adults engaged in all sorts of play – from body and rough-and-tumble play to object-based, imaginative and social play, and compares this to a recent magazine cover on play that exclusively features pictures of children. In the Indian culture as well, reminiscences of the past often include enactment of folktales, storytelling, songs and drama, or adults and children getting together in the evening to play a '*gully*'[2] sport like cricket, *gilli-danda*,[3] or *kabaddi*;[4] a trend that is rarer now.

The other reason for the gradual decline of play from adult life and discourse seems to be the narrative of work-versus-play, which are often seen as opposites of each other. In a culture that glorifies productivity and busyness, many of us seem to feel that playfulness may be seen as slacking off, or not being serious enough about our work. Finally, the time that many of us as adults are making for 'play' in our lives, and more importantly the way we are approaching it (e.g., "let me plug in a game of tennis between these two meetings"), ends up feeling more like work than play! As adults, "we are going round and round in the journey of our life, missing out on what matters the most – playfulness" (Sen, 2019).

Having said that, it is also important to point out that play is very much present in adult life – it's just that its forms vary as we grow older, as do our preferences. In fact, much of our cultural life in terms of drama, film, music, and the more recently popular stand-up comedy, is an outcome of this neotenous drive to (continue to) play. Many people agree, from personal experience, that the play drive is very much alive in adults, just a little jaded and hidden under the piles of complexity and adult responsibility we seem to walk around with, but waiting just under the surface to be tickled.

K: I think it's an innate part of us actually, and we strive to and are taught to suppress it – which is effortful. Letting it out keeps wonder alive – and makes us feel more "present", which creates joy for us, and for others around us.

Bringing back play

There is an emerging narrative that is recognising the evidence and value of holding on to, and bringing back, play into the adult world, especially in the workplace. This movement is based on the innumerable documented and experienced benefits of play and playfulness. First of all, play is simply pleasurable! It energises and relaxes us at the same time. It refreshes us by lifting us out of the mundane. In play, we explore, experiment, adapt, problem solve, take risks, and make discoveries – all skills highly valued and often used by us as working adults as well. Play adds a touch of lightness and humour to an otherwise dull day. And most importantly, the lightness and enthusiasm it generates is infectious! It is one of the very few things we do in life without an agenda, simply for its own sake. It is also known to keep both our bodies and minds active, keep our brains flexible, sprout new neural connections, and to slow down the aging process.

Brown (2010) advocates that play is central to creativity, innovation, and wholehearted living, and Brian Sutton Smith goes as far as to say that, "The opposite of play is not work; it's depression".

One place in which play is emerging as a trend is in the area of corporate innovation. This is because science and research clearly indicate that play leads to a heightened ability to handle tasks that require complexity, fosters cognitive flexibility and adaptability, and leads to higher levels of creativity. When faced with a dead-end while problem solving, play allows for just enough emotional distance from the problem for us to approach it in a novel way, often leading to breakthroughs and creative solutions. These are all core demands in the workplace of today.

To this end, there has been a rise in the concept of the playful workplace in recent years. Companies are bringing in DJ turntables, ping-pong and mini football tables, and also encouraging opportunities for social play and exchange through co-working and shared work spaces. Google®, of course, is well known for its diverse array of playful offices around the world, each with its unique 'play culture' – including razor scooters, Lego™ stations, pets, and whimsical furniture, to name just a few of its features.

Is there more to it?

The question is whether this is all it takes. Is providing opportunities and props for play enough for us to be able to reap its many benefits, or are we missing something here? There were two themes that stood out for me in the literature, as well as in my experience of being part of what we consider a workplace that is "high in PQ" (The term PQ, short for 'playfulness quotient' was itself coined by our co-founder Dr Shelja Sen and first documented in her book *All You Need Is Love*.)

Firstly, to make any of these ideas effective, what is far more important than the ways provided to play, is the culture and ethics that supports play-based practices, i.e., the substrate in which playfulness is encouraged and held. Which brings us to the second point – this culture needs to be able to believe in, and look at, ways in which we are in touch with our playfulness quotient (PQ) on a daily and hourly basis, such that play literally infuses the days of our lives. This cannot happen if we are still stuck in a work-play differential that tells us to take 'play breaks' between work.

In this sense then, a true ethic and culture of playfulness would essentially be based on the premise that:

- Play is a biological innate drive that we are all born with, in other words, play flows through our veins.
- All of us are experts in childhood and therefore in play, as lived experience.
- Continuing to play means continuing to reap its many, many body, brain, social, and emotional benefits for the rest of our lives.

So where do we start?

As promised, I will attempt to round off this chapter by sharing some anecdotal examples, illustrations, and quotes gathered from my observations, as well as an informal survey, at our workplace, Children First®. Here I have tried to weave together the themes and ideas that we believe roots our workplace in a culture of playfulness, and how we come together as a community to create and sustain it.

Culture of playfulness

One of the themes that came through in the survey was that of a culture and ethic of playfulness being far more important and fundamental than an abundance of playful ideas and opportunities alone.

T: I feel like treating each other with respect and as equals creates a space where playfulness can grow and where whimsy is appreciated.

A: Everyone has their own quirks and is able to act them out. As a team we keep the PQ high by joining into other people's playfulness. So if one person starts to sing an old/funny jingle or dance to a 90s song, eventually everyone joins in and it becomes a collective experience.

A1: What motivates us is the trust that everyone shares, which gives us the freedom to be silly and just be . . . if there were a constant demand to only do the right thing then this might not be. . . . In one of my mentoring sessions I was actually told to explore and make mistakes.

Another important point that came up here was that of emotional safety being a prerequisite for a culture of playfulness:

Y: Spontaneity and flow are important . . . as long as you are not hurt and you don't hurt others in playfulness!

Infusing work with play

We spoke earlier about how moving beyond the work-play differential is central to truly reaping the benefits of play at work. We believe that work and play can not only co-exist, but also that each enriches the other.

H: One of my favourite things about my workplace is how we try to mindfully balance the more serious, gritty nature of our work with playful light-heartedness and humour. The playfulness comes in little ways – and stays with me throughout the day. It reminds me that taking yourself lightly is the best way to move forward and bounce back from a difficult session, or an unexpected/unpleasant outcome. Playfulness is woven into the fabric of our "official" rituals as well – whether that be our team meetings, supervision, or even our interactions with the front desk. What helps to sustain this is that it is a shared value, and it is treated as such. People both give and expect playfulness which helps in creating this kind of communication.

Celebrations, festivals and milestones

For centuries, celebrations and festivals have provide a culturally sanctioned avenue for reconnecting with playfulness. At our workplace, we give it a few creative twists, such as celebrating major milestones of team members by making funny movies about them, enacting little skits, doing mimicry, and participating in get-to-know-your-team games or a treasure hunt that involves seeking out corners of the workplace you wouldn't usually visit. Our team day-away usually involves going away to a naturalistic setting where we sing, dance, joke together, and also play actual sport. It is moments like these that cut across hierarchies; I vividly remember a particular game of cricket in which our senior director got bowled out by the most junior member of the team, leading to whoops and much laughter on all sides!

T: All the jokes, humour, and pranks during the team meetings! All the parties and celebrations with games where everyone is an equal!

All of us are wired and inspired differently

One of the core principles of our work is that each child is wired and inspired differently (Sen, 2015), and so is each one of us. That also means that the kind of play breaks that might work for one person may not work for another, or one person's idea of playfulness may not vibe with another's, and that's okay. We constantly strive to create a non-judgemental and inclusive space for everyone to 'play in their own way'. This could mean listening to music while working on a report, taking a walking break, or brainstorming a concept by watching a YouTube™ video about it.

H: The permissiveness and trust that despite me browsing Reddit® or listening to music, I'm going to be committed to working, is something that's helped me be accountable to myself and measure and monitor my own time. It's also helped me to figure out the best working configuration so that I can sustain myself for longer periods of time without feeling burnt out.

Beating stress and staying together

Most importantly, we feel that play fosters connections and allows us an opportunity to connect in a way we wouldn't otherwise. What's more, we feel sharing playful times with each other also makes it easier for us to be there for each other during the more difficult times, while also keeping stress and burnout at bay. In fact, there is a huge body of evidence indicating that playfulness brings down the cortisol levels (stress chemicals) and enhances happiness chemicals like endorphins and oxytocin.

A: We co-create moments of joy and togetherness by teasing each other on the one hand (you will often see us running after each other in the staff room or playing harmless pranks) but also being caring, empathetic and giving space when it's a difficult phase

Play above all

Play is a great leveller – it cuts across hierarchies and differentials in power like nothing else can. When playing, we are all equal – equally vulnerable, equally silly, and equally together. There are no seniors and juniors, bosses and employees, when we are tossing a ball together. As we engage in a lot of playful banter and verbal repartee at our team meetings, it is heartening to see how the best joke can be made by the junior most member and appreciated by the most senior one. As one team member put it, "It is great to see our seniors showing their playful side and being involved in the craziness . . . it tells us we can do it too and won't be judged for it".

To wrap it all up, possibilities for play are all around us, inside of us as well as on the outside. Being aware of our own, and responding to others' 'play signals' (Brown, 2008) – the twinkle of an eye, the cocking of an eyebrow, the hum of a familiar song, is the true honouring of these possibilities. So come, let us not resist the urge to be silly and playful; after all it is a murmur bubbling just under the surface, waiting to be rediscovered. And once it is, it has the power to sweep us all into its contagious swirl of joy, making the days of our lives so much more memorable and worthwhile.

Notes

1 Attributed to George Bernard Shaw.
2 *Gully*: In Hindi, this means a narrow street or alley. *Gully* sports in India typically include cricket, *gilli-danda, kabaddi*, etc. (the last two explained next).
3 *Gilli-danda*: A common *gully* sport played in the rural and small towns of India and some other parts of the world; the game is played with two sticks – a large one called a *danda*, which is used to hit a smaller one, the *gilli*.
4 *Kabaddi*: Another popular game in India; it is a contact team sport involving two teams of seven players each. Often played at a competitive level.

References

Brown, B. (2010). *The Gifts of Imperfection*. Center City, MN: Hazelden.
Brown, S. (2008, May). *Stuart Brown: Play Is More Than Just Fun* [Video file]. Retrieved from www.ted.com/talks/stuart_brown_says_play_is_more_than_fun_it_s_vital.
Brown, S. (2009). Discovering the Importance of Play through Personal Histories and Brain Images: An Interview with Stuart L. Brown. *American Journal of Play*, Spring, pp. 399–412.
Brown, S. and Vaughan, C. (2010). *Play: How It Shapes the Brain, Opens the Imagination, and Invigorates the Soul*. New York: Avery.
Sen, S. (2015). *All You Need Is Love*. New York: Harper Collins.
Sen, S. (2019). Forget IQ & EQ, It Is All About PQ. *The Indian Express*, August 24.

4

PLAYING MEMORIES

Exploring the notion of play amongst participants of participatory theatre in Cameroon

Paul AnimbomNgong

Introduction

Theatrical playing is an activity with intense pleasure that allows the participant to discover unsuspected potentials. It is learning how to explore different states of mind, the nuances of the whole palette of human feelings. By its very history and its device, theatre can offer qualities of openness to the world, imagination and appetite for communication. Through play, theatre becomes a complete art form because it involves the body, voice, soul and space in a collective device.

In a therapeutic area, theatrical play helps to develop self-confidence and to assert self-expression. For this reason, theatre is considered as a form of expression wherein the body, space, words and situations are explored in a playful manner for therapeutic purposes. Based on this premise, it is considered that the use of theatrical play in therapeutic settings was born from Vladimir Iljine (Jones 1996), though this can be argued since the notion of therapeutic theatre based on play can be traced back to African and Greek performances which existed long before Iljine. Despite Iljine's first modernisation attempts, it was Moreno (1889–1974) who, a few years later, formalized the practice. Influenced by Stanislavski, Moreno is at the origin of the practices and terms like "psychodrama", "sociometry", "group psychotherapy" and "theatre of spontaneity" (Kellerman 1992; Bannister 1997). Among these practices, psychodrama exploits the theatrical device the most.

The goal of psychodrama is presented as the true organisation of a form, a creative self-realisation in action, a structuring of space, an understanding, a seizure of human relations during the action of the stage. The theatrical play in this context, occupies a place of choice. It is used in the same way as in many other therapeutic practices. In a therapeutic space according to Didier Anzieu (1975, p. 53), the theatrical play allows the actor/participant to abandon all learning of a pre-established text and any search for an effect on the public, to play his own character or a fictional character whose result is on his own account, so that he can better understand himself and make himself better understood by those with whom he lives or works, where he agrees to change at the same time as knowing who he is.

From this point of view, it is noted that the challenge of linking "therapy and theatre" goes beyond the possible use of theatre in therapeutic practices. It is about the very use of theatre as a therapeutic process. The theatrical play in this case uses psychological mechanisms capable of impacting the personality and operating in its transformations. The case is even more appalling when it comes to working with participants in a precise context. Working in a community setting, following a participatory method of performance, theatrical play can carry participants into a therapeutic journey whose results contribute to a positive change of behaviours. A number of questions can be asked to better understand the issue of theatrical play in a therapeutic setting amongst participants.

What is play and what is its relationship with the theatre? Can a performance in which participants are involved in playmaking be considered therapeutic? How is such a performance conceived, structured and performed? How do participants undergo the self-liberation process? These questions, formulated in their broadest sense, would direct reflections to return to the old debate on catharsis (of which we know the medical etymology) in the theatre. Without seeking to dig deeper into this debate, this write-up aims to highlight the issues and perspectives of theatrical play in a therapeutic setting while taking into account its therapeutic capacity. It draws examples from two participatory theatre workshops organised in Mbalmayo and the University of Bamenda Campus, Cameroon, and portrays how participants reconstruct their past through playing in order to achieve self-purgation. It looks at the use of play and how it is integrated into a theatre production and how this latter can then become therapeutic though this might not have been its primary objective.

Play

The notion of play is wide and has varied implications attributed to it. It goes without doubt that playing is one of the most meaningful aspects of human life from childhood. This is because playful activities are essential for development, permit the adaptation to new situations and can help someone to maintain and/or improve health. It provides pleasure, relaxation, and favours spontaneity. In this light, it can be considered as any activity in which participants spontaneously engage and find pleasure (Vessey and Mahon 1990).

Play can be classified as a form of recreation, entertainment and distraction; stimulation towards the development of cognitive, social, creative, sensorimotor skills; socialisation in the society; and as a therapeutic or cathartic instrument. Used in relation to theatrical play, therapeutic play is a notion that refers to specialised activities that are developmentally supportive and facilitate the emotional wellbeing of participants. It is a form of language that facilitates expression or communication in the therapeutic space that is created. Play gives the participants the possibility to understand their situations and interact better with the context. When participants involve in play, they are given the possibility to express themselves, develop their creative skills, interrelate, interact and use their imagination to dramatize situations.

Imagination and improvisation are vital in theatrical play development. As such, when involved in activities of dramatization or theatricalising, participants are in no way mechanically imitating or reproducing situations. They are engaged in activities of imagination and improvisation wherein they actively and immediately reshape the communication they are producing and receiving, thoroughly transforming them in a manner that tends to invalidate the idea of a neutral encoded message being decoded by a receiver. By this, I argue that the participants in theatrical play are not simply involved in the decoding process but in a somewhat making or re-making of communication through improvisation. This implies that participants need to have a particular focus or objective when embarking on a therapeutic journey. As seen in the under-studied workshops, the conditions the participants are in caused them to set particular objectives which they intend to achieve through playmaking. The use of play here permits the relief of the anxiety caused by their various psycho-social conditions. The results of this intervention are therapeutic, and discussing the therapeutic effects of theatrical play inevitably requires a look into the place of catharsis in the theatrical form.

Participatory theatre and catharsis

Catharsis employed within the domain of therapy intends to purify and purge emotions through art or to evoke a change in emotions that results in renewal and restoration. It is traced from Aristotle in the *Poetics* when he used catharsis to describe the effects of tragedy on the spectator. From the *Poetics* though, there is no clear definition of what Aristotle meant by catharsis and this has allowed theories and definitions emanating from theorists to be different and sometimes contrasting.

Inspired from Aristotle, Freudian psychoanalysis grew to employ the concept and understandings of catharsis. It focuses on the idea that insight into troubling emotional issues and healing occurs only after the patient has achieved catharsis though this is disputed by cognitive-behavioural therapists and others. To them, catharsis and insight are not enough to induce healing or change and therefore should engage new thoughts and behaviours to replace the old. This is not an immediate process with immediate effects but it is a gradual process towards change. It is this process that participatory theatre advocates in its application.

Cathartic feelings in a theatre production are experienced empathetically for the characters in a play by the audience. This theatrical or cathartic experience is shared by the entire audience-participants and by so doing, magnifying release and allowing for an adjustment or change in the entire community's attitude as a whole. According to Thomas Scheff (1979), catharsis is also likely to be derived from other forms of theatre or artistic forms other than tragedy as held by Aristotle. Scheff's attribution of catharsis to other forms of theatre is of interest to me as it points to other areas in which cathartic effects can observed. Catharsis in participatory theatre therefore is not unidirectional but bidirectional because it touches both the onlookers and the looked.

The interpretation and approach of catharsis within therapy-related or therapeutic activities is different. This started with Josef Breuer, who developed a cathartic treatment using hypnosis for persons suffering from hysterical symptoms (Zangwill 1987, p. 118). In this treatment, patients recalled traumatic experiences and were relieved through expressing the original emotions long repressed and forgotten. Moreno on his part developed two forms, a catharsis of abreaction – a purging of the emotions closely related to the Aristotelian model – and "a catharsis of integration, a moment of insight allowing the individual to process intense feeling and find a way to relate it back to the group they are working with, and their own circumstances in everyday life" (Landy 2008, p. 136). These are just two utilisations and approaches of catharsis amongst many others linking this notion to various domains, one of which is drama or theatre.

Joe Winston in his book *Drama, Narrative and Moral Education* (1998, p. 60) presents a definition of catharsis in Boal's view as "a form of correction and purification" with pity and fear being the means, and not the end of the process. Though at this early stage of his career, Boal seems to give a definition of catharsis based on medical analogy, as a purgative process intended to maintain the emotional balance of the individual, his later development of "*The Boal Method of Theatre and Therapy*" gives a more broader comprehension to this notion with regards to theatre and precisely participatory forms of theatre when he states:

> In theatric psychotherapies what is important is not the simple entry of the human body, but the effects of the dichotomy which the aesthetic space brings to bear on that

> body and the consciousness of the protagonist, who becomes, on stage, subject and object, conscious of himself and his actions.
>
> *(Boal 1995, p. 25)*

From this statement, the former broad notion of looking at catharsis from the medical analogy only, is considered to be different as Boal lays emphasis on theatre terminology such as stage, subject, object and aesthetic space, which are all elements used within theatre or artistic contexts well elaborated by Holmwood (2014).

Studies elaborating the understandings of catharsis are many and it is not the intention of this chapter to focus on these. Rather, I have used catharsis here to show the relationship that it has with the understanding of therapeutic effects in theatre in particular and participatory theatre in general. When combined with theatre techniques and elaborated into a complete theatrical form, catharsis stops being regarded as a release of emotions and works towards a complete therapeutic modality termed in various ways by their practitioners to fulfil their objectives and fit the context in which it is practised. From these utilisations emanate forms like dramatherapy (drama therapy), psychodrama, playback theatre or theatre for social change, most of which are participatory and in which theatrical play takes centre stage.

Participatory theatre as employed here covers practices referred to variously as applied theatre or drama, community theatre, theatre of the oppressed, theatre for development, role play etc. The practice ranges from work with performance focus to process-based work aimed at personal, group and/or social development. Participatory theatre can take place in a variety of settings like employment, political, social and communities. The practitioners are also from diverse backgrounds ranging from professional theatre performers and directors, dedicated trained facilitators, or professionals from social work or education. For the purpose of this study, I discuss on participatory theatre practised under the guise of theatre of the oppressed by a professional and dedicated trained facilitator.

Data presentation and analysis

This study is based on two participatory theatre workshops organised in two distinct locations by the same facilitator. The first workshop termed The SOS Village Mbalmayo[1] workshop was carried out with children some of whom had been abandoned by their parents at the mercy of this home and some who have lost either one or all parents. Leaving ones family setting to live in a completely different setting with people who are like "strangers" is likely to cause psychological repercussions to a human being in general and to a child in particular. However, for an event to be considered traumatising, it must be uncontrollable, sudden and exceptionally negative. The change of environment to these children is not only sudden but is also uncontrollable because their opinion is not sought before the decision for them to be taken to a closed home is made. The child's biological aspects and social context prior to the separation is what causes him/her to be affected by the severity of the trauma.

The workshop took the form of a participatory type of theatre for social change combined with methods drawn from theatre of the oppressed. It was carried out under the leadership of Emelda Ngufor Samba, a theatre for development specialist who uses theatre as a tool to sensitise and raise awareness in communities and as a therapeutic tool. Revealed by Emelda Ngufor Samba[2] in an interview and confirmed in the video recording of the entire pre-production, production and post-production process, the therapeutic intervention consisted of two major phases.

In the first phase, theatre practitioners (who were to act as assistants in the workshop) came together in the University of Yaoundé I, Cameroon, to discuss and brief themselves on the project. Detail information on/about the SOS Village was given to these assistants so as to enable them to go with some knowledge of the site. Their role as theatre facilitators was also highlighted and they were called upon to maintain that position. They were not to judge or influence the ideas, decisions or opinions of the participants in any way.

The second phase was field-based. The theatre activist (therapist) with her team of assistants (facilitators) went to the SOS Village Mbalmayo, some 60 kms from Yaoundé to spend a period of three days with the participants. During this period, they met with the coordinators of the village, the mothers and the children. Important information was gathered at both levels and their intention to live and work with the children was made known to both parties and the parties concerned accepted to collaborate, participate or permit the good running of the workshop. This notwithstanding, it was noted by one of the facilitators that during the workshop the children will become silent whenever a mother or worker of the village was passing by for fear of being scolded at.[3] The atmosphere of fear reigns amongst the children as they continue to work their way towards accepting their fate in the hands of people they know little or nothing about.

The children/participants of the SOS Village Mbalmayo workshop were a combination of street children and children from poor families who were living in the centre. While in the village, they do not live in isolation. They live in a family situation in houses containing about eight children each. A mother is designated to take care of these children. In the centre, there is a primary school that has been set up for the children to attend. At the secondary level, the children leave the centre/village in order to pursue their education in town. Despite the presence of these, the children are still not satisfied for they know that the people with whom they live are not their parents or relatives. In addition to this is the fact that the children live in constant fear of expressing their grievances and are victims of frequent insults from the mothers who always remind them of their past or of the fact that they have no parents or had been abandoned by these latter.

It is in this light that the children were taken on a therapeutic journey calling upon their imaginative and improvisational power to recreate scenes which they live in their daily lives as well as those scenes that might have marked them positively and negatively. This is in a way comparable with the process of therapy which psychodrama uses to treat its patients. In this workshop, the children, guided by the team of facilitators, underwent the therapeutic process using theatre activities and games. These techniques tap into the appeal that play has for young people. It assists them to overcome feelings of isolation and gain mastery over conflict (internal/external) and anxieties.

The second workshop, dubbed the University of Bamenda workshop, was organised in 2018 within the framework of the Arts Week of the Department of Performing and Visual Arts of the Faculty of Arts at the University of Bamenda. The main participants were students of the department who, like the children in the SOS Village, are undergoing trauma caused by the on-going Anglophone crisis that broke out in 2016.[4]

This workshop with the students was organised with the hope that, through a playmaking scenario, the participants would get to present their opinions, feelings, fears, anger and frustration as regards the management of the crisis by both stakeholders to an audience whose responsibility would be to alienate itself from the problems and give it a more critical and objective reflection. The audience who were not simply spectators, but spect-actors, would be called upon to invade the stage and propose solutions to the problems facing Cameroon/Cameroonians.

The University of Bamenda workshop was facilitated under the theme "Theatre and Peaceful Cohabitation" by Emelda Ngufor Samba, assisted by Tume Fondzeyuf Kenneth. The main purpose of this workshop was to break the deadly silence that characterised the Anglophone crisis, a situation similar to that of the SOS Village where the children were unable to voice out their worries. It was generally known that a vast majority of the inhabitants of the North West and South West Regions were choking in silence: they had the urge to express their opinions in support of or against the struggle but out of fear of grave consequences that could be meted on them either by the government or the liberation fighters, they had been bound to keep their opinions to themselves or discuss them in secret thereby gradually developing mental health problems.

During the workshop, the participants were encouraged to share the different manifestations of the current crisis and lived consequences. Their attention was drawn to the fact that the goal of the process was not to enable them to take sides with any of the parties in the crisis but rather to act as a therapeutic modality in which they would liberate themselves from the trauma in which they are. At the end of the therapeutic journey, they would articulate the facts and propose a way forward in an open-ended discussion with the audience. There was a great need for the students to critically analyse the opinions of both the Ambazonians and La Republique[5] with Paulo Freire's concept of PRAXIS: reflection – action – dialogue, as the guiding principle. Throughout the workshop process, there was a noticeable insistence on *This is what the people say*. There was a deliberate effort by the participants to separate themselves from the declarations, the thought patterns and arguments that were presented. Even in that 'safe space' at the university where one would imagine that students and lecturers would freely discuss troubling issues, it was very clear from the beginning to the end of the workshop that that was not the case. The frustration, stress, insecurity, forced silence and the toil of the virtual imprisonment of the participants were manifested in the final play that was presented to the audience animated by the notion of playing memories. To obtain the desired effects and affects, participatory theatre just like all other forms of therapeutic interventions, takes place in therapeutic spaces.

Therapeutic space

It is important that therapeutic work should take place in a well-defined environment and setting. The therapist/facilitator chooses the ideal setting for the smooth running of the therapeutic activity. This framework is a fixed and structured landmark. Added to this is the workspace, time and accessories that would be used. It should be noted at this point that all therapeutic activities should take place in the same workspace set aside on the same day of the week and at the same time throughout the process. This is to acquaint the participants with the healing process that is engaged through active participation in the theatrical activities.

Neither the therapist/facilitator nor the participant who engages in such an approach can know at the beginning where it will lead the participant. The transformational power thus of theatrical play depends largely on the ambitions of the participants; talents and means available to the facilitator/therapist; and various spatial constraints. The symbolic dimension of theatrical play reinforces the therapeutic effects sought by the animator/therapist.

This workspace is used to define a transitional area, the scenic space that is conducive for playmaking (Attigui 2012). The therapeutic space is an area wherein all the dimensions of the personality of the participant are invested. The latter lives on an emotional level and feelings associated with the interpretation of the incarnated persona and on a social level, s/he maintains multiple relationships with other participants and the facilitator.

According to Attigui (2012), participants must be considered as real actors. Although the theatrical activities that take place in this safe place are taken from a dramatic text or based on improvisation, the emotions expressed are those of the participant. This is a framework which, from the beginning to the end of the therapeutic process, must be clearly defined to maintain its reliability. In order for the activity to remain playful, the relationship between the character and the participant's life is explored only indirectly. All acts produced in this space by the participants take the value of interpretation. For the facilitator/therapist, the important thing is to incorporate into the theatrical play, clues coming from the participant in order to stimulate his contact with the real world. In this way, participating in theatrical play serves as a factor for the achievement of therapeutic effects.

Theatrical play in participatory theatre and its therapeutic potentials

In every participatory theatre session involving theatrical play, the group needs to identify a particular problem to work on, gather, analyse and prioritise all necessary information on/around the phenomenon. The step that follows is to select the participants that will actively partake in the therapeutic playmaking process. Thereafter, the group discusses their objectives, chooses their working methodology, reviews ethical considerations, and selects the therapeutic activities which will be employed. Amongst the list of playmaking activities to be selected are role-play, dramatization, stories, movements, dance, music, theatre games and improvisation. At this stage, the improvisational acts, if chosen, are rehearsed and enacted. Thereafter, the group will have a post-mortem and closure. This scheme is somehow similar to systematic communication described by Shirley Bach and Alec Grant (2009) wherein five stages including assessment, planning, decision making, review and evaluation, and ending and closure are identified.

The understanding of this scheme enables one to note that "dramatic play is neither a style nor a theatrical form, but a theatre pedagogy" (Bernard Grosjean 2010). It is the activity most naturally associated with theatre. Used in therapeutic space, play has a variable number of therapeutic potentials that can be triggered in the participants. Play can be considered as the exercise of freedom in the face of a constraint or an obstacle. In the course of playing, the facilitator gives the appropriate constraints to the group he is dealing with to navigate the various themes which are under consideration in the workshop.

As the participants execute the instructions given to them by the facilitator, they are involved in playmaking which falls under kinesics in semiotics. Kinesics, Besbes (2007, pp. 52–53) holds, is generally defined as "the study of the human body as a means of communication". He continues that "it comprises all the movements that are the result of muscular and skeletal shift including physical or physiological acts, gestures, automatic reflexes, and changes in the physiognomy that occur in the repertory of human behaviour". Understood as such, kinesics can comprise a variety of components amongst which are facial expression, gestures and proxemics. According to Besbes, these components are identical to paralinguistic signs, tightly correlated with their functions in the communicative process: informative, directive, emotive or emphatic but not random or arbitrary. Kinesics are bound-up with cultural determinants and the task for the audience to read into the actor's body and face is a particularly significant exercises, since it involves a process whereby the spectator is exposed to a constant flux of body messages that become meaningful only in the sense that they are taken as a network of interacting signs.

The SOS Village and the University of Bamenda workshops applied kinesics in a variety of ways. In the SOS Village workshop for instance, the participants made use of their eyes in the playing process. While the children/participants were performing, the use of eye contact with the audience was noted. As I observed, some of the participants were avoiding any eye contact with the mothers in scenes which depicted the manner in which they were treated by these latter. The case of one participant, incarnating the role of the mother, who missed her lines because her eyes came in contact with those of a mother sitting in the audience, is just one example. From her look, she was full of fear probably because she was playing the "bad side" of the mother, and as such, thought she might be detested by the mother. At the end of the enactments, the facilitator made it clear to the participants (actors and audience) that is was just a play, though some people might have identified with the characters portrayed therein. This was a way of eliminating any possible threat to the children/participants by their mothers who might feel that the actions were depicting their (mother's) characters.

Gestures are also part of kinesic signs that can be used to enhance meaning in therapeutic or theatrical play. In theatrical exchange, Besbes (2007) states that gestural signs define themselves in relation to deictic ostension. To him, on the one hand, it serves to specify the position of the actor in connection with the context of the dramatic conversation, with his interlocutor, and with the objects of the discourse. And on the other hand, it plays a cardinal role in relating the 'I' of the dramatis persona and the 'here' and 'now' of the dramatic situation to the actor's body and to the stage context. In one of the scenes presented in one skit in the SOS Village workshop, a child steals food from the pot and is punished by the parents. To portray the punishment, which is through correctional beating, miming is used. The gestures portray the lifting of the arm and pretending as if the child is beaten. The image is symbolically represented and the audience decodes the message therein. In yet another mimetic function, the characters represented in some of the skits are old as compared to the participants playing the roles. For their actions to be true, their movements and gestures oscillate between slow and fast movements, upright standing positions and slightly inclined positions. In all or most of these cases, gesture is represented in its highly mimetic aspects, body movement and positioning.

In the University of Bamenda workshop, gestural signs were used in another dimension. This was following Boal's method of image theatre. According to Samba and Tume (2019, p. 107), the participants were able to move through the three stages of *Real Image* (the exact situation of the crisis at that moment), *Ideal Image* (how the participants would like the crisis to be handled by both the liberation forces and the government) and *Transitional Image* (what procedure and method to use to bring both parties to work towards the ideal situation). Because the theme was very sensitive and the discussions had brought to the lime light very strong emotions, the participants felt they would communicate more effectively through images (gestural signs) and songs. To make the play culturally relevant and to enhance its political and therapeutic undertones, it was performed in English and French, Cameroon's two official languages, while the songs communicated messages in some of the numerous national languages the country boasts of. Participants were hereby taken into a therapeutic journey through involvement in the play-making process as spect-actors.

Following this light, Fidler and Fidler (1978) hold that engaging in meaningful activity for a person is all the more beneficial to one's health, as it truly and symbolically responds to one's needs. Warm-up exercises, such as those engaged in the SOS Village and the University of Bamenda workshops, fulfil this therapeutic quest. This playing activity used in both workshops, thus favours the development and the adaptation of the ego whose consequences are observed in the capacities of execution of the individual participant. This view is similar to that of Vandal and Leahey (1995) who state that "the artistic activity offers to meet reality, to deposit a tangible mark and encourages communication". If we consider the learning process that any involvement in theatrical play can engender, one can only highlight the integration of psychosocial practice in this act.

The theatrical play then allows a participant to incarnate a character in a process wherein he represents someone else but uses his own resources to support the interpretation of the character according to the structure of the theatrical framework. The participant is called to bring to life a fictional or imaginary person inspired by his own life. In the improvised dominatrix processes in the SOS Village and the University of Bamenda workshops, the participants expressed tensions and questions that tormented them and, as Patricia Attigui (2012) so aptly puts it, really live the emotions of the fictional characters. In the same vein, Attigui notes that theatrical play has an intrinsic therapeutic value that promotes health and interpersonal relationships. Thus the theatrical play will allow the recovery of self-confidence, long repressed by the participants following their mental or traumatic conditions.

In the therapeutic space created, theatrical play has the gift of creating lasting communication links. The participants who take part in this activity have the opportunity to represent, in practical actions with others and/or for others, significant situations, whether drawn from life, inspired by stories or myths of life, cultural tradition, or that they are invented by the free exercise of imagination or improvisation. Although it is an exercise, the participant, through his replicas and actions, expresses this something in him, long repressed.

Amongst all activities in a therapeutic or theatrical playmaking process, improvisation exercises are the most difficult as they require vigilance, attendance, quick wit and deep reservoirs of knowledge. This notwithstanding, they are also the most enriching as they carry in them various qualities of interpersonal development and wellbeing. By embarking participants in this journey of improvisation, they are directed towards a process of healing. Improvisation in theatrical play hereby permits a participant to transform in his relation with the other. There is an exigency of listening to be able to discern the intention of the protagonist either to accept or to refuse.

The improvisation that takes place in theatrical play is also spontaneous. This evokes spontaneity, which according to Park-Fuller, is "not rigid but ritualistic – not rote but rite". Park-Fuller further quotes Fox (1986) who holds that spontaneity does not only involve "'not thinking' . . . [but] also involves thinking of the highest order, where the narrational and the rational are comprehended in an understanding which surpasses the limitations of each". In these sessions, the participants are expected to make a clear distinction between self and the acted other. In the course of improvisation, if the participants and the impersonated characters are merged (if there is no distinction between the actor and the character), the play or exercise is annulled. An improvisation is not an exercise to act out self, but one to act with self and develop communication and interpersonal skills. The character is drawn from life and not a character of life.

Improvisation in theatrical play can also constitute a channel of re/educating participants who might have gone through any form of psychological trauma or illness. This is because improvisations carried out in a therapeutic space, as in the SOS Village workshop or the University of Bamenda workshop, participate in at least three learning domains as described by Bloom, Mesia and Krathwohl (1964) known as "cognitive, affective, and psychomotor" activities respectively. Within the cognitive range for instance, during introductory sessions, participants are introduced with a self-introductory exercise which enables them to know each other. In this exercise, they are called upon to choose pet names or workshop names with which they will like to be identified. This permits them to feel safe and equal with other participants. According to Marvin Carlson (1990), "names given to characters potentially provide a powerful communicative device for the dramatist seeking to orient his audience as quickly as possible in his fictive world". These names express the participants' feelings in the workshop. These workshop names can also be seen from the direction of roles which participants incarnate in the improvisation process. Commenting on this, Landy (1994) states that

> in choosing a name for a role, a client [participant] dares to look at the connection between a feeling state and a behavioral state. [. . .] The naming, then, forces the issue of examining contradictions between appearance and reality [. . .]; this leads to a search for connection between the ideal and the real.
>
> *(p. 48)*

Engaging in therapeutic improvisation activities also permits participants to develop psychomotor skills at different levels. Participants are constantly ready to learn how to focus, how to give focus to someone else and also how to finalise a scene when begun. They are called upon to be able to detect through somatic thinking and bodies where to move, take note of what is happening around them, and to talk while paying attention to what others say. These are non-verbal, psychomotor skills that have a great value to interpersonal, group, cultural and rhetoric communication development.

However, theatrical play has the potential of revealing the use of communication in all its dynamic potential but does not discard the possibility of risk by engaging in the use of arts within a psychological context (Park-Fuller, 2012). In this context, the participants utilise improvisation as a sort of purgation of passions and transform these into a new mode. The whole idea about using improvisation in theatrical play now becomes that which seeks to give the participants the opportunity and safe place of hiding behind the identity of an "other" to better reveal the hidden parts of "self". Combining the two – art and psychology – is delicate. This therefore needs to be done in safety under the supervision of a therapist, who can control any scene or manifestation of violence or regret by a participant. As such, the improvisational acts do not only enhance wellbeing but also foster the spirit of self-control in the participants involved. These are considered important qualities in creating a sense of unity in the communities of SOS Village and the University of Bamenda (and by extension, the entire nation of Cameroon), which are predetermined by an increase in the interpersonal relationships of the members. In this way, improvisation stands as one of the playmaking activities through which participatory theatre fulfils a therapeutic quest so highly desired by the practitioners.

Conclusion

As it stands, play is seen as an important facet of human development. Being most of the times voluntary and intrinsically motivated, play has various outcomes, like recreation, pleasure and, in the case of theatre, therapeutic. In this latter, it takes the notion of theatrical play and is applied in various settings by practitioners to respond to specific needs that might arise amongst participants working together. Most children partake in play activities, but the manner of engagement in each varies from culture to culture and from an artistic form to another.

As illustrated in this chapter, the form of play that is mostly used in participatory theatre in Cameroon is improvisation. Theatrical play, through improvisation enables participants to develop means of communicating feelings, misunderstandings and concerns in a language that is not limited to a particular linguistic setting. In the quest for a therapeutic outcome, theatrical play provides the avenue for participants to obtain personal development and wellbeing. In the case of the SOS Village workshop and the University of Bamenda workshop, theatrical play acted like a vehicle for eliciting information from participants while also sharing information about what they are experiencing in their various settings.

Notes

1 The SOS Village situated in Mbalmayo, at the outskirts of Yaoundé, is a home in which children, some of whom their parents cannot take care of them and others who have either lost their parents or have been abandoned by these latter, are kept. Women live in this village and are called "mothers". To each mother is given a certain number of children to care for. She is now considered the mother to these children though the children could be heard complaining that they are usually insulted by the mothers as orphans or abandoned children.

2 Emelda Ngufor Samba, in a personal interview with this researcher in 2011.

3 Ngefor Sheila, extracted from the workshop production video.

4 "The Anglophone Crisis" is a development from "The Anglophone Problem" embedded in the marginalisation of the Anglophones in Cameroon and alternatives towards their self-determination, self-reliance, transparent local governance, strong institutions for a democratic statehood. From various ways of calling for attention and seeking recognition, "The Anglophone Crisis" has transformed into an open-armed conflict between Cameroon's military and armed forces and restoration forces dubbed Ambazonian Defence Force-ADF, as those clamouring for cessation and those insisting on a one and indivisible Cameroon are still playing the blame game. Worse still, both parties have denied citizens the right to express their thoughts about the crisis and people feel imprisoned in this situation of forced silence.

5 The term Ambazonia and La Republique are used by the secessionists to make a distinction between The North West and South West Regions on the one hand and the French-speaking part of Cameroon on the other.

Bibliography

Anzieu, D. (1975) *Le Groupe et l'inconscient*, Paris: Bordas.

Attigui, P. (2012) *Jeu, Transfert et Psychose: De l'illusion théâtrale à l'espace thérapeutique*, Paris: Dunod.

Bach, S. and Grant, A. (2009) *Communication and Interpersonal Skills for Nurses*, Exeter: Learning Matters.

Bannister, A. (1997) *The Healing Drama: Psychodrama and Dramatherapy with Abused Children*, London: Free Association Books.

Besbes, K. (2007) *The Semiotics of Beckett's Theatre: A Semiotic Study of the Complete Dramatic Works of Samuel Beckett*, Boca Raton, FL: Universal Publishers.

Bloom, B. S., Mesia, B. B. and Krathwohl, D. R. (1964) *Taxonomy of Educational Objectives* (two vols: The affective domain and The cognitive domain), New York: David McKay.

Boal, A. (1995) *The Rainbow of Desire, the Boal Method of Theatre and Therapy*, London: Routledge.

Carlson, M. (1990) The Semiotics of Character Names in Drama. In *Theatre Semiotics: Signs of Life*, Bloomington and Indianapolis: Indiana University Press.

Fidler, G. S. and Fidler, J. (1978) *Doing and Becoming: Purposeful Action and Self-Actualization*, Bethesda, MD: The American Journal of Occupational Therapy Association.

Fox, J. (1986) *Acts of Service: Spontaneity, Commitment, Tradition in the Non-Scripted Theatre,* New Paltz, NY: Tusitala.

Grosjean, B. (2010) De l'art de la consigne dans l'atelier de pratique théâtrale. In *Cahiers pédagogiques* No. 483 (Septembre–Octobre 2010) – 51, extracted from http://www.cahiers-pedagogiques.com/

Holmwood, C. (2014) *Drama Education & Dramatherapy: Exploring the Space Between Disciplines,* London: Routledge.

Jones, P. (1996) *Drama as Therapy: Theatre as Living*, London: Routledge.

Kellerman, P. (1992) *Focus on Psychodrama: The Therapeutic Aspects of Psychodrama,* London: Jessica Kingsley Publishers.

Landy, R. (1994) *Drama Therapy. Concepts and Practices*, 2nd ed., Springfield, IL: Charles Thomas Publisher.

Landy, R. (2008) *The Couch and the Stage: Integrating Words and Action in Psychotherapy*, New York: Jason Aronson.

Park-Fuller, L. M. (2012) *Playback Theatre, Communication Pedagogy, and Community Engagement: Improvising Third Spaces and Mutable Selves Through Narrative Performance,* extracted from www.playbackcentre.org.

Samba, E. N. and Tume, F. K. (2019) Prison Theatre in Search of an Effective Methodology. In Animbom, N. P. (ed.) *Participatory Theatre and Therapy*, 97–124, Beau Bassin: EUE.

Scheff, T. (1979) *Catharsis in Healing, Ritual, and Drama*, Berkeley: University of California.

Vandal, G. and Leahey, D. (1995) Vers une conception thérapeu-tique de la relation médiatisée par l'activité artistique. *Revue québécoise d'ergothérapie*, 4, 138–144.
Vessey, J. and Mahon, M. (1990) Therapeutic Play and the Hospital Child. *Journal of Pediatric Nursing*, 5, 328–333.
Winston, J. (1998) *Drama, Narrative and Moral Education,* London: Falmer Press.
Zangwill, O. L. and Gregory, R. (eds.) (1987) *The Oxford Companion to the Mind*, Oxford: Oxford University Press.

5

THE *NESS* OF BEING PLAYFUL

Sylwyn Guilbaud

Many years ago, when I was still in my twenties, I wrote down these words: *Play makes the world and the world makes play*. I wrote these words on a piece of painting paper, along a blue wax-crayon horizon. I imagined these words sitting between the visible and the invisible, the present and the as yet to be, the living and the decaying into newness. I tried to fit these words into watercolour brushstrokes of vortexes, but they were, despite their palindromic sensibility, still too clumsy. I spun on an empty night-time beach breathing to feel the way of the dervish, spinning life into matter; yearning to let my body remember the feeling of a dizzying made up *wee-am* dance, which my three-year-old being circled in the garden every day.

The feeling of play's integrity grew inside me, infecting my noticing. Inside myself I turned over the relationship between play and existence, and this thinking turned me inside out and returned me to myself. Recently, rediscovering my original notes from those years ago, I realised that that wondering became the resurgent sap of and for my explorations with play in the following two decades. Here, as I settle to write about the *ness* of being playful, I find myself acutely aware of my sense of play-full-ness, of being full of play's *ness*, of the myriad interweaving instances and unending threads of play that came before my being here.

As I write the words *before my being here* I reach into my metaphoric back pocket, which is filled with collected expressions in which I sense my heart reflected. There my hand finds and draws out Thoreau's perception that 'The maker of this world but patented a leaf' (Thoreau & Crammer, 2004, p. 329). The *ness* that I hope to explore here is a *ness* that runs to and from this poiesis, through the being of the man who wrote these words, weaving amongst the awareness that we call language, philosophy, science, art, and into my heart as a reader; who was once a child, unaware of conceptual description, tracing the pattern of a rose leaf.

This chapter is about my sense of the *ness* of being playful; as such it is about self and the possible role of play within being. It is about the experience of finding reflections of one's own comprehension in that of others'; as such it is a sharing of the treasures I have collected over the years and secreted in my metaphoric pockets. It is about the affect of these explorations on my perception of and with play and perhaps most importantly it is a questioning of the role which the nature of our questioning, our probing, our view of play may have on what it might be. This chapter is about a sense of responsibility towards the potential that *the world makes play*.

By juxtaposing and interweaving the vast unknowns, unavoidably drawn out by a questioning of the *ness* of being playful, with tangible lived details of playfulness; this chapter aims to create an atmosphere akin to that moment of noticing and responding to playfulness itself. I feel an importance of being part of that representation of play that makes me feel wonder, that makes me feel myself, that touches that part of me that recognises the feeling of playing. I feel this is important – for myself, for others involved in work for play, and for play itself in its absorbent flexibility – to this (un)end, my offering.

From just before the time when I started playing with the possibility that *play makes the world*, one of my old holiday sketchbooks contains notes in reflection of the splash of fat raindrops on hot decking. At that time I mused as to whether the raindrops were playing or whether my perception was playing. I was still trying to separate myself from the rain, still erring towards sureness. Today I conceive the inseparability, and wonder instead as to what connects the water drops' movement, or the moth in twilight, with human experience of play and with our questioning of the relationship.

Today my little girl asks to go into the garden in the cold late afternoon rain. She tells me, as I help her into waterproof overall, that the rain is her friend, and then that the rain is her home. She talks about sleeping in the rain, while her sisters sleep in their beds upstairs. Out the window I watch her with the rain, watering the flowers alongside the rain, in imitation, in parallel play. In the silence of my window-paned distance, I wonder why I might assume to know more of her playing experience than of that of the rain.

Later in the evening, I find myself dipping into Schiller's deceptively little book *On the Aesthetic Education of Man* and open to this:

> The insect swarms with joyous life in the sunbeam. . . . Even in mindless Nature there is revealed a similar luxury of powers and a laxity of determination which in that natural context might well be called play. The tree puts forth innumerable buds which perish without developing, and stretches out for nourishment many more roots, branches and leaves than are used for maintenance of itself and its species. What the tree returns from its lavish profusion unused and unenjoyed to the kingdom of the elements, the living creature may squander in joyous movement. So Nature gives us even in her material realm a prelude to the infinite, and even here partly removes the chains which she casts away entirely in the realm of form. From the sanctions of need, or *physical seriousness*, she makes her way through the sanctions of superfluity, or *physical play*, to aesthetic play; and before she soars in the lofty freedom of the Beautiful above the fetters of the every purposed end, she is already approaching this independence, at least from a distance, in the *free movement* which is itself ends and means.
>
> *(2004, pp. 133–134, first published in 1795)*

This passage is part of a Schiller's description of the many ways in which 'The lawless leap of joy becomes a dance . . . noises of perception unfold themselves, begin to obey a rhythm and weld themselves into song'.

Schiller notices in the tree a fore-echo of the surplus energy, which he believed play serves, disperse. Schiller's suggestion of an organism's surplus energy as the basis for playing was to become married with the somewhat similar views of Herbert Spencer (1820–1903) who saw play as a superfluous activity. The combined thinking of these two philosophers became encapsulated in what is now viewed as one of the classical theories of play and named simply *surplus energy theory* (Brown 2008; Burghardt 2005; Rubin 1982; Saracho & Spodek 2006; Torkildsen 2005). When in my late teens I first studied play and learnt about such theories I did not learn much about Schiller's philosophy or poetry, and I did not encounter anything about his perception of swarming insects or of trees. Now, in my backwards and forwards tracing of the process of this theory, I come across something curious. In the short summation of classical play theories provided by Saracho and Spodek (2006, pp. 12–13). I notice that the authors note play's role in *balancing* energy. While for me the word energy is synonymous with life's force, and such balancing thereby indicative of a slight carrying forwards of Schiller's perception, from the text I doubt this was the authors' deliberate intent. An image of fluttering book pages arises in me, a timeline of references between this book and Schiller's original notes, between this book and the texts that drew from it.

I go back and read the introduction to *On the Aesthetic Education of Man*. Here, serendipitously reflecting the image of the communication between book pages that had arisen in me, Reginal Snell asks the reader to go back through time to Plato and forward to Herbert Read (1943), from whose work *Education Through Art* he quotes. These chosen words beautifully connect with that which is vital to me in this text of Schiller's – namely, the flow of 'harmony and rhythm which enters into the constitution of all living bodies and plants, which is the formal basis of all works of art' (Read, p. 70, in Snell in Schiller, p. 18).

I sit gazing to find a sense of sense in the corner of the room, for the how and the why of the not unavoidable[1] distillation of a theory of surplus energy, from the flow of life in the buds of trees and the buzz of bees via play and towards the personal freedom and equality that was Schiller's frame. How and why did play as an energetic conduit of being become a contributor to the view of play as add on to the non-play of 'real' life?

Today I draw into my brooding Cobb's (1993) lament, in the beginning of her deeply resonant testament to the relationship between the natural world and human imagination that 'Unfortunately, the language of conquest still maintains a supreme hold on our social and political theory, our medical policies, and most serious of all, our teaching of ideas about nature and man' (p. 24). I consider this alongside Huizinga's fascinating tracing of the differentiation of play from non-play in tandem to the naming of play (1970, p. 48). To these I add an old favourite, Jung's description of his visit with the Pueblo Indians in 1924–25. I re-sense his finding that direct questioning was a fruitless means of inquiry and his discoveries, which were made with patience and sensitive watchfulness. I remember and re-read his perception that the survival of the Pueblo culture depended on their guarded secrecy, that it was vital to survival that 'their mysteries were not desecrated' (Jung 1995, p. 278). I hold his words of being entrusted by Mountain Lake who spoke of Father Sun, provider of life and of the vitalness of the kiva rituals which helped the sun cross the sky. I feel the wonder which is conveyed when he told of being told that "If we were to cease practicing our religion, in ten years the sun would no longer rise" (Jung 1984, p. 42). I am grateful once more for this story which since I first encountered it has stayed with me as one of the most beautiful instructions as to the reciprocal relationship between our view finders and what is, and the play between ourselves and the world – of making with belief. Yet now as I write, this story gives me a sense of tenuousness, a feeling of almost loss. I feel the weight of Joseph Campbell (2008) cautioning that 'whenever the poetry of myth is interpreted as biology, history or science it is killed' (p. 212). Perhaps the lens of ***real*** life is truly growing ever away from the way of play, but if so, how? If it grew from play, did it grow that way?

I am also aware, as I ponder, that my questioning of the why and the direction/contextual shifts of distillation is ongoing and reoccurring. When, nearing ten years ago, I found myself reviewing playwork's theoretical framework in order to situate my own research, I questioned the short-handing, and the list-making of complex descriptive perspectives into tick boxes. I wondered at the potential for misrepresentation, in that which was extrapolated and in the act of extracting, and in the encounter of play without its substance. I felt and pointed to the immense responsibility of the playworker to bring their experiences of play to enliven these multi-fit descriptors (Guilbaud 2011, p. 35). I came to express my concerns as a question 'does play's slippery malleability, its way of being in every aspect of life in which we find it,[2] make it vulnerable to the constructs that we create or do not create for it?' (ibid., p. 93).

More recently I found myself musing aloud, in further writing (Guilbaud 2018), as to the reification of play. I probed at the paradoxes, juxtaposing the playing child's creation of new life in detritus with the possibility of atrophy being caused by an arbitrary provision of neophilic opportunities within a play setting. I wondered, if the urge to extract, distil and set in certainty, might, in its most hopeful permutation, serve a necessary rhythmic companion to reinvention. Perhaps providing calcification in the life cycle of play before disintegration – necessary discards and fodder for the resurgent liminal dwellers to have reason to say 'No it shan't be so!' In this meaning making I drew from the inspirational insights offered by Spariosu's (1989, 2013) discussions of play and liminality and Edensor's (2005) re-envisioning of debris. Yet a sense of unease with Laissez-faire, arguably perhaps contradictory to an expansive view of play, prevailed in me, for there seemed a risk that 'recognition of such unspecified potential might rely on perception having been formed by and ever encountering its like, in air, in play, in the wonder of possibility' (Guilbaud 2018, pp. 119–120).

Even more presently three things happen which rub together augmenting a sense of disquiet, a sense of protectiveness, an edging, in me, towards emergency.

Firstly – As I come to the end of the first chapter of the recent tome *The Cambridge Handbook of Play* (Burghardt and Pellis, 2018, pp. 11–29) I read these words: 'Progress has been made in taking some of the mystery out of the origins and evolutions of play in recent decades, but much more needs to be done' (p. 24). Then in the closing sentence at the end of a list of disciplinary techniques, or lenses of investigation, focusing on clarity in identification, types and components, empirical evidence of play benefits and lacks, there is this: 'we envision that play, delightful to watch, may become less of an evolutionary mystery' (p. 25). My whole being resists these words, resists the envisioning of the striping away of play's mystery. Again, I come up against a sense of antithesis to the *ness* of play, in a top down anthropocentric dissection. Though I recognise that the synthesis of thought presented here comes perhaps from enthralled observation, I feel the acute displacement of awe. Within the text the authors state and reflect on the premise that 'we now know play is a heterogeneous category' (p. 12). Yet, such knowledge doesn't lead to a questioning of perspective, or position. There is no reference to *Umwelt* with which Nicolson (2018) drew me in to *The Sea Bird's Cry*; there is none of the diffidence with which Godfrey-Smith (2018) situates the reader's relationship to the octopus in *Other Minds*. Burghardt and Pellis clearly frame the specific scope of their explication, addressing 'the origin of play from non-playful behaviour', with the hope of wetting the reader's appetite to access other studies. In contextualising their approach, they ask us to recall that 'Darwin (1859) never actually discussed the origin of the species (or life from nonlife), instead devoting his efforts to the transmutation process of one (or groups of) species to a different one' (p. 12). However, despite their aim, my appetite finds no attempt to entice, no reflection of the nature of the subject matter, nothing for me to re-enliven through my experience of play,[3] no play cue[4] at all. Furthermore, as I read, the uncomfortable questioning reconfigures in me, as to the role such oft lauded texts play, in the existence, or the life, the cycle, or the play, of play.

Secondly – A text picture moves up the screen of my social media feed, a photograph of children in smiling conversation round a table, raising their tea-set cups, accompanied by the quote:

> Perhaps play would be more respected if we called it something like "self-motivated practice of life skills," but that would remove the lightheartedness from it and thereby reduce its effectiveness. So we are stuck with the paradox. We must accept play's triviality in order to realize its profundity.
>
> *(Gray 2013)*[5]

Perhaps these words, and their soundbite format, flitting through, are jarringly caught by the tone of my already present unease. They seem, in that weird synchronistic overreach of the unknown all-knowing search engine algorithms to arrive in quasi response to my current disquieted questioning. Who is cutting the wood to fashion the building blocks implicated in this value and validity demanding paradox for the gatekeepers of the human child?

Thirdly – Someone asks me if I know of any evidence of insects' play? I find myself grappling around my mind, for if and where I might have seen a study, some data. Suddenly, I feel a sense of revulsion. Who are we to judge, by which narrow amalgamated indicators and to what ends? I mention this to my 15-year-old daughter, who played so strong and long through childhood, and now plays on through her poetry, 'of course insects play,' she laughs, 'it's obvious!'

I recall the bees on our garden lavender in the summer, and this sensory memory somehow brings with it all the feelings from the first time I heard Shane Koyczan's (2007) poem *The Crickets Have Arthritis* – sharing one set of headphones with my beautiful friend Nora Bateson, whose life and work is about increasing perception of the interrelationality that creates life. On that day, we meshed inspirations and wonder-filled conversation in the café of Kew Gardens. Then, together, perhaps illicitly, we climbed a tree.

I will not contribute to the crickets' arthritis. I want to honour the still present human capacity to comprehend the metaphors in Koyczan's poetic lines and feel the importance of healing the crickets. I feel the reflective communications between the crickets' music and the very capacity I have as a human being, who has played, who still plays, to experience metaphor.

Conjoined within a short time frame of encounter these three happenings cause a sense of urgency to rise in me towards the responsibility which we have to continue to be awed, to continue to wonder in life. What other source is there for the continued flourishing of this tendency within us, but play? How far are we willing to alter the frame of play's nature and potential through our perception of such.

It is perhaps unavoidable still, at present, that anyone who studies play, who asks it questions, will be awed in their way. Richard Feynman (Horizon 1981–82) communicates with a twinkle in his eye, as he explains that for him knowing the parts of the flower and how and why of whence they came renders it more rather than less beautiful! Yet when speaking of walks with his father he also tells, with palpable love, of how his father did not teach him the names of birds as the fathers of his friends did, but rather asked him to notice the way the birds were. Feynman traces his approach to physics, his 'playing with ideas' to his upbringing. The propagated fixtures of science that arguably give cause to Joseph Campbell (2008) cautioning discussed earlier, are not the means of this scientist. Nor of one whose explorations and discoveries poeticise a world encircled and embraced by imagination.[6] Nor of she who seeks to comprehend an ever more silent spring (Carson 2000) and also expresses the crucialness of accompanying the young child in their noticing of small moss (Carson, Kelsh, & Lear 1998). 'A scientist in his laboratory is not a mere technician: he is also a child confronting natural phenomena that impress him as though they were fairy tales' (Marie Curie in Leiter & Leiter 2003, p. 70). Clearly, it is not the play-filled pleasure of finding things out that is disjointing us.

Yet still, play is being abstracted from its *ness*. When we speak of it, when we communicate about it, play seems to be fitting into smaller and smaller and neater and more quantifiable packages: We supersede the flotsam and jetsam of a naturally occurring bits and bobs with *loose parts kits*. Acceptance of boredom of the child's need to find their own way to their own play is considered radical and something to be scheduled into a highly prescribed day. Over and over I watch parents involving their children in play on the playground apparatus, their full attention given as they encourage their little ones towards overt play illustrative animation. Then once the phone-photo has been taken, the parent withdraws into their own social media play, from which the child must tug and wheedle to be pushed again on the swing.

Thus play is positioned further and further as something to make time for, something important, the value of which must be quantified, a commodity even in its frivolity. How much further play can be abstracted from its genesis before the weight of reification overwhelms. Do we still have the wisdom, or will, to do as Rumi bids and to sell our cleverness and buy bewilderment. Or might in our incorporation to commodification, such literal purchasing be an all likely catastrophe.

> If all the world were apple pie,
> And all the sea were ink,
> And all the trees were bread and cheese,
> What would we have to drink?
> It's enough to make an old man
> Scratch his head and think.
> *(From* Harry's Ladder to Learning *1850)*

I feel myself step a small, but perhaps more deliberate than ever, step in counterbalance. I will do what I can towards the *ness* of play which in all uncertainty reformed stardust into me, in partnership with the playing child rebuilding from the mess of destruction.

If the *ness* of play is being eroded by dysmorphic representation, might it also be sustained through its communication? To such a connective collective I add my voice and therefore share with you some heartbeats of my playfulness.

I am swept into playfulness as, when cleaning up, I accidentally flick a sliver of stone on my dressing table into the air, and seeing it fly remember the full detail of where and when I found it. Musingly I re-find the words I wrote in that long-ago time.

> I stand in shallow shifting sea, seeing slips of slate stone, that are at once specifically made by and for turning over in the waves as the waves pull back. It feels wrong to steal, but I do so, for I have an urge to leave myself there, to be one with these slips of stone in the sea, to leave myself to merge, but as I cannot become them, I take a few of them to become part of me. I walk back up the beach, as soon as my fingers touch a stone sliver in my pocket they begin turning it over and over.

My sense of play's *ness* is nourished in talking with others about play's detail. For instance, by the recurrent conversations I have had with Bob Hughes over the years about his grandson's invisible yet structurally discernible house.[7] I have felt privileged by each revisit to this magical house, through this grandfather's enchantment, and by the conversational questioning within a heightened sense of awe which this sharing always invited between Bob and myself, as to play's relationship to the uncertainty principle, to the nature of existence and everything.

I feel the validation of kindredness as I read Joel Seath's description of the playground as palimpsest. He writes of the convergence of the *baroque* and the *marvellous real,*

> where the extraordinary is ordinary and the ordinary is extraordinary.
>
> Here the children buried a bird because it was dead; over here, a volcano rose up; there, the children floated on the breeze with open umbrellas; this is the route of the funeral procession following the death of a cardboard box; this is the murder scene of a teddy-bear, spiked through and through and bleeding red; this is the slide

where, once, there was an incidence of surfing on paint, amongst other once-nesses of experiment; . . .

(Seath 2018, pp. 150–151)

I am recognised in my playfulness, in my essence, when someone tells me, because they believe I would like to know, of a tree climbing child who explained that the tree was telling them where to put their feet. I feel again the context of this truth when I came across these words when reading to write here.

To directly perceive any phenomenon is to enter into relation with it, to feel oneself in a living interaction with another being. To define the phenomenon as an inert object, to deny the ability of a tree to inform and even instruct one's awareness, is to have turned one's senses away form that phenomenon. It is to ponder the tree from outside of its world, or, rather from outside of the world in which both oneself and the tree are active participants.

(Abram 1996, p. 117)

I relive the feeling of letting the form of my knowing about play grow from play-full-ness and am thankful for the deep patterning that happened and my continuing capacity to be able to access that perceptual frame.

As a playworking researcher

It was as if my way of being as I researched, which I described as peripheral noticing and drifting, came about through resonance with the children's way of interacting with the bits and bobs available to them. They seemed to find exactly what they were looking for, even though they did not know it before they found it. Their resonance with the play possibility of things seemed to occur in an atmosphere of merging with that which they were exploring for potential. . . . The emulation in my way of researching was not deliberate; at first I was not conceptually conscious of it happening. However as this way of observing became more familiar and ever more responsively refined, my resonance with, or recognition of, what I was observing increased.

(Guilbaud 2011, p. 330)

Letting myself be altered by the play that was happening enabled the organic play-full growth of a way of working with my observations, to share and communicate meaning making, in fulfilment of requirements data presentation and analysis. This way of working trusted play to create not only what I found, but my journey of discovery.

When I remember to hold my sense of the *ness* of play, feeling that play is a home to return to, the place of near sleep, of kin, I am aware of my internal shift towards empathy as my little girl gets caught up in one thing after another, needing to play and then just play this and then that. I see myself trying to get out, to get somewhere, to do something with her, of value. I smile at myself and I pause, and breathe, and wait, and neither coerce nor wrench her out of her need to be within play.

Today, after writing, I sit painting, in my creation time. I reflect on the simplicity of these small gathered personal reflections and the vast-ness of play which they touch, ever so lightly. I whisper a simple hope towards integrity, towards humility, towards negative capability (Keats 2004). I paint a tiny blue butterfly on a card for spring, and then a larger blue butterfly wing on a big piece of painting paper, I let the water drip from my brush, loosening the boundaries of the edges, patterning like raindrops. Then I fold the paper across, so that the wing reaches over the horizon, I feel the metaphor or metamorphosis, of butterflies, with the language of play in my heart I return the ness to itself, when I open the paper again, there is a slight blurry impression.

Notes

1 Such distillation is arguable not unavoidable in that Snell finds reflection of Schiller's view in Read's work. Similar, though arguably more direct lineage can be seen in Cobb's (1993) bringing together of the reverence for nature of many poets in her creation of a research frame and subsequent communication of the complexity of comprehending 'an ecology of imagination in childhood'. A way of understanding which, though she altered the parameters and offered critique, Chawla (1994) carried forward fully.

2 In contextualising my understanding of play arising in every aspect of life I referred particularly there to the perception and perspectives presented by Huizinga in *Homo Ludens* (1970) and Sutton-Smith in *The Ambiguity of Play* (1997).

3 Here I pull on an earlier exploration (Guilbaud 2011) of Gadamer's suggestion that it is only when play exists, continues to exist beyond the player, transformed into structure that it becomes real, as opposed to the rhythmic open ended playing-itself-out of play in the elements, or in children's play (Gadamer, Weinsheimer, and Marshall, 2004, pp. 102–110). Within this proposition Gadamer recognises that such play, even in becoming real, is never completely separated from its origin, for though it exists transformed into something in itself, it has to be experienced/enlivened as play for it to really exist.

4 I use the term *play cue* presented by Sturrock and Else 'The play cue is the lure or invitation from the child to the surrounding environment to join in play productions of one sort or another' (1998, p. 13). This term as part of the *play cycle* has since been incorporated into playwork theory and practice, and incorporated into the many layers of playworkers' recognition and understanding of play. For Instance, in her exploration of sound recordings as a tool to reflective practice, Wilson (2018) describes her realisation of the play cues communicated by the sounds of the playground.

5 This text post is identified as having been created by *Sunflower Creative Arts.org*. The quote can be found on page 156, of *Free to Learn*, by Peter Gray.

6 Einstein's famous view of knowledge and imagination was expressed as: 'Imagination is more important than knowledge. Knowledge is limited. Imagination encircles the world' in 1929 October 26, The Saturday Evening Post, What Life Means to Einstein: An Interview by George Sylvester Viereck, Saturday Evening Post Society, Indianapolis, Indiana.

And

'Imagination is more important than knowledge. For knowledge is limited, whereas imagination embraces the entire world, stimulating progress, giving birth to evolution' in his 1931 book *Cosmic Religion and Other Opinions and Aphorisms*, Covici-Friede, Inc., New York (Einstein & Shaw 2012).

7 This magical house and these conversations are discussed in greater detail in *Sharing Playwork Identities: Research across the UK's Field of Playwork* (Guilbaud 2015) and *The Might of Play as Possibility and Power* (Guilbaud 2018).

References

Abram, D. (1996). *The spell of the sensuous: Perception and language in a more-than-human world.* New York: Pantheon Books.

Brown, F. (2008). The playwork principles: A critique. In F. Brown & C. Taylor (Eds.), *Foundations of playwork* (pp. 123–127). Maidenhead: Open University Press.

Burghardt, G. (2005). *The genesis of animal play.* Cambridge, MA: MIT Press.

Burghardt, G., & Pellis, S. (2018). New directions in studying the evolution of play. In P. Smith & J. Roopnarine (Eds.), *The Cambridge handbook of play: Developmental and disciplinary perspectives* (Cambridge Handbooks in Psychology, pp. 11–29). Cambridge: Cambridge University Press.

Campbell, J. (2008). *The hero with a thousand faces.* Novato, CA: New World Library.

Carson, R. (2000.) *Silent Spring.* London: Penguin Books.

Carson, R., Kelsh, N., & Lear, L. (1998). *The sense of wonder.* New York: Harper Collins.

Chawla, L. (1994). *In the first country of places: Nature, poetry, and childhood memory*. New York: State University of New York Press.
Cobb, E. (1993). *The ecology of imagination in childhood*. Dallas, TX: Spring Publications.
Edensor, T. (2005). Waste matter – the debris of industrial ruins and the disordering of the material world. *Journal of Material Culture*, 10(3), pp. 311–332.
Einstein, A., & Shaw, G. B. (2012). *Einstein on cosmic religion and other opinions and aphorisms*. New York: Dover Publications.
Feynman, R. (1981–82). Horizon – the pleasure of finding things out. *BBC* [Internet]. Available from: www.bbc.co.uk/iplayer/episode/p018dvyg/horizon-19811982-the-pleasure-of-finding-things-out [Accessed 6 March 2019].
Gadamer, H. G., Weinsheimer, J., & Marshall, D. G. (2004). *Truth and method*. London: Continuum.
Godfrey-Smith, P. (2018). *Other minds, the octopus and the evolution of intelligent life*. London: Harper Collins.
Gray, P. (2013). *Free to learn*. New York: Basic Books.
Guilbaud, S. (2011). *A phenomelogical inquiry into the possiblity of played-with-ness in experiences with things*. Doctor of Philosophy, Leeds Metropolitan University.
Guilbaud, S. (2015). Sharing playwork identities: Research across the UK's field of playwork. *International Journal of Play Special Issue, Play and Playwork: Notes and Reflections in a Time of Austerity*, 4(3), 299–313.
Guilbaud, S. (2018). The might of play as possibility and power. In F. Brown & B. Hughes (Eds.), *Aspects of playwork: Play & culture studies, volume 14* (pp. 107–123). New York: Hamilton Books.
Harry's ladder to learning. (1850). London: David Bogue [Internet]. Available from: https://archive.org/details/harrysladdertole00londiala [Accessed 6 March 2019].
Huizinga, J. (1970). *Homo Ludens: A study of the play element in culture*. London: Paladin.
Jung, C. G. (1984). The Pueblo Indians. In T. Hillerman (Ed.), *The spell of New Mexico* (pp. 37–44). Albuquerque: UNM Press.
Jung, C. G. (1995). *Memories, dreams, reflections*. London: Fontana Press.
Keats, J., & Forman, H. B. (2004). *The letters of John Keats*. Whitefish, MT: Kessinger Pub. Co.
Koyczan, S. (2007). *Words aloud 4 spoken word festival in Durham, Ontario, Canada*, November [Internet]. Available from: www.youtube.com/watch?feature=player_embedded&v=6VrZE8MCnIA [Accessed 6 March 2019].
Leiter, D., & Leiter, S. (2003) *A to Z of physicists (Notable scientists)*. New York: Facts on File Inc.
Nicolson, A. (2018). *The seabird's cry*. London: Harper Collins.
Read, H. (1943). *Education through art*. London: Faber.
Rubin, K. H. (1982). Early play theories revisited: Contributions to contemporary research and theory. In D. J. Pepler & K. H. Rubin (Eds.), *The play of children: Current theory and research, contributions to human development* (pp. 4–14). Basel: Karger.
Saracho, O., & Spodek, B. (2006). Understanding play and its theories. In R. Parker-Rees & J. Willian (Eds.), *Early years education: Curriculum issues in early childhood education, volume 2* (pp. 9–27). London: Routledge.
Schiller, F. (2004). *On the aesthetic education of man*. Translated and Introduced by R. Snell. New York: Dover Publications Inc. (Originally published in 1795).
Seath, J. (2018). The playground as palimpsest. In F. Brown & B. Hughes (Eds.), *Aspects of playwork: Play & culture studies, volume 14* (pp. 145–156). New York: Hamilton books.
Spariosu, M. (1989). *Dionysus reborn: Play and the aesthetic dimension in modern philosophical and scientific discourse*. New York: Cornell University Press.
Spariosu, M. (2013). Exile and utopia as liminal play: A cultural theoretical approach. *Philosophy at Play*, 13, April 9–10, University of Gloucestershire.
Sturrock, G., & Else, P. (1998). *The playground as therapeutic space playwork as healing*. IPA/USA Triennial National Conference, Play in a Changing Society Research, Design, Application, June, Boulder, CO.
Sutton-Smith, B. (1997). *The ambiguity of play*. London: Harvard University Press.
Thoreau, H., & Cramer, J. (2004). *Walden*. New Haven CT: Yale University Press.
Torkildsen, G. (2005). *Leisure and recreation management*. London: Routledge.
Wilson, S. (2018). Three reflective tools for playwork practice. In F. Brown & B. Hughes (Eds.), *Aspects of playwork: Play & culture studies, volume 14* (pp. 1–12). New York: Hamilton Books.

6

STAGES NOT AGES

The emerging Developmental Play approach for differently abled children[1]

Caroline Essame

Thirty years' experience – and one child

The Developmental Play approach is rooted in training in occupational therapy, creative arts therapy and education, and over 30 years of clinical practice with clients across the lifespan, including elderly clients with dementia, in acute adult mental health settings, early childhood and special needs work. Consistently, my work has been inspired by the power of creativity and play.

Developmental Play began to crystallise in a small special-needs school in rural Kerala, in India. Deepti Special School (www.deepti.com) was set up by Dr Susan Mathew, a parent of a child with cerebral palsy. She had left India to pursue her postgraduate studies in the UK, where she saw what a difference the right kind of care made to her child. She returned to India with her family and set up the school in 2009, starting with just one child and a vision to make a difference in her community. By 2011 when I first visited, it was clear that there was a great need to develop the capacity of the parents, teachers and the few trained clinicians to work with the growing number of children who were flooding into Deepti. There was no other local provision and there had been no early intervention, so many of the children in the school had quite profound developmental delays.

The challenge was to distil 30 years' worth of experience in an accessible, condensed and meaningful way, and with concepts that could easily be translated into the local language, Malayalam. This was the starting point for Developmental Play, which is now running in India, Singapore and the Philippines. The idea is to build a community of adults with the core skills to help children by giving them clear frameworks to understand child development through different types of play.

There are 600 million people with disabilities worldwide; 80% of them are in developing countries, and 82% live below the poverty line. The World Report on Disability (WHO, 2011) says that more than a billion people are affected by disability. The prevalence of childhood disability in low- and middle-income countries is 15%, double that of developed countries. India alone has an estimated 52 million children with disabilities. The need for intervention and support is immense.

Developmental Play started small, at just one school. The name Deepti means 'light' and one aim is to be exactly that, given the darkness of the situation faced by so many children. Since

2011, we have trained staff in Developmental Play to inculcate play into the children's day at the school and also to inspire parents to engage in play with their children at home. It is the first Developmental Play-infused school in India; the second school is in Lemery, in Batangas in the Philippines.

Firm foundations first

Developmental Play uses play to give children the foundations for cognitive, physical, social and emotional growth. Many differently abled children struggle because they do not have these foundations in place. The Developmental Play approach is like a pyramid (illustrated next) where each level builds on an earlier one. Ensuring that each level is solid – which may involve going back to rebuild it – is the key to the Developmental Play approach. Play is a cornerstone of human development. It helps children learn about their physical and emotional world. It teaches them that they can have impact. It shows them how to collaborate. It is the language of childhood. And best of all, it's fun.

In this chapter, I offer examples and stories to illustrate the power of this process. It's helpful to think in terms of developmental milestones. For example, a child who can't sit still may have missed out on the first Developmental Play level of sensory-body play. Revisit that stage; build that level. Once they are confident of where their body is in space, they can sit better and their time in class becomes more productive. Subsequent levels can be built more confidently once this first, foundational level is solid.

A child who reacts badly to change, meanwhile, can go back to the second Developmental Play level of messy play, to learn about cause-and-effect and how to tolerate unpredictability. It allows them to create change on their own terms, to work with it and gain a sense of control over it. Similarly, a child who is awkward around other children will benefit from engagement at the third Developmental Play level of meaning-making, so that they learn to understand feelings and navigate them. Or they can go back to the very first level, to explore the control and confidence stages of early sensory play and attachment safety play.

Simply put, mastery of each level brings confidence and overcomes many problems that hold children back. This model is rooted in principles of development and neurobiology, understanding how neuron pathways are laid down and built up through a series of relationships and external experiences (Gerhardt, 2004; Siegel, 2012). What makes Developmental Play unique is that it combines earlier interventional strategies to give children holistic support, pulling together many frameworks. It is particularly effective for children with special needs, learning difficulties and those who have experienced trauma. It also focuses on pre-verbal skills and laying the foundations for higher symbolic play (Stagnitti & Cooper, 2009; Stagnitti, 1998; Vygotsky, 1967).

Developmental Play is therefore an approach for teachers, parents, carers and therapists to identify if a child has missed out on a level of play and would benefit from going back to it. It is specifically *not* called 'play therapy'. The word 'therapy' denotes a completely different level of expertise, hours of supervised clinical practice and training and this is often a luxury inaccessible to many children most at risk, particularly in the developing parts of Asia. This approach should be for all adults who work with children – not just therapists – so that *all* children will get a chance to reach their fullest potential through play. Having said that, many therapists in Singapore and Philippines are training as Developmental Play practitioners as an addition to their skills; but it is an approach for their toolbox and should not be termed 'therapy'.

Developmental Play is also part of a bigger, global drive to place play at the heart of childhood. The Real Play Coalition launched in 2018 by the LEGO Foundation, IKEA, National

Geographic and Unilever seeks to do exactly that (www.realplaycoalition.com), while a recent report from the World Economic Forum concluded that:

> We must equip our children with the tools to address and embrace the new realities of tomorrow. To do this, we must allow them to develop through play. Fewer play moments mean fewer opportunities to develop the skills children need to thrive in the dynamic, challenging economies of tomorrow.
>
> *(WEF, 2019)*

Developmental Play joins this global initiative to take play seriously for all children, building on research coming out of PEDAL (Play, Education, Development and Learning) at the University of Cambridge and PlayFutures, a play-based community set up by the LEGO Foundation.

The secrets of the pyramid

The Developmental Play Pyramid is a framework to explain how children learn through play, building foundational skills all the way up to higher play (Figure 6.1). It suggests ideas on how to recognise what level a child is playing at and what play experiences they might be missing or seeking, so that the adults who work with them can support their developmental needs. For most children, the foundation level of sensory body play and attachment is laid down within the first six months. The second level happens once the child can manipulate objects, and crawl around and explore the world around them; they shake a rattle and see that it makes a sound. They smear their food across a table and see that they can make a pattern. The third level is once they begin to make meaning with the objects they play with. They compare objects and differentiate them, seeing that some are larger and some are smaller, that they are different colours, and so on. They use the objects to create sequences and patterns – making meaning out of the objects. This can be as early as 18 months old, seen in a child building a tower of bricks. At its highest level, it can be seen in back-and-forth social play described in the DIR Floortime approach (Greenspan, 1999) as 'circles of communication', in which the child and the carer develop a relationship as a result of emotional exchanges, both verbal and physical.

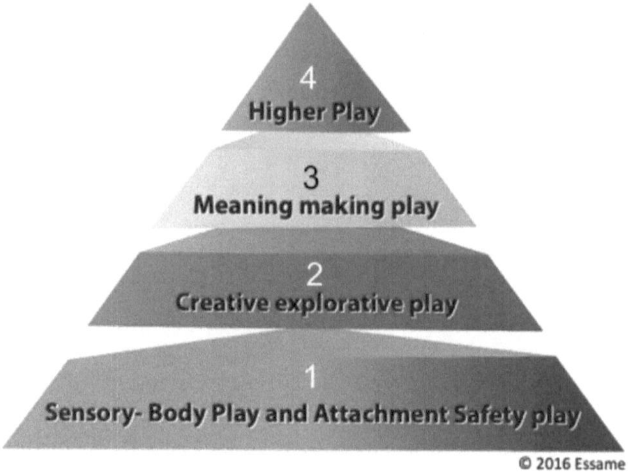

Figure 6.1 The four levels of play

Once all these foundations are in place, we see the emergence of the fourth level of higher play (Metcalfe & Shimamura, 1994). This is when they have imagination, the ability to play collaboratively with other children, sharing and taking turns. An element of this is metacognition, when a child has developed an understanding of their own thoughts and feelings, and what is often described as emotional intelligence (Gardner, 2000), that is, they can describe feelings, express their needs and tell stories with complex narrative. They have developed imagination, a concept of time, object permanence (Huitt & Hummel, 2003) and an understanding of others.

Many children with cognitive challenges struggle with the fourth level of higher play, however, especially children on the autistic spectrum (Baron-Cohen, 1995), and this is where Developmental Play is particularly effective. Giving children more play experiences at the lower levels of the Developmental Play Pyramid, particularly playing with concrete objects, helps them develop the foundational skills they need for higher play. For example, a puppet (level three) can help a child begin to explore roles and expression (level four), which builds empathy. Think about it as building a bridge that links a child understanding her body (level one) and then understanding objects (level two and three) to developing inner thoughts (level four). To take one simple example, developing a concept of object permanence in a small child through playing hide-and-seek or peekaboo (which is a kind of explorative play seen at level two) helps develop the foundations to build a bridge for imagination (level four) in later life. That is the big picture; we now consider the four levels of the Developmental Play Pyramid in more detail.

Level one: sensory body play

Sensory body play is a key foundation stage for all learning and development (Williams & Shellenberger, 1996). It looks at how children learn first through their bodies, through smell, sound and touch, through the movements of their bodies and exploring their physicality, and it ties in with the theory of sensory integration (Ayres, 1974; Ayres & Robbins, 2005). Using this, occupational therapists look at how all the eight senses work together to help children make sense of their worlds:

1. Visual (sight)
2. Tactile (touch)
3. Auditory (hearing)
4. Gustatory (taste)
5. Olfactory (smell)
6. Vestibular (balance)
7. Proprioception (a sense of self-movement and body position)
8. Interoception (a sense of one's body and what is happening inside)

(Ayres & Robbins, 2005; Allen, 2016)

Children take in information from one sense, process it in the brain and then respond often through several different senses, hence the phrase 'sensory integration'. When there is an imbalance in the senses, however, and it is harder to integrate the senses, then the child struggles with learning, with behaviour or with self-concept (Ayres & Robbins, 2005).

Problems in sensory integration often occur across a range. Children may be over-responsive to one sense; for example they may not tolerate loud noises if their auditory sense is over-sensitive. Or they may be under-responsive to one sense; for example they may constantly need to jump and crash onto cushions to feel their body if their tactile or proprioception is under-sensitive. One of the first-rank symptoms of autism is when a child displays sensory issues such as these, giving evidence that sensory integration is important and has huge impact on development at the higher levels of play.

Developmental Play therefore looks at how children play sensorily, what their bodies are telling us and what opportunities we need to give them in order to support them. While children may have many skills at higher levels, if there are major sensory issues, then they will not have a solid foundation, and this may cause problems. This short story illustrates how supporting a child to develop foundation skills in sensory body play can transform higher skills.

> Poppy is a five-year-old girl who is socially withdrawn and nervous. She is startled by loud noises, fiddles a lot with small items like beads and tissues and hates having her hair washed in the bath. So much so, her parents dread bath-time when she will have major meltdowns and will kick and bite anyone who tries to get her in the bath.
>
> Poppy does not have much speech and her play is very limited for her age. Her parents worry about her. She is isolated at school and so unlike her younger sister who is cheerful, boisterous and fearless.
>
> Poppy is sent to see a therapist as her language isn't developing. She barely connects socially, and her parents hope that with extra support she'll catch up with her sister and be able to join in more with family and school life. Poppy sees the therapist once a week for a year in a creative playroom with a colourful sensory gym.
>
> The therapist allows Poppy to lead the sessions and watches and supports her as she plays in the gym and the playroom. The therapist doesn't structure or teach Poppy, but instead allows her to explore things at her own pace.
>
> This delights Poppy. Since she was at pre-school there has always been so much going on, and so much structured play that she couldn't keep up with, so she preferred just to shut down and stay silent. Here, though, she is allowed to play with sand and make a mess with water and shaving foam. She explores creating worlds in the playroom and she swings, jumps and learns to understand her body more in the sensory gym.
>
> Within six months Poppy is more confident and starting to form more words and communicate better. Once her parents understand how sensitive she is and how disturbed she is by the sensation of water on her head, Poppy feels she is in control of some of her inner world, and bath-time becomes easier.
>
> After one year Poppy is transformed. She'll always be sensitive and a little introverted, but now she's a more confident, happier child, more ready to move into the demands of primary school.
>
> *(Essame, 2016)*

Sensory body play sounds natural and automatic for children, but it is in decline. With increasing urbanisation, children don't engage as much in early sensory play and this impacts on how they develop. To take one quite extreme example, in Singapore children do not need to use or develop their vestibular skills much as most surfaces they walk on are smooth, flat, concrete or tile; even many of the parks have flat wooden boardwalks. Physical activities such as tree-climbing are not generally encouraged in public parks, and the sensory profiles of children are altering with some people suggesting that there may be neurobiological changes (Vickerius & Sandberg, 2006; Archer & Siraj, 2014). One study estimates that 5–15% of children have some kind of sensory challenge (Ahn, Miller, Milberger & McIntosh, 2004), so this first level of the Developmental Play Pyramid is fundamental to understanding and supporting children.

Often in the clinic where I work, children come in for social and emotional work and are referred for 'play or art therapy'. In many cases, what they need is to move and feel their bodies; often, that in itself is transformative. The therapy aspect is about learning to recognise the developmental gaps which have been missed, and filling them in. When children feel good in their bodies, their emotions change, which leads on to the other part of level one of the Pyramid – attachment safety play.

Children need a positive-attachment relationship with an adult in which they feel safe when they play. They need to be seen, they need to be loved, they need to be held (Gerhardt, 2004). Recent research from the PEDAL Institute at the University of Cambridge (Leong et al., 2017) shows that if a mother sings to a child at about four months old while making eye contact, the child's brain activity is significantly higher than if a mother sings *without* making eye contact. This helps make sense of why so many children we work with just want to be watched, to be *seen*. That relationship, that attachment or what professionals often called the therapeutic alliance, is crucial to a child. The fact of being seen, the therapeutic or parental gaze, makes them feel safe and validated.

This is even more important for children who have sensory issues. Their brain biochemistry makes them anxious and fearful, and they need a key adult to help them to co-regulate and to feel calm. They need support to navigate their world and to feel that it is a safe place. Most research has been done in situations where this has *not* happened, which indicates how important it is. Sunderland (2006) explains how fear impacts a child's developing brain and how vital it is to help a child wire the brain to feel safe and connected. The significance of play and attachment is also well covered in the book *Healthy Attachments and Neuro-Dramatic-Play* (Jennings, 2010) and builds on the early work of Bowlby (1969) and Rutter (1972) who identified the negative impact of not having a primary care relationship. We have also seen how lack of early attachment can be detrimental way into adulthood and this was evident in work coming out of Eastern Europe and children with early attachment disorders in which the social and emotional parts of the brain simply did not develop (Brown & Webb, 2005).

This foundation level of development is preverbal, and its absence is felt by the child long before it can be articulated. For this reason, talk therapies are often not effective. The child feels this absence, this confusion deeply in their body (for more on how experiences are embodied, see Van der Kolk, 2015). It cannot be spoken through the language of higher play, higher cognition and language, but needs to be expressed through sensory play in the context of a safe attachment relationship. Time and again, if you can return with children to this fundamental, first level of Developmental Play, the results are transformative.

Level two: creative explorative play

If a child is scared, they will freeze, fly or fight; but if they feel safe, they will start to explore the world around them. So, creative explorative play grows from the foundation of feeling safe and being able to integrate the senses, which happens at level one of the Pyramid. At this second level, the child might kick a mobile hanging above their cot, they may splash in water, they will smear their food or, if you're unlucky, the contents of their nappy. It feels great for them to explore the world around them, to begin to experience cause-and-effect, and to expand into control and ordering those sensory explorations. Here, the child starts to create form, to build motor skills, and to hardwire more complex circuits of understanding. This is where the engagement with the outside world begins. Herein lies the essence of playfulness: joy, spontaneity, flexibility.

In neuro-typical development, a child often moves quite quickly through level two and into the higher level of meaning-making (described in the following section). But some children stay a long time in creative explorative play before they're ready to move on to more cognitive, construction play, such as building a tower of blocks. The important thing in Developmental Play is to follow the child's lead, to remember that play builds neural pathways and that often you may need to revisit lower levels of play or stay with them for quite a while before the child is able to move on. This will be illustrated in Jamie's case story, later in the chapter.

Level three: meaning-making play

Meaning-making is patterning, sequencing, ordering. It's the beginning of children creating their own languages from play. It's when the child begins to see problems and solve them. It's intrinsically linked to the development of language, the understanding of symbols and the laying down of memory. In order to make meaning of their play, children need to have an idea of object permanence – to know that things still exist even when you cannot see them. Understanding object permanence reduces anxiety; if they know that a toy is still there even if it's under a blanket, then they know that their primary carer still exists even if they have left the room.

Here, they're also laying down foundation skills needed for later use of symbols and being able to understand metaphor which is crucial in higher play. They explore shape and form in their drawings, they repeat sequences to really understand them, often they play the same game again and again or watch the same cartoon so that sequence, pattern and meaning becomes hard-wired. Children who cannot see this patterning and recreate it may struggle later with creative thought and even language development and self-regulation. If they do not have this ability, you need to take them back down the pyramid to level two, and do more creative exploration, messy cause-and-effect play, hide-and-seek and concrete object play where the child uses her hands to move or manipulate something.

Level four: higher play

Most educators, when they talk about play, talk about this level, which is all about imagination, narrative, 'as if' and pretend play (Bodrova, 2008; Bergen, 2013), as well as complex problem solving and being able to understand roles and rules (Vygotsky, 1980). This level of play is highly linked with language and social learning, and for many of the children I work with this level is a challenge. And yet, as adults, this level is often how we understand play, simply because it's the only kind of play we remember. This is why understanding the foundation levels of the Developmental Play Pyramid is so important, because if children have not reached this fourth level, we need to be able to see where the gaps are, and often to revisit those earlier levels and strengthen the child's foundations as the following case story shows.

Case story: Jamie moves up

Jamie is a seven-year-old boy with cerebral palsy and developmental delays who was referred for play work by a speech therapist because he was unhappy, constantly having meltdowns and his development had plateaued. Although he could read and write, his engagement, understanding and regulation did not mirror his cognitive skills. He was home-schooled and had been in therapy for years to develop the functional skills he needed for daily life. After assessment, and discussion with his mother, it was agreed that he needed more child play which *he* could direct and control. He worked with therapists twice a week in child-led play.

On first meeting Jamie showed limited engagement and reciprocity, until he climbed into the sand tray and began to cover himself with sand and play hide-and-seek. That was when the engagement began. For many weeks, he played with sand, shaving foam and messy paint at the first level of sensory body play, and often at the end of each session returned to writing words to make meaning of his activities, moving up to level three. Joy and engagement began to emerge as he was given complete control over his sessions and not directed to learn letters or numbers or complete tasks. He began to create his own safe world, and his behaviour began to change.

Out of this sensory, messy and explorative play, Jamie moved into lining up paints by colour, ordering, sequencing again and again. We recognised that it had meaning for him and worked with it, such as sometimes expanding it into more playfulness by hiding the paint bottles on the shelves so he had to look for them and engage more with the environment. This went on for some months, until we started to wonder if we had got stuck. But if there is one thing that children teach you, it's that repetitive play is an important part of development. Children need to do the same things again and again until they are ready to move on, and what we were seeing was Jamie making his meaning and feeling control over his world. That's a very important stage for a child who'd had so much done *to* him throughout his early years of therapy. Sometimes, we would revisit early messy and sensory play (level one and two) together, and that helped him expand on the meaning-making stage (level three).

After several months, Jamie moved away from the playroom and into a sensory gym. He began to engage more with big-body play, and his imagination began to develop. For example, a big tube became a lift he could climb in and get stuck in, and we played going up and down in lifts. He would roar with laughter as I wobbled the tube saying, "Oh no, we're stuck in the lift!" Spontaneity in play and roles was beginning, and we began to engage socially in play far more often. Having initially been very rigid and not open to my engagement with him, once he was able to use his body more in play, Jamie developed an openness to others in his space. We played spontaneously and together with much laughter.

Throughout these sessions, Jamie moved up and down the Developmental Play Pyramid starting with sensory body play and moving up into creative exploration and then meaning-making and finally beginning to engage with higher play, imagination and social engagement. Whenever he got stuck, we found that revisiting lower levels of the Pyramid helped. By creating a safe space where we saw him and validated his play even if we did not always understand it, he found safety to develop and learn.

After two-and-half years of weekly therapy, Jamie was referred on to more social group therapy as he needed to develop higher-play skills with his peers. He had reached beyond the potential that had been expected of him when he was first referred, and he continues to develop and thrive. He will always have challenges, but the Developmental Play work has helped fill in some gaps, and has built the foundation skills needed for further development.

Enter the world of the child

We talk of the levels as if they are separate, but they have a lot of natural overlap. If you watch neuro-typical children playing freely, you'll often see multiple levels of play going on at the same time. They will run, feeling their bodies in space (level one); manipulate and play with objects such as toys (level two); repeat the same kind of play to create patterns (level three); and tell a narrative involving rules and roles (level four). Children also have cycles of play. As we saw with Jamie, they may move up a level, but then they may go back to sensory play again to cement one process of learning, for example. While the pyramid is about levels of development it should not be rigidly applied. After all, life is not a linear journey and the complexity of the human brain is like the branches of a tree that go in all directions and make what often seems like random connections. That's the wonder of play, or as Loris Malaguzzi, founder of Reggio Emilia called it, "the hundred languages of childhood" (Edwards, Gandini & Forman, 2011). Developmental Play gives a clear view of how children play at different stages and what their play tells us. Its focus is on supporting children to be the navigators of their own journey, and to support and scaffold that journey where needed. The more we can enter the world of the child, the more we will be able to work successfully with them. The aim of this approach is to help us revisit and understand how and why we played before we could put it into words – or write it in a chapter.

Note

1 Differently abled is the term used in India for children with disabilities and special needs and we have chosen to use this terminology to reflect the roots of Developmental Play in India and also the inclusive nature of the term reflecting diversity and difference rather than a disadvantage or inability.

References

Ahn, R.R., Miller, L.J., Milberger, S. & McIntosh, D.N. (2004). Prevalence of parents' perceptions of sensory processing disorders among kindergarten children. *American Journal of Occupational Therapy*, 58(3), 287–293.

Allen, S. (2016). *Let Me Tell You About Sensory Processing*. London: Jessica Kingsley Publishers.

Archer, C. & Siraj, I. (2014). Measuring the quality of movement-play in early childhood education settings: Linking movement-play and neuroscience. *European Early Childhood Research Journal*, 23(1), 21–42.

Ayres, A.J. (1974). *The Development of Sensory Integrative Theory and Practice: A Collection of the Works of A.J. Ayres*. Dubuque, IA: Kendall, Hunt.

Ayres, A.J. & Robbins, J. (2005). *Sensory Integration and the Child: Understanding Hidden Sensory Challenges*. Los Angeles, CA: Western Psychological Services.

Baron-Cohen, S. (1995). *Mindblindness: An Essay on Autism and Theory of Mind*. Cambridge, MA: MIT Press, Bradford Books.

Bergen, D. (2013). Does pretend play matter? Searching for evidence: Comment on Lillard et al. *Psychological Bulletin*, 139(1), 45–48.

Bodrova, E. (2008). Make-belive play versus academic skills: A Vygotskian approach to todays dilemma of early childhood education. *European Early Chidlhood Education Research Journal*, 16, 357–369.

Bowlby, J. (1969). *Attachment*. NewYork: Basic Books.

Brown, F. & Webb, S. (2005). Children without play. *Journal of Education*, 35(1), 139–158.

Edwards, C., Gandini, L. & Forman, G. (Eds.). (2011). *The Hundred Languages of Children: The Reggio Emilia Experience in Transformation: The Reggio Emilia Experience in Transformation*. Santa Barbara, CA: ABC-CLIO.

Essame, C. (2016). *Fighting the Dragon, Finding the Self: Why Art and Play Matter in Early Childhood*. Singapore and London: CreateCATT.

Gardner, H. (2000). *Intelligence Reframed: Multiple Intelligences for the 21st Century*. New York: Basic Books.

Gerhardt, S. (2004). *Why Love Matters: How Affection Shapes a Baby's Brain*. Hove: Brunner-Routledge.

Greenspan, S. (1999). *Building Healthy Minds*. New York: Da Capo Press.

Huitt, W. & Hummel, J. (2003). Piaget's theory of cognitive development. *Educational Psychology Interactive*, 3(2), 1–5.

Jennings, S. (2010). *Healthy Attachments and Neuro-Dramatic-Play*. London: Jessica Kingsley Publishers.

Leong, V., Byrne, E., Clackson, K., Georgieva, S., Lam, S. & Wass, S. (2017). Speaker gaze increases information coupling between infant and adult brains. *Proceedings of the National Academy of Sciences Dec 2017*, 114(50), 13290–13295.

Metcalfe, J. & Shimamura, A.P. (Eds.). (1994). *Metacognition: Knowing About Knowing*. Cambridge, MA: MIT Press.

Rutter, M. (1972). *Maternal Deprivation Reassessed*. Baltimore, MD: Penguin Books.

Siegel, D.J. (2012). *Developing Mind: How Relationships and the Brain Interact to Shape Who We Are*. New York: Guilford Press.

Stagnitti, K. (1998) *Learn to Play: A Practical Program to Develop a Child's Imaginative Play Skills*. West Brunswick: Co-Ordinates Publications.

Stagnitti, K. & Cooper, R. (2009). *Play as Therapy: Assessment and Therapeutic Interventions*. London: Jessica Kingsley Publishers.

Sunderland, M. (2006). *What Every Parent Needs to Know: Love, Nurture and Play with Your Child*. London: Dorling Kindersley.

Van der Kolk, B.A. (2015). *The Body Keeps the Score: Brain, Mind, and Body in the Healing of Trauma*. New York: Penguin Books.

Vickerius, M. & Sandberg, A. (2006). The significance of play and the environment around play. *Early Child Development and Care*, 176, 207–217.

Vygotsky, L.S. (1967). Play and its role in the mental development of the child. *Journal of Russian and East European Psychology*, 5(3), 6–18.

Vygotsky, L.S. (1980). *Mind in Society: The Development of Higher Psychological Processes*. Cambridge, MA: Harvard University Press.

Williams, M.S. & Shellenberger, S. (1996). *How Does Your Engine Run? A Leader's Guide to the Alert Program for Self-Regulation*. Albuquerque, NM: TherapyWorks.

World Economic Forum. (2019). *What the Global 'Play Gap' Means for Our Children's Futures*. Available at www.weforum.org/agenda/2019/01/play-gap-hurting-childrens-skills-futures/.

World Health Organization. (2011). *World Report on Disability*. Available at www.who.int/disabilities/world_report/2011.

7

PLAYFUL WORKSHOP TOOLKIT FOR PROFESSIONALS WORKING WITH CHILDREN

Irina Calota

This chapter focuses on offering practical and creative solutions for counsellors, when working with teachers or other staff members who interact with children, using the embodiment-projection-role paradigm.

EPR was developed by dramatherapist Dr Sue Jennings (2011) following her observations of the key stages of dramatic play from birth to reaching the age of seven. It is therefore a developmental paradigm that uniquely describes the progression of play in direct contact with the physical, cognitive, emotional, and social development in children.

The child's first social play experiences and their development of secure attachment are essential for the overall development of sensory and brain functions.

By following the stages of EPR and using EPR play techniques, counsellors and teachers can really get in touch with the child and with her or his true needs. Using play, sometimes sensorial play, other times projective play or role play, leads to a very special place, where the child can feel safe, seen by others, connected, powerful and belonging. And when a child is offered the chance to experience these feelings, we usually have a higher chance of creating a proper environment for academic learning.

The three developmental stages of dramatic play are:

- Embodiment: birth – twelve months (everything is experienced through the body)
- Projection: thirteen months – three years (toys and media beyond the body are explored)
- Role: three years – seven years (roles and stories are developed in dramatic play)

The embodiment stage is about messy play, communication by touch, massage, clapping and singing, peep – boo. All these playful interactions are helping both the brain and the body to develop. So, embodiment is all about sensory development, body self, body image, and body movement. The projection stage is about the child responding to the world beyond her or his body: sand and water, paints, bricks, puzzles, drawings, stories etc. The Role stage is about the child taking on roles, using masks, costumes, objects. In the dramatic reality everything is possible, and the child can experience and express different voices, emotions, sounds and roles. Sometimes the child's EPR development can be distorted by different factors, like unhealthy attachments, overprotection, under-protection and abuse (Jennings, 2015). That is why EPR is an extraordinary instrument that can guide and offer support to counsellors and teachers, especially when working with children dealing with challenges like bullying, aggression, anxiety, unsecure attachment, loss, neglect or abuse.

The observer tools

Many times, in schools, adults tend to address behavioural aspects first, focusing on ways to decrease or stop the "unpleasant behaviour" of the child or teenager, creating different behavioural and thinking strategies to eliminate that targeted behaviour. And these can work on a very limited time frame, but after a while, a different "bad behaviour" appears. This happens because the behaviour is just a symptom, a way in which the child is telling us something is wrong.

Another tendency is to focus too much attention on the child that has the disruptive behaviour. The attention itself is not the issue, but the manner in which it's presented, oftentimes by telling the child to stop, by putting different labels on them or describing what she/he does, like "You talk too much/You interrupted me again/You are not nice to. . ./You . . .". In this way the adult is just putting the behaviour in words, adding some judgmental labels, suggesting what the child does is not right.

At the other end of the line is the child, who will most probably get ever more attached to this behaviour because it delivers what they want or need. Maybe it was a need of excessive attention (I don't feel I belong, or I am not connected and I need to be seen), maybe the child was in a power struggle (I don't feel strong or capable enough and I need to show my power), maybe they had a revenge goal (I feel like I don't count and I don't trust others, so I will hurt back), maybe they just gave up and simply refused to try anymore (I feel so inferior, I have no courage and I will fail again). All these are called mistaken goals in Adlerian theory (Dreikurs, 1984), and the child is using them when her or his psychological needs are not met. They are feeling discouraged most of the time and will therefore cope through misbehaviour (Bettner, 2006). The child will feel disconnected and discouraged by the judgmental adult's attitude, so her or his energy and resources to cope will be stifled because they must cope with other emotions like guilt, anger, shame etc. All this makes the job of adjusting the child's behaviour much more difficult for the teacher or counsellor.

That is why it is crucial to address and see what is beneath that behaviour and create those play experiences that can offer the child precisely what they need. It can be as simple as nurture and support. Sometimes, the child only needs more space to express themselves, more time to be seen and heard, or validation and appreciation. Other times the child needs structure and healthy limits because these are missing in her or his day to day life. By using play strategies, we can create that safe space within which the child can freely explore.

How we see, hear and respond to the child is an essential tool for those working with groups or individuals in schools. It is important to see deeper and further, and ask ourselves what the child needs, what they want to obtain with that behaviour. When we see how the child's body moves, we gather crucial info about her or his emotional state and current challenges. When we listen closely, we catch the tone and modulations of her or his voice. Keeping the observer alive, the child's messages are never ignored and represent her or his way of communicating deeper needs.

By maintaining this mental state, the adult is less likely to get caught in her or his own emotions and becomes less reactive. Sometimes in schools, teachers or other staff may very well wind up feeling attacked and taking things very personal when relating with misbehaviours. In these moments it is crucial to Breathe and tell yourself it is not about you, to put on the empathy glasses and breathe some more. And get in contact with your creative and playful resources, they are a big help when dealing with difficult situations.

When the adult is calm, contained and still warm, the child will most probably also feel contained, safe, respected and protected. And this is a good ground for starting to plant the seeds of change and new coping strategies.

And how can we activate all these resources in schools or other organizations working with children?

Playful workshops using EPR

A very useful way is for the counsellor to create **special play workshops for teachers**, where they have the chance to experience the three play stages once more: embodiment, projection and role.

In these workshops, teachers have the chance to stimulate their right hemisphere, the one that is responsible for creativity, imagination, flexibility, intuition and emotions. It can help them more naturally tune to the children's worlds. Another benefit is that they get in touch with the energy of their inner child and activate creative and playful resources. We all know we need those creative resources when working with challenging situations, or with big groups of children.

The workshops can have different main topics, related to playfulness, nurture, exploration, community spirit. For example, we can work and focus on themes like the inner child; breathe in and out; welcome the newcomers; new year, good vibes; let's play together; relax and smile etc.

A big advantage of these workshops is that they offer teachers different tools they can use for themselves and in their work in class. The beauty of the method is that it allows teachers to first experience these tools directly by having them used for their own relaxation and stress relief. They can feel the benefits of using body work and sensorial games, playing with their voices and with rhythms, drawing and painting, improvising or creating scenes. This will naturally help them better relate with children and their needs. They make new connections to help them better understand different emotional responses or misbehaviours. Playing with their colleagues gives teachers the opportunity to become more open, relaxed, flexible, joyful and optimistic, both more sociable and realistic about how we all work.

It is the active and experiential way of learning, a dynamic workshop where the teachers are in action, learning by doing and feeling. These workshops are not therapy or counseling and it is important to explain that to the participants, they are in a workshop for their wellbeing and the activation of playful and creative resources that can help them in class and in their day to day lives.

The timing of the workshop is very important. A good moment is at the beginning of the school year, for any newcomers to feel welcome and for the older staff to reconnect with their new school year goals, intentions, resources etc. Another good moment is after the end of a semester when teachers usually feel tired, overworked and stressed and therefore need some "fresh energy" and to give themselves time to rest and reset. Another useful time is at the beginning of the year, for example creating a 'new year, good vibes' workshop, focusing on new year resolutions and what energy they want to activate in the new year. The end of school is the time to look back and celebrate all the achievements of the past year, so another naturally good moment to bring teachers together in a playful context.

By using this method, we are accessing the ritualistic dimension of play therapy. As we know, social and cultural rituals have the power of making others feel that they belong, create solidarity and increase the connectivity between others. As Sue Jennings emphasizes "Ritual and risk form the basis of children feeling safe in the world, as well as a desire to explore the world" (Jennings, 2012: 4). The balance between the two is very important because the context of the workshops makes the participants feel safe enough that they can get in easier contact with their exploratory dimension. They can risk more in a good way, try new things, open more, engage in new experiences. That is why every workshop has a specific structure, a warm-up, middle and closure. This well-defined structure offers containment.

The warm-ups, middle and closure games or activities are chosen to sustain the main objective of the workshop, either to offer new ways of relaxation, or connecting with their inner child resources, maybe creating the mission of the group, or cooperation and teamwork etc.

The counsellor can choose the length of the workshop, depending on the number of participants, so it can be from two hours to six hours. The time frame is a flexible one; each counsellor or play therapist would take account of the needs of the group, the internal policies of the school, well known group dynamics, different impacting factors. The maximum number of participants is sixteen to eighteen, this way all participants will have the chance to express their ideas and feelings, as well as get closer to colleagues.

The power of circle – the safe place where you can express yourself and explore the new

The first short introductory discussion and the first games should be done in a circle, the same with the closure part. The group starts and finishes in a circle, again the ritualistic dimension proves to be very important in creating the safety and stability of the group. The circle is also a nurturing element, offers a high level of containment and gets people together, close to one another, having the support of others. When in a circle you can see everybody else and you are seen by all others. A lot of human needs are covered by standing in a circle – to belong, to be seen, to count, to be with others – so it is a wonderful way to start and to finish a group session, and can also be used for some of the middle activities.

Circle games and activities examples

A lot of embodiment, projection and role techniques are, or can be done in a circle. Here are just a few ideas of activities that can be used in playful workshops with teachers or others.

Embodiment

- ***Body movement exercises***: Stretch and shake every part of the body. Each participant will make a gesture showing how they feel, and the others will repeat it; change places when you catch somebody else look; move a plastic bottle or a balloon to the person next to you without using hands.
- ***Sound and voice exercises***: Each participant will say his name using different tones: loud, kind voice, whisper. The others will repeat after, all together. Playing with vowels and consonants. Throw a soft ball to another person and say your name or something that helps you relax. The game finishes when all from the circle got the ball.
- ***Breathing exercises***: breath of the lion, breath of fire, breathe in calm and relaxation and breathe out stress.
- ***Rhythmic exercises***: clapping, using drums, sticks, awareness of heart beats, creating rhythms on the floor using the feet or palms.

- ***Massage***: self-massage, tapping, massage others – each participant turns to the right, they get closer and massage the person in front of them. In this way everybody gets a massage and it can be used both as a warm-up, energizing the person in front, and in the closing part when each participant can say a word suggesting how they feel or what they want to give to the group after the time spent together. That word will be represented by some massage moves on the back of the person in front. When using massage it's very important to respect the massage safe zone, starting from the base of the neck until the waistline.
- ***The shower game***: a person is asked to stay in the middle of the circle. All circle around and at the same time place their hands on top of the head of the person in the middle. All at once, they will start to descend their palms, doing a soft massage to the person in the middle. The person is asked to protect their private parts, a woman to cross her arms around her breasts and if a man, to put his palms over his private area. The exercise can be repeated three times. Also, the person can choose if they want the "hands shower" with touch or not. With no touch, the participants will place her or his hands near and do the same movement but without touch. This exercise is a very helpful tool to create a connection, to give and receive nurture and soft care, to yourself or others, and gets participants in a good state of mind. It works very well as one of the closure activities with a group.

All these embodiment exercises help participants be more active, get in touch with their body, create space for change, not only at the rational and mental level, but also at the physical level. I remember that during the first two workshops with teachers, a lot of them felt a little uneasy to stand up and start moving and playing using their bodies, but after a few activities, many started to become more open and active, most carrying a smile and growing ever more curious. The very next year, the very same colleagues asked when we would meet again to play some more several times throughout the year. It seems getting in touch with your body is a great way of reconnecting with your self-image, it grounds the person more. It relaxes and activates good energy. As Sue Jennings suggests:

> The body self is the first step towards the growth of identity. Once children have a strong body self, they are able to develop a balanced sense of body image. I cannot have an image of my body unless I feel that I inhabit my body.
>
> *(Jennings, 2006: 36)*

So, using embodiment techniques offers a great wealth of ideas for professionals to use in their work with children. These help the entire educational process go smoother and improve the development of children in a holistic way.

Projection

- ***Drawing, painting, modeling with clay, plasticine***. The participants can stay in a circle on the floor, using cushions or on chairs if they need it.

There are a lot of projective activities that can be used, such as:

Draw what relaxes you: each participant will draw on a piece of paper things that relax them. Of course, we can use the circle again, so we ask the participants to draw a big circle on the page and to draw within it, creating the containment needed.

Draw your name: each participant will choose a colored paper and then will draw their name, being as creative as they like, thinking of their own personality, what they like, and expressing these through the way they draw the letters in their name. This exercise works very well at the beginning of the year when newcomers arrive. It really helps people get in contact with their identity and to show parts of themselves to others. After they finish drawing, they can put the drawings themselves in a circle and then walk around it and see other's pieces. After everyone has gone around once, they can each stop behind their own paper, saying their name as they would like to hear it from others (tone, vibration of the voice, fast or slow, kind voice or firm). This exercise therefore combines a projective technique with an embodiment one.

Draw your favorite toy from childhood: First draw your favorite toy. Participants can then write on the same paper some of the qualities and resources drawing the toy brought back from childhood. The connection with their favorite toy and the good memories from childhood it brings, facilitate the rediscovery of strong points, qualities, resources, abilities and help bring all these into sharp, present focus, so the participant can use them straight away.

Paint your safe place: Participants will paint a representation of their safe place, where they feel secure, good, calm and relaxed, contained and in-touch with themselves. This exercise is useful because it awakens their inner resources and creates a helpful anchor for any situations that can have a higher level of anxiety or distress at work.

These last two projective exercises are more efficient and have more benefits for participants if done after a *guided relaxation and creative visualization*, another projective technique. The participants are asked to sit in a comfortable position and close their eyes if they feel like it. They then relax their entire body, from head to toes. The next step is to start the creative visualization process.

For the favorite toy activity, they could imagine a travel machine that brings them back when they were little, seeing their favorites games, toys and they could then choose a toy that they like a lot, then imagine holding it, exploring it sensorially, remembering it's smell, how it feels to the touch, if it makes any sounds, how they play with it. Participants should try to pay attention to their feelings while holding the toy. Once the toy is fully explored, slowly return their attention to here and now and finally open their eyes.

For the safe place activity, we ask them to picture a road in front of them, as they would like it to be. They start walking on it, paying attention to the things around the road, nature, animals maybe, putting a focus on the senses, what they see, how it smells, what they hear. At the end of the road they will find a place that can be as they want, a house, a shelter, a cave, a tree, an imaginary place that they will explore. And in this place, they find a spot where they can sit and relax, it's her or his safe place, where they feel welcomed, relaxed, secure. Again, we focus on all senses, so we create strong anchors within the body. We let people take some time in that safe place, to rest, breathe and calm down. Afterwards they can choose an element from this place to take with them, a reminder of the feelings experienced there. They can hold it, feel its sensorial properties. You can then let them know that whenever they feel anxious, stressed, uneasy, they can imagine holding that element. Lastly, we ask people to return on the road and then to start being aware of the sounds in the room, the smells, and finally, gently open their eyes.

- ***Writing***

Work on common vision: each participant writes on a piece of paper the professional goals and values they would like to adhere to and see around them in the coming year. From the list, they then choose one or two and write the chosen ones on a green piece of paper.

On another piece of paper, each person will write their resources and qualities, especially those used and showed at work. They then choose a couple, those they feel can best sustain the values written on the green paper. They then write these resources on a light brown piece of paper. They are given the instruction to write holding the paper vertically.

Afterwards, they receive another paper, a dark brown one this time, on which to write an encouragement message that they would want or like to hear throughout this year, about them, or their work.

All the green and brown papers will then be arranged in the shape of a big tree on the floor, the green papers are the leaves of the tree, the light brown the trunk, while the dark ones represent the roots. Each participant will place their own papers, so the tree is created by them. It is important to give the instruction that they can place their own papers but cannot move or take other's papers from the tree.

The result for each participant is a visual representation of their goals and values, the qualities needed to support them, and a strong personal mantra that can in turn support these qualities. While individual, each of these is a part of the tree that, in its entirety, represents the goals, values, and qualities of the whole group. Therefore, this activity offers a lot of clarity while empowering participants and creating stronger connections within the group, it provides the feeling of working together to get to where we want to be.

- ***Confidence list***

Each participant writes his name on top of a paper. All the papers are ordered in a circle on a big table. Everyone receives the instruction to write a personalized message for each individual on the other's pieces of paper. This message would contain some positive feedback, what they notice and appreciate regarding the other person, what inspires them or a feature they identify as a model in the other person, what they validate or encourage for the future. So, at the end of the exercise, each member of the group will be left with a single piece of paper full of positive messages about the way they are perceived by their colleagues. This supports the stabilization of their professional identity and helps a lot in empowering them and awakening pleasant emotions.

Of course, there are more projective techniques that can be used, playing with cards, pictures, using stories – we can read a story to the participants, or they can create their own, using small toys and figurines, or by creating masks. We can use a large range of projective techniques so we can help the process of exploring and expressing much of the internal material and personal meanings of the participants.

Role

- ***Individual or group sculptures***: To sculpt with their own body one of the things drawn earlier, for example, whatever helped the participant to relax, so each participant will express that relaxation symbol by using their body, gestures and mimicry. The participant will freeze for a few seconds in that posture like a sculpture. A variation of this exercise is to ask everybody else to repeat/mirror the posture. Sculptures can be created to express different emotions, animals, values and attitudes (kindness, honesty, patience, active, strong).
- ***Dynamic sculptures*** are a variation that creates more movement. The participants receive the instruction to create all together in the middle of the circle one big dynamic sculpture that will represent a seaside picture or an extraterrestrial ship or a giant vacuum, or any complex setting. One by one, the participants will choose to become an element of the

given structure and will enter the middle. They will stay in a spot and from there they will find a repetitive movement to express that element. They can also use a repetitive sound. The game stops when all participants are in the structure doing her or his movement and voice.

- ***Improvisation games***
  What are you doing?: One person sits in the centre of the circle and gets the instruction to mime an action, after they mime it for 20–30 seconds, another person enters the circle and asks: What are you doing? The person in the middle will answer with another action, not the one they were doing up to that moment and then returns in the circle. So, the person who asked remains in the middle and starts to mime the new action, so on and so forth until all participants get to be in the middle.
  What is this?: A simple object, like a pencil, a stapler, a brew will be passed from one participant to another. Each participant will transform the object into something else by miming an action with the object. For example, the pencil can be a comb, a microphone, a wand and so on.

There are a lot of improvisation techniques that can be used in a circle. All these offer participants the chance to become engaged with different roles, ones that are more or less comfortable for them, to explore different situations, to get out of their comfort zone, to activate their playfulness, flexibility and ability of putting themselves in other's shoes. And this exploration is made in a safe way because of the distance created by the power of playing a role.

The power of exploring out of the circle (the nest)

During play therapy or counseling workshops for the teaching staff it is essential to create a dance, a dynamic between the two dimensions, ritual and risk so that the participants can benefit from both and develop and activate their full resources.

For example, after creating a safe atmosphere by doing introductory embodiment games in the circle, we can go further and explore with the participants that space. We can ask them to start walking around, owning the entire space of the room. From this basic random movement, a lot of variation can be achieved, various embodiment and role explorations. Walking in different ways: faster, slower, with heavy steps. Walking on different imaginative surfaces: ice, tall grass, water and, why not, mud. Walking and having different shapes of the body: round, thin, long, square. Walking like different characters or animals: ballerina, giant, monster, dwarf, zombie, king and so on. Walking and interacting with others: when you encounter another participant, say 'hello' using different voices or saying 'hello' by using hands, shoulders, elbows.

Another great way of activating playful resources, helping participants get in better contact with themselves and others, and facilitating closer interactions, is by using dyads exercises that can better sustain the goals of the workshop. For example, we can use the classical Mirror game to help people tune in more with one another and explore different postures. There is a great set of dyads exercises for exploring different roles and attitudes in relations. They help finding balance in interactions, confronting in an assertive way, putting limits.

Balance exercises: Participants find a partner. Facing them, they will hold each other's wrists and slowly lean themselves on the back of the other person, so they could find their balance while sustaining each other.

Confronting exercises: facing each other and putting her or his hands in front and touching the other person hands, they will have to push the other person, to gain more space, of course, while taking care of the safety of themselves and their partner. Holding one hand of their partner, they will have to grab the other person towards them. Yes and no dialogue: one person says 'yes' and the other responds 'no' and after a few times they chance the role. The idea is to explore more ways of saying 'yes' and 'no'.

Nurture exercises: in a pair, one will massage the back of their partner following an imaginative support offered by the facilitator. For example, 'pizza' or 'apple pie' massage or 'seasons' massage. They then change the roles.

These are just a few examples of activities that can be done without using a circle. There are so many more, thinking of role and dramatic techniques, like acting parts of a story that was read, creating scenes by using masks or puppets, and many more – really, it's all up to your imagination.

The essential point is to let ourselves as professionals open to the new, to be flexible and in contact with the needs of the group we work with. This creates a structure that simply fits better.

In conclusion, playful workshops for teachers or other members of organizations focus on activating their strengths and offer the chance to interact in a very fun and playful manner. They bring about contact with the participant's own emotions, but most importantly, they *empower* them. And it's like a domino effect, we want teachers or other adults to focus more on the strengths of the children they work with, to empower more, to put limits in a healthy way, to inspire and be role models. We therefore need to create plentiful opportunities for adults to re-experience all these attitudes and mindsets in a safe, free, enjoyable and nurturing way.

References

Bettner, B. L. (2006) *The 'Creative Force': How Children Create Their Personality*, Media, PA: Connexions Press.

Dreikurs, R. and Soltz, V. (1984) *Happy Children: A Challenge to Parents*, New York: Harper Collins.

Jennings, S. (2006) *Creative Play with Children at Risk*, London: Speechmark.

Jennings, S. (2011) *Healthy Attachments & Neuro-Dramatic Play*, London: JKP.

Jennings, S. (2012) *Neuro-Dramatic-Play and Trauma 'Towards Healing and Hope'*, Malaysia: B. Braun Medical Industries Penang.

Jennings, S. (2015) *Neuro-Dramatic-Play & Embodiment-Projection-Role Part Two*, London: UK: A Play Book for Adults of Theory and Practice, Sue Jennings.

8
LILALOKA
A place of play

Anna Oijevaar, Jitta Skoles and Marrianne Dubouchet

Introduction

Lilaloka is a self-learning resource centre for the child, big or small, having special needs or not. A place where they feel safe and accepted and are unique and happy with themselves amongst others. The Lilaloka project is situated in Auroville, an international township in Tamil Nadu, Southern India. Auroville was founded by Mira Alfasa, known as 'the Mother', and Sri Aurobindo and based on their teachings and guidelines concerning integral education. Integral education, as envisioned by the Mother and Sri Aurobindo, regards the child as a growing soul and helps him to bring out all that is best, most powerful, most innate and living in his nature. The Mother believes that each individual manifests specially in the Universe and that each person's true path has to be completely unique.

The focus of our work is the integral development of the whole human being and is especially aligned with guidelines from the Charter of Auroville (Majumdar, 2017), which wishes the town to be truly universal, open to all men and women and a place of education, growth and peace for all. Research in education is thus of prime importance; children open the way to the future, to develop new educational approaches and places for all, based on free choice, creativity and non-judgment. It is a demand expressed by many families in many parts of the world, where such initiatives are seen as the beginning of an answer to the major human crises of today. Our project comes from a great demand for the new approach to which integral education provides for our new age children.

Auroville has amazing resources and unique educators. Our centre is a space dedicated to the child, parallel to the schools, open to all ages and types of children, attending school or not, with or without special needs, beyond all categories. Here, all will meet and benefit from the energy born from these interactions, each individual being enriched through the differences of the others; where joy shows the path and play is the work. It is Lilaloka's dream to become a resource centre dedicated to this, a place where educators can meet and share their research and experience.

Lilaloka's site covers one and a half acres and is located next to an indigenous forest. A building, no doors or windows, surrounded by a magical world of nature influenced very much by our way of working. In group sessions we are providing several facilitators to support each individuals need. One of us is always available when a child needs some individual, quiet time.

This individual space is a very important aspect of the soul of our work. Most children who come to our sessions are in the age range 5 to 15. We work with what we call free flow, free choice, follow the children. These expressions are used all over the world manifesting different aspects of this way of working which for us means that we create an atmosphere which is so set up that it provides a stimulation to explore different possibilities: a table with art supplies, drawing material; a corner with books, dolls, long pieces of different coloured cloth; a table of musical instruments; garden tools. Children are full of magic and curiosity and love to play and explore. They come in and are immediately intrigued.

The children are allowed to flow freely where they want. As facilitators we observe and are available but we do not give direction or impose. Children are full of ideas, when they are able to flow freely through the diversity of possibilities they feel at ease to reveal and discover their inner being. "Where are my interests?" What am I good at ?" "What can I learn"? These interactions with the children guide us towards the journey we will embark on together.

With each group you go on a different journey. We aim to create a place where play transforms, where joy shows the path. All activities are centred on the self, the other, and the world around us. Each child should be able to be the creator of their own world and learn that this knowledge can benefit the bigger world. To experience that one's talents are recognized and made useful for the whole gives a feeling of belonging. For children this means playing happily together. The need for a safe personal space, and harmonious, creative, stimulating social interactions is an atmosphere needed by us all to be able to reconnect with the inner child which lives within us. The team of facilitators works as one but within the frame of their own specialty and provides that aspect within the group. We have the following specialties in our team:

- Nature's theatre
- Discovering medicinal plants
- Communication through sound
- Storytelling, puppets and games
- Individual play sessions
- Shiatsu and reflex therapy

Nature's theatre

Nature's theatre wants to create a safe and inspiring environment. The first sessions are based on getting to know each other. They always start with the proof that we are all stars. We use the story, '1, 2, 3 and a Star We-Be', developed by the facilitator. A star comes to visit a child to prove that we are all just like a star. The star takes us on a journey where we meet three other stars. Each star offers us a part of a magic sacred tool. They explain to us the essence of that part and what to do with it. They let us experience the magic of a first point which contains an infinite sea of possibilities. Whatever we see, whatever is created, starts with a first point as do we, in the belly of our mother. With the use of our own body we take measurements to construct our star in our own size. The last star shows us how, with the use of this tool, we can pull a line where the beginning meets the end and just goes on forever.

Our first construction, the circle, called the one, which is just our size, is the space we take in the world, the space in which we all have the right to feel safe. Within that line our star is born. We are uniform in our shape but no point is the same and each star has its own path to follow. Together we make a constellation. All the instructions the stars give us and actions we have to take are accompanied by 19 square illustrated flashcards [55cm by 55cm] which makes it visible and easy to understand. Playing with shapes, relating them to our self, our surroundings and our language helps to prepare the mind for understanding.

Many of the games and exercises are developed to help the child to recognize patterns of behaviour which help to reconnect with their inner being and happiness.

The circle, gives many possibilities to play, create, explore, and learn to recognize principals which stay in harmony with our self and with others.

Working together exercise – cut out two circles. Those two circles represent the two, in this case two children working together. They are busy and move around, sometimes partly overlapping each other. Suddenly one circle totally covers the other one. Followed and covered wherever it moves. What would you do if you are that circle not being able to show yourself and let your star within shine? They understand, the one under will start to push and pull, scream to be seen or cry and feel squeezed.

Energies of lines and colors, verbal social interaction exercise. Put some circles on a piece of paper. Play within that space with lines and colours. Fill one circle with round flowing lines and soft pastel colours, another with pointy lines and dark, hard colours. What feeling does each one provoke? Which ones make you happy, which ones sad? If those circles are made of thick white carton paper you can cut them out, stick a small pencil through the centre of the circle and you have a spinning top. Each spinning top will reveal a different surprise but they have all one thing in common. When turning we see many circles within the circle. Geometry is used in our language, especially in expressions. For example, "You go around in circles, there is no point in what you do. It's pointless. We have different points of view". When we can relate those words to our constructions we make it visible, easy to understand without many words. They show us the path towards harmony within and in the collective so we can shine on our own path and together.

Two examples follow.

> One of the participants, a girl age seven, was restless and jumping around, not happy or interested in anything that was going on. I approached her and smiled asking her, "What are you doing? You go around in circles and there is no point in what you do, you are not happy with it". She took a deep breath and puffed it out. "Yeah", she said, "Yeah". So I suggested she should make a point of herself and feel her star and see where it wants to shine. Less than five minutes later she came back full of light and sure about what she wanted to do. She wanted to be a writer and write stories. We created a Writers' Corner. At first she asked for each word to be spelled out. When I answered that the spelling did not matter right now, it is all about the story, she wrote continuously for 45 minutes.

> My group entered the room arguing, pushing, an unhappy, irritated, conflicted bunch of children age seven to ten. So I suggested that we play a game I call 'Points of View'. I put a big pot on the floor filled with pens, sticks, and flowers and told them that's the (centre)? point of this game. Around it we make our point so we form together the never ending line, the circle. I asked everybody to describe what they see. When everybody had their turn, I started to pretend-play an argument with the person who is sitting opposite me. I say that it is not possible that they understand what each other is saying and doing. At this moment the children start to feel uncomfortable and restless, "How can you say that Anna, you cannot see what is there, but it is there?" So I made my point and showed them that only together we can see the full picture; together we see more. We can only see things from one point of view unless we move.

After that there is always some sharing and the conflicts and irritations are totally forgotten. We started to play and create in harmony. My experience is that once we have constructed shapes, played with them and made them, using different techniques and material, the participants are inspired to design and create. With children this mostly results in creating a world together and playing within it. Creation and play are the foundation of nature's theatre.

While we create we build a world, while we play in that world we create the objects needed to manifest that world. In a playful natural way we become a community where all have their part to play by sharing their talent in the service of the development of this community. Some become the painters, builders, care takers, farmers etc. This results in a feeling of belonging and joy which gives positive life energy. This creates harmony and mellows down the competitive attitude. When the struggle reduces, our abilities will be nourished and have space to develop.

Figure 8.1 Preparing the plant creations

Natures theatre is an ongoing exploration, creation and research project. All material, games, and storybook '123 and a Star We-Be" are developed by our facilitator.

Discovering medicinal plants

We learn the use of mainly local medicinal plants in a playful way. The activity is based on team-work. Together we honour Mother Earth, each other and the world around us. We explore and learn how we can support each other. We do this by exploring the land around us and learning to recognize medical plants that can help our body and mind, to be healthy and comfortable.

We learn how to prepare the remedies, [the children call them magic potions]. Simple home remedies for rashes, burns, dehydration, common cold, body coolers, anxiety, nightmares, anger, shyness and make a spa or one's own new potion. (This includes making wood fires for our teas, mixing and squashing with our hands, cutting, pouring, stitching and delegating tasks to each other so all get a turn.) The preparation of our plant creations and their application invites communication, sharing, and team work (see Figure 8.1).

We share stories of our experiences related to the plants healing abilities. This builds trust among the children; we become aware of our inner self, the other and the world around us. For example, one child spoke about having nightmares which began a sharing of dreams and resulted in the making of a dream pillow. It also helps to develop the senses by experiencing different textures, smells, and tastes together with the joy of discovering the interrelation of all things.

To support the natural curiosity in the child. To provide a safe space to freely express themselves and feel comfortable to ask questions. We start our first activity by putting our hands in the earth to make an offering to Mother Earth and show her our love and gratitude. We use clay and natural materials found in Auroville (like feathers, dead insect wings, seeds, shells, crab claws, porcupine quills, wild cotton) to create our personal unique creation to offer. We arrange our first creations with flowers, water, fire and do a small ceremony together or alone. We leave it and through time see how it's becoming one with the earth again. Starting in this way gives the opportunity to silently see where each child is in their development and social interactions. These observations will guide towards the choice of plants to introduce. The main objective is to build confidence, respect and ownership of oneself and ones feelings. To help the child to learn to express and respect themselves and the world they live in. This is done with the help of medicinal plants and how they live in coexistence with other living things. So also with us.

An example follows.

> A child, six years old, naturally shy and introverted and with a small social circle, became withdrawn and wasn't socialising very much with others after her one close friend had moved away. She was struggling with issues of fear, having nightmares, not making new friends easily, and was very shy. She showed a lot of interest in the sessions and loved to make potions and dream pillows. She bonded strongly with the facilitator and this safe place allowed her to slowly build her confidence and trust in other people. This change was noticed by her parents and school teachers. Over the course of six months the child gained a lot of self confidence and strength within herself. Her parents informed us that the child has now started first grade with a new outlook and confidence to grow new friendships and interact with her classmates. She wanted to continue this year with individual sessions, not because she needs them but for the joy of contact with the facilitator and being in Lilaloka.

Communication through sound

Music and movement activities are used for self-discovery and self-expression. The individual and group activities are focused on sound dialogues during one hour when the child can freely explore sounds, instruments, and elements. At first the adult is the witness, and then begins an imitation/mirroring process to start a dialogue, based on sounds.

In the middle of the mirroring process, some sound/rhythmic variations are added by the adult, and the reactions of the child are observed and transformed in a playful communication through sound and movement. These activities are mainly based on the awakening and stimulation of the mirror neurons (Lacoboni, 2008) which reacts both when a particular action is performed and when it is only observed. They underlie imitative action, awareness, and understanding of another person's act, intention, or emotion. I offer a series of different instruments and children can choose which one to take. We start by simply listening to the different sounds of the instruments and explore what we can do with each different instrument. After that we include our voice and movement. We become tigers, dogs, wind, rain etc. Sometimes we turn into a group of musicians and dancers.

The main aim is to give more self-confidence and awareness to children, let them freely explore themselves through sounds and movements, so that they will have a more free and clear way of expressing themselves; to discover, understand and change their patterns and keep their imagination and dreaming world awake. This will lead to the development of intuition and empathy.

Storytelling, puppets, games, and role play

We offer books, hand puppets, games, and dressing up props. We read, create stories, draw, and play games. The activity is integrated in most sessions and aims to encourage communication and verbal expression. For the individual sessions, which take place mainly in the mornings, Lilaloka is working with children and parents in collaboration with schools and the team of Supportive Respite for Children who need Supportive Learning. The individual sessions mostly host children who have difficulties to participate in groups and to connect with other children. This affects their social, emotional, and intellectual skills.

Individual play sessions

The sessions aim to provide the opportunity for children to explore themselves, to be themselves, to find their joy which connects to a deeper layer inside, in the presence of an adult; to contribute to a balance enfolding of children's potentials by allowing them to face challenges and find their own way of dealing with them.

The non-directive individual play sessions are centred on the child, which means he/she is the one who leads the play. There is an atmosphere created for the child to feel safe and accepted as possible, feel as comfortable and free as possible to express what they need to. We are there to acknowledge and reflect to the child some of the facts and feelings they express so that they can gain insight into their behaviour, feel accepted and free to change or continue. We do not direct the play or conversation.

The room is designed in such a way that it is one with a small walled garden. A small pond provides the possibility to play with water, dig, build etc. There are garden toys like spades, watering cans, plastic animals, boats, play-mobile people and mermaids. The room is small, protected and invokes a feeling of privacy. Various toys are available like hand puppets, masks for play acting, story books, drawing material, clay, construction toys, games, doll house, doctor set, baby doll, kitchen set etc. The only limitation set is to make the child aware of their responsibility in relationship to the adult so, no destruction of play material, room furniture, or violence towards the adult or him/herself.

An example follows.

A child, nine years old, who was being homeschooled and did not speak until five years of age had a tendency to withdraw into his own world. He started individual play sessions where he was mainly interested in the construction toys, Lego blocks, and play mobile. The child was especially fascinated by the propellers; turning them around and looking to the world behind them. In the beginning he got very upset when he would arrive and did not find the world around Lilaloka the same as it was when he left it in the previous session. Slowly he understood that there are more children coming to Lilaloka and he got curious. So we decided to do the sessions parallel to group activities to give the child the chance to integrate in his own time from a safe place. His curiosity overcame his shyness. He made friends and the parents observed that his communication skills were improving. In a letter to Lilaloka the parents wrote:

> Lilaloka has also been a blessing for him. Since it is essentially a "place of play" there is no other guiding demand how a child should be other than he grows in his own unique way through play. There are no labels of "retarded", "special needs", or "on the spectrum" since each child is regarded as an unfolding. There is no mental image of perfection that the children are being pushed towards by the teachers. Our view is that if you give children the right place and people, there is never anything to "worry" about.

Shiatsu and reflexo therapy

Following a two-year course from Bernard Bouheret 'l'art et la voie du shiatsu familial' and through the teaching of reflexo therapy for children, I decided to combine shiatsu and reflexo therapy as ways of massage and as a tool to connect with the children and gain their trust through non-verbal communication.

We start with a small meditation/concentration based on Holo Energetic, developed by Laskow "Healing with Love" where we align our self through different visualisations. The session takes between 5 and 30 minutes. It can take five or six classes with hyperactive children to be able to receive a 15-minute relaxation massage. The aim is to release the body of the day to day tensions. Once this tension is released, the child is able to find a connection with the self, which opens the way to increase confidence. Once the child feels more confidence they open up to receive the tools which can help them to better cope with their emotions. This results in better social skills, which leads to a more joyful interaction with other children and grown-ups. This relaxing massage is applied in all sessions.

After the relaxation, the child receives full attention to their needs, and together we find the best way through play, games, drawing etc. for the child to regain their self-confidence. Some children need only two sessions to start overcoming their emotional knots. The children come once a week, during the school year in collaboration with the schools, parents, and the team or supported respite for children who need supportive learning. The children are aged between 5 and 12 years old. Most of them, whatever their diagnosis (hyperactivity, autistic spectrum disorder, or emotional damage), show improvement in their behaviour. The trust gained through the relaxing massage allows both the child and the facilitator to communicate with each other on a non-verbal level. Once we gain trust, we can start moving forward.

An example follows.

A five-year-old boy was insecure in his body, moving and fiddling all the time, having difficulty to draw or write or stay concentrated for a while. It took four sessions of relaxation before he could manage to close his eyes without flickering and not moving a single part of his body. Then slowly we passed from 30 seconds of stillness to one and two minutes. After the relaxation, I spent the rest of the session playing play mobile and also in playing specific exercises with a ball.

As a premature child, he has been over cocooned by his parents who treated him like a baby. His speech was that of a toddler. Throughout the year, I started to encourage him to talk as a five-year-old by having him articulate his needs. I could share a lot with the father and he understood the necessity to have his son treated as a boy and not as a baby. He also bought some painting materials so the child could express his artistic side at home as he is full of imagination. As a team we decided to integrate him in the play session of a mixed age group. Being with older and younger children exposed him to many different levels of language and communication skills, which allowed him to develop his speech, gain confidence in his own capacities, and to become more independent at home; eat by himself, go to bathroom on his own, share a story of the day. His active social participation increased inside and outside school.

Conclusion

It is important to us to preserve the innocence and wonder of childhood, to practise and encourage wholesome living, and to provide emotional and environmental safety. Creativity, artistic expression, and play are essential for a child's healthy growth and development, and are fully integrated into our play sessions.

For the last six years we have worked on the Lilaloka compound, a place dedicated to the child and the inner child in all of us. We have regular groups of children who attend the two-hour play sessions for a full year once or twice weekly. The fact that the groups are mixed in age showed some very interesting group dynamics. We observed that there was a similarity in the journeys we embarked on together. When you offer the children what they need and let them freely flow to explore those possibilities they create a natural hierarchy within the group. In a very natural way they find their space and express freely who they are in this particular stage in their life. Rolls are being played and artists, architects, scientists, teachers, caretakers, actors etc. reveal themselves.

We manage to manifest a lot of these ideas. We made a gate, swing ropes, arranged fallen trees to balance on. They made clay sculptures from earth in the garden, little dwellings, fairy houses, and dragon mountains. In the art corner they made little beings to live in those dwellings. They made a fire place. They constructed a gate together. We collected and made through the years a lot of items to play in the garden, a bit like the idea of the sandbox only not in a box but at any place of their choice. We used plastic sheets, earth and stones to make ponds for the boats and mermaids and a place where the animals can drink. And there is always the magic of the fireplace where we can brew magic potions, herb teas, cooling drinks for when it is very hot. The demand is bigger than what we can provide right now. Through our observations, feedback from parents and schools we could conclude that the work we do develops social and communication skills. This influences their emotional and social well-being.

The interest and need for places like Lilaloka is growing all over the world. Many educators and parents who do homeschooling, from India and abroad, visit Lilaloka. People who believe the child is a soul in evolution. They search for new ways and places which can answer to the needs of our fast evolving children. There is a great desire to share our work and ideas. We all touch different aspects, come from different social and cultural backgrounds, and use different material, games, and techniques. We all run into obstacles but to enrich each other benefits the quality of our work, which benefits the children. Lilaloka wants to be a place of active and concrete research and sharing in this field.

Our aim is to finish the main building and further develop the compound with a Watsu pool, sound cave, collective hall, separate library including reception and consultation office, two staff quarters, store room, kitchen and toilets, all surrounded by a magical garden full of elements to discover together with an herb and vegetable garden.

Gardening and outdoor activities can take place there. Nature and architecture has to blend in together, a oneness which is an extremely important element in the manifestation of the concept of Lilaloka.

In the adjoining forest we are allowed to use we hope make a space for a garden of creation where we can build and play with natural local materials. We want to provide an accessible

Figure 8.2 The future vision of Lilaloka

building, garden, and infrastructure for children with different abilities to be able to play alongside their peers within a rich framework of activities accommodated to their needs. These children are sometimes misunderstood and many have needs and capacities which have not yet been recognized. The garden will be involved and made accessible to allow them to experience the senses, the elements and healing, and the magical call of nature.

We hope to provide an outreach facility so that Lilaloka is accessible to teachers, parents, and children from India and guests from abroad to share knowledge, offer training, and demonstrate new visions and educational tools. We hope our space will always be occupied by a team of professionals in the field of Supportive Respite for Children who need Supportive Learning and team member for a parents' support group. We envisage this space in the same building as the library to have easy access to books and games. We hope to continue to laugh and play!

References

Atkinson, Mary (2017). *Healing Touch for Children*, London: Albert Bridge Books.
Bouheret, B. (2006). *L'art et la voie du shiatsu familial*, France: Editions Quintessence.
Joshi, Kireet (2012). *Sri Aurobindo's Philosophy of Education*, New Delhi: Popular Media.
Lacoboni, M. (2008). *Mirroring People: The New Science of How We Connect with Others,* New York: Farrar, Strauss and Giroux.
Laskow, Dr. L. (2016). *Healing with Love*, Los Angeles: Jesse Krieger.
Majumdar, Anuradha (2017). *Auroville, City for the Future*, New Delhi: Element.

9
PLAY WITH CHILDREN IN HOSPITALS

The situation in Japan

Chika Matsudaira

Introduction

I am writing this chapter with the intention of creating an awareness of the importance of play for children in the hospital. I hope to change the model that medical care is the only means of treating children who are unwell to one that also includes play as an important part of their treatment. I suggest that all children under all circumstances need play, but children with complex medical needs need a professional staff to contribute to help them to play.

I will introduce the stories and certain children and their families that I have gathered from professionals working with children and play in hospitals. I will also describe a boy who has complex medical needs which prevented him from playing for several years. Through learning to play he has restored his 'power to live' and his 'power to connect' – his playing has served as a medium for him to reconnect with the world.

I work for the University of Shizuoka, Junior College where a decade ago I created an education course to train working adults to become specialists in play. Their role was to support sick children and children with special needs through play. In the UK this role is referred to as hospital play specialist (HPS), and is based on attachment theory and the recognition of children's rights. I was commissioned by the Japanese Ministry of Education to develop an HPS curriculum based on Japanese culture and education. This took three years of initial research from 2009.

During the past decade the HPS has been introduced in Japan as a trained professional whose main role is to prepare children for medical procedures and treatment. It enables children to independently participate in their treatment programmes. It should be noted that the word 'preparation' has been introduced into paediatric nursing textbooks as an independent medical term. However at the core of what is termed 'play preparation' is play itself, nothing more or less. If the HPS does not conceptualise these roots of play, the hospitalised children will find it hard to access their independent participation in treatment. As a general rule, HPS will need to find a medium appropriate for each child, and perhaps will need to make one when it is not available. In relation to what has been discussed here, the knowledge and skill to find or make a medium which links the world on this side and the other side is the core expertise of HPS that works face to face with children who need medical care.

Play-based theoretical constructs in Japan

Before we move onto case studies I would like to contextualise Japanese constructs on play. Professor Satoji Yano (1995, 2006) explains play in two different categories. Namely, play which links to a world where things are evaluated by its usability is known as *kocchi-no-sekai* (a world of practicalities) and play in which children experience a multidimensional world or a world that transcends the exact meaning of words and links to something beyond *acchi-no-sekai* (a world on the other side; world of spirit; world of fairies and goblins). Play which links to the latter world, *acchi-no-sekai*, holds immense power.

Yano tells us that since the beginning of time humans acquired the ability to play. This was how culture was born. Science, technology, religion, art, and philosophy were born. Play requires an extremely high level of competence. Animals know pleasure, but humans live their lives at a more profound level beyond mere pleasure. Humans have spent a long time pursuing this fact, deepening it, and passing it down through generations. Play or various forms of art as an extended form of play can deepen the dimension in which humans live. Such power of play acts on *kocchi-no-sekai* where things are assessed by usability, and adds depth to and activates the taken-for-granted ordinary world. That is why adults sometimes need to go back and forth between the ordinary world and *acchi-no-sekai* (the world on the other side) so that they can live more fruitfully. Infants and children tend to put more weight on the world on the other side, and they try to deepen that world by asking guidance on how they can find the path to lead them there. Thus, play needs to be considered from both developmental and biological points of view. And the most important is that the adults as parents are the first people who will teach the child to play, which means that parents are giving to the child that there is a spiritual world that you see, which is the meaning of life.

Considering how Yano categorizes play, the techniques of hospital play seem to pull play back into the world symbolised by usability. However, skills which work on children's imagination such as guided imagery, drama and storytelling, for example, enable children with no legs to play soccer, or children with severe paralysis to jump on the trampoline with a straight back. These are key skills that allow children to play in a world their psyches/minds have created. No matter how dependent they may be on medical care to sustain their lives, a human being is defined as a being who possesses a spirit. So children as human beings are in need of play in order to keep their spirits alive, or to heal, remonstrate, and stir their souls. Yano also explains that play is an act which best demonstrates the nature of children.

First time admission to hospital for a child

Everyone understands the fear of medical procedures for children if there is a lack of preparation and information. The unfamiliar atmosphere at medical facilities, including straight corridors, medical instruments, uniforms and disinfectant smells, do not convey an atmosphere of kindness. These fears can continue into adulthood. In gathering of stories from grown-ups about their childhood experiences having surgery or hospital care, I have found it can affect their relationships. The negative feelings can be passed down generations. The following story from a young woman, aged 20, shares the reason she does not tell others about her early hospital experience:

> I went to buy sweets in the hospital store; I was in my pyjamas as I was told to wear them. In the shop a young child asked 'Mum, why is that girl wearing her pyjamas in the daytime?' The mother

stopped the child saying 'Shush, you shouldn't say something like that'. I was shocked to hear that conversation and felt that I shouldn't be here. But I have to be here because I'm sick and not complete. I worry about how my friends outside think about me and whether they will accept an incomplete person and if they will think I am a nuisance. I feel I will never catch up with anybody. Being sick and hospitalised is an experience I cannot share with anyone.

A 23-year-old college student looks back on his experience of being admitted to the hospital at the age of four and five.

I don't think I will live long, because I refuse to go to hospital. Logically I do understand that I should if I am sick, but I will not. The hospital is not a place I can go. It's because of my first memory that I had as a child in the treatment room. No one was listening and I was shouting and screaming.

I was lying on my face, being pushed down. I was being forced and could feel many hands pushing me. I was really scared because my parents weren't by my side. I remember until some kind of needle pierced my back. This memory comes back in my dreams when I am ill. I know I have a high temperature when I dream about the hospital.

When I was discharged from the hospital and started infant school, the teacher told my parents that my linguistic development was slow. Although my parents started worrying again, I knew what caused the delay. Conversations at the hospital were unnecessary: nurses let you take medicines, brought you food, doctors asked one-side questions and I would answer yes or no; you don't need words when you are lying on a bed all day! My parents then became stricter from the age of ten, thinking my illness was cured and I had to catch up educationally. My mother seems to have been condemned by her relatives for giving birth to a weak child. One day when my mother was angry she shouted 'Something like this would not have happened if only your body was right!' I still can't forget what my mother said, it's not my fault, and neither is it my mothers.

This young man also shared that everything he needed was taken away from him as a child. When I asked him what was taken away, he answered after some reflection, "play and family relationships".

The following story is of a father who refused to allow his child to have surgery because of his own experiences in primary school. He suffered trauma and was too embarrassed to talk about it, and therefore asked his wife to be interviewed instead. The mother talked about the impact the hospital experience had made on two generations.

Since we are both working full-time our son has attended day-care since he was one year old. He developed colds so we frequently took him to a paediatric clinic and eventually he was diagnosed with adenoid hyperplasia and enlarged tonsils.

Because my husband had such a frightening experience, he tried to postpone the surgery and I felt a conflict as a wife and mother. By three years old he had breathing difficulties and snored badly and the ontologist and paediatrician at the nearby hospital diagnosed him with sleep apnoea and strongly recommended surgery. My husband and I initially just ran away from the decision, not knowing how to explain to our son. Finally we met with the HPS and everything changed.

I was very moved when I met the HPS on the ward and saw all the preparations so my son's life would not be disrupted. I realised the importance of not lying to a child and explaining everything. We went as a family and had a very positive experience. My child enjoyed being an inpatient with lots of play, painting, and water play. If his throat hurt after the operation he would say that he wanted to go to the playroom and his tears soon stopped.

On the day of the surgery I was anxious about how my son would respond to the procedure; he wasn't even allowed a drink of water! However when he was playing with the toys he looked happy and relaxed. The HPS showed him the preparation book which had pictures of the clothes he would wear, and pointed to his throat and explained that was going to be fixed. I believe that the HPS led my son's surgical success and he is now recovering well. His speech has improved and he has a better appetite. Meeting the HPS has been a blessing for my son and us as a couple. We sincerely look forward to the spread of creating an environment where people involved in the medical field, and the parents, can understand the feelings of the child.

What can we learn from these three stories? Children who will be hospitalised need to first and foremost be children before being sick children. Listening to the voice of the child is as important as the medical care they receive. I summarise these points as follows:

1. Play in the hospital is an essential activity for children who receive medical treatment and for children with special needs.
2. Play in the hospital is a method to protect children emotionally and create a sense of security and confidence. This will allow the child to approach their treatment in a more positive way.
3. Play in the hospital provides information for all hospital staff who treat children, to understand their emotional needs.
4. Play in the hospital defends children's right to play. This can be impeded in a medical environment which prioritises medical aspects of the child's life.

Paediatric practice and play in Japan

When I started the hospital play specialist training project ten years ago, a Japanese medical system researcher told me:

> when you go to the children's hospital there is a place at the back of the corridor where it is very quiet. When you enter this dim room, all you can hear is the sound of the breathing machines. The children on the beds move their eyes but nothing more. If you ask the nurse in charge, 'when do these children play', she will look surprised and not have an answer. Medical procedures are advanced now, and children like this are increasingly kept alive. But are they alive? Children who have never played, can you say that they are alive?

I pictured what the researcher said but I did not know how to respond to this situation as I did not then have a true understanding of the value of play for ill children.

Meeting Taiga, the boy with complex medical needs

Taiga had hypoxic encephalopathy due to the surgery he had when he was six. As a result he lost most of his mobility including his speech. Since all possible hospital treatment was over, home-based care and rehab had begun. Taiga needed care 24 hours a day, and regular posture changes and aspiration of sputum were performed by his parents. It was difficult for Taiga to take a standing position so he spent most of his time on the bed. A heart rate monitor was attached to Taiga, his hands and feet were paralysed, and he also had a tracheotomy.

The first time I met Taiga was about two years after he went home. The HPS in Japan had started a project to provide hospital play for children in the community, partnering with a paediatrician who worked for a children's rehab centre. She knew many children who could have a much better life if they had support from an HPS.

I arranged to support Taiga. I clearly remember the day I first met him. When I entered the room Taiga was lying on a care bed installed at the far end of the room. Even though I was far away from him, I felt from the atmosphere that he was slightly nervous, wondering what this person does. There was a tense atmosphere throughout the room, including his parents. I went to the side of Taiga's bed, quietly introduced myself, and kneeled down to meet him. Taiga cannot voice words as his trachea is cut. However, his eyes stared back at me. After a while, he began to work hard to move his lips. I spoke a lot to Taiga, using meta-communication techniques. Just like I would talk to a child who could speak, I asked what he liked to do when playing. I also asked him about study at school, and asked questions about relationships with family and friends. Of course, Taiga did not answer using his voice. However, every time he moved his lips and blinked, I would respond to him saying things like 'Really? I see. That's how you feel' and enjoyed conversation with him. As I had the impression that Taiga was a very talkative boy, I told this to his mother. The mother's face became instantly bright, and said, 'Yes. He's a child who used to talk a lot'. Every time I talked to Taiga, his dull facial expressions became increasingly excited. And his face became more shaped, and personality came back to him. The soundless conversation was also fun. When his parents came to look at Taiga's face, it was funny to see Taiga showing through his facial expression, "Don't interrupt". After chatting for about 20 minutes, Taiga had started to doze, which might be because he used all his energy to engage. Before leaving the house, I told his parents that I had fun playing with Taiga, and I could feel that he has many excited feelings inside, and I would like to draw on this. I also told them that although there is medical care, I would like to do things a ten-year-boy would do. What I had shared with his parents seemed to be extremely fresh, since it was completely different from what medical practitioners would discuss. I felt an atmosphere of relief from his parents.

I later heard that Taiga was looking for me with his eyes after waking up from his nap. From this time, I believe the journey with Taiga, reconnecting his spirit and body through our conversation, helped him to begin to make sense of his situation. His parents began to connect more with him.

The second visit, igniting the fire of the power to live

On my second visit, I decided to create a story about Taiga having an adventure. Taiga was feeling persistent pain because of the rigidity and strain in his body. I wanted him to be free from the pain and start an adventure using his brain and imagination. This was because when I imagined the feelings of a ten-year-old who could not move his body for several years, I guessed that he would surely want to engage in play involving physical expressions, such as running, jumping, shouting, or throwing. Following is the therapeutic story I actually created for Taiga:

A day of Taiga
One afternoon when the end of summer could be felt, Taiga left the house and started walking, invited by the sun.
It was a hot day but he enjoyed walking and kept on going.
Let's walk, lets walk, it's a fine day lets walk.
Taiga started to have so much fun, he started to also sing a song.
What a beautiful day to have an adventure, what kind of adventure will be in front of me?

A shopkeeper opening a store by the side of a road smiled at Taiga,
Looking at him singing happily.

A butterfly was also flying.
The chirping sounds of the birds also echoing.
It seemed as if everyone was cheering for Taiga's departure.
Taiga felt like he was starting a new adventure and was excited.

When he was walking on a paved road, his face became hot because of the reflected heat.
'It's hot. Was it better to stay at home after all and watch TV?'
Taiga thought to himself. He started to also feel thirsty.
No, no, an adventure is more fun than anything, it's a challenge.

When he came to a fork in the road and saw a large camphor tree.
Taiga thought to himself, I know, I'll cool down under that camphor tree.
There is a shadow below the camphor tree. It is cool and very comfortable.
The wind that can be felt on his face is also very comfortable.
The feeling of grass that can be felt from the tip of his toes is also cold and feels very good.
Taiga was satisfied that he left his room with air conditioning.
Suddenly, he noticed that there was a small path on his right.
Although he could not see well as weeds were growing on both sides of the road, a narrow pathway was surely there.

'A hidden road, that looks interesting. I'll try walking on that road', Taiga thought and stood up. Taiga brushed off the leaves on his bottom with his hand. He courageously went into the small path.

The path was perfect for Taiga's adventure.
A grasshopper hopped high, surprised by Taiga.
A frog hurried to dive into a rice field with a splash.
There was also a grass snake.

The grass snake was relaxing and sleeping, coiling itself.
'This is great. The path is perfect for me', thought Taiga as he kept walking.
Just then, he started hearing voices of children from afar.
What? What? Is there a park ahead of this path?

The voices became louder and louder the more he walked.
'Yes, that's right'. 'Go for it!' 'That's good, that's good'. 'Ah, that was close'. Large and powerful voices became louder and louder.
Taiga started to run, wanting to know what was there.
The voices also became louder and louder.
Suddenly, he reached an open field.
The field was surrounded by a fence of trees. Over the fence, children were playing with a ball.
Taiga decided to see the children through the fence.
He wondered what kind of game they were playing.
The children are absorbed in kicking a ball.
They cheer every time there's a goal.
Nine children are playing divided into two teams.

Then, one child wearing a white shirt noticed Taiga watching.
And, with a loud voice, he called out beckoning him, 'We're missing one person. Come and join'.
Then, the rest of the children said, 'Come', 'Let's play together'.
Taiga hesitated just for one second but opened the wooden fence and ran to everyone,
And started playing with everyone.
He immediately could understand the rules.
'Taiga, this way'.
'Turn it around, be careful'.
'You did it, that's great'.
Lots of voices come up.
Taiga plays with all his might and with all his body.
His face is serious.
It's so much fun he can't help it.

The wind slightly changed.
Taiga shook himself.
When he realised the sun had come to quite a low position.
The evening time is approaching.
Yes, it's time to go home.

Taiga said 'Goodbye' to everyone.
The children said, 'Come back here any time and let's play again'.
Yes, I will certainly do so, said Taiga.

Taiga started returning home.
The wind feels cooler than when he first came.

Frogs had also started to make noises.
Taiga walked, thinking to himself, 'Wow, today was so much fun'.
He then came near to his house in no time.
He would arrive home at the next corner.

Taiga thought to himself. 'I'll go to another adventure'.

Every specialist involved in childcare knows that in the world of imagination, children can be anything, and do anything they want to do. It is also understood that narratives are essential for the growth and development of children. Unfortunately, there are not many attempts in Japan to actively read stories to children who do not have an ability to move. Taiga was very focussed when listening to the story. When the story featured running, he ran hard, his heart beat fast (it is clear as his heartbeat rate became faster), and at the scene of going home, he became tired and his facial expression looked dreamy. After reading the story and checking how Taiga was doing, I noticed he seemed dissatisfied. I asked him if he wanted me to read it again, and he blinked his eyes several times saying yes.

I told his parents that Taiga could do anything in his imagination and that play could increase the courageous power to challenge, which is essential for children to live. The parents had watched Taiga's situation playing in his head, and sensitively perceived the changes in their child's feelings that had been born. The power of play is easier to understand if you actually show it. They took videos to show the sight of Taiga playing to other professionals.

After that, my visits continued twice a month. The relationship between his parents and Taiga and his younger sister also started to change, along with the deepening of the relationship between me and Taiga. The change could also be seen from his living environment. Toys continued to increase in the room, whereas before it was all about cleanliness. His mother and father started to talk to Taiga more. They learnt to have a conversation that could consist of the movements of the eyes and slight facial movements.

In the past two years from the first visit, Taiga was hospitalised and had to go through surgery. Therefore, depending on the situation, I gave him play preparation and breathing methods that would lead to relieving the pain. Other things we worked on included dressing-up play, flower arrangement, snack tasting, sand play, water play, storytelling, play dough play, drama, slime making, reading cartoons, making Kiwanis dolls, medical passport making, and video games. Based on the idea that all play was possible.

His parents recorded all the sessions in a video and gave it to me on a CD. The changes in Taiga's emotions, spirit, and body were clearly recorded in the CD, including how Taiga's facial expressions changed. Depending on the state of his play, he became more confident and would sit on a chair to take a sitting position to play. It was a video that truly conveyed the feelings and joy of his parents.

Meeting the practical needs of ill children with hospital play

Whilst receiving medical care, children, including Taiga, are surrounded by many professionals. Various adult staff surround a single child, including multiple specialists (such as doctors to visit when having a mild cold, orthopaedists, paediatricians in internal medicine, surgeons), nurses, facility rehabilitation staff, and teachers from the special education school. In the United

Kingdom, I feel that the barriers between institutions and facilities are lower than in Japan. For example, in the UK there are many cases in which HPS working in the community follow the children who are hospitalised and prepare them for hospital admission. The concept of multi-sectoral collaboration is not merely a philosophy but is also in practise. Professions will always gather together with family and children. They will seek ways to cooperate by utilising their respective specialities. In Japan, unfortunately, there is no such flexibility. There are very clear lines between professions.

Taiga's parents openly expressed their opinions and feelings to the medical staff. The hospital guardians believed that other specialists had something to learn from the methods used by HPS. They believed that if they could share the information, they could create a richer circle of support for Taiga. The guardians had continued rehabilitation from the desire of wanting Taiga to be able to stand again.

Hospital guardians had shared that Taiga would sleep when they took him to rehabilitation, which was why the rehabilitation was not progressing so much. I had noticed this in other countries. I was attending a training at Children's Trust, a medical intermediate facility in Tadworth in the United Kingdom that specialises in the rehabilitation of children with brain disabilities. I was told by a nurse specialist that 'These children cannot run away even if they experience something they don't like. That's why they sleep'. When I shared this information with Taiga, he blinked his eyes and agreed with me. He told me that he wanted me to follow him to rehabilitation. It seemed that Taiga wanted to communicate to the rehabilitation staff that he was sleeping because he could not communicate well and they did not understand him. It is not easy to advise other professionals in other institutions. However, the enthusiasm of the hospital guardians and the rehabilitation physical therapists wanting to have a smoother communication with Taiga led to permission for me to accompany him.

When I accompanied Taiga I noticed the rehabilitation physical therapists were relating to Taiga in a very kind way; they did not know the importance of waiting for Taiga's own timing and lacked an ability in waiting for him to respond. During the rehabilitation, I observed Taiga and the therapist's involvement, and advised them, such as 'I think Taiga wants you to wait now'. 'I don't think he would dislike it if you move him now'. In order to tell the therapist that medical instruments can also be used for play, I suggested that he put a stethoscope around Taiga's neck before listening to the heartbeat of the mother or the therapist. As a result, Taiga experienced a happy surprise of listening to the heartbeat, and the relationship with the therapist became closer. The natural way I talked to Taiga also seemed to surprise the therapist. After saying 'You talk to him so normally', she apologised to me for saying 'normally'. However, this indicated how the therapist was so careful on a daily basis when relating to the patients.

Children who receive medical care are often overprotected. There are cases where they do not receive age-appropriate responses, and sometimes they are deprived of opportunities to be independent because they are children who need medical treatment. Collaboration with other professionals is crucial for practising child-centred support. By knowing what techniques each other has in contact with children, you can adjust your role and relationship. In addition, by sharing information, it is also possible to confirm the degree of achievement of support. Taiga never slept during his rehabilitation when I accompanied him. Moreover, Taiga has never slept during rehabilitation ever again. After that, Taiga would ask for me to accompany him when he accomplished something, such as when he took a sitting position.

The ending of Taigas story

The quality of Taiga's life changed because 'play' had returned into his life again. It is also true that play stimulated Taiga's growth and development. Taiga's mother summarised the following comments in writing:

> I have had strange emotions since my son was bedridden and returned home. Although it is difficult to put into words, it's a strange, fluffy feeling that you do not know if the child in front of you is really your own child. It's a feeling that he is your son but is not your son. However, after an HPS came, Taiga's eyes became brighter and brighter, stronger and stronger. I could understand clearly what he liked and disliked. My son started stating his own opinions, and I came to have confidence to say yes, this surely is my son. My son came back to me. I felt that my son had come home.

This was the moment I could feel that the connection between Taiga and his parents, which seemed to have been cut off through medical intervention, was reunited. Today, daily family interactions such as Taiga and his mother quarrelling and his younger sister intervening can also be seen. Of course, Taiga must continue to receive medical care from now on. He will continue to receive daily phlegm suction, feeding from the stomach, injections to relieve tension, plastic surgery to correct crotch deformity, and highly invasive medical care. Although it is an extremely simple thing to say, I would like to suggest that all children in medical care, including Taiga, need the 'joy of living' to meet the many painful treatments and medical care. Sick children need to explore the meaning of life through play.

I have introduced the relationship with Taiga as a case example in which the relationship between parents and a sick child could be strengthened by play. It is because children are sick that adults must play with their children.

Changes occurring in Japan

The path to develop professionals who will support children with medical needs using play in Japan has never been easy. I faced severe criticism. At the beginning of the HPS training activities, there were the following four things that I considered necessary to change in Japanese paediatric care.

1. Visiting hours for parents
2. Not have the parents leave the treatment room during treatment situations such as blood tests
3. Restraining children during procedures
4. Not offering play to children

I also noticed that although social work and medical care have similar goals, medical care does not always share the same values as social work. There are significant differences regarding the perception of children's rights despite the medical profession being a profession that deals with children. For example, at a Children's Hospital, parents can only see their children eight hours a day, while a therapy dog can stay in the ward for 24 hours. There are also situations in which children being hospitalised are called 'dear patients' while they are being restrained in treatment rooms. Medical staff may restrict children's play by stating 'this is a hospital' or 'because this child is sick'. An example of this is as follows:

> My child had been hospitalised for nearly a year. The hospital where my child had been hospitalised would give children a pacifier and tape their mouths with a tape in the evening because of the lack of staff. I think they had explained that it would help the children sleep longer. At first there was a sense of incongruity. I was not able to meet my child at night. However, surprisingly, I got used to it. After leaving the hospital, it became a daily routine to take my sleepy child on a stroller and take a walk. At that time, I gave my child a dummy. I also taped my child's mouth with a tape after giving my child a dummy because it was convenient. Then, a neighbour whom I heard was a former kindergarten teacher later told me,
>
>> You shouldn't do something like that. If the dummy falls from the mouth, the child will cry and inform you. But it's important to pick it up again and give it to the child. At that time, the child will interact with the mother. When the child cries, the mother responds. That is very important for the child.
>
> This is when I realised that we had been in a very odd place. The way people think at the hospital had affected me. Thanks to the medical treatment, the child grew up safely and was able to leave the hospital. However, I feel that there are many things that I have to regain from now on.

I noticed differences in how doctors and nurses perceived expertise. For example, when talking to a hospital manager, I was often told, 'Well, this is the place to cure disease', or 'We only have severely sick children so not many can move and play'. One play specialist told me that she was decorating a treatment room to make it a child friendly environment. When a doctor came in he immediately said 'That's not science, is it?' I used to have phone calls asking me to convince paediatricians who could not understand the importance of distracting a child with play during blood tests. It showed us that many medical professionals were not interested in play and did not believe that HPS techniques could help because they had not been proven. If preparation is done effectively, children will be less likely to refuse treatment, making it easy to treat them. Therefore, many would then listen to me when I said 'preparation is the role of HPS' instead of saying play was the role. As a strategy, I initially introduced the system saying, 'When hospital play is introduced, everyone (including medical staff) will benefit'. However, I felt a sense of discomfort, as the more I recommended it to medical professionals that 'you will receive benefits', we were moving further away from the value of play itself. I began to think the more I was selling the effects of preparation, I was trivialising the sight of children playing and distancing play for children in the hospital. I wanted to establish play as not only a benefit but play as a gift from adults to children who are sick and in the hospital.

Concluding thoughts

Earlier I stated that doctors and nurses described play as 'not science' and 'What you are doing is grandma's wisdom'. This expresses an aspect of medicine that treats numbers as the only evidence. It is not completely possible to quantify the effects of play and to quantify the effectiveness of HPS. In fact, I have travelled abroad to study various play therapies to conceptualise my practise for sick children. I now practise an Adlerian Play Therapy approach to conceptualize and evaluate my work. The theory to connect, count, be capable, and develop courage is central for HPS practice. On the other hand, it is important to convey the value of play as a philosophy. It will be difficult for play to take root in current Japanese paediatric medicine unless there is a multidimensional way for it to be evaluated. When looking back at HPS training, I believe that there is a need to focus on finding demonstrability and persuasiveness in the "dialogue" with different fields.

In Japan people who use the word 'play' are not only people who are involved with children. Those in the field of engineering also use the word 'play' in Japan. It is called *asobi*. For example, the 'play (*asobi*)' of a steering wheel or brake is incorporated into a mechanical device as a safety system built in at the design stage. This 'play (*asobi*)' is incorporated in the operable range of the system, but it is not reflected in the movement of the device. In short, play is an essential quality built into the system from the beginning in order to manage the shock caused by unexpected incidents or impacts from bumpy roads. Play for children may have a similar meaning.

In Japan, there is a famous Buddhist priest by the name of Ryoukan (1758–1831). This priest did not belong to a particular sect, and he led a poor life although he had a high status background so he was called the beggar priest. He is also well known for having dedicated himself to playing with children. In drawings and statues, Ryoukan is always depicted with children holding a *temari* (Japanese ball) in his hand.

Seeing him play, many adults at the time seemed to have considered him a lazy man. Why did Ryoukan devote himself solely to playing with children? One answer was because 'it was in children that he saw the image of Buddha'. When people condemned him and asked why he did not stop playing with children, he made no excuses and simply replied, 'It is no use explaining to those who cannot understand'.

However, when we turn our attention to the children, there seems to be an answer to why he dedicated himself to play with them. The children were all from farming families in the poor Tohoku region. In that area boys would either be given hard labour on the farm or sent out to apprenticeships, girls were destined to work as maids or, at worst, as prostitutes. Ryoukan must have known the severe destiny of these children. That is why he played with them with all his heart. Ryoukan saw Buddha in each child playing.

> Deprived of play, the child is a prisoner, shut off from all that makes life real and meaningful.
>
> *(Susan Harvey, Cofounder of HPS, 1972)*

References

Harvey, S. (1972) *Play in Hospital*. London: Faber and Faber.
Yano, S. (1995) 子どもという思想 *Kodomotoiushisou*. Tokyo: Tamagawa syutpanbu.
Yano, S. (2006) *Towards the Élan Vital: Philosophical Study on Playing Children*. Tokyo: Seori Shoten.

PART II

Therapeutic play

Introduction

Section 2 specifically moves away from generic approaches to play, not that Section 1 was wholly generic; however, Part II could be considered to be slightly more "applied" in that practitioners are aiming towards specific uses of play that are more specifically focused around the "therapeutic" nature of play. We again see some international social and cultural perspectives, including play-based therapeutic work centred in Malaysia, Ireland, the UK and the Czech Republic.

We begin with Professor Fraser Brown's consideration of "playwork" as a therapeutic tool. Brown himself is the first Professor of Playwork in the UK and brings a breadth of knowledge and experience to this debate. In Chapter 11 Sue Jennings revisits theatrical connections to play and otherworldliness, connecting her experiences as an anthropologist in the Forests of Malaysia with the otherworldliness of the characters in Shakespeare's *A Midsummer's Night's Dream*. The notion of "altered states" which occurs in the Temiar she lived with as an anthropologist are not dissimilar to the altered states of otherworldliness that children enter into when they play.

In the next chapter we consider, from an international school in Prague, Leong Min See's work attempting to use play as a way of bringing together both able bodied and less abled children, using play as a mediator between levels of physical and learning ability, in order to build trust and cooperation. In Chapter 13 Sue Gascoyne develops this notion of the necessity of play being able to connect children to feelings and emotions by considering the importance of "messy play", however she also brings considerable academic rigour in thinking about the theoretical ideas underlying something that seems so simple. Messy play is further enhanced in Chapter 14 from Prendiville and Casey's very practical approaches to using messy play. They offer a range of practical ideas and activities to further enhance this notion of playful messiness.

Finally, Chapter 15 considers the use of aspects of Jennings' Neuro-Dramatic-Play (NDP) (2011) working with children who are beginning to struggle within confidence and other issues in a middle school in the UK. Holmwood, though himself a dramatherapist, chooses to work not as a therapist but as a Neuro-Dramatic-Play practitioner. Making links between story, character and drama and the latent aspects of play which he considers to be inherently endowed with "therapeutic" potential. He particularly focuses on the use of large puppets being a playful transition between drama and play-based approaches to support general anxiety and lack of confidence. This moves us towards Section 3 where we will specifically consider play in a therapy context.

Reference

Jennings, S. (2011) *Healthy Attachments & Neuro-Dramatic-Play*. London: Jessica Kingsley Publishers.

10

PLAYWORK AS A THERAPEUTIC TOOL

Fraser Brown

This chapter explores the way in which playwork may be used as a therapeutic tool. Playwork is rooted in an understanding that children learn and develop while they are playing, and that there are many instances in modern society where that process is interrupted or impaired. The playwork profession accepts that adults have a responsibility to address this deficit. Playwork involves identifying and removing barriers to the play process, and enriching the child's play environment (Brown 2003a). In most cases this involves nothing more complicated than providing a staffed adventure playground, or an after-school club, for children who would otherwise have nowhere to play. However, at the extreme it can be as complex as creating an environment to assist the recovery of children who have suffered severe play deprivation (Brown 2018).

In *Aspects of Playwork* (Brown 2018, pp. 104–105), I summarised the main characteristics of therapeutic playwork. While not claiming that each characteristic was unique to playwork, I nevertheless suggested that taken together they represent a unique way of working with troubled children. Thus:

In its underpinning thinking, therapeutic playwork:

- Draws on elements from most of the major strands of psychology
- Understands that play is a basic human drive
- Believes in the uniqueness of every child
- Respects the child's ability to solve their own problems through playing
- Accepts that children may use play as a reconciliation mechanism
- Accepts that children's play has elements of both 'being' and 'becoming'
- Aims for the development of the potential inherent in all children

In their practice, therapeutic playworkers:

- Work with groups of children, while paying close attention to the needs of individual children in the group
- Create an atmosphere of safety and security
- Create a resource-full play environment
- Create an environment characterised by fun, freedom and flexibility
- Establish a permissive environment of openness and honesty
- Adopt an attitude of unconditional positive regard towards the children

- Work as far as possible to the child's agenda, in a non-directive way
- Emphasise the development of the child-playworker relationship
- Internalise negative capability as a creative approach, and thus adopt an open-minded attitude towards the children
- Use the technique of 'joining' where appropriate
- Encourage the development of fundamental life skills, e.g. sympathy, empathy, mimesis
- Recognise the fundamental role of rhythm in early child development
- May sometimes act in an advocacy role for a child

In order to illustrate some of these points, this chapter draws extensively on my own personal experiences, and those of various playwork colleagues, as recounted in the book *Play and Playwork: 101 Stories of Children Playing* (Brown 2014). I have also referred substantially to work undertaken by myself and my colleague Sophie Webb (2014) in a Romanian paediatric hospital with a group of children who were abandoned at birth. Perhaps it is worth enlarging on this unusual situation. The children, ranging in age from one to ten years old, had suffered chronic neglect and institutionalised abuse. They had been tied in a cot for most of their lives; they were fed on meagre rations, and their nappies were changed no more than once a day. They could see and hear each other, but could not leave their cots, and so experienced hardly any social interaction. The extent of their developmental delay was shocking. It has generally been accepted that children who suffer chronic abuse during their early years, have very little chance of recovery. Our work challenged that pessimistic view.

The focus of the work, and accompanying research study, was the children's play development, which was assessed using an instrument developed in connection with an earlier study (Brown 2003b). During the 18-month study nothing changed in the children's lives, other than their introduction to the playwork project. Nevertheless, the change in the children was remarkable. Their social interactions became more complex; physical activity showed an identifiable move from gross to fine motor skills; their understanding of, and interaction with, the world around them was improved; and they began to play in increasingly creative ways. They no longer sat rocking, staring vacantly into space. Instead they became fully engaged active human beings. The study's conclusion was simple, but nonetheless powerful; i.e. the children's developmental progress was clearly identifiable, and apparently made possible through their experience of the playwork project.

The fundamentals of therapeutic playwork will now be discussed under six significant subheadings, extracted from the preceding list.

Play as a reconciliation mechanism

After the most wide-ranging study ever undertaken of researchers and theorists who focus on the subject of children's play, the developmental psychologist Brian Sutton-Smith (1997) concluded there is overwhelming evidence that play supports a child's general growth and development. For Freud (1900) this development takes three forms:

- reconciliation: coming to terms with traumatic events
- gratification: satisfying libidinous desires
- aspiration: enabling the achievement of wish fulfilment

He suggested that, in all three cases, this is achieved by altering the circumstances in play to how we might wish them to be in real life. Indeed, Kuschner (2019) suggests that the 'suspension of reality' is a factor common to most forms of play. In general, this is merely a mechanism to enable children to explore their thoughts and feelings. However, when children 'play out' painful experiences in their fantasy play, it helps them come to terms with their own feelings (and possibly those of other people) more effectively (Dwivedi 1993). This reflects the earlier ideas of Axline (1969), who suggested that, in the long run, through playing, children can reconcile themselves to events that have troubled them. A powerful example of this is provided in the following vignette, which is taken from Sophie Webb's reflections while working at a specialist therapeutic recovery centre in the south of England, in the early part of this century (Brown 2014):

> 6th August – Charlie's mother died two days ago. Today was his birthday. He is three years old.
>
> Charlie has always enjoyed pushing things and transporting objects. He'll push other children's wheelchairs and buggies if he can. Today he decided to use his own buggy – but this was different. He wanted to push a baby doll around in it. We chose a doll together and Charlie even wanted to dress her appropriately for the hot weather. I asked Charlie what his baby was called. He replied, 'Mum'.
>
> At this point I realised that this meant more to him than a normal 'home corner' activity. This was his way of attaching an object to his loss. That afternoon the doll called 'Mum' went out with us to the park and had to sit on his lap in the taxi. When we came back he needed the toilet (a big thing for him at the moment as we're potty training), but he wouldn't do anything until 'Mum' was watching him. He sat the doll on the bath opposite him.
>
> I then suggested we give the doll a bath. I did this because I wanted to give Charlie the opportunity to use the doll as a symbol for his mum if he wanted to, or he could simply play at dressing and bathing dollies. Charlie was really pleased with this idea and helped to run the bath. 'Lots of soap' he told me, and dunked the doll in the water. He also wanted to wash her hair. While doing this Charlie was calm, thoughtful and gentle.
>
> This was an activity that I hadn't played with Charlie before and he really enjoyed caring for the doll, all the while calling her 'Mum'. This day has highlighted a number of vital messages to me; the most obvious being, he still needs his mum around.
>
> I learnt the next day that about an hour after I left, Charlie displayed extremely aggressive behaviour and was very difficult to manage. This little boy is still grieving, and we need to find structured and unstructured ways of helping him. He is trying to speak to us through his play and I am concerned that there is no continuity in the way we are helping him.

Webb is highlighting the issue raised by Sturrock and Else (1998) that if play is not allowed to flow, the child is likely to develop some form of neurosis. This is confirmed in the classic animal studies of Suomi and Harlow, who found that "no play makes for a very socially disturbed monkey" (1971, p. 492). It is further confirmed in the study of abused and abandoned children in Romania (referred to earlier) – in particular by Sophie Webb's description of the children at the beginning of the study:

> When I observed the children in the playroom, they were unaware of each other, fixed on their own activities – barely communicating. Some just sat and seemed bewildered and vacant.
>
> *(Webb 2014, p. 139)*

Creating a resource-full environment

Simon Nicholson's *Theory of Loose Parts* holds that:

> In any environment both the degree of inventiveness and creativity and the possibility of discovery are directly proportional to the number and kinds of variables in it.
>
> *(Nicholson 1971, p. 30)*

In other words, the more flexibility offered by the environment, the more creative we are likely to be. In the Romania example (Brown & Webb 2012), we have always been keen to stress that the remarkable progress made by these severely neglected and abused children, was substantially due to the fact that they were at last able to play with each other, and interact with a 'loose parts' environment. It is true that we engaged in some very focused one-to-one work with individual children, but we have no doubt that our most significant contribution to their development came from the resource-full environment that we created for the children. Hughes (2012) supports this idea, in as much as he suggests a child's interaction with the environment is fundamental to their future development. However, he largely focuses on the ancient Aristotelian elements of the environment (earth, wind, water, fire). While not dismissing the importance of those elements, we would also wish to emphasise the significance of children being able to play with each other, i.e. the social aspects of children's play. In that sense, even the children who are played with become loose parts in the wider loose parts environment.

On the face of it the following example is not about the therapeutic benefits of children's play. However, it is included here in order to reinforce the point that play has wide-ranging benefits; and when taken together, those benefits lead the child towards a state of emotional equilibrium – a kind of homeostasis. The example, which was reported by my colleague Ben Tawil, relates to an adventure playground where a lorry has helpfully left a lot of old furniture just inside the gates of the playground (Brown 2014).

> As the children arrived for the evening session, they took immediate interest in the furniture. A group of about seven children, aged eight to twelve, both boys and girls, started to sift through it. At first their search seemed indiscriminate, almost chaotic, with very little communication between them . . . the children seemed to have concurrent ideas that stemmed from one person's initial placing down of a piece of furniture. Two leaders emerged – the eldest girl of about twelve, and one of the younger boys of about eight. They seemed to be taking on the role of interior designers – telling the rest of the group where to position the furniture. These instructions were followed to the letter with great seriousness. Together they created a home environment. . .
>
> Straight away a boy of about eleven sat down at the bureau and exclaimed, "Can you keep the noise down? I'm trying to write a letter to the council", and without question or hesitation the oldest girl (until this point the chief interior designer) addressed the other children in a sharp authoritarian voice, "Your Dad's told you to keep the noise down. Now go and play quietly". Immediately the rest of the group took on roles as brothers and sisters, grandparents, daughter and visiting boyfriend. . .
>
> This play . . . continued for two weeks – every evening and for full eight-hour days at the weekend. Different groups of children used the materials and altered the environment and the narrative to suit their needs. . . . Eventually the children's interest waned:

perhaps they had played out their need for this type of play for the time being, they had certainly worn out the already dilapidated furniture. The play began to morph once more as the children found uses for panels from the furniture in construction play, and the remnants were put to good use fuelling our nightly campfire.

Table 10.1 summarises the benefits of this 'furniture play'.

It is not hard to see these elements, and more, in Tawil's story, and that would be the case during most instances of children's play. Sadly, we tend to trivialise play with phrases like 'it's child's play' and 'just playing around', whereas play is actually very complex, and the benefits are multifaceted.

The artifacts in Tawil's example are basically recycled scrap. Clearly, playing with 'junk' in this way has its minor risks (splinters, sharp edges, etc.). However, it also means the children have all manner of unusual opportunities to combine materials that would not be available to their counterparts in a more sanitised environment. Sylva says "the essence of play lies in . . . combinatorial flexibility" (1977, p. 60), which Bruner suggests is the initial building block for human creativity. He says:

> Play and playfulness free us from the narrowing immediacy of pressing demands, enabling us to better explore the combinatorial possibilities that are opened to us by our cognitive powers.
>
> *(2006, p. 5)*

My argument would be that it is only when experiencing such social, physical and cognitive satisfaction that we can be expected to achieve any sort of emotional release from the trials and tribulations of everyday life.

Establishing a permissive environment, characterised by fun, freedom and flexibility

When observing children at play it is often impossible to tell what is going on in their minds. They frequently surprise us, and as has already been suggested, their play is often far more complex than we realise. For example, something that looks like a fight may in truth be a 'play-fight'. To the adult mind there is nothing very unusual about the story that follows. However, if the children's actions were taken too literally, we might actually find them frightening. The fact that most people would

Table 10.1 Furniture play

Nature of the experience:	*Leading to:*
Interaction with the environment	Deeper understanding of specific materials and the potential they offer
Social interaction	Forging and/or reinforcing friendships
Societal behaviours	Internalising the value of co-operation
Physical activity	Health, fitness and the development of motor skills
Cognitive challenge	Stimulation of the imagination, creativity and problem solving skills
Being in control	Self-discovery and a recognition that it is possible to influence the world around you

simply find the actions amusing is explained away by Tinbergen (1975), who suggests that children's play has three clearly identifiable characteristics (that are also seen in adult behaviour). He says play is:

- Clumsily executed
- Often chaotically arranged
- Generally performed with abandon

The following observation, which happened during a playwork student's practice placement, is a good example of this (Brown 2014):

The Party – Rachel Kelly

A group of five children decide to play families. The girls choose to wear long dresses and two have aprons. The boy chooses a baseball shirt and a cap. There is much discussion about who should play what role. The group finally finishes up as Mum, Dad (the boy), two sisters and a baby. They also add a soft doll as another baby. They decide that they are going to cook a meal for a birthday.

Linda starts to give orders about the setting of the table but mostly she seems to be talking to herself because no one takes any notice but she doesn't seem to mind. Jane and James set the table talking together excitedly. Lottie crawls around the floor making baby noises. Jane is nursing the doll. Lottie says "I hate babies".

Linda telephones her mummy and asks for party lights. Jane leaves the group and goes to the computer. Mary is busy cooking, James is helping. There is much discussion as to whether the carrots are ready. Mary shakes the cooking pans furiously. She seems fascinated by the sound and the plastic food swirling round the pan. Linda announces she has put up the decorations. Mary says food is ready, but James looks at the timer and says there are ten minutes left. He then counts down from ten to zero and the others join in.

Lottie retires to bed and says let's pretend Dad has cut my arm off. James obliges and soon they are all having bits cut off by Dad with a plastic knife. They are all laughing and there is no hurt done. Jane rejoins the group as a doctor and examines Lottie's glands and listens to her chest. Suddenly the meal is remembered and they return to the table.

In this example, no one is interfering with the flow of the children's play. Lottie is not chastised for 'hating babies'; Jane is not told how to 'play nicely'; the children are not told that ten minutes is a longer time than counting from ten to one; James is not told to stop 'cutting bits off' the children. The fact that the children are playing in a permissive environment means they can explore and experiment with ideas and actions at their own pace. In other words this is an environment characterised by fun, freedom and flexibility. Taken together these things give children control over their play. Indeed, Panksepp (2004) suggests that this form of play is an extremely positive form of activity, because it engages children in over half of the brain's genetically ingrained emotional systems: caring, social bonding, playfulness and explorative urges. As such, it leads to the sort of emotional equilibrium that is at the heart of therapeutic resolution.

Working to the child's agenda

It is a fundamental tenet of playwork (not just the therapeutic variety) that playworkers work to the child's agenda (Brown et al., 2018). Indeed, playworkers would argue that they are the only adults who do this. Most other adults who work with children are bringing an adult agenda to their work. In most cases their motives are well-intentioned, but nevertheless they have adult priorities, and work towards an adult goal. The child-playworker relationship is different, and it is the nature of that relationship that aligns very readily to the powerful therapeutic technique of *joining* (Kaufman 2019). The Autism Treatment Centre of America developed this technique largely for working with autistic children. Kaufman argues that autistic children are actually keen to communicate; they just lack the ability to do so in a way that the non-autistic world can easily understand. He also rejects the widespread view that autistic children have no imagination, and therefore cannot play. Kaufman argues that it is simply hard for us to recognise an autistic child's play, because we have such a restricted view of what is 'acceptable' play. He argues that instead of trying to force the child to conform to our 'normal' world – a world which they struggle to understand – we should instead 'join' the child in their world. In so doing we establish a relationship with the child who recognises us as someone whom they can trust. Once the child begins to let us in, then we can begin to help that child to learn and grow (Kaufman 2019).

This is also true of non-autistic children. A child who has experienced neglect or abuse will often exhibit unusual, and sometimes repetitive, behaviour patterns. Perhaps they are subconsciously asking for help. By 'joining' the child in their repetitive behaviour, the playworker shows the child something they can understand, which in turn will help the child to begin to trust the playworker. Once the initial trusting relationship is established, then a deeper bond may follow, often with formidable results. This was certainly true of the children in the therapeutic playwork project in Romania. The following examples are taken from Sophie Webb's research diary, which was kept during her work on that project. In the first example she is using 'joining' to encourage a ten-year-old boy to walk unaided for the first time. He had previously played with shoes and gloves, to the exclusion of anything else.

> I made sure I spent some time with Nicolae today. I have played 'shoes' with him for the past two weeks and that appears to have led him to trust me. Today, after playing 'shoes' yet again, I stood him in the middle of the room, about four steps away from me. Usually he just sits down, but this time he walked towards me with his arms stretched out for a hug. I think these may have been his first independent steps (after 10 years!).
>
> *(Webb 2014, Day 18)*

The second example is of a 6-year-old girl who would not play with the other children, and resisted all contact with the playworkers. She seemed at her happiest when standing in a corner by herself, making the same repetitive sound – a bit like a melodic ambulance siren. We were keen to find a way for her to be confident of beginning to make social contact with us and the rest of the children in the group.

> Today I started to repeat the noises that Elena makes 'waaooo, waaooo' and her reaction was amazing! The look on her face was just like someone had spoken her language. It felt like a little breakthrough as you can rarely communicate with her. . . . I

> started to repeat this noise back to her and she responded by instigating a sequence . . . exploring my face and trying to decide where the noise was coming from. By offering myself as the play environment Elena was comfortable to allow herself the freedom to communicate and investigate.
>
> *(Webb 2014, Day 29)*

Clearly, these examples relate to a set of very unusual circumstances, but the same approach can be applied in any setting where a child shows signs of emotional turmoil. In most cases therapeutic playwork would occur by default, because a child is in urgent need and has turned to the playworker for help. It would not normally be within the remit of most playworkers to undertake extensive work with a troubled child. Playworkers do not see themselves as qualified play therapists, but sometimes there is really no alternative to stepping in, because the child is in urgent need, and it is likely that the playworker is the person who has the strongest relationship with the child. Once the need has been clearly identified, the child would normally be referred to an appropriate specialist. In most cases this process can happen with more urgency because of the strength of the child-playworker relationship, and the advocacy role of the playworker.

Unconditional positive regard and the child-playworker relationship

The American psychologist Carl Rogers (1961) highlights the importance of unconditional positive regard, and this is a fundamental tenet for both play therapists (Axline 1969) and therapeutic playworkers (Brown 2018). It is essentially an approach that supports and accepts the child for who they are, no matter what that person says or does. This is not always easy, as this extract from Sophie Webb's diary shows. She is reflecting on a visit from some potential sponsors.

> The American sponsors came to visit today. They left almost as soon as they arrived – all that preparation by Edit and me, for a ten minute inspection! Without being too judgmental or stereotypical they looked and behaved exactly how I imagined they would. I noticed the children they approached and touched, like Ion and Virgil. They look cute and have all their teeth and so on. But when a hand was held out by Olympia or Nicolae (who don't have teeth and usually have dribble, snot or vomit down their faces) they literally turned the other way. It really broke my heart and I felt like screaming at them, "You don't know how special these children are".
>
> I can't imagine what this must do to Edit. She just smiled the whole time not saying a thing, and I kept smiling and answering their questions politely, when they realised I was English. When they asked how old the children were, the doctor didn't even know! She said, "They are kept here until they are five". In fact, Alexandru is eight and Nicolae is ten! It's so very sad that this was only done for show, and that these wonderful children are not treated with such care and respect by the hospital staff every day; instead of just for ten minutes.
>
> *(Webb, 3rd March 2000)*

Setting aside Sophie's understandable anger, it is clear how important it is for her that the children are respected for who they are. If these children were not treated with respect, then it would not be reasonable to expect them to develop any regard for themselves. No matter how difficult the child may make it, that positive regard must be unconditional – even when dealing with a child such as Nicolae, who in the early days took great delight in hurling snot at us, or Elena who sat on my lap and peed all over me.

The fundamental role of rhythm

The final illustration has been recounted in several texts, but I make no apology for including it here, verbatim. The vignette relates to an incident in the summer of 2005, when I had returned to the Romanian paediatric hospital to offer some training to the Romanian playworkers who were now employed on the project.

> On my final day I was introduced to a very agitated 4 year old girl who had been left in a ward totally alone. She stood at the bars of her cot rocking back and forth, making strange hooting noises. Every so often she walked rapidly round the cot, before settling back into her rocking.
>
> Her doctors said she was "blind and mentally retarded". I felt uncomfortable with this diagnosis, as she was clearly aware of my presence, and appeared to be reacting to my movements (albeit not in a very positive fashion). An added complication was her fear of men's voices. This was confirmed when I called her name, 'Liliana'. Straight away she retreated to the back of the cot.
>
> The playworkers were wondering how they could work with her. How could they get beyond the obstacle of her poor sight?
>
> I started singing to Liliana quietly: 'Twinkle, Twinkle, Little Star'. She calmed down immediately, moving her head to locate the sound. At the end of the song she made a noise in the back of her throat, which I interpreted as a request to sing again – a kind of play cue. I did this three times, and each time she moved closer to the sound.
>
> Then I started to clap gently in time to the rhythm of the song. When I stopped, she reached for my hands and put them together – another cue for me to sing. I repeated the song three more times, and each time she gave the same cue. On the last occasion she not only took my hands, but also started clapping them together in time to the song. Finally she picked up the rhythm of the song in her own hand movements, and I sang in time to the rhythm of her clapping.

This whole sequence took no more than five minutes. In that short space of time I was able to show the Romanian playworkers how to start making a relationship with Liliana by using rhythm and music.

Later that afternoon I went back into her ward, to find her rocking and hooting again. I called her name, "Liliana". She came across the cot, and felt for my hands. Clasping them together in hers, she started to clap our hands together in a rhythm that I recognised – 'Twinkle, Twinkle, Little Star'. This was truly a magic moment, but should not surprise us, because babies enter the World equipped with an understanding of the concept of rhythm. Indeed Trevarthen (2005) suggests that babies use that understanding as one of the basic building blocks of development.

Conclusion

By way of a summary, there is a strong argument that most playworkers will, at some time during their career, become involved in therapeutic work – sometimes without even realising it has happened. This is almost inevitable, because of a combination of factors. Firstly, children experience play as the only time when they are truly in control of their own life. Secondly, children tend to see adults as the people who have the answers to their day-to-day problems. Thirdly, playworkers work to the child's agenda, i.e. they do not take away the child's sense of control. Taken together, these factors mean a child is likely to form a trusting relationship with a playworker quite quickly. I would suggest the child's subconscious thought process goes something like this: I have a problem – adults can usually solve my problems – here is an adult who is playing *with* me – I can trust her, because she is not trying to take away my control. In these circumstances, it is not surprising that a troubled child would be likely to approach their playworker, before anyone else for a resolution to their problem. Therefore, playworkers have a responsibility to make themselves aware of certain therapeutic techniques, while at the same time being aware of local sources of specialist help.

Bibliography

Axline, V. (1969) *Play Therapy*. Rev. ed. New York: Ballantine Books.

Brown, F. (2003a) *Playwork Theory and Practice*. Buckingham: Open University Press.

Brown, F. (2003b) *An Evaluation of the Concept of Play Value and Its Application to Children's Fixed Equipment Playgrounds*. Unpublished PhD thesis, Leeds Beckett University.

Brown, F. (2014) *Play and Playwork: 101 Stories of Children Playing*. Maidenhead: Open University Press.

Brown, F. (2018) Therapeutic Playwork: Theory and Practice. In: Brown, F. & Hughes, B. (eds) *Aspects of Playwork: Play and Culture Studies*. Vol. 14. Lanham, MD: University Press of America.

Brown, F., Long, A. & Wragg, M. (2018) Playwork: A Unique Way of Working with Children. In: Smith, P. & Roopnarine, J. (eds) *The Cambridge Handbook of Play: Developmental and Disciplinary Perspectives*. Cambridge: Cambridge University Press.

Brown, F. & Webb, S. (2012) Children Without Play. In: Johnson, R. & Maguire, N. (eds) *Complex Trauma and Its Effects: Perspectives on Creating an Environment for Recovery*. Brighton: Pavilion Publishing Ltd.

Bruner, J. (2006) *In Search of Pedagogy*. London: Routledge.

Dwivedi, K.N. (1993) *Group Work with Children and Adolescents*. London: Jessica Kingsley Publishers.

Freud, S. (1900) The Interpretation of Dreams. In: Freud, S. (ed) *The Standard Edition of the Complete Psychological Works of Sigmund Freud*. 24 Vols. Translated from the German Under the General Editorship of James Strachey, in Collaboration with Anna Freud; Assisted by Alix Strachey and Alan Tyson in 1974. London: Hogarth Press, Institute of Psycho-Analysis.

Hughes, B. (2012) *Evolutionary Playwork*. London: Routledge.

Kaufman, R. (2019) *The Son-Rise Program Versus ABA*. The Option Institute [Internet]. Available from: <www.autismtreatmentcenter.org/contents/other_sections/aba-vs-son-rise-program.php> [Accessed 12 October 2019].

Kuschner, D. (2019) *As if: The Importance of Make-Believe for Children (and Other Human Beings)*. Exton, PA: Sticky Earth Books.

Nicholson, S. (1971) How Not to Cheat Children: The Theory of Loose Parts. *Landscape Architecture Quarterly*. Vol. 62, No. 1, October, pp. 30–34. Also In: *Bulletin for Environmental Education*. No. 12, April 1972, London: Town & Country Planning Association.

Panksepp, J. (2004) *Affective Neuroscience: The Foundations of Human and Animal Emotions*. New York: Oxford University Press.

Rogers, C. (1961) *On Becoming a Person*. Boston: Houghton Mifflin.

Sturrock, G. & Else, P. (1998) The Playground as Therapeutic Space: Playwork as Healing. In: *Proceedings of the IPA/USA Triennial National Conference, Play in a Changing Society: Research, Design, Application,* June, Boulder, CO.

Suomi, S.J. &Harlow, H.F. (1971/1976) Monkeys Without Play. In: Bruner, J.S., Jolly, A. & Sylva, K. (eds) *Play: Its Role in Development and Evolution*. New York: Basic Books.

Sutton-Smith, B. (1997) *The Ambiguity of Play*. Cambridge, MA: Harvard University Press

Sylva, K. (1977) Play and Learning. In: Tizard, B. & Harvey, D. (eds) *Biology of Play*. Ch. 6. London: Heinemann.

Tinbergen, N. (1975) The Importance of Being Playful. *Times Education Supplement*, January

Trevarthen, C. (2005) First Things First: Infants Make Good Use of the Sympathetic Rhythm of Imitation, Without Reason or Language. *Journal of Child Psychotherapy*. Vol. 31, No. 1, pp. 91–113.

Webb, S. (2000) *Therapeutic Playwork Project: Research Diary*. Unpublished.

Webb, S. (2014) Therapeutic Playwork Project: Extracts from a Reflective Diary. In: Brown, F. (ed) *Play and Playwork: 101 Stories of Children Playing*. Maidenhead: Open University Press.

11
PLAYFULNESS AND OTHER WORLDS

Re-visiting *A Midsummer Night's Dream*

Sue Jennings

Introduction

Otherworldliness is not often explored in therapeutic work. Rather there seems to be more of an emphasis on 'worldliness'. Worldliness is supposed to be a sign of maturity, having one's feet on the ground and imagination in check!

However in creative work, and in arts and play therapy specifically, we are actually inviting our participants to enter another world, the world of their imagination, and the world of their dreams. I refer to this as 'dramatic reality' (Jennings 1990), as contrasted with everyday reality. When we encourage people to use their creativity and imagination, we are encouraging them into another world experience. It is usually so powerful that we need to make sure that they are 'out of role' and 'out of landscape' before they finish the therapeutic experience.

The play *A Midsummer Night's Dream* has a central focus on journeys to other worlds, whether it is the dangerous and magical world of the forest or the trance states brought on by magic potions. The forest is a place where everything is turned upside down!

During my research with the tribal peoples in Malaysia, known as the Temiar, I discovered their understanding and practice of entering an altered state of consciousness during séances for healing or prevention. This understanding enabled me to see the importance of otherworldliness in therapeutic work and indeed in health maintenance as a whole. The substance of this chapter will be to explore *Dream* as a means of enabling therapeutic journeys, and also drawing on examples from the lives of the Temiar peoples.

The Temiars

The Temiars are a tribal people who live in the rain forest of Peninsular Malaysia, either side of the hilly divide between the states of Perak and Kelantan (Jennings 1995). Recently their lands have been ravaged by the encroachment of logging companies and the rivers have become polluted by industrial waste. They are a riverine people who gain their sense of space and direction by living in close proximity to the main river, Sungei Nengiri. Not only do they find their spatial security through the river, it is their main means of transport to other villages. As well as providing fresh fish, it provides water for their daily needs, as well as bathing, clothes washing and loo facilities.

Villages are usually built around a sibling group, and families join together to plant tapioca root and hill rice. Underlying their total way of life is a basic belief in sharing, and to be called mean is the greatest insult a Temiar can inflict. They live in bamboo stilt houses which are raised off the ground, although bamboo has given way to wooden planks in some places.

The tribe stage their own 'dance-dramas' with the shaman going into trance to effect healing. The shaman speaks of his soul going on a journey to bring back healing (or other discoveries such as a dance or song). The head soul is able to leave the body not just in the shaman, but also in ordinary people when they enter a trance. The Temiars say that it is like dreaming but it is a public event rather than a private one. Nevertheless they were always very attentive to reports of dream states, as the content could be important for others. They also said that you must not wake someone suddenly as their head soul would not have time to return to their body.

With the Temiar, the shaman is not only healer but entertainer, philosopher, social leader and counsellor. After a healing séance, performers transform into themselves again, often very bleary and disorientated. Everyone stays until dawn and they emerge into daylight to return home.

While I was there I witnessed several variations of their established shamanic tradition. The shaman is of various grades from minor to major and very rarely, the great shaman. The great shaman, it is believed, can actually metamorphose into a tiger, which is the dangerous animal most feared by the Temiar. Such shaman can embody the extremes of danger and during the entranced period becomes a very powerful healer. A special hut is constructed in a domestic house inside which the shaman will turn into tiger. The group of women are playing the bamboo stompers and echoing the shaman's assistant as he sings to call the tiger spirit closer to the house. We are alerted to the transformation when there is a scratching on the branches that form the shelter; it is believed to be the tiger's claws, accompanied by a soft growling. People who are unwell are placed near the shelter as the tiger healing is supposed to be the most powerful of all. It is a very deep emotional experience that has a profound effect on the well-being of everyone present.

The tribe often refer to a playful trance experience using the expression 'to forget'. Ordinary people as well as minor shaman 'forget': the younger people dancing ecstatically and the older men gently rocking. People who are ill or parents of new-borns do not go into trance as it is believed it will endanger their head-soul. It is believed that when in trance, the head-soul leaves the body and goes on a journey: sometimes to play with other head souls or to discover a new dance or song or hunting destination or cure. The trance state is one of creativity and discovery, which in turn is shared with the community. An ethos of sharing permeates both the other world and the day-to-day mundane events around eating and hunting.

A Midsummer Night's Dream

The Temiars do not see themselves as forest people, rather they live in villages that happen to be in forests! And there are strict rules governing their journeys into the forest. In the *Dream* the forest is the place of the 'otherworld' people, the fairies, who will interfere with or annoy or even assist their human counterparts. The *Dream* takes us on a journey from the rigidity of the court into the otherworldliness of the forest and back again to a more fluid and compromising court and market place. Whereas in the court there are strict rules governing parental hierarchy, once in the forest all the limits are released. People change partners, change roles, challenge the status quo. The magical world encourages new thinking and shifts in perspectives. There is chaos in the forest. *Dream* is also a play of shadows and darkness. By allowing us to enter 'other world experience' we are able to experience a greater understanding of our own shadow self.

However in the *Dream*, there is additional chaos as the main fairy leaders, Titania and Oberon, have had an enormous disagreement, so loud that it frightened the other fairies! The disagreement is about an adopted child who Titania promised to care for when his human mother died in childbirth. Oberon is claiming the child to be one of his helpers. Titania gives an extraordinary description of how their row has caused havoc in the human world (very reminiscent of the descriptions of climate chaos!).

Titania addresses this very succinctly when she is challenging Oberon on their disagreement about the Indian boy; the scene well illustrates the type of domestic conflict when a child is in the middle of two warring parents. Titania points out that there is chaos for mortals when the fairy characters are at war; when there is dissent in the 'other world', humans are insecure and disorientated.

> These are the forgeries of jealousy:
> And never, since the middle summer's spring,
> Met we on hill, in dale, forest, or mead,
> By paved fountain, or by rushy brook,
> Or in the beached margent of the sea,
> To dance our ringlets to the whistling wind,
> But with thy brawls thou hast disturb'd our sport.
> Therefore the winds, piping to us in vain,
> As in revenge, have suck'd up from the sea
> Contagious fogs which, falling in the land,
> Hath every pelting river made so proud
> That they have overborne their continents:
> Midsummer Night's Dream 11.1. 81–92
> *(Shakespeare 2015)*

She goes on to describe how the crops have rotted, sheep are dying and the crows feed well on the corpses, and the humans need their winter cheer. There is dissent in fairy land which creates destruction on earth. And it is not just dissent. She describes a lack of playfulness, dance and music that brings about 'contagious fogs' and floods.

I find it remarkable that Shakespeare is able to anticipate so much of the conflict that exists in modern times, including the difficulties of adopting children, the rigidity of parents and the chaos of climate disruption.

Interactive play and child development

The ethos of the Temiar tribe and the structure of the *Dream* came together in giving me inspiration to create Neuro-Dramatic-Play (NDP). Having lived with the tribe, together with my own children, I was able to gain more than a passing insight into their beliefs and practices.

NDP is the developmental paradigm that I have developed for therapy, education and parenting (Jennings 2011) which includes sensory play, messy play, rhythmic play and dramatic play. There is also a sense of performance and heightened dramatic expression between mothers and babies.

Through sensory play all our senses are developed; we don't just see the world and others, we hear, taste, smell and sense the totality of our experiences. Our senses make us alert and we are fine tuned to the world around us. It may be dangerous or it may be creative but our antennae pick up the important messages. I was sitting in my house in the Temiar village and there were various children playing inside and out including my own three children. Suddenly the Temiar children en masse rushed away – they had all heard the fall of a durian fruit drop off the tree some two hundred yards away! I and the children could hear nothing at all. However the Temiar brought the fruit back to share with everyone, whether or not we had heard it fall! Of course an acute sense of hearing can be crucial to survival where the crack of a twig could mean an approaching predator.

Rhythmic play continues the rhythmic experience of the baby before he or she was born, when babies are aware of their mother's heartbeat and the rhythms of her day-to-day life: walking, sleeping, dancing, and rocking and so on. Rhythmic play can be through physical movement, the spoken word through poetry including Shakespeare's verse, as well as drumming.

However, the Temiars are imbued with rhythm as well as routine throughout their lives. The flooring of houses is of sprung bamboo, which gives a rhythmic sensation when walking. Children are carried in slings first to the front and then on their mother's back until they can walk independently. Then they accompany the adults to the séances and usually sit with the women who are playing the bamboo stompers that accompany the shaman singing the invocations to the spirits.

Their 'rhythm of life' is governed both by the monsoons as well as their own rules regarding dancing and singing, storytelling and serious healing séances. When there are major thunderstorms these rhythms are disrupted as senior people in the community try to appease the anger of thunder. The Temiar believe that a major thunderstorm is a result of being careless with food and spatial rules. A common practice is to place a burning log in the rear doorway of the house and burn haircuts from the individual who is thought to have broken some rule. If there is any doubt then hair is cut from everyone present. The senior person, in my village, it was usually the midwife, blows the smoke from the burning hair in the direction of the thunder while muttering various apologies.

So for the Temiar, time and energy is spent in keeping other world entities at bay, thunder and tiger. The only exception is the rare example of the tiger séance. However the 'other world spirits' that are positive and assist the healing process are encouraged to enter the human world through special songs and decorations. During a healing séance they are believed to reside on the roof of the house and the central decoration which is made from leaves and flowers. These spirits are said to be the head souls of flowers who are small and shy and need to be coaxed to come down. The Temiar see themselves as rather like these sprites, shy and contained. It is only in recent years that they have been galvanised into action as their land is being stolen by multi-national companies. They have been able to engage pro-bono lawyers to represent them in court and they have built blockades to try and prevent the loggers entering the forest areas. This is a complete change of being for them: change of life's rhythms and rituals but it has been necessary for survival.

It strikes me as similar to the seismic changes that occur for children who have been suddenly removed from their families and placed in a setting that has different values, strange food and strange people. They will often feel 'lost in the landscape' and confused about their identity and expectations placed on them.

The *Dream* also allows us to explore difference: court characters, work people and fairy folk; young people and old people; male and female. We are also able to address reality and fantasy; chaos and order; this world and other worlds; dreams and every day; strangeness and truth; trance and reality: differences in time, space, consciousness, role and imagination. The play enables us to take risks while in the safety of our building. We can enter the magic forest and see what happens and then make informed choices about future decisions.

However the *most* important thing is the direct experience of the play itself. We gain much from witnessing it but we will gain much more from direct participation.

Therapeutic application of the *Dream*

This therapeutic story with actions was created by a group of children aged five to six during an NDP training practice. The children were very firm that the fairies should not be arguing about the baby and kept making sure it was safe (a very Temiar response!).

This story was enacted beginning with a soundscape and then a tree narrating the story while the action took place.

Joy and happiness in fairy land

In a green lush forest with beautiful big trees there were colourful flowers and butterflies, many fierce and gentle animals and also fairies making the forest their wonderland.

Action: *personify the animals, enact with sound to give impression of forest*

There was a fairy queen Tatiana with special fairies Blossom, Naughty, and Tinkerbell to attend her; and a fairy king Oberon, with special fairies, lion king, cheetah king and Puck to attend Him.

Action: *Tatiana with her fairies goes to their place and Oberon moves to his place with his fairies. The fairies attend to their respective king or queen.*

Tatiana has a human baby and Oberon wants the baby. So Oberon and his fairies go to Tatiana's place and ask for the baby. The queen and her fairies do not want to give the baby and ask them all to leave her place.

Action: *The preceding is displayed with a stomping of feet and hand gestures and verbal sentences by all.*

The following scenes are also enacted.

Oberon is very angry. He asks Puck along with lion and cheetah kings to bring a magic flower from the forest. When they bring back the magic flower, they make a magic potion, and Oberon asks them to go and put a few drops of the magic potion on the fairy queen Tatiana's eyelids while she sleeps. They went stealthily and did so. When the queen wakes up she falls madly in love with a donkey who was passing by. Oberon and his fairies have a good vengeful laugh but Oberon also feels bad. They go to the fairy queen's place and sprinkle the magic potion again. When the queen sees Oberon she is happy. Oberon and his fairies ask to share the baby. All were full of love with the magic potion and agreed to share the baby. They all danced with Joy singing 'Let us share the baby'.

Action: *All held hands and the baby was held by some of the children while everybody sang, dancing in a circle with Joy and Happiness.*

This is a delightful sequence for younger children. It enables full participation and exploration of important themes. I witnessed the sharing, with the children dressed in colourful costumes and wings. The overriding preoccupation was with keeping the baby safe.

A Therapeutic Story based on the Lovers of 'A Midsummer Night's Dream'

This original story is designed for older children and young teens but could be explored with older teens and possibly with parents. It addresses so many issues and conflicts that are common to teenagers.

Hi my name is Hermia and I am madly, deeply, blissfully and totally in love with a man named Lysander. I love everything about him. He is kind, generous, thoughtful, helpful, funny, AND guess what? He loves me too! He loves me as intensely as I love him. Fantastic, right? I should be blissfully happy, right? And I am NOT! You want to know why? For one thing, my dearest friend since childhood – Helena – is also totally in Love with a man named Demetrius. That shouldn't be a problem, right? Well guess what? Demetrius is in Love with me! I hate to see my friend so torn. She suffers in her love. Demetrius does not return her love and always sends her away. He can be so cruel to her. And I feel awkward when Demetrius tells me how much he loves me. I just don't feel it for him. I only have eyes for Lysander and him for me. I would like nothing more than to marry Lysander.

And guess what? My story gets even more complicated. My father – he is a very important man in the court of the Duke – has always been good and generous with me. Since forever, he has given me whatever I wanted. He has taken care of me, kept me safe, secure and loved me. BUT now he wants me to marry Demetrius! Can you believe this? He said I must obey him! If I do not, he insists I will **never** marry and must live like a nun. What am I to do? I don't want to disobey my father but how can I marry someone I do not love? How can I marry one man when I deeply love another? My heart is so open with love for Lysander yet I feel pain. I see my friends Helena suffering. I also see Demetrius is suffering and torn. UHHHHH! I don't know what to do! Do you have any ideas?

I have decided, at this moment that I can't move on in matters of the heart. My life is at a standstill. How can I be happy with Lysander when I will cause so much pain for others? How can I be happy with my decision? There is so much going on and so many people's feelings to consider.

I just went and spoke to a dear, wise old woman named Nadia. I have known her all my life, I love and trust her. She uses few words, yet when she speaks her words hold a deep wisdom. When I asked if she would help us through this difficulty, she just nodded her head. So Helena, Lysander, Demetrius and I will go together and share our feelings. Everyone has agreed to be part of this process.

I know Nadia will listen, and this will help. I want to share what I am feeling and listen to what others have to say – what are they going through? And maybe through Nadia's supportive listening, we can resolve these conflicting feelings. For instance, you might think that I would feel good knowing I have not one man but two that love me. But it actually feels terrible.

So we are going to start talking and sharing. It gives me hope that everyone is willing to at least talk. And I trust that she will be able to help us come to a place where we can ALL feel ok; that we can stop hurting each other. I am willing to take this journey of self-discovery with the support of Nadia. I am ready to be open and honest with everyone.

I know we will come to a harmonious resolution in this difficult situation. I have deep faith and hope in this process.

I will let you know what happens.

Think good outcomes for us all.

Until later. . .

oxo Hermia

This story is written like a letter and raises so many issues for young people: parental conflict, loyalty, jealousy and much more. It could be used as a basis for improvisation, and other characters in the play can be explored in similar ways. For example, what do the workmen talk about when they go home? Especially when Bottom is transformed into the ass! There are boundless opportunities to stimulate the imagination, explore roles and relationships and test outcomes (Jennings 1992). And of course to speak the beautiful rhythms of Shakespeare's poetry – another part of rhythmic playing.

But a very useful caution!

As Theseus says to his newly married wife Hippolyta:

The lunatic, the lover, and the poet,
Are of imagination all compact.
IV.2. 7–8

End piece

It is this very nebulous state within 'dramatic reality' that allows people of all descriptions, whether the walking well or those with unwell being or social disadvantage or disability or those people restrained or contained in secure settings, to challenge their ideas of self and the world.

The Temiars are a contemporary example of a culture that has built both prevention and cure through their creative and artistic experience. It includes trance and performance. In Western Europe we need therapists and theatre to do this for us!

A Midsummer Night's Dream is a journey from the rigid rules of the court to the chaos of the forest. The dangerous forest is aided by otherworld fairy characters that lead the mortals into unsafe places through transformational potions! The transition back to the safety of a now more human court with flexible borders, and market folk with their entertainment, completes the journey.

Everyone is blessed by the other world fairy folk and the audience is reassured that their shadow selves, portrayed by the actors need not give offence.

All will be well

In his penultimate speech Puck, who is a rule maker as well as a rule breaker, describes his role of ultimately tidying everything away! But in this speech he also reminds us of the darkness and the shadows. The house he is referring to is not the Duke's palace but the theatre which is often referred to as 'The House' (notices outside will say 'House full'). In this speech and the final one, Puck is reassuring the audience that it can be safe from disturbance (and mice!). He also says that it is a hallowed space, like the shamanic sacred space.

Now the hungry lion roars,
And the wolf behowls the moon;
Whilst the heavy ploughman snores,
All with weary task fordone.
Now the wasted brands do glow,
Whilst the screech-owl, screeching loud,
Puts the wretch that lies in woe
In remembrance of a shroud.

Now it is the time of night
That the graves, all gaping wide,
Every one lets forth his sprite,
In the church-way paths to glide:
And we fairies, that do run
By the triple Hecate's team
From the presence of the sun,
Following darkness like a dream,
Now are frolic; not a mouse
Shall disturb this hallowed house;
I am sent with broom before,
To sweep the dust behind the door.
Midsummer Nights Dream
V.1 361–380 (Shakespeare 2015)

All is well and all manner of things are well.

Acknowledgements

To the Senoi Temiar for their welcome and honesty; to the dramatherapists who use Shakespeare!

References

Jennings, S. 1990 *Dramatherapy with Families Groups and Individuals*. London: Jessica Kingsley Publishers.
Jennings, S. 1992 'Theatre of Healing: Nature and Scope of Dramatherapy', in *Shakespeare Comes to Broadmoor*. Ed. Cox M. London: Jessica Kingsley Publishers.
Jennings, S. 1995 *Theatre, Ritual and Transformation*. London: Routledge.
Jennings, S. 2011 *Healthy Attachments and Neuro-Dramatic-Play*. London: Jessica Kingsley Publishers.
Shakespeare, W. 2015 *A Midsummer Night's Dream*. London: New Penguin Classic.

12

'THE MAGIC DRUM' IN ACTION

The use of therapeutic play group intervention to foster positive peer interaction among children with diverse abilities and needs in an international primary school in Prague

Leong Min See

Introduction

All children have abilities, including SEN children. It is our collective responsibility as a society to focus on children's diverse abilities and facilitate their self-growth to bring out the best in them. In order to cultivate positive self-identity among SEN children, we need to advocate positive terms such as 'children with diverse abilities and needs' which acknowledges the diverse abilities inherent in SEN children while recognizing their individual needs.

In this chapter, 'children with diverse abilities and needs' refers to all children including disabled or SEN children. 'Disabled' or 'SEN' children refer to children whose access to equal education opportunities are restricted due to physical and social barriers.

Research evidence indicates that children with various Special Education Needs (SEN) across different countries experience peer rejection, victimization and bullying which could lead to long-term negative social and psychological consequences. For inclusion to be put into good practice and to yield effective results, early and effective interventions to promote positive relationship among SEN and non-SEN children is needed. This research aims to investigate the use of therapeutic play group intervention in fostering positive peer interaction among SEN and non-SEN children in a primary school in Prague.

The relationships between SEN and non-SEN children can be a barrier and an enabler to the successful inclusion of SEN children in mainstream schools. A three-year systematic literature review indicates that social engagement is vital in enhancing the academic and social inclusion of SEN children (Rix *et al.*, 2009). Studies in Cyprus (Koutrouba *et al.*, 2006), UK (Frederickson *et al.*, 2007) and Malaysia (Jelas, 2000) indicate that the success of inclusion depends to a great extent on the attitude of non-SEN children towards their SEN classmates.

Conceptual framework for therapeutic play group intervention

Stigmatization and lack of awareness among non-SEN children towards their SEN peers tend to lead to peer rejection. This study used therapeutic play group intervention to emphasize the strengths of SEN children and foster empathetic understanding of non-SEN peers to support the difficulties encountered by SEN children.

The conceptual framework (Figure 12.1) shows the focus of this study to investigate the learning experiences of children participants in the therapeutic play group intervention and the factors that determine the successful implementation of the therapeutic play group intervention.

In this study, a class with SEN and non-SEN children, with an issue of peer rejection is 'purposively' selected in order to implement the therapeutic play group intervention. A class of seventeen children aged ten in an international school was selected. The rationale to choose children in this age group is that they are more able to reflect on their learning experiences and communicate their insights compared to younger children in the primary level. Sixteen children participated in this study as one child was absent during the intervention session.

The research school is an international school based on the English National Curriculum in Prague with students from both the local and international communities. The participants are sixteen children, eight boys and eight girls, aged ten to eleven from the same class. A majority of these children have been in the same class for five years. Five newcomers join the class a year ago. The participants are from different nationalities, including Czech/Indian, Czech/Chinese, Czech/Israeli, Spanish/British, Italian, French, Danish, Russian, Afghan, British and South African.

Three children are diagnosed by educational psychologist as having special educational needs. According to the SENCO, one child is diagnosed with dyslexia, another child with attention deficit hyperactive disorder (ADHD) and another child with semantic pragmatic

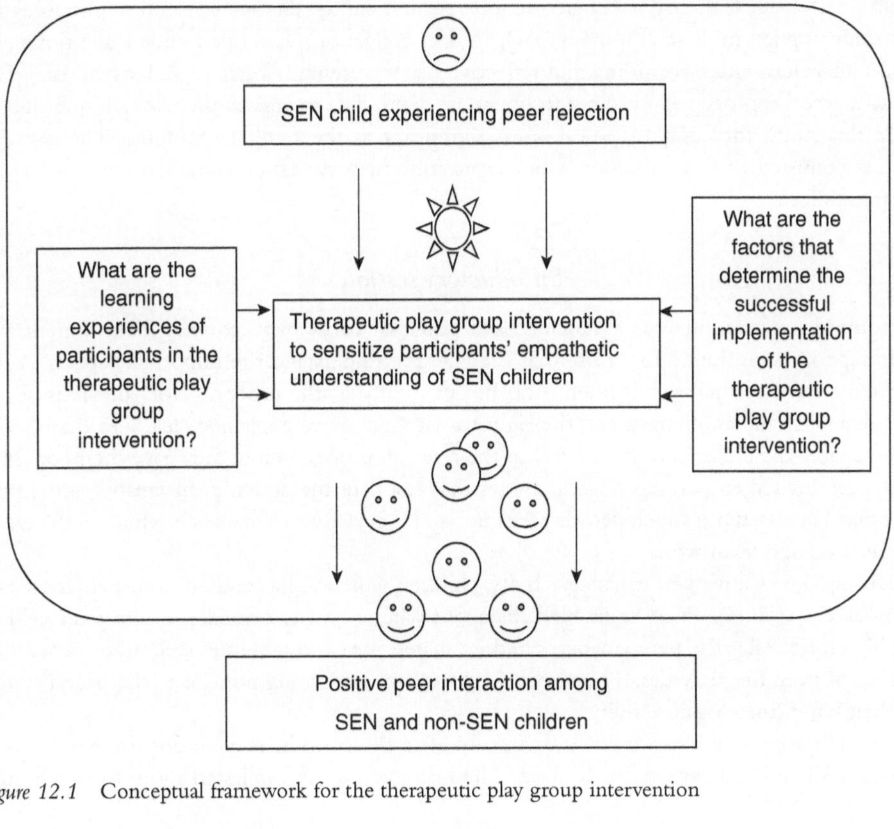

Figure 12.1 Conceptual framework for the therapeutic play group intervention

disorder. These children receive learning support from SENCO twice a week for literacy and mathematic subjects. The sessions consist of one individual session and one group session for three of them. Each session lasts for an hour. Two SEN children who are diagnosed with dyslexia and ADHD also receive learning support in the class from teacher assistant for four hours a week.

Two SEN children are native English speakers whereas for one of them, English is a second language. Two SEN children have just joined the class a year ago. Two SEN children join after school clubs and sports whereas one SEN child does not join any after school activities.

Interviews with the head teacher, drama teacher and SENCO reveal that the child with semantic pragmatic disorder seems to experience peer rejection. The child is a native English speaker and has been in the same class for four years. The child joins after school activities and has friends who are much younger. According to the teachers, the child appears sensitive, withdrawn and gets frustrated easily. The child usually plays alone and does not socialize with classroom peers during play time. The child does not receive any psycho-social program or counseling program.

My observation during introductory session, intervention session and closing session was consistent with teachers' reports regarding the issue of peer rejection. For example, during introductory session, the child did not join her group during group activity. During closing session, the child appeared sensitive and angry when her peer described one of her strengths as 'funny' during the activity 'the positive things about me'.

The research process

In line with critical theory paradigm, action research and qualitative research methodologies, this study applied multi-method approach in data collection. The open ended questionnaire, group interview, video recording and reflective diary presented concrete and visual data. The multi-method approach allowed me to obtain in-depth data to answer the research questions.

In this study, the research process is as significant as the resultant findings. The research process consisted of three sessions which were introductory session, intervention session and closing session.

Introductory session

An introductory session was conducted with potential participants to seek their consent and build rapport with them. To ensure that the children understand the important aspects of the research related to their involvement, the aim of the study, the process involved, the issues of confidentiality and anonymity, and the right to withdraw were explained clearly to them.

To enable the children to have a feel of the research process before they gave their consents, they were invited to participate in activities that resonate the activities in creative arts intervention. These activities included three games and a creative performance which emphasized cooperation and team work.

During the creative performance activity, children worked in a group of four people to create a vehicle. Children enacted disabled characters when they performed; they created a vehicle as a big group. After the performance, children shared their thoughts and feelings on what they had learnt from the activities. They were invited to say something positive to the friend sitting on their left in the closure activity.

Children's consent was sought individually after the introductory session to avoid group pressure. All children gave verbal consent. The introductory session lasted about forty minutes.

Intervention session

The intervention session was conducted two days after the introductory session. The intervention session consisted of greeting, narration of the story 'The Magic Drum', creation of a thick forest and rehearsal, performance of the journey to search for 'The Magic Drum'. It also included narration of the ending of the story 'The Magic Drum', debriefing, completing questionnaires, a closing activity, and group interview. The intervention session lasted about two hours.

1 GREETING

The intervention session began with greeting and introducing the camera person to the participants. Participants were reminded that they would do some creative activities together and share their thoughts and feelings after the activities.

2 THE STORY OF 'THE MAGIC DRUM'

Once upon a time, there was a special village in a faraway country. What made it special was there were quite a number of villagers who were different from the others. They had different abilities and needs. Among them were children. Some of these children were very good at singing and acting although they had difficulties with spelling or mathematics. Some of them were very good swimmers although they were very active. Some of them were very helpful and caring although they were disabled.

These children went to school just like all of you. But they faced a number of challenges in schools because they were different. Sometimes, they felt very frustrated and upset when they were struggling to learn and make friends in school.

One night, one of the children had a dream. She dreamt of her grandmother who had passed away. The grandmother was a very wise and well-respected woman in the village. Her grandmother told her in the dream,

"My beloved granddaughter, I know you are struggling with your schooling because you are different. At the eastern part of the village, there is a thick forest. There is a magic drum in the forest which has special powers. If all the children from the village hit it together at the same time a miracle will happen.

It is a very challenging journey. You have to go through the thick forest where unknown danger and adventure await you. You cannot do it alone. You can invite all the children in the village to go with you to search for the magic drum. You have to work as a team to help each other in this journey."

The child woke up in excitement. She quickly spread the news and invited all the children to join her in the journey. All of them decided that they were going to take up the challenge.

After the story was told participants were invited to role play children and villagers with different abilities and the need to search for the magic drum.

Participants formed two groups with four girls and four boys in each group. Each participant chose to enact a disabled character with strengths. The strengths included very helpful and caring, good at skating, swimming, acting, singing, mathematic, playing basketball and playing guitar. The disabilities were blindness, mute or physical impairment with only one hand that could move.

3 CREATION OF A THICK FOREST AND REHEARSAL

Participants transformed the classroom into a thick forest. One group created a narrow cave whereas another group created a big rock using tables and chairs. After that, participants used cloths to blind fold their eyes or tie their hands to disable themselves. Each group practised how they would crawl in the narrow cave and climb on the big rock to reach the magic drum. Each group decided what would happen when they hit the magic drum together at the same time and how the story ended.

4 PERFORMANCE OF THE JOURNEY TO SEARCH FOR 'THE MAGIC DRUM'

After several rehearsals, each group took a turn to perform their journey to search for the magic drum. When one group performed, the other group was the audience. Participants were reminded to work as a team to help each other manage the challenges they face in their journey to search for the magic drum.

5 NARRATION OF THE ENDING OF THE STORY 'THE MAGIC DRUM'

After the performance, the ending of the story of 'The Magic Drum' was narrated to the participants as follows:

> After going through all the challenges, the children from the village found the magic drum in the forest. They made a circle and hit the magic drum together at the same time. And what happened? (Paused) They created wonderful music.
>
> Their difficulties and disabilities did not disappear. But the children realized that everybody has his or her own strengths and difficulties. The children realized that by working together as a team, helping each other, they can manage a lot of challenges in life. And these children become very good friends forever. The end.

6 DEBRIEFING

Participants were invited to come out of their characters by shaking their bodies. They were reminded that they were not the children villagers in the far away country and they were back to their own self.

7 COMPLETING QUESTIONNAIRE

Participants were invited to reflect on their learning experiences by responding to seven open-ended questions in the questionnaire after debriefing.

8 CLOSING ROUND

After all the participants have handed in the questionnaires, they were invited to sit in a circle to do a closing round. Everybody was invited to take turn to share what they would like to say.

They were reminded that it was ok to say pass if they felt they have nothing to say. After the closing round, participants turned the room back into the classroom.

9 GROUP INTERVIEW

After participants took their seats, they were invited to participate in the group interview to express their views on five questions drawn from the questionnaire.

Closing session

The closing session was conducted five days after the intervention session to round up. It lasted about forty minutes. The first activity was 'the positive things about me'. The activity began with a participant telling the group his or her own strengths, followed by the group members telling the group the strengths they observed in that particular participant. Everybody took turn to participate in this activity. After that, the video recording of 'The Magic Drum' lasted about thirty minutes was presented to the participants on the LCD screen.

Play and social learning

Social development theory (Vygotsky, 1978), collaborative learning theory (Johnson & Johnson, 1989) and situated learning theory (Lave & Wenger, 1990) argue that learning involves social processes at many levels and social interaction is fundamental to the progression of learning. According to social development theory, collaborative learning theory and situated learning theory, learning is best achieved through social interaction rather than mere teaching.

Therapeutic play groups use social interactions to stimulate children's development and embrace the support of others through collaborative work. Activities such as dramatic play provide ample opportunities for collaboration. Children learn to be responsible team players. They learn to share, take turns and negotiate with group members.

Therapeutic play groups engage participants in social and collaborative learning. Collaboration made the learning process reciprocal and effective. Participants took ownership of and responsibility for their own learning in the reciprocal relationship when they worked as a team to create the forest and plan the journey to search for the magic drum. Participants' comments reflected their enthusiasm in working as a team:

- I like working as a team (P13).
- It was fun when we work as a team (AG).
- It was fun and I like fun and work with others so I would lik (like) to do it again (P9).

According to Vygotsky (1978), effective learning experiences are organized within the zone of proximal development (ZPD) where learners start from participating in a range of tasks they can do independently to tasks which require help from others. In this study, therapeutic play group was designed according to the zone of proximal development (ZPD) of participants. Participants engaged in problem solving activities which they could achieve independently and learnt new strategies with the help from peers and researcher.

The findings of this study are consistent with the argument of social development theory (Vygotsky, 1978), collaborative learning theory (Johnson & Johnson, 1989) and situated learning theory (Lave & Wenger, 1990) that social and collaborative learning provide positive role modeling and increase feelings of support. Participants had multiple opportunities to learn from peer modeling when they created the forest, climbed the big rock and crawled in the narrow cave as disabled people.

Play and reflective learning

Constructivist learning theory (Bruner, 1966) argues that the interaction of new stimuli with pre-existing understandings enables learners to acquire new knowledge. According to constructivist learning theory, meta-cognition plays significant roles in learning process (ibid.). Meta-cognition involves reflective learning where participants actively reflect and monitor their learning and thinking processes (ibid.).

Therapeutic play group engages children in reflective learning. Dramatic play for example requires children to invent character, reflect on the character and reinvent the character. Children engage actively in reflection and revision of ideas by planning how to enact the character, rehearsing, stepping out of the character and reflecting on the process.

Engaging in reflective learning through therapeutic play group, children generate new understanding of the problems occurred in the learning process and improvise and create different strategies to solve the problems. Children engage in deep learning when they step back regularly to reflect and review on the learning progress.

Reflective learning helps children to gain deeper understanding of themselves and others. Reflecting on their thoughts, feelings and actions, children are more aware of their own emotions and behaviour and the effects of their emotions and behaviour on the learning process. Reflective learning enables children to recognize their strengths and weaknesses, and transform their learning experiences into insights.

Constructivist learning theory argues that meta-cognition is crucial in learning process (Bruner, 1966). The therapeutic play group engaged participants in reflective learning when participants enacted the disabled characters, rehearsed, stepped out of the characters, reflected and revised ideas.

Reflective learning enabled participants to generate new understanding of the problems, improvise and create different strategies to solve the problems. This is reflected in participants' efforts to rehearse several times in order to find different strategies to climb the big rock and crawl in the narrow cave as disabled people.

Reflective learning enhances the development of observation skills. Participants' comments on what they could do to create a better environment for disabled people showed that they were observant of the existing facilities that accommodated the needs of disabled people. Their awareness was reflected in their comments:

- They have er like a lift . . . they have like the number, the little dots that they can touches them and know which number it is (HB).
- Like we go to the tram, it tells where the station are, and then next station, so the blind people know where, when they will have to get off (AG).
- And also when at zebra crossing, it makes the sound like when to go (MG).
- Like people who can't hear, ya if they are at tram, they are like say which is the next place . . . for people who can't hear, there are like writing on this kind of board, it helps them (BG).

Children are observant and have a rich understanding of their learning processes. Deep learning occurred when participants stepped back regularly to reflect and review on the learning progress. The questionnaires and group interview enabled participants to reflect on their learning experiences and gain deeper understanding of disabilities issues. Participants realized that:

- It doesn't mater (matter) if you are disabled you are still good at something (P1).
- You have to work as a team to achieve something if you have a disability (P6).
- That is not funny having disabilities and you must be gentle with disabilities (disabled) people (P12).
- That it doesn't matter if a person has a disability because everybody is good a (at) something (P13).
- When people have a disability, they are also good at other things. Everybody has his disability and everybody has his strengths (EG).

Engaging in reflective learning, participants gained deeper understanding of themselves and others. When participants reflected on their thoughts, feelings and actions in the therapeutic play group, they were more aware of their own emotions and behavior. Participants learnt more about themselves and their peers through self-reflection that emerged from their observation of others during creative and playful exchanges. They realized that:

- I have disability, when eating like using chopsticks (CB).
- I felt good on the start but I know that if I had the disability forever that won't be nice (P12).
- I felt great with all the other people helping me and I liked helping them (P13).
- I felt good because it is good when somebody was helping you (P14).

Play and emotional literacy

Emotional literacy refers to the ability to manage a full range of positive and negative emotions experienced by ourselves and others (Gardner, 1993; Goleman, 1996). Emotional literacy enhances concentration, memory, problem solving and learning skills, creativity, innovation and leadership (ibid.). Emotional literacy is essential to children's holistic development and an important preventive strategy in promoting children's mental health (Coppock, 2007).

Children need to develop necessary emotional literacy in order to engage meaningfully with peer. Therapeutic play group provides multiple opportunities for children to express themselves and have their voice heard. Participants learnt to manage a full range of positive and negative emotions experienced by themselves and others in therapeutic play group. They learn to accept and express these emotions constructively.

Imagination is crucial in developing children's empathy. Imagination enhances empathetic understanding and enables children to feel the suffering of others although the intensity of suffering is different for every individual. By enacting roles in dramatic play, children identify with the suffering of the characters and learn to understand the complexity of the characters' inner world, their fears, their desires and their hopes. Children learn to feel the pain and the impact of the suffering on another person who is different from them. Enacting the roles of disabled characters enabled participants to experience how it felt to be disabled. The ability to feel becomes the foundation for participants to develop their capacity to care for others. Participants' comments reflected that they realized that life was hard for disabled people:

- Different and I realized it is hard to have disability (P2).
- It wasn't the best feeling and it was hard (P3).

- I felt really weird and like I shouldn't be in this world (P5).
- I feelt (felt) different (P6).
- I felt bad because I couldent (couldn't) move my hands (P9).
- I feel sad for them and there (they) are not lucky (P10).

Therapeutic play groups have the potential to cultivate in children a capacity for empathetic understanding that will enable them to comprehend the motives and choices of people different from themselves, seeing them not as alien but as human beings who share many problems and possibilities with them (Nussbaum, 1997). Enacting roles in dramatic play offers boundless opportunities to portray the human condition and enables participants to acknowledge their own vulnerability to misfortune. Participants were able to imagine that they might go through the same suffering experienced by disabled people who were different from them:

- I felt good when I was blind but I really didn't, I really don't want to be blind (GB).
- It was fun having my hands tied (cross both hands as if tied) but I don't want to be like that for the rest of my life (HB).
- When I was being blind, I felt good but if I will be blind forever then it won't be nice (CB).
- It was fun, play but not for all of my life (IB).

Dramatic play enables children to enter others' world in a form of 'as if' to better understand other people. Participants were not limited by their restricted life experiences when their imagination opened the doors for them to experience alternative realities. Being in someone else's shoes allowed participants to feel the suffering of others and look at things at different levels and in different ways as expressed by participants:

- It doesn't matter if you are disabled you are still good at something (P1).
- That helping can be full of fun and friendship (P5).
- We learn how life is like when you can't do things . . . sometimes it just feels bad (BG).
- When people have a disability, they are also good at other things. Everybody has his disability and everybody has his strengths (EG).

The ability to identify and express one's own emotions and to read another's emotions correctly and comprehensively helps children negotiate multifaceted social web, engage meaningfully in relationships and reap the necessary social support crucial to healthy human development (Brouillette, 2010).

Children imagine, recreate images and ideas from their previous experiences through play (Vygotsky, 1978). The storytelling, interactive games and dramatic play emphasize affirmation, communication and cooperative problem solving. Participants learnt to share, take turns and negotiate with each other when choosing the disabled characters, deciding as a group how the story ended and going through the journey to search for the magic drum.

Cattanach (1994) asserts that dramatic play enables children to transform their life experiences through active engagement in a safe and symbolic space. Jennings (1993) asserts that children develop trust and a sense of autonomy and belonging to the group though dramatic play. Participants learnt to stretch themselves, explore possibilities, and take risks when their emotional and imaginative needs were met in the therapeutic play group. They played and learnt together in creative, investigative and problem-solving manners. By engaging in an imaginative world, participants acted out problems and experimented with possible solutions through peer collaboration when rehearsing their adventure of climbing the big rock and crawling in the narrow cave as disabled people.

Empathetic understanding facilitates the capacities of children to look at their world with a more complex and critical lens. By nurturing children's empathetic understanding, children develop the potential to become creative thinkers and problem-solvers with the capabilities to explore difficult issues in creative and innovative ways. Participants were able to make some suggestions when asked what they could do to create a better environment for disabled people:

- May be you could design a club wich (which) will every day go somwhe (somewhere) to help the disability (disabled) people (P5).
- To be kind and helful (helpful) (P7).
- Help them to move and say things (P8).
- Some hous (houses) for people without (with) disability (P9).
- Try to help him and be friend with him (P12).
- I would say to everybody to help people with disability (P14).
- Build a school for children with disability so they can have extra care (KG).

Dramatic play enables children to develop empathy and conscience which form the moral stance in the world. Children learn through 'being the other' rather than receiving a moral code 'from the other' (Jennings, 1999). Therapeutic play group has the potential to cultivate in children the ability to form moral decisions towards creating an inclusive society. The ability to feel for others' pain and suffering enabled participants to form a particular stand point:

- That is not funny having disabilities and you must be gentle with disabilities (disabled) people (P12).
- . . . it is mean to be to be staring or laughing on (at) people with disabilities (P13).
- Er . . . like people with disability need like, extra help, when you see them, you should really like, try to help them in the best way possible, because like, if blind people, if they are like, shopping or delicate things, you should like, show them the way so that they don't get into trouble. . . . Tell him show him, like hold his hands, like say go here, go there (FB).

Therapeutic play groups have the potential to cultivate in children the ability to take moral action. Of the sixteen participants, five explained in the questionnaire that they would interact differently with disabled people after participating in this project as they knew how it felt to be disabled. Another five participants stated that they would help disabled people without indicating whether there were changes in their interaction with disabled people after participating in this project. Their comments include:

- Yes because I know now how hard it is for them (P4).
- Yes I would beacause (because) I would know how they feel. So Surely I would Help them) (P5).
- Yes because now I know how hard it is for them, and I will help them (P7).
- I would help him because he would need help from somebody for example if somebody was blind or couldn't walk or couldn't talk! (P14).

Education needs to play a proactive role to nurture the development of compassion, creativity, autonomy and the ability to take moral action (Davies, 2006). Empathetic understanding is crucial for the development of compassion. Compassion involves the ability to acknowledge one's own vulnerability to misfortune (Nussbaum, 1997). The ability to feel the pain of others and imagine that they might go through the same suffering enables children to form moral decisions towards creating an inclusive society.

Appropriate matching of the intervention to meet the specific needs of participants

The finding of this study is consistent with previous studies which emphasized the appropriate matching of the intervention strategy with specific needs of the targeted children (Gibb *et al.*, 2007; Spence, 2003; Gresham *et al.*, 2001). Data analysis showed that by meeting the specific needs of participants, the therapeutic play group successfully engaged participants in co-operative and collaborative creative work which enhanced their development of empathetic understanding of disabled people. Participants' comments reflected that they understood how it felt to be disabled:

- Different and I realized it is hard to have disability (P2).
- It felt hard (P4).
- I felt really weird and like I shouldn't be in this world (P5).
- I feel (felt) different (P6).

Active engagement

Active engagement with SEN and non-SEN participants through therapeutic play group activities made the learning process enjoyable and challenging, thus encouraged participants to be autonomous, energetic and imaginative. Participants expressed their joy of engaging in therapeutic play group activities by commenting that:

- I liked being blind a (and) the obsticles (obstacles) (P4).
- It was fun because we were acting (P9).
- It was fun being blind (P11).
- I enjoy being without eyes (GB).
- I enjoy the adventure (CB).
- I enjoy building the cave (JB).
- I enjoy building the rock and being blind (KG).
- I enjoy when we were building the rock, er . . . the cave (HB).
- I enjoy the part where we have to build the. . . (FB).

Majority of non-SEN children do not understand the nature of the special needs and difficulties encountered by their SEN peers. Some non-SEN children avoid their SEN peers because of the fear and insecure feelings they experience when they are with SEN peers and the 'mysterious' behavior shown by their SEN peers (Nicolaidou *et al.*, 2006). Lack of awareness tends to lead to peer rejection (ibid.).

Research findings suggest that non-SEN children hold negative attitudes towards SEN children (Hodkinson, 2007; Laws & Kelly, 2005; Swaim *et al.*, 2001). A multi-modal intervention with 285 ADHD children in US showed that even though the behavior of ADHD children improved with treatment, peers may often resistant to change their perceptions and feelings toward a child whom they reject (Hoza *et al.*, 2005).

Children interpret peer relationships and respond differently to SEN peers based on their belief and values (Hay *et al.*, 2004). Non-SEN peers may response positively or negatively toward SEN children depending on the extent to which SEN children are perceived as being responsible for their non-normative characteristics (Hennessy *et al.*, 2007; Frederickson *et al.*, 2007; Coplan *et al.*, 2007; Frederickson *et al.*, 2001).

It is therefore crucial to engage with non-SEN children to educate them about the nature of SEN to diffuse their anxieties, fear and stigmatization. Non-SEN children tend to empathize and accept their SEN peers when they understand that their SEN peers cannot control some of their own behavior (Nicolaidou *et al.*, 2006). Effective interventions highlight the strengths of SEN children and cultivate empathetic understanding to support the difficulties encountered by SEN children (Frederickson *et al.*, 2007). Non-SEN peers are informed about the strengths of SEN children, their favourite sports, leisure activities and hobbies to emphasize the similarities shared among SEN and non-SEN children (Gibb *et al.*, 2007).

Non-SEN peers play vital roles in promoting social inclusion for SEN children. Non-SEN peers can become the buddy of SEN children by staying, playing and talking with SEN children. The peer buddy approach has contributed to increased social interactions among SEN and non-SEN children in UK (Jones, 2007), Cyprus (Nicolaidou *et al.*, 2006) and Malaysia (Jelas, 2000). Study in UK indicated that peer buddies of autistic children achieved a better understanding of autism and learnt about individual differences (Jones, 2007).

Active engagement enabled SEN and non-SEN participants to interact in meaningful and enjoyable ways. Participants expressed their enjoyment in peer companionship and support by commenting that:

- That helping can be full of fun and friendship (P5).
- I like working as a team (P13).
- I felt good because it is good when somebody was helping you (P14).
- It was fun when we work as a team (AG).
- I enjoy the part where we have to build the . . . you give us, and I enjoy the part where we have to help like people who couldn't see or people who needed help (FB).

Therapeutic play group activities encouraged participants to be autonomous, energetic and imaginative. Video analysis revealed that participants were enthusiastic to enact different disabled characters and they asked my permission to change their disabled characters. I encouraged them to negotiate among themselves. Some participants managed to exchange their disabled characters with their friends. Video analysis showed that some participants climbed the big rock and crawled in the narrow cave in different disabled characters. Indirectly, this has enriched their experiences in the learning process.

Active engagement in role playing allowed participants to step into disabled people's shoes to better understand causal characteristics of disabled people. Participants' comments showed that they understood the possible limitations and frustrations experienced by disabled people:

- I felt bad because you cannot sometimes do things (P8).
- I felt bad because I couldent (couldn't) move my hands (P9).
- It was hard when we like, it was really hard when you have like disability (JB).
- Er . . . I learn that it is hard for the people with disabilities, that it is hard like move thing. . . (HB).
- We learn how life is like when you can't do things . . . sometimes it just feel bad (BG).

Pupils need to be perceived as co-educators and be actively involved in the efforts to foster the values of inclusion. The findings of this study are consistent with previous studies that indicate the importance of active engagement with SEN and non-SEN participants to foster positive peer interactions among them.

Conclusion

The findings in this study are theoretically coherent with contemporary learning theories, reflective learning and emotional literacy. The findings suggest that the therapeutic play group intervention engaged participants in peer collaborative learning and reflective learning which contributed to the development of participants' emotional literacy. The findings also suggest that the intervention has the potential to cultivate in participants the capacity of empathetic understanding which is essential for fostering positive peer interaction among SEN and non-SEN children.

Upon reflection, I have identified areas of improvement for the therapeutic play group intervention. The findings from this study suggest that some participants held negative views on disabilities issues and the creative arts intervention failed to change their negative views. These negative views included perceptions that associated disability with 'something wrong' and 'not feeling normal':

- I learned how hard it is to have something wrong with you (P7).
- I felt that people with disability are not feeling normal, so we try to be like them, so we know now how to help them to overcome it (LG).

Due to time constraints, participants did not have sufficient time to engage in deeper reflective learning and discussion to explore these negative perceptions. In future, I would facilitate one group of participants at a time with maximum eight participants in a group. By doing so, I would have more time to engage participants in deeper discussion on the themes of empathetic understanding, strengths and difficulties, and similarities and differences between SEN and non-SEN children. This would enable participants to develop a more balance and positive perceptions on disabilities issues.

Moreover, I would conduct video screening to present the positive characteristics of disabled people around the world and their success in different areas of achievement. I would enrich the disabled characters in the dramatic play by emphasizing their strengths. I would invite participants to demonstrate their strengths when they enact the disabled characters. For example, the blind participants could sing to motivate others whereas the mute participants could be knowledgeable in identifying edible plants.

In addition, I would enrich the adventure by inviting participants to create some happenings in their journey. For example, participants encounter a snake and one of the participants directs the group to react appropriately to avoid the danger. Another happening could be that one participant falls down and gets injured when climbing the big rock and the participants work collaboratively to carry him to continue the journey.

I would enrich the creative elements by bringing some props, cloths or boxes for participants to create the forest. This would enhance participants' creativity and collaboration. I would also use musical instrument to draw participants' attention and take greater precautions to ensure the safety of participants when they go through the adventure as disabled people.

Peer rejection, victimization and bullying can be a lifelong process of frustration and alienation for SEN children. The negative long term social and psychological consequences of peer rejection, victimization and bullying on SEN children further asserts the need for early and effective interventions to foster positive peer interactions among SEN and non-SEN children.

This study has identified four important factors that determine the successful implementation of the therapeutic play group intervention in fostering positive peer interaction among SEN and non-SEN children. The four important factors are appropriate matching of the intervention to meet the specific needs of participants, active engagement with SEN and non-SEN participants, reflexivity of researcher and collaboration with school teachers.

Human behavior is influenced by consistent reciprocal interactions among behavioral, cognitive and environmental factors (Bandura, 1977). A child's individual skills and temperament, factors in the family and greater social environment affect the child's ability to engage positively with peers at different developmental stages (Hay *et al.*, 2004).

Inclusive school settings are ideal situations to teach social skills as these settings provide naturalistic spaces where SEN children can learn through modeling from their non-SEN peers and teacher-led direct instruction. Teaching non-SEN children appropriate strategies to interact with SEN children can contribute to the development of positive peer relationships among SEN and non-SEN children over time.

Interventions that enable non-SEN children to understand and accept differentiated behavior exhibited by their SEN peers help to build positive and collaborative relationships among SEN and non-SEN children. Most importantly, these interventions enable SEN children to develop friendships with non-SEN peers and join the social network of school, community and greater society.

Social skill intervention is most successful when the intervention is implemented using multimodal approaches, naturalistic settings, positive peer models and socially valid skills (Spence, 2003). Social skills intervention engaging non-SEN peers should involve non-SEN peers at every stage of the design, delivery and evaluation of the intervention to promote a sense of ownership of the intervention.

Inclusive practices need to be underpinned by inclusive policies and the creation of inclusive communities and cultures (Booth *et al.*, 2006). Strategies to foster inclusion of SEN children are likely to be successful if they are underpinned by core values and attitudes of respecting and valuing diversity (Humphrey *et al.*, 2006).

Research studies in UK (Fox *et al.*, 2004) reveal that friendships among SEN and non-SEN children exist in schools where there are opportunities for SEN and non-SEN children to work together in all aspects of school life. School culture that encourages cooperative learning and understanding for all children promotes more positive attitudes and behavior toward SEN peers than school culture that emphasizes social comparison and competition on school-related tasks (McDougall *et al.*, 2004).

Study in UK identified effective school management and inclusive school ethos as vital elements to ensure successful inclusion for ASD children (Gibb *et al.*, 2007). Another study engaging thirty-five primary teachers from seven countries across Europe revealed the need to perceive diversity as an enriching factor at the whole school level in order to promote inclusive education (Humphrey *et al.*, 2006).

The school's commitment to create an inclusive school ethos with active involvement of teachers and pupils is crucial in fostering friendships among SEN and non-SEN children. The values of inclusion and solidarity can bring all members of a school community together to form a supportive social network in school to promote inclusion for all children.

The findings of this study demonstrate the potential of therapeutic play group intervention in enabling children with diverse abilities and needs to form positive peer relationship. Future study would contribute to the development of a more inclusive society in which children with diverse abilities and needs could truly enjoy equal education opportunities to develop their personality, talents, mental and physical abilities to the fullest potential.

References

Bandura, A. (1977). *Social Learning Theory*. Englewood Cliffs, NJ: Prentice Hall.

Booth, T., Ainscow, M., & Kingston, D. (2006). *Index for Inclusion: Developing Play, Learning and Participation in Early Years and Childcare*. Bristol: CSIE.

Brouillette, L. (2010). How the Arts Help Children to Create Healthy Social Scripts: Exploring the Perceptions of Elementary Teachers. *Arts Education Policy Review, 111*(1), 16–24.

Bruner, J. (1966). *Toward a Theory of Instruction*. Cambridge, MA: Harvard University Press.

Cattanach, A. (1994). Dramatic Play with Children, the Interface of Dramatherapy and Play Therapy. In Jennings, S., Cattanach, A., Mitchell, S., Chesner, A., & Meldrum, B. (Eds.), *The Handbook of Dramatherapy* (pp. 133–144). London and New York: Routledge.

Coplan, R., Girardi, A., Findlay, L., & Frohlick, S. (2007). Understanding Solitude: Young Children's Attitudes and Responses Toward Hypothetical Socially Withdrawn Peers. *Social Development, 16*(3), 390–409.

Coppock, V. (2007). It's Good to Talk! A Multidimensional Qualitative Study of the Effectiveness of Emotional Literacy Work in Schools. *Children & Society, 21*(6), 405–419.

Davies, T. (2006). Creative Teaching and Learning in Europe: Promoting a New Paradigm. *Curriculum Journal, 17*(1), 37–57.

Fox, S., Farrell, P., & Davis, P. (2004). Factors Associated with the Effective Inclusion of Primary-Aged Pupils with Down's Syndrome. *British Journal of Special Education, 31*(4), 184–190.

Frederickson, N., & Furnham, A. (2001). The Long-Term Stability of Sociometric Status Classification: A Longitudinal Study of Included Pupils Who Have Moderate Learning Difficulties and Their Mainstream Peers. *Journal of Child Psychology & Psychiatry & Allied Disciplines, 42*(5), 581–592.

Frederickson, N., Simmonds, E., Evans, L., & Soulsby, C. (2007). Assessing the Social and Affective Outcomes of Inclusion. *British Journal of Special Education, 34*(2), 105–115.

Gardner, H. (1993). *Multiple Intelligences: The Theory in Practice*. New York: Basic Books.

Gibb, K., Tunbridge, D., Chua, A., & Frederickson, N. (2007). Pathways to Inclusion: Moving from Special School to Mainstream. *Educational Psychology in Practice, 23*(2), 109–127.

Goleman, D. (1996). *Emotional Intelligence*. London: Bloomsbury.

Gresham, F., Sugai, G., & Horner, R. (2001). Interpreting Outcomes of Social Skills Training for Students with High-Incidence Disabilities. *Exceptional Children, 67*(3), 331–344.

Hay, D., Payne, A., & Chadwick, A. (2004). Peer Relations in Childhood. *Journal of Child Psychology & Psychiatry, 45*(1), 84–108.

Hennessy, E., Swords, L., & Heary, C. (2007). Children's Understanding of Psychological Problems Displayed by Their Peers: A Review of the Literature. *Child: Care, Health & Development, 34*(1), 4–9.

Hodkinson, A. (2007). Inclusive Education and the Cultural Representation of Disability and Disabled People: Recipe for Disaster or Catalyst for Change? An Examination of Non-Disabled Primary School Children's Attitudes to Children with Disabilities. *Research in Education*, 77, 56–76.

Hoza, B., Gerdes, A., Mrug, S., Hinshaw, S., Bukowski, W., Gold, J. *et al.* (2005). Peer-Assessed Outcomes in the Multimodal Treatment Study of Children with Attention Deficit Hyperactivity Disorder. *Journal of Clinical Child & Adolescent Psychology, 34*(1), 74–86.

Humphrey, N., Bartolo, P., Ale, P., Calleja, C., Hofsaess, T., Janikova, V. *et al.* (2006). Understanding and Responding to Diversity in the Primary Classroom: An International Study. *European Journal of Teacher Education, 29*(3), 305–318.

Jelas, Z. (2000). Perceptions of Inclusive Practices: The Malaysian perspective. *Educational Review, 52*(2), 187–196.

Jennings, S. (1993). *Play Therapy with Children: A Practitioner's Guide*. London: Blackwell Scientific Publications.

Jennings, S. (1999). *Introduction to Developmental Play Therapy*. London: Jessica Kingsley Publishers.

Johnson, D. W., & Johnson, R. T. (1989). *Cooperation and Competition: Theory and Research*. Edina, MN: Interaction Book.

Jones, V. (2007). 'I Felt Like I Did Something Good' – the Impact on Mainstream Pupils of a Peer Tutoring Programme for Children with Autism. *British Journal of Special Education, 34*(1), 3–9.

Koutrouba, K., Vamvakari, M., & Steliou, M. (2006). Factors Correlated with Teachers' Attitudes Towards the Inclusion of Students with Special Educational Needs in Cyprus. *European Journal of Special Needs Education, 21*(4), 381–394.

Lave, J., & Wenger, E. (1990). *Situated Learning: Legitimate Peripheral Participation*. Cambridge: Cambridge University Press.

Laws, G., & Kelly, E. (2005). The Attitudes and Friendship Intentions of Children in United Kingdom Mainstream Schools Towards Peers with Physical or Intellectual Disabilities. *International Journal of Disability, Development & Education*, *52*(2), 79–99.

McDougall, J., Dewit, D., King, G., Mille, L., & Steve, K. (2004). High School-Aged Youths' Attitudes Toward Their Peers with Disabilities: The Role of School and Student Interpersonal Factors. *International Journal of Disability, Development & Education*, *51*(3), 287–313.

Nicolaidou, M., Sophocleous, A., & Phtiaka, H. (2006). Promoting Inclusive Practices in Primary Schools in Cyprus: Empowering Pupils to Build Supportive Networks. *European Journal of Special Needs Education*, *21*(3), 251–267.

Nussbaum, M. C. (1997). *Cultivating Humanity: A Classical Defense of Reform in Liberal Education*. Cambridge, MA and London: Harvard University Press.

Rix, J., Hall, K., Nind, M., Sheehy, K., & Wearmouth, J. (2009). What Pedagogical Approaches Can Effectively Include Children with Special Educational Needs in Mainstream Classrooms? A Systematic Literature Review. *Support for Learning*, *24*(2), 86–94.

Spence, S. (2003). Social Skills Training with Children and Young People: Theory, Evidence and Practice. *Child & Adolescent Mental Health*, *8*(2), 84–96.

Swaim, K., Morgan, S., Lenhart, J., Hyder, K., Zimmerman, A., & Pevsner, J. (2001). Children's Attitudes and Behavioral Intentions Toward a Peer with Autistic Behaviors: Does a Brief Educational Intervention Have an Effect? *Journal of Autism & Developmental Disorders*, *31*(2), 195–205.

Vygotsky, L. S. (1978). *Mind in Society: The Development of Higher Psychological*. Cambridge, MA: Harvard University Press.

13

PLANNING SUCCESSFUL THERAPEUTIC MATERIAL ENGAGEMENTS

Charting a course from danger to the safe zone

Sue Gascoyne

Being a passionate advocate of messy play doesn't make me immune to experiencing its stresses and strains – far from it! Working therapeutically with children can be physically and emotionally demanding, without even considering offering messy play opportunities with short gaps in between sessions for clearing-up. Understanding the value, benefits, characteristics, stages and themes of messy play (all areas explored in *Messy Play in the Early Years*, Gascoyne, 2019) will provide a solid foundation for preparing the environment and yourself, but this knowledge and experience are no guarantee that messy encounters won't feel discombobulating and out of control! Children's therapeutic material engagements are a heady mix of alchemy, transformation, un/containment, agency and control, which is why I place Deci and Ryan's *Theory of Self-Determination* (2001) and three psychological needs of autonomy (A), relatedness (R) and competence (C) at the centre of my own practice and self-care.

Offered as a 'useful framework for considering messy play' interactions (Gascoyne, 2019, p. 9) this tool for focussing on children's and adult's psychological needs is, I believe, central to the success of children's therapeutic material engagements since messy play has the power to elicit extreme and sometimes uncomfortable reactions, so sustainable and authentic interactions need to accommodate adult's psychological needs, as well as those of the child.

Collectively termed **ARC** (Gascoyne, 2019), the need for Autonomy, Relatedness and Competence, set out in Figure 13.1, can be applied to messy play as follows:

Autonomy – Better understood as a child's ability to explore and use their body to test ideas rather than necessarily taking control, autonomy is described by Morris (2015) as 'being me'. Although at first sight counterintuitive, structure and the setting of limits provides children with 'security of exploration' (Grossmann et al., 2008, p. 10) to discover more about themselves and the world around them. As well as enabling children 'to feel safe to be curious and explore' (Gascoyne, 2019, p. 17) it will help therapists and the adults around children feel safe and in control.

Relatedness – With messy play a trigger for extreme love/hate reactions we cannot ignore the importance of a child's relatedness to the resources, environment and caregivers, nor,

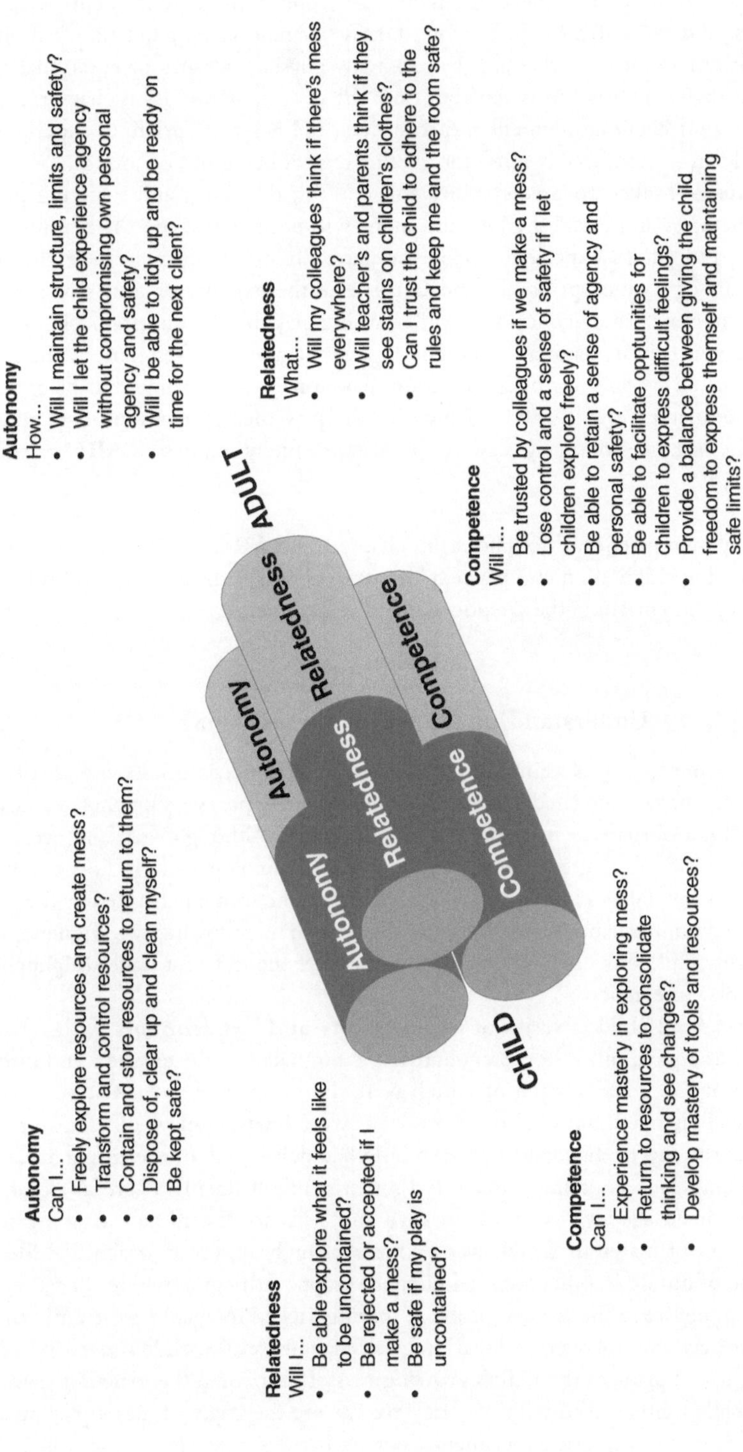

Figure 13.1 Acknowledging typical messy play doubts

Source: Adapted from Gascoyne (2019).

equally, the adult's own positive or negative experiences of messy play materials and interactions. A vital component in children developing strong attachments with adults, resources, and the self through growing bodily awareness, the adult therapist also will bring with them into the play therapy room any physical, sensory or emotional associations that they may have. As is apparent from the play vignettes in this chapter, the quality of therapist client attunement and trust lie at the heart of enriching experiences, or conversely an uncomfortable sense for the therapist of being out of control.

Competence – Linked to predictability and control, the limits and structure provided by the therapist and established environment will positively shape the quality of both the child and adult's experience of messy play. Child-centred practice rightfully lies at the heart of Axline's principles, but if this is at the expense of adult's own need for safety, competence and agency, then the resulting play may be unsustainable, inauthentic and potentially damaging to the child's and therapist's self-esteem, as the message communicated by anxious body language may convey fear, shame and blame. As we will explore later, several simple changes to the play therapy environment can greatly reduce the negatives of messy play and in doing so finely retune the **ARC** balance (see Figure 13.1).

Mirroring my own practice, throughout this chapter I use **ARC** as a lens for unpicking and better understanding children's messy play explorations, as well as shaping my own role in supporting satisfying and enriching therapeutic material engagements.

Understanding the stages of messy play

Far from viewing messy play as a simplistic and homogenous process, Gascoyne (2019) asserts that messy play can be characterised by its varied qualities, energies, stages, benefits and appeal, as well as a plethora of spatial, temporal and visual patterns. Although each and every child's interaction will be unique, that does not mean that patterns aren't evident in how children use materials. In fact, several stages in messy play encounters can be seen if permitted by the physical and emotional environment put in place by the therapist. The following play vignette, typical no doubt of many children's explorations, neatly describes the different stages (highlighted in bold) of messy play interactions:

For one 7-year-old child, a series of **explorations and experiments** with a range of materials, including glue, oil, water, paint and salt, culminated in the creation and **containment** of numerous substances, each of which were duly inspected at each session to see if and how they had changed. Some of these creations were **harnessed** as 'magic paint', while others were **cleared-up** or **discarded** when mould had developed, or the aroma became too pungent! Checking each container, much as a scientist might during a careful experiment, the child experienced agency over what to **save** and what to discard and in doing so over the transformation of materials which was a source of highs and lows much like life itself. The appearance of mould in one container elicited disappointment at where the experiment had gone wrong, before resilience and problem solving kicked in enabling the child to move forward. Conversely the discovery of sparkling salt crystals on the child's saved mess was a source of delight and pride as the child's growing mastery mirrored the growing crystals.

In this one child's interactions with materials we can see the stages of messy play possible if children are permitted to explore, experiment, contain, harness, save, clear-up and discard their creations. The associated benefits to **ARC** are also apparent for this young child:

Autonomy

- Enjoyed the freedom to shape their own investigations and test their ideas within the safely set limits.
- Unpicked why the mixtures might have gone off (too little salt?) and experimented with remedying this.

Relatedness

- Developed a connection to their creations.
- Increased their connectivity to the therapist from their acceptance of the child's need to explore.
- Gained a greater understanding of themself and their abilities.

Competence

- Was given permission by the therapist and the physical and emotional environment provided to masterfully experiment with adding ingredients.
- Developed resilience and gratification delay as they weathered the disappointments of making foul smelling creations.
- Developed a sense of agency, self-esteem and mastery as they grew their own crystals!

A focus on the exploring, experimenting and containment stages of messy play

Sometimes a child's exploration of materials appears to mirror their exploration of their own difficult feelings or the mess of their world and life. For a child with limited ability to shape much of their everyday life, experimenting with and transforming materials can give them a rare opportunity to experience autonomy, agency and change. Pacini-Ketchabaw, Kind and Kocher highlighted the potential of materials 'as not just resources for shaping children's thinking but resources for thinking with' (2017) and this reminds us of the value and significance of children's explorations and the important role played by play therapists in supporting children on this journey of self-discovery. As the following vignettes reveal, children's ability to experience the different characteristics and qualities of each of these stages affords opportunities for the development of **ARC** too. Where this is not at the expense of the adult's own **ARC** needs, then play falls within the safe zone (Gascoyne, 2019).

For some children or stages of messy play the ability to contain creations is key, while at other times there is a blurring of boundaries between their body and materials:

> Having rolled out a length of wallpaper on the floor, one child proceeded to **explore and experiment** as they spooned, SLOPPED, poured and dropped an array of materials onto the paper from a standing position. Starting with sunflower oil, the child then poured from a 5-litre container of glue, watching with evident satisfaction as a growing puddle of gloopy white glue formed. (Needing the limits gently restating to bring them back to reality and the need to keep the resources safe, the child was supported without shame in needing lots, but not all the resources.) To this pool was added a generous sprinkling of pungent turmeric spice and several scoops of an enticing concoction

of dried lentils, glitter, gems and dried rice, all provided in small containers (to provide **containment**) for the child to access. Several more glugs of oil were added to the pool of glue (which took over a week to dry), before topping with some large sparkly gems. They watched intently as these settled, some part submerged in the goo, some standing proud, then stood back pausing to admire their finishing touches.

Dragging this to one side to **save** it and fetching another length of wallpaper, their energy and focus seemed to shift as they started **experimenting** with flicking paint from a fat paintbrush onto the paper, at first tentatively, and then gradually with more energy and delight. As you will discover if you try paint flicking, there is a definite knack to getting more paint on your target than yourself! With a plastic sheet laid on the floor, but the flicked paint rapidly spreading in all directions, I noticed my normally calm state starting to gradually tense, as myself, the radiator, wall and soft furnishings behind me also had become targets. Clearly surprised that more paint seemed to have flicked on ourselves and the room behind us than the carefully placed cover, the child momentarily paused then continued with heightened energy and gusto. This gave me the dilemma of how best to support this child in **containing** their exploration, while also maintaining the limits of keeping the room and resources safe. With my mind bouncing back and forth with the arguments for leaving or further containing this child, I wrestled with feelings of disappointment that I, a messy play advocate, should be a 'killjoy' more concerned about the mess than their process!

Reminding myself of the dimensions of **ARC** and the importance of limits to keep the child, myself and the room safe, I reflected that they were enjoying flicking the paint and seeing how far it had spread and that if they wanted to continue, we needed to go outside. Far from limiting and interfering with their experiments, dragging their work in progress to a grassy area outside gave them permission to flick to their hearts content, quickly staining the grass in the process. Pausing finally to inspect their finished creation, we both discovered with delight that not only had they created a Jackson-Pollock style mural, but their face, apron, arms and legs were also a bespeckled part of this artform! The child visibly bubbled with pride at the 3D picture created on their legs and was adamant that this should not be washed off!

As Figure 13.2 shows, the very characteristics, qualities, attributes and benefits of these different stages (which make them so appealing to different children in different situations) have the potential to present the therapist with either opportunities or threats.

Agency in messy play

In the preceding play vignette the child was clearly drawn to lots of wet, sticky and undeniably messy resources, but this is not always the case which is why it is important to consciously consider which types of resources we make available for children to enable them to pick those which best speak to them and suit their current energy, focus or mood. Clearly as therapists our focus is on the child, but ignoring our own emotional state, needs and reactions to different resources is short-sighted, as this will also impact upon the child's experience and potentially risk shaming the child if the encounter becomes stressful and overwhelming. To clarify, I am not suggesting that therapists only offer the types of resources and opportunities that they favour. Nor am I advocating that the main focus is on the adult's needs. All resourcing and provision

	CHILD perceived opportunities	*STAGES OF MESSY PLAY*	*ADULT perceived 'Threats'*	*ADULT perceived 'Opportunities'*
A	**Using body to test ideas & discover more about self**	**EXPLORING and EXPERIMENTING**	**Lots of mess to dear-up. limited time between sessions, time & peer pressures, therapist insecurities**	**Lots of communication**
R	**Discovering about own body resources & environment**			**languages to consider**
C	**Experiencing cause & effect, transformation & mastery orientation**			
A	**Ability to manage, change & control creations**		**Challenge of f indi ng enough space for safe storage of wet resources**	**Minimising mess through**
R	**Ability to develop a connection, manage & find out how things change**	**CONTAINING**		**containment**
C	**Ability to be inspired/be creative with containers**			
A	**Ability to transform, experience agency &**	**HARNESSING**	**Concern in case this creates even more mess**	**Indicator of creativity, problem solving & positive changes**
R	**self-esteem Developing greater connection to resources, ideas & self**			
C	**Greater sense of competence & mastery**			
A	**Evidence of being kept in mind, valued & process kept safe**		**Challenge of f indi ng enough space for safe storage of wet resources**	**Opportunities for child to**
R	**Greater understanding of resources & self**	**SAVING**		**revisit, galvanise thinking, develop agency & mastery, transform resources**
C	**Ability to experience gratification delay**			
A	**Taking responsibility for self & self-care**		**Stresses & time pressures of dearing-up**	**Opportunity for developing**
R	**Developing self-nurture & connection to resources & others**	**CLEARING UP**		**mindfulness & self-care – therapist & child**
C	**Developing mastery & agency over own mess**			
A	**Agency of deciding what to throw away as not needed.**		**Therapist may be more attached to creations than child**	**Cathartic moment indicating**
R	**Greater knowledge of self & needs**	**DISCARDING**		**client no longer needs it. Reduced storage pressures.**
C	**Ability to make choices & decide what is important**			

Figure 13.2 Opportunities and threats

decisions that I make are based on and shaped by my knowledge and sense of a child's state and needs and the potential for the messy play material to become a communication tool and means of self-expression for the child. That said, in order to keep children safe, remain authentic and best support their process, practitioners need to be self-aware, and I have found the **ARC** lens to be an excellent tool for planning enriching and affirming messy play experiences.

Achieving this balancing act requires us to question whose need we are fulfilling when we resource messy play provision? For many practitioners offering children porridge oats to explore, it is the adults rather than the children who quickly tire of this resource in its dry (and let's face it much less messy) state, suggesting the addition of water. In Godfrey's experiment in a mainstream setting to replace ready-mixed paints with yellow powder paint for an array of child-led explorations and experiments, interestingly it was also the staff not the children, who felt the need to change the colour focus (Gascoyne, 2019).

> Watching a four-year-old deeply engrossed in mixing up meals from a tray of rock salt and sand in a play therapy session, adding sprinklings of dried lavender petals, dried lentils, sequins and coloured stones, their agency and mastery were apparent. The child was happily absorbed for over thirty minutes of their forty-five-minute session, mixing, pouring, combining and dishing-up their dry concoctions. Using the sink of a wooden play kitchen to 'wash up' in, I found myself resisting the urge to suggest that they could use the large bottle of water, if wished, to wash-up for real, reasoning that this was not necessary as they knew there was a bowl of soapy water for washing hands which could easily have been used. Inviting me to wash-up, while they 'ate' their meal, I pretended to use the tap to add water and clean the plates and cups, considering as I did so whether the cathartic effect of water would be helpful? I was swiftly reminded by this expert in play that this was neither necessary nor appropriate when they exclaimed 'where's the cloth to wash up, we need a cloth' at which I uncharacteristically replied, 'we don't have a cloth but I can see if I can get one for another time'. Turning to me smiling this young child remarked 'We don't need a cloth as we can use a pretend cloth!'

Mess as an adornment

Reactions to messy adornment are as varied as the spectrum of messy play materials themselves. For some children (and adults) sporting the signs of mess on our clothes can feel like a trophy emblematic of a satisfying session, while for others this represents a cause for consternation, fear or shame. Of course, this greatly oversimplifies reality as for the same individual and even materials, a change to the emotional or physical context can rapidly transform the child or adult's experience. Equally, the same actions, such as mixing messy play materials in a play kitchen, may be deemed acceptable, with a different child or on a different day when the therapist feels secure in the child's acceptance of the limits. The colour, coverage and whether a gradual or sudden exposure to the mess can also significantly contribute to how safe or unsafe an experience feels.

> For one child who had been purposefully yet sloppily transferring milky cornflour from one container to another (much like his emotional needs were roughly passed from one parental home to the other), the discovery that stubborn spots of white had sprayed onto his school trousers

and shoes, sparked his concerted scrubbing. Each time the fabric dried, and the spots reappeared, another frenzy of cleaning ensued. Watching this child taking steps to clear up his own mess I was struck by the fact that this was somehow now visible to him, as in an earlier session he had been unaware that he'd flooded the floor!

For another primary-aged child, much like a toddler perhaps his school top told the story of his daily meals and interactions, with smears of food, dirt, phlegm and messy play marks creating an abstract painting on his once white top. Declining the offer of an apron, his experiment with adding pink powder paint to a take-away tub of gelly play, invited exploration by his fingers and an automatic wiping of these on his top. Seemingly not noticing the obvious and war paint-like marks he'd created across his chest, or the fact that paper towelling was closer to hand, it is through careful noticing of children's material encounters that we are afforded an insight into their world.

In another session, I too experienced the joy of being mess splattered as a water fight with home-made water guns left my dress dark spotted with water droplets. Comparing how dark each of our clothes were at the end of the battle, the sense of freedom rather than lack of control was evident. Crucially the client's listening to the limits I set of not hurting me (by squirting in my face or drenching me from close range shots) shaped the messy play experience for me also as it successfully navigated the tightrope of feeling liberated yet safe. The child's ability to listen to and work within the limits ensured **ARC** for both client and therapist as play remained in the safe zone. Mindful not to destroy the cathartic sense of freedom and release that can be experienced when you flood a lino floor, but aware of the need for some containment as the child had been referred for their uncontained and unsafe behaviours, I permitted the child to visibly wet the floor, then introduced structure by offering a large washing-up bowl on the floor as a target for the water gun. Spotting an even more fulfilling alternative, I was instructed by the child to put down my water gun and shield and replace these with the washing-up bowl repurposed as a shield for protecting me and the room from the jets of water! The result was an uplifting session and positive example of messy adornment, as the client and therapist's needs for safety and agency were met and the coverage of wet mess was manageable, and being water, I knew it would disappear when dry.

This compares starkly to a very messy episode with another primary-aged child in the same week. Following a physical outburst within and outside the room in the previous session, in which neither myself nor the resources were kept safe, I was naturally a little guarded. Conscious of the importance of providing containment, I tried to establish as many limits and structure as I felt necessary for supporting this child whose behaviours could instantaneously switch from calm and kind to angry and uncontained. Crucially I recognised that I too needed these limits (set out in Figure 13.3) in order to make me feel safer and in control.

Initially located in the messy play area (on lino and next to the sink) the child began squirting and mixing different coloured paints together in several role play cooking containers. Pivoting to

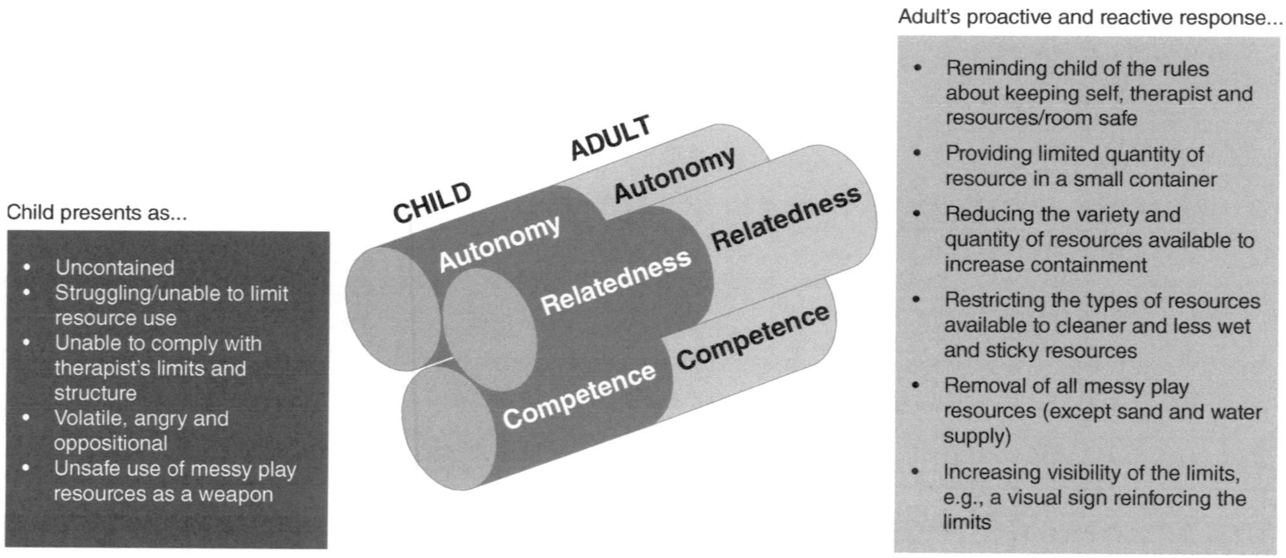

Figure 13.3 Strategies for achieving ARC

take some black sludgy paint to the role play kitchen at the opposite end of the room, (a carpeted area next to beanbags, a sofa, puppets and cushions) I successfully managed to set limits to restrict the spread of messy play explorations into the carpeted and soft furnished part of the room. (This highlights a failing of the room set up, as ideally the role play kitchen would be located next to the messy play area to facilitate children's use of messy materials in context, in this instance as imaginary food.) Stopping at a low table on the lino, yet immediately adjacent to the carpeted area, the child continued to mix and create alchemy, splattering the tables as they did so. They created a thick tarry black paint from their considered additions and mixing of several different paint colours in a takeaway food container.

Given the child's volatility and chosen proximity to the carpet and soft furnishings, the session and their messy play explorations felt fragilely balanced, poised at any moment to spiral out of control and into the danger zone. For a child who had previously demonstrated a preference for order and a natural inclination to always clear up their mess, there was a free and deliberate sloppiness to the way the utensils were used and then slid or 'thrown' onto the surface creating fingers of black paint in the process. Other containers of pale powder paint contrasted with the increasingly thick black paint being created. Each time a squirt of bright yellow, pink, green or blue paint was added like exploding fireworks, these quickly rippled and marbled through the paint before disappearing into the black.

With several containers of dark slimy mess created the child decided that they no longer wanted to stick to the agreed limits of me keeping their creations safe within the room and instead shared their determination to take them out of the room. Lifting the largest container, the child stood poised as if about to throw a mud pie. As the boundary was tested again and again and I restated the agreement and rules, I knew that their messy creation was in danger of being launched at me but persisted with gentle but firmly held limits. Then my intuition was rewarded, as second guessing that the tarry black paint was about to be thrown, I quickly managed to turn my head to the side, narrowly avoiding being hit full in the face. The walls, floor, beanbag, table and entire side and back of my head and clothes were however, caked in gunge at which the child legged it from the room. Grateful for not actually having been hurt, it took running my head under the tap followed by a shower to shift the sticky mess! After much reflection on what actions I needed to take to keep myself and the child safe whilst also being authentic from feeling competent and safe, I decided to further reduce the toolkit by removing the paint and other messy play resources (except water and sand) from the room to increase containment and structure.

As Figure 13.4 illustrates, for each of these three vignettes the child's use of the materials creates a tension between the therapist and child marking a turning point in the therapeutic session. If the restating of limits and provision of therapist structure and containment enable the child to accept the limits, as in the examples of the child flicking paint and child spraying water, then the interaction moves out of the danger zone back into the safe zone, where both the child's and adult's three psychological needs are met. If, however, the child is unable to accept the therapist's limits, play passes from the safe zone through the turning point to the danger zone, where the adult's need for **ARC** are disregarded and the session will consequently feel unsafe and out of

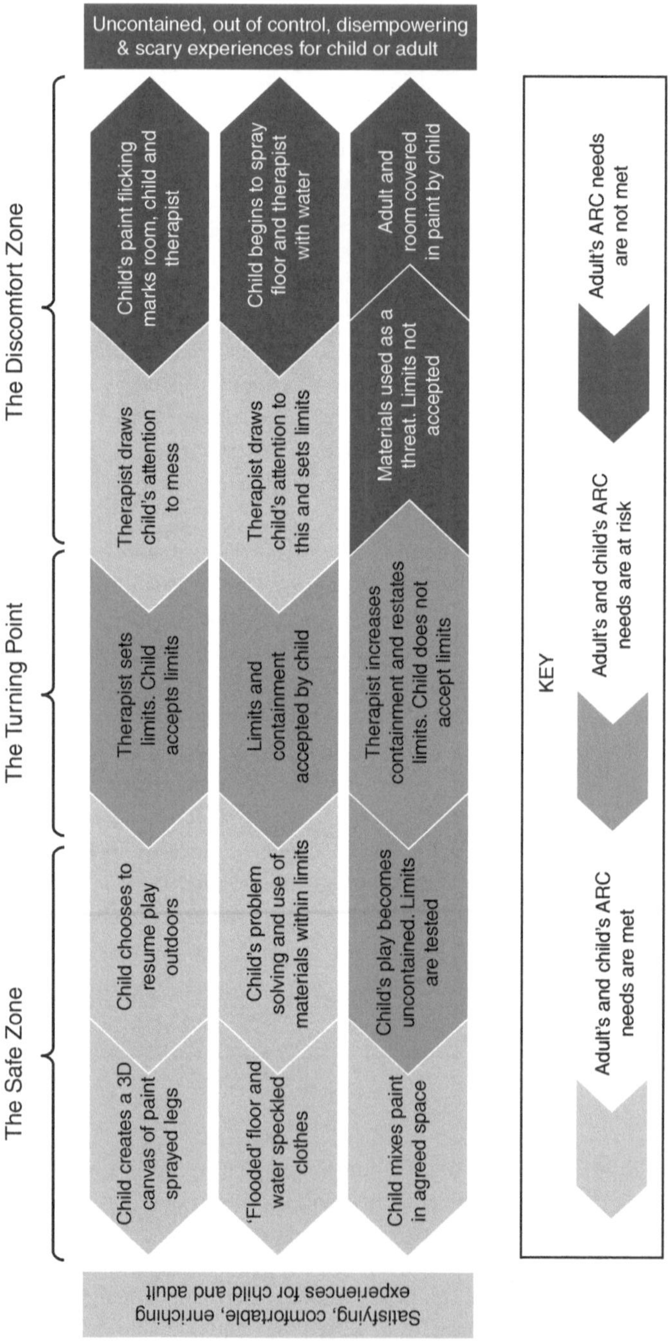

Figure 13.4 Tensions in therapeutic messy play reflections on autonomy

control. Being receptive to messy play I have not personally encountered an example where the adult's overly rigid and unfair restrictions limit the child's experience of **ARC**, however, examples of this are cited in Gascoyne, 2019.

Structure and limit setting

As Figure 13.4 demonstrates, setting appropriate limits in therapeutic (and mainstream) messy play is central to providing safe and satisfying material encounters because it lies at the heart of supporting (rather than restricting) children's and adult's autonomy. In the preceding vignettes, structure and limits were created both through the physical and emotional environment provided by the therapist. Since structure and limit setting determine a child's experience of autonomy, relatedness and competence, it is to this area that we now turn as we explore what this structure might look like in practice. Using a series of simple and multi-layered approaches, the intention is to enhance not diminish the quality of children's experiences:

Increasing structure and limiting mess by. . .

- Restricting messy play use to outdoors or the wipe clean/non-carpeted areas of the room.
- Providing visually discreet protective table/floor covering and aprons.
- Providing a variety of different sized and shaped containers (some with lids and some without), for children to mix and create in; transport and transfer resources; inspire role play or ideas; transform and return to materials; as well as for saving creations to return to.
- Providing extra visual guides such as a place mat, length of wallpaper, sheet of A4 paper or recycled packaging tray on a table or floor to support appropriate positioning and further contain a child.
- Limiting which materials are made available, e.g., bars of dry soap or playdough for cleaner and more sensory focussed mess rather than gloop, clay, wet sand or a paint concoction.
- Providing a bowl of soapy water or access to a tap or container of water for clearing-up.
- Providing appropriately sized clearing-up tools, which empower children to take responsibility.
- Offering some resources on trays 'Montessori style' for children to freely access for self-directed exploration. These sets of complementary resources, such as a bowl of dried rice, tongs, a spoon, funnel or sieve and chocolate packaging tray, enable children to engage independently with the resource, either following the activities suggested by the tools, or re-purposing them if desired. By containing the activity on a tray this provides much needed containment and structure for children, as well as recreating the equivalent of Lowenfield's 'world'. A simple example of this in a therapeutic context was the provision of a dry bar of soap within a large recycled packaging tray, and the arrangement of some appropriate tools next to these for the child to use or ignore if wished.

Not only do limits such as these provide containment and reassurance for adult's basic **ARC** needs, but they also support children in keeping them safe, as without any limits, material engagements can quickly feel scary, chaotic, stressful and out of control.

Reflections on relatedness

As is evident from the examples described, there can be a very thin line between therapeutic mess which is experienced as positive and comfortable by the practitioner and mess which feels unsafe and out of control. Central to this is the relationship and trust between the therapist and child as this enables the therapist to feel sufficiently relaxed to be authentic and able to accept the child's mess. Equally the child needs to trust the adult that they will still be accepted and not shamed by their 'mess'.

Limit setting – The setting of clear limits helps anchor the child to reality (Axline's principles) and clearly sets out what is acceptable and what is not. Each time a child is able to adjust their actions to accommodate the therapist's reminder of the limits, this reinforces the therapists trust in the therapist/child relationship and safety. Conversely when the limits set by the therapist are ignored, this understandably impacts upon their ability to trust in the child at that time to keep themselves and the resources safe.

Visual signs to reinforce verbal messages – In one setting where I had justified concerns about myself, the resources and room being safe I decided to put up an A4 poster on both sides of the therapy room door outlining the need for the child to be kept safe, and myself and the resources not hurt. By pointing this out to children before they entered the room, and on occasions when within the room, I was hoping to further reinforce the verbal limits and 'contract' governing special time.

Self-care – As I discovered first-hand, repeated clearing-up of extensive mess can take a toll on the therapist's hands and be extremely draining. It is important to recognise this and take steps to minimise any discomfort through the provision for example of hand cream, and a drink or snack, as these not only provide valuable opportunities for self-nurture, nourishment and safe touch, but also enable us to be one hundred percent present and authentic in subsequent sessions.

Acceptance – Just as a child has the potential to be shamed by their messy creations, so too the therapist can experience a sense of shame and exposure if the aftermath of sessions is construed by other practitioners as a lack of competence, control or foresight.

Reflections on competence

When working with clients with limited time in between sessions for clearing-up and storing the aftermath of wholesale messy play, I have found several simple environmental factors to be particularly useful in minimising the negative stresses of mess:

Storage Space – Personally I have found a lack of time, rather than the amount and type of mess to be the biggest factor in shaping my own levels of discomfort. For me clearing-up a child's therapeutic mess is an important, caring, mindful and respectful act which should not be rushed. So being able to temporarily store children's creations until you have the time to properly record, clear-up and safely stow these, can significantly reduce time tensions, enabling a calm and unhurried end to the session. An area explored in depth in *Messy Play in the Early Years*, if there isn't adequate time between client sessions to do so properly, I have found the availability of storage space a godsend. Access to plenty of lockable cupboard space, an out of reach shelf, cupboard top or even the space under a table covered with a waterproof cover, can facilitate the safe storage of messy creations between sessions, allowing the therapist to return to these when there is more time and space to reflect upon the child's creations and any hidden messages.

A sink – Access to a sink not only increases opportunities for developing child agency in transforming resources but can also positively support both the child and therapist in clearing-up. To fully appreciate the potential of a child being able to clear-up their own mess, this stage may also need to be re-defined as a calming process of self-care and mindfulness, rather than simply a functional act.

Paper towelling – Having paper towelling that is big and absorbent enough to take on the job, and copious quantities of it, can also help contain and therefore make light weight of wet sticky mess.

A bin – I have also found that having basic but not necessarily available equipment such as a bin or bin bag can make a huge difference to a therapist's sense of competence and wellbeing, especially when the overflowing bin is emptied!

Noticing the patterns and significance of children's play

Within a safe yet permissive therapeutic environment, the child will be empowered to explore and experiment in the safe zone, giving the therapist the space and opportunity to reflect upon the significance of the child's play and any messages that this may be communicating. The types of materials a child picks to use and what they chose to do with them can give us valuable insights into their potential world. A child may interact with materials in a careful, calm, measured, robotic and precise way; be spontaneous, carefree, boisterous and rough; or all manner of stages in between! Careful noticing can highlight potentially significant patterns, points of difference and tools for communication. Watching a primary-aged boy sloppily transferring gloop between two containers it would be easy to dismiss this as unnoteworthy given the child's often boisterous and uncontained state, however, this compared markedly to the same child's noticing and careful scrubbing of a stubborn gloop stain on his clothes within the same session. Similarly for one boy roughly swirling a container of gelly at speed with a spoon, seemingly not bothered by the mixture sloshing in all directions, it would be easy to fail to notice how this contrasted with his delicate mixing, pouring and spooning of 'lotion food' and careful feeding of this to the characters in his role-play. It is not enough therefore to simply notice what resources children choose to use and how, whether for creating potions, experimenting with cause and effect or mixing up imaginary meals, because the colour, texture and appearance of a child's messy creations and energy, care and nature of the child's messy play interactions can, if we notice all potentially communicate 'This is my world!'

References

Gascoyne, S. 2019. *Messy Play in the Early Years – Supporting Learning Through Material Engagements*. Abingdon: Routledge.

Grossmann, K., Grossman, K.E., Kindler, H., & Zimmermann, P. 2008. A wider view of attachment and exploration: The influence of mothers and fathers on the development of psychological security form infancy to young adulthood. In J. Cassidy & P.R. Shaver (Eds.), *Handbook of Attachment: Theory, Research, and Clinical Applications* (2nd ed.). New York: Guilford Press, pp. 348–365.

Morris, K. 2015. *Promoting Positive Behaviour in the Early Years*. Maidenhead: Open University Press.

Pacini-Ketchabaw, V., Kind, S., & Kocher, L.L.M. 2017. *Encounters with Materials in Early Childhood Education (Changing Images of Early Childhood)*. New York: Routledge.

14

LET'S GET MESSY!

Enhancing and enriching children's learning and development through messy play

Leeanne Casey and Siobhán Prendiville

This chapter will explore the use of messy play in early childhood education settings, including both pre-schools and more formal primary schools. Both authors have vast experience of using messy play in education and therapy settings. We will draw on these to present novel, easy to use, messy play ideas. Leeanne's undergraduate degree was a Bachelor of Arts in Early Childhood Care and Education whilst Siobhán's undergraduate degree was a Bachelor of Education. Siobhán also holds a Master of Education in Early Childhood Education. Both Siobhán and Leeanne have completed comprehensive, experiential training in play therapy as part of a fully professionally recognised MA qualification as a psychotherapist with a major emphasis on the practice of play therapy, at the Children's Therapy Centre in Ireland. The Children's Therapy Centre (CTC) is the longest established and foremost provider of professional play therapy training in Ireland and has been delivering high quality specialised training since the 1990s. They train practitioners in the expanding field of psychotherapy with a specialisation in the use of play and creative approaches. Siobhán is now MA Course Leader on their Master of Arts in Creative Psychotherapy (Humanistic & Integrative Modality); this incorporates their Postgraduate Diploma in Play Therapy.

We will introduce Leeanne's master's dissertation research "Let's Get Messy! A Study Investigating Pre-school Practitioners and Primary School Infant Teachers Perceptions on the Benefits and Barriers of using Messy Play" (Casey, 2018). The chapter will initially outline what led Leeanne to this area of research and will then continue to highlight her main research findings. We will continue to explore and present practical ideas on how messy play can be incorporated into early childhood education settings. It will highlight the main benefits of using messy play in education settings and attempt to tackle head on the barriers that were highlighted in Leeanne's study.

Introducing the research

After working in early childhood settings for over seven years with under 5's where messy play equipment was freely available throughout the day for children to engage with, Leeanne made the move to formal primary school education. Whilst messy play equipment was not freely available here, infant teachers were observed using elements of messy play materials as teaching tools. Whilst working in both environments Leeanne met colleagues who exhibited fears, reluctance and anxieties around using messy play materials with children for various reasons. Leeanne's experience of working in the dual environments and observing the therapeutic powers of messy play, inspired her research on messy play "Let's get messy! A study Investing Pre-school Practitioners and Primary School Infant Teachers Perceptions on the Benefits and Barriers of using messy play" (Casey, 2018).

Leeanne's study was based in Ireland and focused on the provision of messy play for children aged between three and six years. International definitions identify early childhood as being from birth to six years therefore in Ireland this age range incorporates children in both pre-schools and infant classrooms, to focus on one setting would give an incomplete picture.

Aistear is the Early Childhood Curriculum Framework for birth to six-year-olds in Ireland, which "complements and extends existing curriculums and materials" (NCCA, 2009, 6). Aistear is applicable in both settings and advocates learning through play and hands on experiences, presenting learning and development through the use of 12 principles and four themes, whilst also having an entire chapter devoted to play. The primary school curriculum (NCCA, 1999) is implemented solely in primary schools supporting children's learning from 4 to 12 years, this curriculum does not overtly focus on the use of play.

Main findings from Leeanne's research

Data was ascertained through the use of questionnaires and interviews with primary school infant class teachers and pre-school practitioners. The overall findings from the study revealed that all participants showed high levels of understanding in relation to messy play. Beckerleg describes how messy play is different to any other type of play: "It smells, has some wonderful textures, looks unusual and has many varying forms and because of this it can produce amazing results that wouldn't be seen elsewhere" (2009, 23).

Evidence drawn from the data analysis showed that all participants viewed messy play as beneficial to child development. Practitioners identified messy play as being 'extremely important' on physical, social and emotional domains, whilst teachers identified messy play as being 'extremely important' on language development. Gascoyne suggests using our senses through play "provides the foundations upon which all subsequent knowledge, thought and creativity are based" (2016, 3).

From analysis it is clear that utilising messy play as a playful learning opportunity in teaching is constrained by 'too many children' which came out on top for both infant class teachers and preschool practitioners, this was followed by 'curricular constraints' for teachers and 'limited space' for practitioners.

However, whilst barriers exist, there was still evidence of the provision of messy play in both the infant classrooms and teachers with just one teacher identified as not providing any messy play. Both sectors revealed they involve themselves in messy play in order to enhance it, however teachers did appear to be more formal and directive than practitioners during messy play. According to French "the adult has a responsibility to provide rich environments where children are able to explore, touch, manipulate, experiment with a variety of real life and diverse materials and where children can ask questions, make hypothesis and develop their thinking" (2007, 3). The majority of participants in both sectors revealed they have received ineffective and insufficient training with regards the implementation of messy play within their rooms, therefore we have compiled practical ideas to support learning through messy play and how to overcome the barriers.

Overcoming the barriers to including messy play in education settings

Unfortunately, due to the word 'messy', often this type of play can go unrecognised and undervalued as a play that enhances and enriches children's learning and development. Leeanne's study identified 15 barriers to providing messy play. We will now tackle these and outline easy and realistic ways to solve each barrier.

Staff

Before introducing messy play to any child, it is vital staff have been educated on how messy play can enhance and enrich children's learning and development. Some professionals have anxieties around mess or the mess getting out of control, whilst others simply do not like getting dirty or putting their hands or feet into different gooey substances. It is essential staff explore and experiment ever before messy play is provided to children as staff attitudes and anxieties around mess and lack of control can be picked up by the children. Adults need to experience messy play materials themselves using their own senses in order to get a true indication of the value and what the experience is like. When providing messy play opportunities to children it is important to be open minded and flexible. What you have planned may not be where the children take it, for example the professional might expect children to put their hands into baked beans, however the child may want to experience putting their feet in, a cereal might fall on the floor the children might choose to do angels on the floor.

Too many children in the room/not enough staff

Messy play does not have to be a big messy experience, for example shaving foam can be contained to a table top area and it is easy to clean up as it rubs off easily with a cloth or towel. Drier messy play materials can be used for whole class experiences as these are easily cleaned up, whilst wetter messy play can be kept for small groups/stations providing a balance through group and individual activities. Some messy play sessions do not have to be adult led; they can be child led, giving the child freedom to explore and experiment and less pressure on the professional.

Space

When space is an issue keep the play contained through the use of containers or designated spaces, tables etc. Keep the activities simply, do not over think them. However, if the space is too small take it outside. It is important to be prepared for the weather getting parents to bring in waterproof jackets and wellies at the beginning of the year.

No end product going home

Educating parents on the benefits of messy play is essential. When educating parents, it is vital to turn this negative into a positive emphasising how critical messy play is in supporting children's emotional well being as there is no end product in messy play there is no pressure on children around strengths and weakness. Using photographs is a great way to capture the children's participation and also explain how particular activities support different developmental domains of the child. This can be done by attaching speech bubbles to the different photos and displaying for parents to see.

Apprehensive parents regarding dirty clothes/cleanliness

Messy play can sometimes be discouraged at home due to exceptionally clean houses and parents who want to eliminate all germs. Similar to the preceding, education of parents around messy play is crucial, it is also vital to reassure parents that the mess can be washed off clothes. Partnership with parents is essential, invite parents in to explore messy play for themselves. Waterproof painting jackets are inexpensive to purchase at your local bargain shop, these will keep the children's clothes protected. Of course, buying these for an entire class can be costly, another idea is to get every child to bring in an old adult sized t-shirt from home to help protect their clothes. Food colouring can be difficult to wash off so consider using paint as an alternative to avoid stained skin/clothes.

Curricular constraints

Why not involve a messy play-based approach to teaching – experiential work allows children to actively learn and understand as they solidify their learning through doing. As most children find messy play enjoyable, using messy play as a teaching mechanism will motivate children to be more focused and eager to learn, for example teaching sink and float through water, colours through coloured rice. Experiential, hands on and active learning through messy play gives children a break from the typical rote learning.

Time

To make the most of your time it is important that your preparation is done. Plan your activity well, ensuring you have all products available. Once planning and the experiential task is complete, it is essential to complete a short reflective review, looking at what worked well and what you would do differently next time, exploring if the main aim was reached. It is important to note paper work can be time consuming and overloading, short, concise bullet points are just as effective and quicker. Sit back and observe whilst children play. Children should also not be rushed or hurried, allowing them ample time to absorb the experience and increase their learning from the hands-on experimentation.

Allergies/intolerances/dietary requirements

It is essential to ensure you are aware of all child allergies in your classroom before you commence any messy play. Common allergies, intolerances and skin conditions, e.g. nut-free diets, lactose-free diets, gluten-free and gelatine free diets and eczema, were highlighted as barriers to messy play. Gelatine-free marshmallows and jelly are now widely available in large chain supermarkets. Egg and milk would not be commonly used in classrooms as part of messy play with the exception of baking; in this case eggs should be avoided and water used as a substitute for milk. Nut allergies can be extremely dangerous, each professional needs to find out what kind of nut allergy is it and avoid all products containing that nut. When using flour or cereals as part of the messy play materials alternative gluten free products can be found in large chain supermarkets. It would also be recommended when using materials such as shaving foam that you would use the sensitive skin.

Use of food

The use of food can be controversial, particularly around views of children losing respect for food and also around poverty. We believe that individual professionals and organisations should use their own autonomy, coupled with their knowledge and experience of the children in their setting when planning for the inclusion of food in messy play. For us we believe using food in the context of play offers so many wonderful benefits for sensory exploration and development. Taste safe messy play recipes allow children to engage all of their senses.

Consideration for the children who do not like engaging in messy play

Children will join in at their own pace and should not be forced to join in. Some children can have difficulty with tactile and/or other sensations. If professionals approach messy play with fun and excitement a child is more likely to partake in the activities presented. However, if staff are anxious about mess etc. the children will pick up on these apprehensions and may cause the anxious child to become more anxious or reluctant to join in in the messy play experience. Having all staff experience the materials will allow them to overcome anxieties or fears prior to the activity being carried out. Starting with drier items rather than wetter items may be easier for the child as they do not get dirty. Model to the child how to use the different materials. Alternatively, materials could be placed in a sealable plastic bag and felt at a safe distance from the outside of the bag. Providing gloves and wellies can also support hesitant children in getting involved.

Too messy

Engaging in messy play does not mean the room or outdoor area has to be destroyed. Large washable sheets can be used on floors and tables, tuff trays are large and offer wonderful containment, on dry days bring the play outside. Most children love jobs; get children to assist in the clean up by providing cloths, child sized dustpan and brush. This will also teach the children important life skills. If you are lucky enough to have the space in your room for a messy play corner keep the amounts of material in each container to a minimum at the start. As this corner will be new for the children it is important to be aware they will do lots of exploring and materials will be found everywhere around the room over the initial introduction period. However, as time passes the children will learn to contain the mess in the designated area once you have put boundaries in place. Remember mess is fun!!!

Safety

Safety is vital particularly around messy play, never leave children unsupervised. If sand is stored outside, ensure it is well covered and free from rodents and animals. Similarly if water is left in a container, inside or outside, ensure to change it regularly, not making it too deep. Examine all containers and play instruments for sharp or broken edges and foreign objects.

Expense

Some materials can be found free in nature, for example water, mud, leaves, stones, twigs, pine cones. Some materials can come from the recycling e.g. shredded paper, plastic bottles for scooping and filling. Playdough can be made easily and in large quantities which when stored in an air tight container lasts a long time. As this is play and not dinner, products do not need to be expensive names, keep costs to a minimum by shopping in your local bargain shop.

Language barriers: e.g. English as an additional language

From our experience play is the universal language of all children, they do not have to have the same vocabulary or spoken language to engage in messy play together. And in fact, engaging in messy play is a wonderful way of enhancing and developing children's communication and language skills.

Practical play ideas on how messy play can be incorporated into ECCE settings

As professionals with experience in the field we believe messy play and teaching do not have to be separate practices. When effectively integrated into a curriculum messy play provides the experimentation and practice of skills needed to support teaching, creating a rich learning environment and a less formal didactic approach for ECCE settings. Whatever form messy play takes it always offers opportunities to connect with our seven senses including tactile, olfactory, gustatory, auditory, visual, kinaesthetic, vestibular or visceral. Gascoyne identifies that learning through our senses and a hands-on approach appeals to all "children with different thinking and learning styles" (2016, 9). Jennings (1999) clearly identifies the need to engage all children in messy play.

We have compiled some practical ideas to make teaching through messy play more fun and exciting for both children and professionals whilst supporting children's various developmental domains. All our practical ideas have many parallel benefits such as supporting children's sensory exploration, curiosity and imagination skills, whilst providing opportunities to develop fine and/or gross motor skills supporting eye and hand coordination and pre-writing skills, growth of social skills such as turn taking skills and eye contact. It allows opportunities for emotional expression and self-soothing as well as self-esteem enhancement as there is no right or wrong as it focuses on the process rather than the end product (Casey, 2018). Messy play also supports cognitive development through the addition of new vocabulary and exploration of new mathematical and scientific concepts such as counting, sorting, more/less, bigger/smaller, prediction, volume, gravity and measurement. Children can also create the materials themselves, enhancing their independence and creativity. Messy play is also critical in children's play development (Jennings, 2014). Messy play is an essential component in both Neuro-Dramatic-Play (NDP) (Jennings, 2011) and Embodiment-Projection-Role (EPR) (Jennings, 1999) competence in both NDP and EPR are critical for a child's overall development and maturation.

Messy play ideas for early childhood education settings

Taste safe playdough (uncooked)

Ingredients

- 1 cup flour
- ½ cup salt
- Teaspoon cream of tartar
- Food colouring
- ½ cup water

Method

1 Add 1 cup of flour to your bowl.
2 Add half a cup of salt to your bowl.
3 Add a teaspoon of cream of tartar.
4 Mix, mix, mix!
5 Add a couple of drops of food colouring to half a cup of water and mix again!
6 Pour this into your flour and salt.
7 Add a couple of drops of food essence.
8 Mix, mix, mix!
9 Take it out of the bowl and knead it with your hands.
10 If it is too wet add extra flour. If it is too dry add extra water.

Taste-safe cooked playdough (makes a large batch)

Ingredients

- 4 cups plain flour
- 2 cups salt
- 2.5 tablespoons of cream of tarter
- 4 cups water (add food colouring here)
- ½ cup cooking oil

Method

1 Mix ingredients in a saucepan and keep stirring over a medium heat.
2 Mixture will thicken and combine into a lump.
3 Tip onto a pastry board, knead with a little flour, until smooth.
4 Allow to cool.

Additional ideas

As it is so versatile, use it in different themes, e.g. construction theme; it can be used as cement to join bricks together or dug up by the construction vehicles.

Add different types of essence to make it more exciting.

Benefits of playdough

- Mathematical concepts: counting, measuring, more/less, longer/shorter, heavy/light prediction, 3D shapes.
- Science: Discover new colours by mixing different coloured playdough together.
- Brings new vocabulary to life such as smooth, lumpy, squeeze, twist, flatten.
- Allows children to release tension and express emotions through hitting, squeezing, pulling, poking and throwing.

Taste-safe slime

Ingredients

- 1 cup marshmallow fluff
- ¼ cup caster sugar
- ¾ cup cornflour
- Food colouring
- Coconut oil

Method

1 Add 1 cup marshmallow fluff into a mixing bowl.
2 Add in ¼ cup caster sugar.
3 Add in approximately ¾ cup of cornflour, a little bit at a time.
4 Add in a few drops of food colouring.
5 Use a spoon to mix this all together; it is quite sticky and hard so you need to mix thoroughly.
6 If it still seems too sticky you can add in little extra of the dry ingredients.
7 Before taking the mixture out with your hands coat them in coconut oil (this stops the slime from sticking to your hands while you stretch and fold it).
8 Take the mixture out and "stretch and fold" it repeatedly to get it to the slime consistency you like!

Benefits of taste-safe slime

- It is non-toxic compared to shop bought slime.
- Can be regulating for children.
- Great for science experiments.
- Children can use all their senses when making and playing with taste-safe slime.

Taste-safe moon sand recipe

Ingredients

- 2 cups flour
- ¼ cup vegetable oil
- Few drops of food colouring

Method

1 Put oil in a cup, add in a few drops of food colouring and mix together.
2 Put the flour in a bowl.
3 Add your coloured oil into your flour.

- Mix with a mixing spoon first and then get your hands stuck in and mix and crumble the moon sand.

Benefits of moon sand

- Supports pre-writing and writing skills.
- Allows children make moulds and 3D shapes.

Additional ideas

Sensory bin with small objects
Sand writing e.g. letter formation/word development
Sand pictures
Can be used to develop themes e.g. construction theme/under the sea

Jelly dig

Ingredients

- Two packs of jelly
- Water
- Sensory bin
- Plastic numbers/letters

Method

1. Scatter your plastic letters/numbers into your sensory bin.
2. Make up the jelly following the instructions on the pack.
3. Pour the jelly into the sensory bin (allowing it to cover the letters/numbers).
4. Allow the jelly to set.
5. Set the children a task to find all the letters/numbers.

Additional ideas

Alternatives hide dinosaurs, animals etc.
Make jelly in a variety of small moulds and have children break them up.

Benefits

- Supports numeracy and phonics recognition.
- Increases children's vocabulary e.g. sticky, slimy, slippery, wet, wobbly.

Pom-pom ice cubes

Materials

- Water
- Ice tray
- Colourful pom-poms

Method

1 To make the ice-cubes, pop a pom pom in each section of an ice-cube tray, then fill it with water.
2 You can save batches in the freezer so you build up a large collection.
3 Pop all the pom-pom ice cubes into a large bowl or sensory bin.
4 Allow the children to play with them.
5 They can squirt them down with extra water to help the melting process if they like!
6 This moves into some lovely water pom-pom play as the ice cubes melt.

Alternatives ideas

- Play around with different temperatures and add a hairdryer to the activity (adult supervision).
- Colour sort: sort the pom-pom ice cubes into colour categories.
- Science experiments, e.g. sink and float, add salt to the ice and observe the changes.
- Allow children to break up the frozen ice with appropriate materials under adult supervision.

Benefits of ice pom-poms

Explore the art melting and freezing.
Explore the weather.
Introduce vocabulary such as liquid, solid, melting, frozen, splashing.
Practise pouring skills.

Coloured and scented rice

Ingredients

- Rice
- Paint/food colouring
- Food essence
- Sealable bag
- Grease proof paper

Rice can become more visually appealing by adding colour to it. This can be done through the following steps:

Method

1 Pour rice into sealable bag.
2 Add a squirt of paint or food colouring and a drop of scented food essence into the bag and seal.
3 Using your fingers gently rub the outside of the bag until all the rice and paint have mixed together.
4 Pour out onto a baking sheet and allow to dry.

Using coloured rice

Dropped into different types of containers, it can provide a good auditory experience, particularly if replicating rain for drama. Putting rice into a container with different scoops, jugs, funnels cups, tweezers will allow children to develop their pincer grip, pre-writing skills, eye-hand co-ordination, fine and gross motor skills as they pour, scoop, mix and pick up single grains of rice.

If adult directed, the adult can introduce vocabulary such as scoop, pour, sprinkle, grains and teach different colours.

Benefits

- Vocabulary, e.g. scoop, sprinkle, grain, pour, heavy/light, fast/slow.
- Assists in teaching different colours.
- Mathematical concepts: counting, sorting, bigger than/smaller than, volume, gravity and measurement.
- Promotes calmness and relaxation.

Additional ideas

- For children who are anxious or find it difficult to sit down, pouring rice over their hands can act as an excellent sensory experience.
- Once children are finished exploring the rice the children can put it into recycled plastic bottles/containers to make musical instruments.
- Pour rice into a balloon and make stress balls to help children in the class with their big feelings.
- Create a treasure hunt by hiding objects in the rice.

Pasta (or spaghetti)

Pasta comes in various shapes and sizes and can be coloured just like rice. Pasta can be painted dry using paint and paintbrushes or cooked and then coloured using the same method as the rice.

Benefits of using pasta

- Dry or painted pasta can be great for maths creating patterns, counting or sorting by shape/size.
- Introduces vocabulary such as crunching, snapping, slippery, slimy, soft, hard, sticky and squishy.
- Supports colour recognition.
- Non-toxic.

Additional ideas

- Give children straws, pasta and play dough and watch their creativity bloom as they make sculptures.
- Added to a string to make a sequence. Encourage children's creativity during art by giving them pasta, rice and beans to create a picture.

Shaving foam

Delight in play with shaving foam!

Benefits

- Develops their hand-eye co-ordination, pre-writing skills, letter formation practice.
- Easy clean up and rubs off clothes easily.
- Support sustained attention and concentration skills.

Additional ideas

- Pre-school: Children can make or draw their own pre-writing patterns or pictures in the shaving foam.
- Primary School: Take a break from the pencil children can practise drawing their letter formation patterns and designs.
- Extended add small world animals, e.g. artic animals to create an artic scene at stations time.
- Add paint and PVA glue to make puffy paint.
- Food colouring can also be added for more creativity.
- If a child is not yet comfortable with playing with the shaving foam directly, squeeze some into a bag and allow the child to explore and play with the shaving foam through the bag.
- We recommend using sensitive skin shaving foam.

Gloop/oobleck

Ingredients

- 2 cups cornflour
- 1 cup water

Method

1 Add 2 cups of cornflour and 1 cup of water and slowly mix together.

Benefits

- Science concepts: When children put their hands into the bowl they experience the wet mixture which turns dry in their hands once out of the bowl.
- Encourages children to dig, mix, squeeze, sprinkle, pour and poke.
- Allows vocabulary to be introduced, such as solid, liquid, wet, slimy, runny.
- Gloop is easy to brush away and rub off clothes.

Additional ideas

- Create prints from hands/feet on coloured paper.

Snow dough

Ingredients

- Cornflour
- Hair conditioner

Method

1 Add 2 cups of cornflour to 1 cup of conditioner.
2 Mix into a soft silky snow dough using your hands.

If it is too sticky just add more cornflour.

Additional ideas

Add Artic animals, shredded paper, coloured rice to create an artic theme.

Taste-safe paint

Ingredients

- 1 cup plain flour
- 2 tablespoons salt
- 1½ cups cold water
- 1½ cups hot water
- Food colouring

Method

1 Add flour and salt into a saucepan.
2 Add the cold water and stir until there are no lumps.
3 Add the hot water.
4 Bring to boil. Once thickened, remove it from the heat.
5 Whisk until lumps are gone.
6 Divide into containers and mix in desired food colouring.

Benefits

- Allows for experimentation: mixing colours.
- Promotes social skills: sharing pots, taking turns, working together.
- Non-toxic.

Laminated sensory pouches

Materials

- A4 laminated pouches
- Iron or hair straightener

- Hair gel
- Small objects to fill your sensory pouch

Methods

1. Using an iron seal around three sides of the laminating sheet, leaving the forth side open so you can fill your pouch.
2. Fill the pouch with a tube of hair gel and small objects.
3. Seal the pouch by running the iron gently along the top of the pouch.
4. Allow it to cool and then squeeze the pouch to make sure it does not leak.

Additional ideas

Glue different tactile materials along the sides of the closed pouch to increase the sensory experience.

Benefits

- Not all children like to get messy or participate in messy play. This practical idea allows children engage in a fun experiment without getting themselves in a mess.

Note on taste-safe recipes

When we say taste safe we are not proposing that these recipes are for eating. If you are looking for messy play ideas for tasting, we suggest getting your hands stuck into some cooking and baking! Think homemade pizzas and meatballs, rice krispie buns, brownies and other treats. However, we do think it is a bonus that these recipes do not have any toxic ingredients in them and therefore if a child did happen to put some into his/her mouth it would be okay. Remember to store all taste safe products in air tight containers.

The great outdoors

Not forgetting of course, the incredible learning opportunities offered in the great outdoors. The endless supply of fun provided free by nature from the production of mud pies to simply jumping in muddy puddles, rolling in grass, throwing leaves in the air, relishing in feeling the frost or the wind on your face, catching and tasting hailstones, rolling and making snowballs, igloos and snow people. Play with natural objects – twigs, cones, mud, leaves, stones; echoes of joy and laughter tend to propel through the air as children in the squelching, splashing and crunching of materials, whilst the non-existent pressure to produce an end product increases the contentment.

Conclusion

This chapter has explored the use of messy play in early childhood education settings, including both pre-schools and more formal primary schools. It has explored and presented a range of novel, yet practical, ideas on how messy play can be incorporated into early childhood education settings. It has also highlighted many of the benefits associated with using messy play in education settings and tackled head on some of the barriers and obstacles that hinder the use of messy play in education settings. We hope that the ideas we have presented have been persuasive in presenting the case for making full use of messy play in education settings.

References

Beckerleg, T. (2009). *Fun with Messy Play: Ideas and Activities for Children with Special Needs*. London: Jessica Kingsley Publishers.

Casey, L. (2018). *Let's Get Messy! A Study Investigating Pre-school Practitioners and Primary School Infant Teachers Perceptions on the Benefits and Barriers of Using Messy Play*, MA, Children's Therapy Centre, Westmeath, Ireland (unpublished).

French, G. (2007). *Children's Early Learning and Development, Aistear: The Early Childhood Curriculum Framework, Research Paper*. Dublin: NCCA.

Gascoyne, S. (2016). *Sensory Play: Play in the EYFS*, 3rd ed. London: Practical Preschool Books.

Jennings, S. (1999). *Introduction to Developmental Playtherapy*. London: Jessica Kingsley Publishers.

Jennings, S. (2011). *Healthy Attachments and Neuro-Dramatic-Play*. London: Jessica Kinsley Publishers.

Jennings, S. (2014). Applying an Embodiment-Projection-Role Framework in Groupwork with Children. In E. Prendiville and J. Howard (eds) *Play Therapy Today: Contemporary Practice for Individuals, Groups and Carers* (pp. 81–96). London: Routledge.

National Council for Curriculum and Assessment. (1999). *Primary School Curriculum: Introduction*. Dublin: The Stationery Office.

National Council for Curriculum and Assessment. (2009). *Aistear: The Early Childhood Curriculum Framework*. Dublin: NCCA.

15

NEURO-DRAMATIC-PLAY AND A HERO'S JOURNEY

A play-based approach in a UK junior school

Clive Holmwood

Introduction

The obvious nervousness was etched onto their young faces as they stood in front of me pensively in the middle school hall. It was a Wednesday afternoon, in September, they stood silently awaiting instructions at our very first meeting. Some children could barely speak they felt so anxious. One or two spoke, 'What are we here for?' asked one; 'To see a play, I thought' said another. A third child played nervously and excitedly with a ball, whilst waiting to start. I was a teacher, I explained, but a different sort of teacher to what they might be used to, I had come to play, make stories and go on an amazing adventure, using drama; through which we could carry out an exciting research project and learn some new ideas about play together.[1]

Over the next 50 minutes, the apprehensive faces of 11 young people aged between 8 and 11 began to ease as we played a series of warm-up and introductory games. We threw balls to each other, created emotions and feelings with the help of the balls, ate giant imaginary pieces of toffee and began to walk through snowstorms, quicksand and forests. The playing of games and standing in circles and natural rhythms of the group's movements and actions seem to allow young people to gain a greater sense of self and body awareness. As Russell Meres states the 'inner life, which we sense as insubstantial, is founded on physical things such as toys and parts of the body' (2005, pg. 3). Through creativity, touch, objects, rhythms, group cohesion and a physical grounding and awareness of the floor beneath their feet, the children, had by the end of the session come alive, their obvious worries had lessened and some of them could even smile.

This chapter aims to consider the principles of Neuro-Dramatic-Play (NDP) (Jennings 2011) as a form of pre-therapy/structured interventional play. By running nine sessions with a group of 11 children (years 4 to 6, 8- to 11-year-olds), all of whom had been handpicked by school staff due to confidence and self-esteem issues, we were to explore the notion of NDP as an effective form of play-based intervention. By allowing the children to build bridges with each other through the play and going on a fantastical and dramatic hero's journey together, we considered the appropriateness of NDP as a way of supporting the confidence and self-esteem of a small group of middle school age children in the UK. I used a combination of a practitioner, action-based research (Whitehead & McNiff 2006) from an empirical perspective, acknowledging the importance of children playing an equal role in the project as co-researchers (Kellett 2005).

Two major theoretical ideas relate to this work. Firstly, as already mentioned, Jennings (2011) *Healthy Attachments and Neuro-Dramatic-Play* (NDP). Secondly, the concept of the 'Hero's Journey' (Campbell 1993) in which he considers the concept of the 'monomyth' – the concept of the hero going on a journey, who undergoes a range of challenges and returns somehow changed, which has similarities to Van Gennep's (1960) 'Rites of Passage.' I will also consider, later, the notion of attachment and Bowlby (1988) and his concept of transitional object's, using large puppets throughout the project and on the journey with the children, which became a key part of the research findings.

NDP

Sue Jennings, one of the early pioneers in play and dramatherapy, describes NDP as 'a new synthesis of several approaches to child development that include our greater understanding through neuroscience of the complexity of early brain development' (2011, pg. 28). Jennings has developed these ideas since the mid-1960s alongside the development of play and dramatherapy. It is based on the notion of sensory, rhythmic and dramatic playfulness, that are part of normal and healthy early childhood development; specifically, 'from conception to six months' (2011, pg. 33). From the womb the baby can feel the warmth of the mother whilst encased in a liquid membrane. The very first rhythm it feels is the beating of the mother's heart alongside that of its own. From birth the baby interacts with the mother (or caregiver) whose face is inches away as it is held, safely, in her arms after birth. So sensory rhythmic and dramatic play and interaction acts as part of a baby's earliest experiences and assists the notion of feeling safe and contained from conception through to birth. Its brain development and neurology are all based around these basic instinctive (hopefully positive) experiences and sensations. This has particular relevance in relation to Porges' Polyvagal Theory (2015), which suggests that children who are in a high state of distress rely on their nervous system and senses to understand the world around them, as thinking, psychological processes shut down. Jennings describes this reliance on the nervous system as being akin to the 'Theatre of the Body or Theatre of Life' (ToB) (Jennings 2011, pg. 16). The natural process of a child's development which she describes as being preventative and (usually) supports positive early attachments between child and caregiver. Equally she describes 'Theatre of Resilience' (ToR) (Ibid), which plays a similar role in early childhood development, but is more specifically therapeutic in that it supports the social development and awareness of the child from birth onwards, especially if the child is anxious or has developed attachment difficulties due to early adverse childhood experiences.

Jennings also parallels the notion of Embodiment Projection Role (EPR) (2011, pg. 17) alongside this, which is a developmental play based paradigm that can be used to assess a child's natural development, from early messy play with slime or flour and water (embodiment), the playing with objects and beginning to make stories (projection), to fully formed characters in dramatic play and actions (role). By the time a child is six or seven, most children unaffected by trauma or negative early childhood experiences will have easily developed through to the 'role' stage. Noticing the child who has not naturally progressed to this level might suggest early experiences have held back this natural process and may suggest a 'stuckness' that requires additional support.

Jennings stresses that NDP is not in itself therapy but might be used by therapists as a form of assessment or as part of their work. Moreover, it acknowledges and echoes the natural, rhythmical development of the child. When a child is stuck and appears unable to develop this might be when more specialist interventions might be required by play or other arts therapists or health care professional. NDP as a process is not therapy, not invasive and not investigatory, it is something that can be used by any competent and skilled practitioner within a general child and adolescent population to assist their natural development.

Monomyth and rites of passage

In his now classic and seminal work 'The Hero with a Thousand Faces' (1993), originally published in 1949, Campbell argues that almost all stories follow a specific universal pattern of events. He describes the hero's journey as a tri-partite progression that he called, departure, initiation and return, beginning with 'a call to action'. He has entitled this notion the 'monomyth'. Campbell openly acknowledges the connection between his work and that of Van Gennep's 'Rites of Passage' (1960). Van Gennep, the celebrated twentieth-century anthropologist also noticed a pattern in behaviours when working with indigenous peoples who would often send a male child/adolescent out on hunting trip who would then return changed and matured (it was almost always males), having killed an animal. Van Gennep described these as three phases: pre-liminal, liminal and post liminal (from the word limen, meaning threshold). These could be more easily described as separation, or death – to the old way, when the 'call to action' comes, the transitional phase is the test or activity the individual is set, then finally incorporation, or re-birth, the individual returns triumphantly and has been changed due to the experience or the journey they have been on.

We can see this tripartite phase in most modern human social activities such as a wedding for example. The groom would go from being a bachelor, 'separation' from being single, the 'transitional' or liminal state – the wedding service, through to the 'incorporation' when he is reborn in a new union as a husband to his new partner.

This tripartite notion of story is not unusual, as famously discussed by Bettelheim (1976), with its safe boundaries of beginning, middle and end, as used in this project with middle school children who all have confidence and mild anxiety issues. Using notions of NDP, Campbells 'Hero's Journey' and Van Gennep's 'Rites of Passage' the children will in one sense, unknowingly, through play be allowed to playfully test out their capabilities, build confidence and overcome challenges through drama to hopefully come out the other end, reincorporated into the school, but also transformed and reinvigorated.

The school

Having worked in schools as a registered dramatherapist for many years (Holmwood 2014), it was challenging for me to return to a school in a slightly different role, as Neuro-Dramatic-Play practitioner, not as a therapist. The boundaries are in one sense looser. There is no clinical application, issues around confidentiality are not as cogent or essential as might be required in therapy – pupils were at liberty to walk back to their classes and freely share with classmates the nature of the journey we had been on that day. I was not there as a therapist, nor was the work therapy, but it had the potential to be therapeutic. The challenge for me as practitioner was to remember the specific role I was to play, which was primarily as instigator of a great adventure!

I was fortunate to be able to run the group with a specialist school mentor, who worked in the school, she knew each child, had experience of working with children with mild to moderate difficulties, and had been involved in handpicking the group on my behalf and organising them for each visit. Having a key person in school to support the work when my own time was very limited was essential, and I was indebted to her for this support.

The school was on the outskirts of a regional city in England. The children themselves came from a range of both white, Asian and mixed Afro-Caribbean heritages. Some of them could be described as coming from disadvantaged backgrounds. The area itself also had major pockets of social deprivation.[2] All the children had some issues that related to their general confidence. Several of the girls, including a small group of Asian girls, were described as very quiet in the classroom, as were some of the boys. One boy appeared to be autistic, this was confirmed to me by the school mentor. He talked a lot, was loud, got frustrated easily, frustrated others, and would go into minute detail about such things as black holes and the stars and the science behind them, even when science was not central to the discussion. For example, when we pretended it was snowing in the introductory session, he would insist on explaining scientifically the shape of snowflakes. Another boy had been described as having odd thought processes, would rarely speak and when he did much of what he would say would make little sense. He was as yet undiagnosed, but school felt there was probably some sort of mental health issue underlying the way he thought and spoke. The group was a cross section of children that might be found in any classroom, anywhere. I was not (in this instance) a therapist, nor was I here to assist with or fully understand the complexity of issues that might also be around for some of them. I was here to facilitate an adventure through which it was hoped they would gel as a group, develop confidence and camaraderie and most importantly do this by using play and drama (from an NDP perspective) whilst having an enjoyable and exciting time.

Fear and monsters as a path to the children's journey

As is the unpredictable nature of schools for our second session we were out of the school hall and in a rather cramped classroom. This happened on numerous occasions throughout the project. The children had been asking all week about my return, which was positive. Week two was 'feelings week,' and I introduced the notion of things that 'frighten' us. Fear is not necessarily a negative experience. In fight and flight mode the adrenaline is released by the body to assists increased heart rate and blood pressure (Van Der Kolk 2014, pg. 46). Children like scary stories. In the average child the adrenaline release is temporary and helpful, it allows them to learn when they should be frightened and run away, and when they should not. Jennings' echoes this by stating that 'the development of our brains and the maturation of our emotions are therefore dependent on early playful attachment' (Jennings 2011, pg. 38). Learning about fear is part of this process. We explored dark haunted houses and graveyards at night. The autistic boy relished in frightening others with stories of monsters who ate flesh and even, if you were not careful, someone's heart.

I also introduced the central aim of the rest of our work, the idea of going on a journey, but more importantly, the companion who was going to go with us – a monster, called Sam. McCarthy (2007) was an advocate for the monster being in the play space and a central focus to Jennings NDP approach is the idea of mess *and* monsters (my italics) (2011). Additionally, the use of large puppets allows children to anchor their attachments, or as Winnicott would describe it use them as transitional objects (2005), the notion of human experience that exists between the inner and outer world of the infant, the me and not me. Sam, luckily in this case, was to be a friendly monster, so giving the children an alternative perspective to notion of fear and being frightened. Sam was a round velvet, soft to touch, androgynous large purple monster with a large mouth and eyes that allowed him/her to squeak as the mouth opened and closed. She/he had a small mop of purple hair with a small neat bow in it (See Figure 15.1). As I pulled Sam out of the bag and introduced her/him to the group there was an audible 'aaah' from around the room. Everyone took turns holding and stroking Sam as he/she was passed around the circle. Not only did Sam fulfil the notion of being an attachment figure, the soft touch of the fur allowed each child to intimately connect through touch with a monster that was neither male or female, so disposing of gender stereotypes, but also was not scary. Sam would be coming on our journey and so was central to the process. Whilst in clinical settings monsters might represent aspects of the child or clients own 'monsters', this is not the intention here from an NDP point of view, as Jennings describes it, Sam represented 'affection and affirmation through sensory methods' (2011, pg. 166).

Figure 15.1 Sam

Sam had been an instant hit with the children. Each week the ritual began by removing him from the bag of toys and allowing him to be passed around the circle for each child to touch and stroke and speak to.[3]

A call to action

During session three I introduced three concepts which we would work on in the remaining sessions:

- Leadership
- Creating a map
- Making props

Firstly, the notion of leadership. We discussed what it meant to be a leader and I asked who might like to lead the group when we went on our adventure. To my surprise almost everyone put their hands up, including the boy who was almost always very quiet and vacant. He was unable to articulate what it meant be a leader or how he might lead, but the point that he had put his hand up was very encouraging.

I had already decided that all the children would be given, if so desired, a chance to lead part of the adventure. I had always been interested in Dorothy Heathcote's notion of 'Mantel of the Expert' (O'Neill 2015). Heathcote was one of the UK's foremost drama in education teachers and very enlightened in the area of giving children the responsibility to manage and run aspects of drama classes themselves. She described this as the notion of the '"Mantle of the Expert" where the class is set a task in such a way that they function as experts' (O'Neill 2015, pg. 44). Heathcote goes onto say that 'good "interior form" can never have a "wrong" external pattern' (2015, pg. 44). In other words, embedding the notion internally of taking empathic responsibility for others around you will always lead to positive external outcomes. This seems akin to Jennings notion of EPR – embodiment, projection, role – the idea that the forming of attachment is a natural developmental process, and that giving children the 'role' of leadership, at certain times assists with this natural internal/external connection. They at once come to see leadership as both caring for others and being cared for by others. There does appear to be strong theoretical connections between Jennings's NDP and EPR and Heathcote's model of 'Mantle of the Expert.'

This notion of leadership or taking the lead was echoed in the next decision made by the group – the idea of planning their own journey by creating a map. The notion of creating maps is nothing new in education, for example it is used in teaching reflection at graduate level when training arts-based practitioners (Parker-Eames 2019). In this case it is used to develop the student's individual creativity and confidence, and to allow them to individually, physically and externally create and embody their ideas whilst working on a project as a group. The very notion of using crayons, pencils, paints, messy play, creating bright colours and shapes is central to NDP. It also developed and heightened their ability to negotiate who would draw what and where, especially when drawing together on one large piece of paper. The group made a collective decision that their journey would be to an island. Once started, the group could not stop their ideas for the island adventure flowed out, they included mermaids, dolphins, dragons, jungles, woods, black holes, volcanoes, stepping stones, rivers, swamps, as well as good and bad monsters to name but a few. There was no end to the creative ideas that the children had in imagining where they might go and what they might do.

It was at this stage I decided to introduce a second attachment monster figure – Alix – the snake. Alix was again neither male or female, but was a five-foot-long soft, green, velvety, friendly snake with a red tongue, a mouth full of teeth and big eyes. The group responded to Alix equally as well as they did Sam. Alix, too, would be coming on the journey with us. I hadn't realised the significance or importance of the roles played by Sam and Alix until much later. They represented what Bowlby described as an 'attachment figure' (1979). He hypothesised throughout his work that for a child to feel secure and confident they needed an appropriate attachment figure, from birth (usually the mother, or parent figure) who was available to them who they could connect with both emotionally and physically. Sam and Alix became the attachment figures for the group. They could hold and touch Sam and Alix, stroke them, speak to them, knowing that Sam and Alix were available and safe whenever a child wanted to pick them up or ask a question and wait silently as an answer was whispered into their ear. They also had to learn to be patient and share them. I also introduced a bright yellow stress ball with an

Figure 15.2 The map

emoji face on it. Group members loved to squeeze it and in one case one child continued to squeeze the staff members arm after he let go of the ball.

The final element to our adventure was the props, the practical objects that would help us on our journey. The group agreed we would need to make things such as torches, not just modern electric torches, but ancient medieval ones, sticks with fire on the end made of orange and yellow paper, so that they could see in the dark. They also needed shields and backpacks and even sandwiches and burgers in case they got hungry on the way. The group spent two weeks making the props. This was an important part of the process, we used cardboard, scissors, glue felt pens and paint. The children found the cutting and manipulating the material, sometimes hard, but the staff member who supported me felt they worked better than some other children in the school at this. The mess of making things, the physical touch, the cooperation between group members, sharing materials, the frustration when it didn't always go right, were all part of the integral physical and embodied experience of preparing to go on the journey. Akin to the idea of actors rehearsing for their first night. All of this is paralleled Jennings' notion of sensory and messy play in which she states 'messy play is a means of discovering order: by making a mess I can learn limits, containment and borders. It is also a means of developing sensory play and may continue in some form into adult life' (2011, pg. 71). Jennings also reminds us that messy play is not necessarily chaotic play. Through the mess and chaos of playing with various materials we can eventually find boundaries and order.

In order to create even more excitement and a sense of adventure the group agreed we would arrive on the island by boat, a boat in which the group, like in a Viking long boat would row towards the island, something we practised several times, with some group members being given the role of captain and helmsman. Finally, and even more excitingly in order to create absolute ecstasy and direction on the adventure the group were going on a hunt – for treasure!

The hero's journey

I made a conscious decision to not rehearse our adventure, other than, rehearse rowing in the boat, deciding who was going to lead each section of the journey and then decide the route of the adventure. I hadn't even told the children where the treasure was to be located. The adventure itself was going to be completely improvised and be a one-off experience (Johnstone 2017). There was the potential for it to be messy but hopefully contained. Though it wasn't a performance, we decided not to open it up to other staff or pupils to watch; the only person present was a member of staff taking a few photographs on behalf of group members. I had provided a structure and hoped that the children would take responsibility for the sections they were to lead, and that the improvisational nature of play would become '*running play*, arms outstretched, faces joyful' (Italics original) (Slade 1995, pg. 74) as Slade, the drama educator once described it.

'Arms outstretched and faces joyful' (ibid.) was, in the end, a very good description of the performance. I played the role of director of the adventure, I moved around with the group and suggested what might come next. Various group members and I held Sam and Alix during the adventure. One of the girls was the captain and began to find her voice as she barked instructions to the group on the boat, others steered the boat and let down the anchor on arrival.

The adventure was an onslaught to our senses. We avoided icebergs and mermaids, walked through gardens filled with poisonous flowers, relaxed briefly on a sun kissed beach with good monsters, crossed a slippery bridge, went over stepping stones, past bad monsters and went through a jungle with an angry monkey in it. We manage to pass an exploding volcano and a fire breathing dragon, who we befriended, crossed a dessert, went through a forest and avoided a black hole and aliens. Finally finding the treasure, a blow-up treasure chest covered with a

Figure 15.3 The treasure chest

cloth. We made our way back to our boat and left the island having survived. Each and *every* child took responsibility for leading a part of the adventure, in itself a major achievement. I gently guided when necessary. On arrival back from our adventure we opened the treasure chest, each child (unknowingly) had an emoji stress ball, chocolate money and sweets to take home, after all it was two weeks before Christmas, and the children deserved the treasure they had so valiantly gone on the adventure for. The stress ball being their own permanent reminder of the adventure, and maybe a help in the future.

The adventure had been a great success. One of the quiet girls who had played the captain on the boat declared afterwards she had found her voice. My colleague from school assisting me said that the quiet boy with the strange thought processes had taken part so well in the project, she was so pleased and surprised. She said, 'in the classroom he is doing nothing; almost nothing at all.'

Conclusions and an unexpected outcome

It was mid-January when I met the children for the final time to carry out a feedback session. At first, they were quiet and withdrawn. It was a struggle to get anything from them as we sat in the school staff room, the only room available for us to meet in. It was as if all the work we had done only a few weeks earlier had evaporated into the air during the Christmas break, which of course it might have done.

In hindsight, I had completely forgotten the structure of our sessions as I had been so focussed on getting raw 'research data' in the form of verbal feedback. At that moment, in desperation, I opened my bag and pulled Sam and Alix out; suddenly everything changed. On seeing Sam and Alix the children came alive they wanted to hold, stroke and talk to them. The noise in the room went from silence to excited chatter in a matter of moments on Sam and Alix's arrival. They had their anchor, their attachment friends, something that was familiar that had held them as a group. This also reminds me of Chasen's view, as discussed earlier on engaging 'mirror neurons' (2011). Even though the original experiences were not in existence the children's own mirror neurons fired as if they were back in the room on their adventure. The children then freely spoke and shared their thoughts and reflections on our work as they passed Sam and Alix around the room and touched, cuddled and chatted with them. Their reflections flowed out and are recorded in Table 15.1.

It is clear from these comments that the project has been successful for these children. In one sense there is nothing groundbreaking about what the children have said. Play is central to children's development and growth, an established fact that needs little quantifying here. It also suggests that a relatively short-term piece of work, which is relatively inexpensive can have positive benefits for children in school, on the margins, who might not otherwise have received additional support due to scant resources.

It also opens the wider debate about what should be central to primary and middle school curriculums in an era when the education system is becoming much more concerned about traditionalist approaches to education, which include the need for rote learning and knowledge and is less concerned about the progressive educational policies and approaches of the past (Holmwood 2014). The children's comments could arguably suggest that a more playful, progressive experiential and embodied approach to learning, as suggested by people such as Heathcote and Slade, would be helpful in today's digital world. It was particularly noted in relation to the boy with strange thoughts, that in the classroom he barely functioned and did nothing at all, yet in the group with peers he somehow appeared to come alive and participate in a way he had not been able to in class.

More central to NDP, the focus of this chapter, is the obvious importance of the large attachment puppets. Even as researcher myself I had forgotten the absolute necessity that the children had in needing to feel grounded, not only in each other, but, more importantly, with their large attachment puppets. Jennings notes this in the importance of 'Theatre of the Body' (2011, pg. 18). She discusses the idea of 'thinking with the body,' 'body intelligence' (and) 'body memory' (2011, pg. 19) and 'that the body is a primary means of learning' (ibid.). One wonders whether the notion of 'mirror neurons' might also play a part here (Chasen 2011), the idea that additional mirror neurons fire when young children observe certain actions, something that is repeatable and held in memory. Does the proximity, size, colour, shape and texture of the attachment puppets, with big eyes, mouths and teeth, create an imbedded memory for each child of what it was like to be cared for as a tiny baby. The ability to touch and interact with the puppets clearly had an impact on the whole group throughout the process, and during the chaotic times of the adventure, they would turn to them for advice. They only came awake and alive again in the final session when the puppets re-appeared. Maybe their mirror neurons were again firing, reminding them of their previous adventures as well as lost early unconscious embodied memories.

This notion of the large attachment puppet is certainly an area worthy of further research and development, and both its place in the classroom as well as less formal play environments. It fits

Table 15.1 The children's responses

Question 1: What was different to what you normally do?
Having fun and learning at the same time. Bigger chance to use our imagination. Meeting new people from different classes. Learning by watching others. Meeting Sam and Alix.
Question 2: What did you learn?
How to be confident. Use my imagination. Facing fears. Sticking together. Learn how to make friends.
Question 3: What did you enjoy?
Spending time with new people. Having fun. Drawing the map. Liked building things. Acting as part of a journey.
Question 4: What will you do differently?
I will not be as quiet as I was. Be brave and not be scared.

alongside the notion of Jennings' 'Theatre of Life' and 'Theatre of Resilience.' This small-scale project has reminded us of the importance of play and that play *is* education and is central to all children's natural development and growth. I very much hope that I can begin to develop research into the notion of attachment puppets in many more classrooms, schools, and educational settings in the future.

I am indebted to the pupils, the specialist mentor and the school for allowing me the privilege of working and researching together with them.

Notes

1 Written ethical agreement and approval had been sought and granted from both the school and the children's parents or guardians, and each pupil had also given verbal consent to attend. They were free to leave at any time and return to their classroom, in fact only one child left throughout the process of the project. Ethical approval had also been sanctioned and approved by the University of Derby where I work as a senior researcher and associate professor in professional practice in the Department of Therapeutic Practice. The project was also part of my training in Neuro-Dramatic-Play – NDP.

2 Specific information does exist on the social deprivation in the area but is held back here to maintain confidentiality.

3 Using a multi-sensory bag would be even better, a bag with shape colours and textures in which the attachment figure could come from and return to would benefit the work further.

References

Campbell, J. (1993) *A Hero's Journey*. London: Fontana Press.

Chasen, L. R. (2011) *Engaging Mirror Neurons: To Inspire and Social Emotional Development in Children and Teens on the Autistic Spectrum*. London: Jessica Kingsley Publishers.

Bettelheim, B. (1976) *The Uses of Enchantment: The Meaning and Importance of Fairy Tales*. New York: Random House.

Bowlby, J. (1979) *The Making and Breaking of Affectional Bonds*. London: Routledge.

Bowlby, J. (1988) *A Secure Base*. London: Routledge.

Holmwood, C. (2014) *Drama Education and Dramatherapy: The Space Between Disciplines*. London: Routledge.

Kellett, M. (2005) *How to Develop Children as Researchers: A Step by Step Guide to Teaching the Research Process*. London: Paul Chapman Publishing.

Jennings, S. (2011) *Healthy Attachments and Neuro-Dramatic-Play*. London: Jessica Kingsley Publishers.

Johnstone, K. (2017) *Improvisation and the Theatre*. London: Bloomsbury.

McCarthy, D. (2007) *If You Turned into a Monster: Transformation Through Play: A Body-Centred to Play Therapy*. London: Jessica Kingsley Publishers.

Meares, R. (2005) *The Metaphor of Play – Origin and Breakdown of Personal Being*, third edition. London: Routledge.

O'Neill, C. (2015) *Dorothy Heathcote on Education and Drama – Essential Writings*. London: Routledge, David Fulton Books.

Parker-Eames, M. (2019) Chapter 2: Teaching Reflection to New Undergraduate Students. In Taylor, J. and Holmwood, C. (eds.), *Learning as a Creative and Developmental Process in Higher Education – A Therapeutic Arts Approach and Its Wider Application*. London: Routledge.

Porges, S. W. (2015) Making the World Safe for Our Children: Down-Regulating Defence and Up-Regulating Social Engagement to 'Optimise' the Human Experience. *Children Australia*, 40:2, pp. 114–123.

Slade, P. (1995). *Child Play*. London: Jessica Kingsley Publishers.

Van Der Kolk, B. (2014) *The Body Keeps the Score: Mind Brain and Body in the Transformation of Trauma*. London: Penguin Books.

Van Gennep, A. (1960) *Rites of Passage*. London: Routledge.

Whitehead, J. and McNiff, J. (2006) *Action Research: Living Theory*. London: Sage.

Winnicott, D. W. (2005) *Playing and Reality*. London: Routledge.

PART III

Play in therapy

Introduction

Section 3 shifts specifically towards using play within specific therapy or counselling contexts. All the writers in this section are qualified therapists or counsellors and use play within their work, which is specifically therapy orientated, though they may not be registered in their country or state as play therapists. We again bring a range of social and cultural perspectives, with therapists sharing their work and experiences with their clients from Israel, Malaysia, Turkey, USA, UK, Israel and Greece.

Our first chapter in this section, Chapter 16, considers the complexity and sensitivity of cultural differences for a counsellor using play-based work in his counselling practice in Malaysia. Gian Yong considers the multiple complexities of language and religion and the impact this has upon his play-based counselling practice in a country where notions of play (outside of the forests, at least; see Jennings Chapter 11) are sometimes approached with suspicion and concern. We then contrast this with similar issues in a very different cultural world with Hayek Damouni using play in a dramatherapy context working with addicted Palestinian men in Israel.

Chapter 18 moves several thousand miles to the west to the UK where Torrance, a highly experienced dance movement therapist, uses play at the core of his work with autistic children and young people. Staying in the UK, we then consider the Sandplay based work of art therapist Emma Allen working in the complex area of a forensic secure unit with people who have not only committed serious crime but have other complex issues around identity. Allen focuses on one case with a very complex client and we travel on their journey together. Similarly, for Chapter 20 we remain in the UK with another art therapist, Bridget Rees, working with a very different client group using 'Dyadic Art Therapy' and play-based approaches with children and their caregivers.

In Chapter 21 we return to Israel and consider Pendzik's six key models of dramatherapy using play as a central focus with work carried out by her colleague Galila Oren with one specific school-aged client in an education context. The following chapter, Chapter 22, considers play from a Filial Therapy perspective, particularly in relation to working with children and families in Turkey.

Chapter 22 moves us to the USA where Neal Brodsky, a licenced family therapist shares his accounts of using play in his therapy work with two enchanting case studies around children and families. We remain in the US with Chapter 24 where Dennis McCarthy, originally a licensed dance movement therapist, who has used play in his therapy work for over four decades, shares his experience of children who panic with some beautifully illustrated case studies.

In the final chapter to this section, Chapter 25, Couroucli-Robertson explores the connection between dramatherapy and play therapy, partly through the setting up of a Panhellenic Professional Association of Dramatherapy and Play therapy in Greece. She then considers the case of "Angela" from this multiple play and drama-based therapy approach. Further blurring the boundaries between professions which learn from and teach each other. This making way for the final section that follows after this, which focuses specifically on play therapy.

16

USING PLAY AS A COUNSELLING TOOL IN MULTICULTURAL SOCIETY

Jaff Choong Gian Yong

Introduction

As a registered and licensed counsellor in Malaysia, I was trained to be highly aware and sensitive to multicultural counselling, particularly with the three main ethnic groups in Malaysia (Malay, Chinese and Indian). I found that using play as a tool in counselling sessions helped to support my professional services, especially with children and teenagers. I trained in Neuro-Dramatic-Play with Professor Sue Jennings in 2010 after I completed my psychology degree. I then started to integrate the practice of Neuro-Dramatic-Play within counselling approaches and skills during my master's degree in counselling.

It is said that "Toys are children's words and play is their language" (Landreth, 2002). Throughout the years of practice using play as a counselling tool I discovered how playfulness strengthens the therapeutic alliance between the counselee and myself. I experienced the power of non-directive play in reinforcing the therapeutic process. I applied mostly play in my counselling sessions during the time when I was attached to the Hospital (Children's Ward) and the Juvenile School.

When I volunteered in the Boys' Juvenile School, though the majority of the residents were Malay and they spoke only Bahasa Paku (Malay's dialect), I was the only counsellor available within the school. Despite language issues I worked well with the boys using the practice of therapeutic play. I speak multiple languages (Malay, English, Mandarin, Hockkien, Teowchew and Cantonese), but still find it challenges me to understand the deeper expression of the languages, especially when dealing with clients who speak the Terengganu accent of Bahasa Paku. Therapeutic play helps to build the therapeutic alliance with the clients and indirectly helps us progress in the therapeutic process, too.

Malaysia, a multicultural society

Malaysia is a harmonious and beautiful multicultural country where the population is made up from different ethnics groups; as reported by the Department of Statistics of Malaysia in the year 2019, Bumiputera (a combination of Malay and Orang Asli aka Indigenous group – 69.3%), Chinese (22.8%), Indian (6.9%), and others (1.0%) (Mahidin, 2019). Since the independence of Malaya in 1957, Malaysians maintain not only their own unique culture, belief system, values and languages among the groups; but moreover, the cultural integration of such cultures as Baba-Nyonya and Mamak which have also risen over the past 60 years (Bee & Leinbach, 2019).

Malaysia's official language is Malay and the secondary language is English. Besides that, people in Malaysia speak Mandarin, Tamil, Baba-Nyonya, Chinese dialects (i.e. Hockien, Hakka, Cantonese, FuZhou and Teochew), Indian dialects (Telugu, Urdu, Punjabi and Malayalam) and Native languages (i.e. Termiar, Iban, Kadazan, Semang, Senoi, Temuan, Jakun, Che Wong, Jahut and Semelai).

Religion or spiritual practice is significant and essential for Malaysia; most Malaysians have a strong religious belief. According to the Malaysia's Constitution, Islam is recognised as an official religion and it also one of the major factors that differentiate Malay and non-Malay (Balakrishnan, 2017). By law, all Malays are Muslim; and non-Malay have freedom of religious choice. According to the codex, every non-Malay in Malaysia has the right to profess and to practise his or her religion and propagate it in peace and harmony (subject to the application of law: to propagate religion to Muslims is prohibited) (Bedford, 2018; Hays, n.d.; Islam in Malaysia – Islamic Tourism Centre of Malaysia, n.d.; Samsudi, n.d.).

Three-fifths of Malaysians are Muslim (mostly Sunni Muslim) (Hays, n.d.). Most of the Chinese in Malaysia do not have a dominant religion, but mostly are Confucian, Taoist, Buddhist (more Mahayana and Tibetan) (Dipananda, n.d.; Nalanda, 2012), Christian (mostly are Baptist, Methodist and Catholic) (Nicholas, n.d.; Wong & Ngu, 2014) and also a small group of Chinese-Muslim. The majority of Indians in Malaysia are Hindu or Buddhist (mostly Theravada) (Dipananda, n.d.) and there are minor groups of Indian-Christians and Indian-Muslims. Nevertheless, most of the Orang Asli in Malaysia follow Paganism and Animism, originally; but more recently, many of them are being converted into Christians and/or Muslims after years of missionary outreach by churches and mosques (Chan, 2016; Nobuta, 2007). Apart from this, there are also small amounts of Punjabis who believe in Sikhism in Malaysia, too (Bedford, 2018; Islam in Malaysia – Islamic Tourism Centre of Malaysia, n.d.; Samsudi, n.d.).

Mental health professions and play therapy in Malaysia

Psychiatrist, clinical psychologist and counsellors are differentiated by the different types of education and training they have completed in order to obtain their credentials. On the whole, there are only three mental health professions recognised by the central government in Malaysia:

1. **Psychiatrist** are registered and certified by the Malaysian Medical Council (Ministry of Health) under the Act 50 – Medical Act 1971.
2. **Clinical psychologists** are registered and certified by the Malaysian Allied Health Professions Council (Ministry of Health) under the Act 774 – Allied Health Professions Act 2016.
3. **Counsellors** are registered and certified by the Board of Counsellors (Ministry of Women, Family and Community Development) under the Act 580 – Counsellor Act 1998.

"Malaysia is grappling with mental healthcare. Affordable and effective treatment exists but is out of reach to many, thanks to a confounding and constricting system" (Foong, 2018). In Malaysia, although mental health professionals (MHP) are highly in demand, the ratio of MHP to public is insufficient, much more mental health awareness is also needed in the general public, especially in the area of identifying the authenticity of the MHP. There is a lot of malpractice; false practitioners and/or religious practitioners who claim they are offering mental health services in this peaceful land. Thus, the educational need for public awareness becomes a much bigger challenge for MHPs in Malaysia, especially in the area of play therapy which has no recognition by the central government.

Multicultural awareness plays an important role for the trained MHP in Malaysia so that they are ready to serve the community better. Although counsellors in Malaysia were trained in multicultural counselling and most of them are culturally sensitive, there are still some barriers in the counselling process, especially when we serve the children or people in rural areas where the smaller communities speak only their mother tongues and limited English or Malay. Therefore, therapeutic play can be an effective counselling tool for counsellors in Malaysia, especially when the counsellors are dealing with children in these small rural communities.

Recently therapeutic play and play therapy has become popular and is being promoted among educators and parents, unfortunately, this is also the beginning of malpractice allowing unqualified practitioners of therapeutic play/play therapy, without any experience or formal training. As the demand for therapeutic play/play therapy in Malaysia is increasing MHP's are facing challenges, especially those who use play as a tool in their service:

1. **Challenge 1: Counselee/patient lost confidence in MHP.** According to my records, 80% of my cases who seek help from are due to their experiences with the unqualified therapist, they are losing their trust towards the MHPs due to their bad experiences in the past. Most of them appear to be involuntary clients or completely resistant to a MHP.
2. **Challenge 2: The public do not seek professional help straight away.** Due to the stigma of mental health, people tend to seek help from a so-called teacher, guidance or motivator who claimed they too provide play therapy services, rather than seeking help from a certified MHP. There is such a phenomenon in Malaysia where the people (especially the business mindset educators) attended a workshop, for example, a play therapy workshop and they then claim themselves to be qualified in doing what they learned in the workshop.
3. **Challenge 3: The public are confused with the so-called diagnosis.** Many unqualified practitioners tend to "diagnose" their customers (I call them patient or counselee, but often they are described as customer for the unqualified practitioner). This "diagnosis" actually causes depression and anxiety to the customer as well as the family members. For instance, there was once a traumatised boy (who developed selective mutism) who was sent to me by the parents due to an unqualified practitioner "diagnosing" him as having ASD, the "teacher" (the teacher is the child's class teacher offering a service) and is selling health care products to the parent and claiming that it is a special medicine for speech. The parents were panicked and confused as they have gone through several "treatments" and "play therapy" sessions with the "teacher" and their child's condition was not improving.
4. **Challenge 4: Public complaints about professional charging.** Once there was a parent who came to me saying that the MHP's fee is high (counsellor's charge a fixed price), so they would rather seek help from a "therapist" who also does the same job as mine but at a cheaper price. Luckily, after a while the parent came back to me and stated that the cheaper charging "therapist" is repeating the same pattern of "play therapy", and they found out the "therapist" did not hold any qualifications or related educational background; the "therapist" had just shown a certificate of attendance for an Introduction to Play Therapy. I was wondering, how many parents and clients out there are facing the same situation as this parent does?

Case study – Naughty Joker

Background of Naughty Joker

Naughty Joker is an 11-year-old (Chinese-Malay) Muslim boy who was suffering from emotional issues. The mother is a single parent whose husband had eloped with his mistress. The boy was brought up by the maternal grandparents since the age of seven.

Naughty Joker's biological father is Chinese, and converted to Islam after he married. According to the maternal grandparents, they had never been in touch with nor met with the paternal family and they don't even know who the paternal grandparents are.

Precautions: *Most typical Malaysian-Chinese and Malaysian-Indian families oppose their children's marriage with Malay or Muslim, as this is strongly related to the law and the belief system. Most of the time, the family members of converted Muslims are suffering from the contradiction of the belief system:*

1 Law: two (2) laws found in Malaysia – Syariah Law specifically for Muslims and Secular Law for everyone else.

a According to the Constitution of Malaysia 1957, Part I – The States, Religion and Law of the Federation, Article 3: "1) Islam is the religion of the Federation; but other religion may be practiced in peace and harmony in any part of the Federation". "Malay means a person who professes the religion of Islam . . ." as stated in the Constitution of Malaysia 1957, Part XII – General and Miscellaneous, Article 160, Clause 2 (Islamic Department of Malaysia, 1984).

b According to Malaysia Islamic Family Law (Act 1984), Section 10 – Persons of Other Religion: "1) No man shall marry a Non-Muslim except a Kitabiyah; 2) No woman shall marry a Non-Muslim".

2 Taoist-Chinese belief: Chinese are concerned about life and death issue, especially the funeral after death. As most of the Taoist Chinese believe that to bury in a Chinese grave, especially together with their parents is a must, or else they will be considered an unfaithful child. Also, it is typical for Taoist-Chinese to be concerned about the funeral ritual.

3 Hindu-Indian belief: Indian belief states that a prayer after death is important for reincarnation. This is same as Taoist-Chinese families, they are so concerned about the ritual after one's death.

However, the dad died when Naughty Joker was three, then the mum remarried when he was six. Unfortunately, the stepfather left them when he was seven because the stepfather had another family and eloped with his third wife (the mum is the second wife of the stepfather).

> ***Precautions:*** *According to Malaysia Marriage and Divorce Act (Act 164) – Reform Act 1976, Monogamous Marriages are applied in Malaysia. However, Polygamy is allowed under Malaysia Islamic Family Law (Act 1984), Section 23 – Polygamy.s*
>
> *(*Law Reform (Marriage and Divorce) Act, 1976: Act 164, *1981)*

Since then, Naughty Joker was being taken care of by the maternal grandparents, and the grandparents were annoyed due to continuous complaints from the school teachers. The teachers complained about his uncontrollable anger, he would shout and scream when he is angry; recently, he even pointed a blade at his friends in school due to bullying.

Taking history

As usual with any counselling cases related to children, I often start with taking a history, whilst the guardians were also asked to fill in the informed consent as well as the Strength and Difficulties Questionnaire (SDQ). I will always show the special room and introduce the categories of toys as well as the usage of toys to the guardian, especially when there are religious miniatures around. The same is offered to Naughty Joker's grandparents, after I shake hands and meet them.

> ***Precautions:*** *Gender and age order – elderly then youngster, male followed by female; gender boundaries – no shaking hands or touching different genders.*

I show them the special room, introduce the toys, taking a history, and filling in the informed consent and SDQ. Naughty Joker's maternal grandpa felt uncomfortable with the religious miniatures (cross and statues) and the figures of dogs and pigs.

> ***Precautions:*** *the Christian cross, dog and pig is non-halal for Muslims.*

Then we discussed through the values and taboos of Malay-Muslims that they wish me to take note of during the sessions. As I explained to the grandparent, the miniatures are just a tool for children to express and there is no missionary intention involved. After reflection, both grandparents allowed me to keep the other religious statues and the dog miniature but not the pig, the cross and the statue of Jesus Christ. The grandpa especially insisted to remove the cross and the statue of Jesus Christ during the sessions with their grandchild.

Forming therapeutic alliance

"*Aku tau ini bilik, untuk main-main dan cikgu akan suruh aku duduk diam-diam lagi, aku memang tak suka bilik ini*" (I know this room is for playing and the teacher asked me to sit still while playing, I don't really like this room). Naughty Joker whispered while exploring the room when he first entered. He graffitied the blackboard in the room, he mixed the sand tray with small stickers then he put all the toys into both wet and dry sand tray, he messed up the special room during the first session. I noticed that he looked at me after each action, thus I chose not to stop him but just retold his behaviour in the Malay language.

> ***Precautions:*** *speaking the mother tongue to a child is valued by most Malay families.*

The mess continued up to the third session and stopped after we discussed and created the "goal of BaHaSKan" (see later in the chapter). Personally, I prefer to set a target goal with my clients instead of introducing the boundaries as ground rules.

Setting boundaries

In the third session, Naughty Joker repeated his messy play for 20 minutes, then he sat on the sofa and watched me and shouted: "*Bosanlah*!" (It's so boring!) I invited him to play a word game with me after he voiced the "*Bosanlah*", and he agreed. First, he was allowed to share a word that he wants to share with me; immediately, he said "*kebebasan*" (freedom). Then, he asked what my word was and I gave him a word of "*kesopanan*" (courtesy) but it was rejected by him. He stated that courtesy is the word that he hates the most, because everyone told him to be polite and courteous. Then I suggested "*menghormati*" (respect), and I explained to him that mutual respect was what we can give to each other, and he agreed with the word. He then followed by his suggestion and wished to have "*seronok*" (fun) whenever he is in the session with me, then I asked is that okay for me to suggest we "*kemaskan*" (clean up) after we have had "*keseronokan*" (fun), and he agreed to that suggestion. Thus, we decided to form a word from "*kebebasan*", "*menghormati*", "*keseronokan*" and "*kemaskan*"; we created BaHaSKan (which means let's talk) as our guideline and common goals throughout the counselling sessions. Naughty Joker requested to write the common goals on the blackboard as a reminder to both of us.

BaHaSKan (Let's talk), included our four common goals:

First goal: **BEBAS** (freedom)
Second goal: **HORMAT** (respect)
Third goal: **SERONOK** (fun)
Fourth goal: **KEMAS** (clean up)

Building trust

Naughty Joker told a story of two men who were falling in love on the fourth session. "These men are in love with each other, they are living happily. These kids are crying because the dad is falling in love with another man, the dad is the one who wears the blue jacket. The lady is crying together with the kids as the husband left her". Naughty Joker kept looking at me when he told the story, he was observing my reaction to his story.

> ***Precautions:*** *Homosexuality is totally prohibited amongst Muslims and is unacceptable in the society.*

"*Kenapa hang tak marah aku?*" (Why don't you scold me?) Naughty Joker asked in a curious way when he saw I did not stop him from discussing the topic, I just stayed calm and listened to him attentively. Naughty Joker created this story and shared it with many people, especially adults; some scolded him for telling an inappropriate story, some stopped him from continuing to tell the story, and some tried to educate him that homosexuality is an inappropriate behaviour, some even threatened him to not tell this story anymore. He was curious as to why I did not respond as the others did, as I explained to him that I remember the common goals – "*bebas*" and "*hormat*", and I said no matter what, I will accept and respect his freedom. "*Aku minta maaf kerana test hang tadi. . . . Aku sangat takut, aku takut hang akan marah aku dan aku jua takut hang akan tipu aku. . . . Maka aku bohonglah tadi*" (I am so sorry for testing you with the story. . . . I am so worried, I worried that you will scold me as others did and I am scared you will cheat me. . . . That's why I am telling a lie); he cried and apologised for not trusting me and testing me with the story.

Release fears, worries and aggression

I introduced an emotional thermometer[1] to Naughty Joker during the fifth session, and then we played an emotional stones game. He admitted he is angry all the time (mostly his thermometer is at seven or eight) when we discussed it. He found that his anger often came from worries. He cried and shared his worries, he said he was worried that he would lose his loved ones, especially his mum as he is trapped by the teasing from the classmate: "*anak terbiar*" (abandoned child) and "*anak kurang ajar*" (rude child). He said he was so angry for being teased as an orphan. "*Aku geram je dengan dia orang, pikir nak gaduh sangat sama dia orang, aku nak fight back! Aku bukan anak terbiar! Tapi, aku tak kuat macam dia orang, maka aku guna lah pisau*" (I feel mad with their words; I wanted to fight with them! I am not an abandoned child! But I am not strong enough like them, so I have to use a blade). Over the session, he shared a lot of his feelings: anger, worries, helplessness and insecurity. Since then, he has claimed himself as "*Pelawak Nakal*" (Naughty Joker). He was happy when I invited him to play with the emotional stones and he shared a lot when we were playing with them.

When the session came near to the end, I decided to do an imagery guided exercise with Naughty Joker so that he was able to formulate and visualise a safe-place to calm down his long-term built-up emotions. The lovely Naughty Joker looked peaceful when he was in his safe place; he smiled when he described his safe place. He imagined he was running and playing with the birds in the cloud, rolling in the clouds, he said he felt he could embrace the clouds too. Naughty Joker was so satisfied at the end of the session, and for the first time, he "*salam*" (which is the action of shaking and kissing the hand of the elderly or respectful people) to me before leaving the special room.

> ***Precautions:*** *"Salam" is a social etiquette as a respectful action in Malay culture. The younger people will shake and kiss the elderly or the respectful person's hand. After the handshake, both palms of elderly and youngsters will be brought to the heart as a gesture of sincerity. Women only shake hands with direct family members.*

Naughty Joker requested to play something active in the next session; thus we played the body movement games in the sixth session. We started with a walking and freezing game; while walking we made facial expressions. We had a good time playing the walking and freezing game. In my reflection I considered that as Naughty Joker was a boy and we did not need to concern ourselves with gender boundaries, or else the outcome may well have been different with a girl.

> ***Precautions:*** *Body boundaries with different genders among Muslims are needed as body touching is discouraged between different genders, including handshakes.*

After the walking and freezing games, we did body tracing. We started this game by picking up our favorite color, I then invited Naughty Joker to trace the body outline by allowing him to lie down on the white paper. He was so excited doing this activity that he suggested I lie down too so he could trace my body outline. We discussed the bodies reaction to happiness, sadness, anger, fear and disgust, then we used the colours to represent the emotions that were discussed. We used yellow to represent joy and happiness, blue for sadness, red for anger, purple for fear and green for disgust. Naughty Joker was laughing and he stated that it was similar to the movie "Inside Out". Naughty Joker jumped on the traced body and wrote a few names on it; then he started to scold and curse the traced body. "*Aku mohon maaf terhadap kelakuan biadab aku*" (I am so sorry for my rudeness) he apologised for his lost control and rudeness.

Precautions: *Activities of cursing are discouraged in the Malay culture as it is seen as a rude behaviour; but I allowed Naughty Joker to do this as it was a way for him to release and express his anger.*

Gaining power

I shared the story of "Lucky Burger" (a story I created by using ABC Model. ABC Model aka as Antecedent-Belief-Consequences Model is a model proposed by Albert Ellis and known as an important part in Rational-Emotive Behavior Therapy, REBT, and commonly used in Cognitive-Behavioral Therapy, CBT (Corey, 2009; Selva, 2018)) with 'Naughty Joker' at the beginning of session seven. Then we discussed the story. Curiosity drives Naughty Joker and he asked and reflected in the session: "Why does the Lucky listen to the witch?", "Why does Lucky still eat the burger bun, as he knew there was no meat patties in the burger?", "It is not delicious without the burger in the bun (only the bread)", "What can Lucky do to make the bun more delicious?"

Story of Lucky Burger

Lucky is an owner of a burger shop near Masjid Jamek, KL. Lucky and his family operate the burger shop for over three decades since his grandfather's era. The burger was famous for its brilliant chewy and tasty barbeque beef as the patties of the burger (**Precautions:** Avoid non-halal food such as pork and alcohol when creating the story for a Muslim.) Every day, Lucky produced barbeque beef by following his ancestral secret recipe; this makes the business good and he made huge profits from it.

Abu, the owner of the fried chicken at the next hallway was so jealous of the business of Lucky, he always tried to find ways to defeat Lucky and take over their business. One day, Abu sought help from a Bomoh (Malay Shaman) in the Gunung Puteri Ledang. Abu requested the Bomoh to cast a spell and cause Lucky to loose his taste buds so that Lucky will no longer produce the fabulous beef.

Next day, the Bomoh went to Lucky's shop and claimed that his burger is disgusting. The Bomoh said, "you shouldn't be proud of your burger, because your burger without the patties is yuck. You are cheating on your customers as you are using the tasteless bun!" The Bomoh continued to say he was here to help Lucky, but the Bomoh cheated Lucky saying a famous royal chef from the istana (palace), has a special burger he would like to share with Lucky. The Bomoh promised Lucky if he tried his burger then Lucky will learn how to create the best burger. The Bomoh then showed a burger without patties to Lucky and told Lucky to finish it . . .

Since then, Lucky lost his sense of taste after one bite of the burger given by the Bomoh. Lucky felt sad and angry after he lost his sense of taste, he blamed himself for no longer being capable of making the best burger patties. He started to produce only burger bubs without patties as 1) he can't accept the fact that he lost his sense of taste, 2) he refused the help from his friends and family members, 3) he no longer trusted in people, 4) he feels insecure.

Day after day, he kept producing the burgers without patties and this affected his business and his life. He cried and grizzled to everyone, including his family members.

> He believed that he would never succeed again and he thought he was a hopeless person. He gave up . . .
>
> What can you do to help Lucky?

Naughty Joker said a burger without patties is not nice. If he were to help Lucky, he said he would ask Lucky to teach him how to make the patties, then he would try to encourage Lucky to produce the burger with him. I drew a burger and a box (I wrote an "M" on top of the box), immediately Naughty Joker shouted happily: "I knew this, it is McDonalds!" Then, I introduced the box and the burger based on ABC Model to him:

- Cover of the box known as Events.
- First bun of the burger known as Feelings.
- Patties of the burger known as Thinking.
- Second bun of the burger known as Action.
- The box known as Consequences.

Naughty Joker agreed with the explanation. He said: "No wonder Lucky will eat the burger without patties, because he never thinks. He acted when he feels angry after being challenged by the Bomoh". Then I explored the belief and thinking with Naughty Joker, I asked how can he help Lucky by using the concept of the "burger model" we discussed. Without thinking too much, Naughty Joker answered: I will help Lucky in changing the way – he thinks or he will never succeed again, and Lucky also cannot believe that no one can help him, he is not hopeless. . . . We discussed about Lucky and the burger, then I tried to guide Naughty Joker by using the burger in reflecting his own issues. At the beginning, Naughty Joker stated that he didn't see the relation between those two, but he whispered by saying "if there is a similarity, then sometimes I am the same as Lucky. I listen to people's comments, then I get angry". We explored his situation more by using the ABC model, then we did a Progressive Muscle Relaxation (PMR) technique at the end of the session.

In session 8, Naughty Joker came with anger. He said his friend had called him names again, and he doesn't like it. He punched and kicked the tumbler when he entered the room, then he cried and shouted "*aku bukan anak terbiar!*" (I am not the abandoned child!) I allowed Naughty Joker to express himself in the room without any interrupting. When Naughty Joker was ready, we practised PMR techniques and Naughty Joker requested me to imagine his safe place.

Facing challenges and reclaiming future

In session 9, I told the story of Monster, then I invited Naughty Joker to play the character of the monster. We shouted as a monster, walked like a monster; then I invited Naughty Joker to draw the monster in his mind. We discussed the monster, the character, the natural self, the influencer, the guidance, the temper, the strength and the weakness of the monster. We wrote what we discussed into a "monster mandala" (the personal mandala was developed by Professor Sue Jennings and I used the monster as a metaphor). Naughty Joker requested we play with clay dough; Naughty Joker used the clay to create the monster, then he played with the clay as if the monster was walking towards him. He said the monster is angry all the time; sometimes the monster even lost control with his own aggressiveness. Then, Naughty Joker

Figure 16.1 "There is a little good deed in the evil; there is a little devil in a good deed"

transformed the monster into a rose. He said: "*Kalau jadi ros, dia tak kan marah lagi, dia akan jadi peace*" (If the monster becomes a rose, he will no longer be angry and he will become peaceful).

In the last session, I planned to invite Naughty Joker to free draw reclaiming his future. The Naughty Joker drew a Tai-Chi Dragon where he wrote "There is a little good deed in the evil; there is a little devil in a good deed".

Conclusion

Over all the sessions with Naughty Joker I felt his ability to be cooperative and tolerant improved. Indirectly, the sessions taught me that rapport and trust can make a difference to both me and the client. "When rapport is there, trust is there, then hope follows" this is the greatest thing I have learnt from these sessions with Naughty Joker.

As stated in the beginning, counsellors and practitioners in Malaysia need to be more culturally sensitive, and undoubtedly using play is helpful in counselling sessions. Nevertheless, training and supervision for using play as a therapeutic tool in Malaysia is still in high demand and in its infancy. Hopefully, one day there will be more trained play therapists or play practitioners and there will be complete lawful protection for all the mental health professions in Malaysia.

Note

1 An essential evidence-based tool for identifying feelings in different intensities, it allows users to explore the different strength of feelings through the scale 0–10.

References

Baharudin, D. F. (2008). Counselor trainees perception and understanding on multicultural counseling in Malaysia. *PsycExtra Dataset*. doi:10.1037/e501382012-025

Balakrishnan, N. (2017, May 29). *What the Malaysian constitution says about your religious rights and freedom*. Retrieved from https://says.com/my/lifestyle/religious-rights-you-have-as-a-malaysian

Bedford, S. (2018, February 26). *A guide to religion in Malaysia*. Retrieved from https://theculturetrip.com/asia/malaysia/articles/a-guide-to-religion-in-malaysia/

Bee, O. J., & Leinbach, T. R. (2019, June 13). *Malaysia*. Retrieved from www.britannica.com/place/Malaysia/Religion

Carlson, R., & Arthur, N. (1999). Play therapy and the therapeutic use of story. *Canadian Journal of Counselling, 33*(3). Retrieved from https://files.eric.ed.gov/fulltext/EJ603020.pdf

Chan, M. (2016, October 17). *Don't let religion stand in the way of helping Orang Asli*. Retrieved from www.freemalaysiatoday.com/category/opinion/2016/10/17/dont-let-religion-stand-in-the-way-of-helping-orang-asli/

Constitution of Malaysia 1957. (n.d.). Retrieved from www.commonlii.org/my/legis/const/1957/

Corey, G. (2009). *Theory and practice of counseling and psychotherapy*. Belmont: Thomson Brooks, Cole.

Dipananda, B. D. (n.d.). *Features | themes | art and archaeology*. Retrieved from www.buddhistdoor.net/features/reflections-on-buddhist-sites-in-malaysia

Foong, L. M. (2018). *Malaysia's failing mental healthcare system*. Retrieved from http://sea-globe.com/failing-mental-healthcare-in-malaysia/

Hall, T. M., Kaduson, H. G., & Schaefer, C. E. (2002). Fifteen effective play therapy techniques. *Professional Psychology: Research and Practice, 33*(6), 515–522. doi:10.1037/0735-7028.33.6.515

Hays, J. (n.d.). *Islam in Malaysia*. Retrieved from http://factsanddetails.com/southeast-asia/Malaysia/sub5_4c/entry-3633.html

Islam in Malaysia – Islamic Tourism Centre of Malaysia. (n.d.). Retrieved from https://itc.gov.my/tourists/discover-the-muslim-friendly-malaysia/islam-in-malaysia/

Islamic Department of Malaysia. (1984). *Akta undang-undang keluarga Islam, wilayah persekutuan, 1984: Act 303 = Islamic family law, federal territory, act 1984*. Kuala Lumpur: MDC.

Jaladin, R. A. (2013). Barriers and challenges in the practice of multicultural counselling in Malaysia: A qualitative interview study. *Counselling Psychology Quarterly, 26*(2), 174–189. doi:10.1080/09515070.2013.793046

Jennings, S. (2004). *Introduction to developmental playtherapy playing and health*. London: Jessica Kingsley Publishers.

Jennings, S. (2010). *Storybuilding: 100 ideas for developing story & narrative skills*. Brackley: Hinton House.

Jennings, S. (2011a). *Healthy attachments and neuro-dramatic-play*. London: Jessica Kingsley Publishers.

Jennings, S. (2011b). *The anger management toolkit: Understanding & transforming anger in children & adolescents*. Brackley: Hinton House.

Jennings, S. (2013a). *101 activities for social & emotional resilience*. Buckingham: Hinton House.

Jennings, S. (2013b). *101 ideas for managing challenging behaviour*. Brackley: Hinton House.

Jennings, S. (2017a). *Creative play with children at risk*. London: Routledge.

Jennings, S. (2017b). *Creative storytelling with children at risk*. London: Routledge, Taylor & Francis Group.

Landreth, G. L. (2002). *Play therapy: The art of the relationship*. New York: Brunner-Routledge.

Law Reform (Marriage and Divorce) Act, 1976: Act 164. (1981). Kuala Lumpur: Jabatan Percetakan Negara.

Mahidin, M. U. (2019, July 15). *Department of statistics Malaysia*. Current Populations Estimates, Malaysia, 2018–2019. Retrieved November 19, 2019, from www.dosm.gov.my/v1/index.php?r=column/cthemeByCat&cat=155&bul_id=aWJZRkJ4UEdKcUZpT2tVT090Snpydz09&menu_id=L0pheU43NWJwRWVSZklWdzQ4TlhUUT09

McLeod, S. A. (2019, January 11). *Cognitive behavioral therapy*. Retrieved from www.simplypsychology.org/cognitive-therapy.html

Nalanda. (2012, July 2). *Buddhism in Malaysia*. Retrieved from www.nalanda.org.my/buddhism-in-malaysia/

Nicholas, C. (n.d.). *Similarities we often overlook: Christianity and Indigenous religion compared*. Retrieved from www.coac.org.my/main.php?section=articles&article_id=36

Nobuta, T. (2007). Islamization Policy toward the Orang Asli in Malaysia. *Bulletin of the National Museum of Ethnology*, *31*(4), 479–495. Retrieved from http://citeseerx.ist.psu.edu/viewdoc/download;jsessionid=24CCBCA8B76BAE4ACF5042B7B5773DCB?doi=10.1.1.625.6689&rep=rep1&type=pdf

Prendiville, E., & Howard, J. (2014). *Play therapy today: Contemporary practice with individuals, groups and careers*. Abingdon: Routledge.

Raja Mohan, V. S. B., & Sorooshian, S. (2012). Counseling in the context of Malaysian living style. *Australian Journal of Basic and Applied Sciences*, *6*(10), 292–296. Retrieved from http://ajbasweb.com/old/ajbas/2012/October/292-296.pdf

Samsudi, S. M. (n.d.). *Understanding Islam in Malaysia*. Retrieved from www.ikim.gov.my/index.php/2019/01/28/understanding-islam-in-malaysia/

Selva, J. (2018, March 8). *Albert Ellis' ABC model in the cognitive behavioral spotlight*. Retrieved from https://positivepsychology.com/albert-ellis-abc-model-rebt-cbt/

Sue, D. W., & Sue, D. (2003). *Counselling the culturally diverse: Theory and practice* (4th ed.). New York: John Wiley & Sons.

Tekin, G., & Sezer, Ö. (2010). Applicability of play therapy in Turkish early childhood education system: Today and future. *Procedia – Social and Behavioral Sciences*, *5*, 50–54. doi:10.1016/j.sbspro.2010.07.049

Wong, D., & Ngu, I. T. (2014). A "double alienation": The vernacular Chinese Church in Malaysia. *Asian Journal of Social Science*, *42*, 262–290. Retrieved from https://umexpert.um.edu.my/file/publication/00012892_123613.pdf

17

THE ISLE OF SILLINESS

Play and dramatherapy with addicted Palestinian men in Israel as a way of balancing in patriarchal society

Natalie Hayek Damouni

Introduction

In Arabic, there are seven different terms to refer to "play", which are more likely to include the word "game" than be a reference to "theater". Play therapy is often considered childish, and thus requires specific adaptations to make it approachable for adult participants. This is especially the case with Palestinian men suffering from substance addiction.

This chapter aims to shed light on addicted Palestinian men in Israel and the complex reality they live in. It presents a unique process of dramatherapy in the contexts of different unstable realities: the Israeli-Palestinian conflict; Palestinian society's shift from traditionalism to modernity; and the challenges to masculinity and patriarchy within Palestinian society. The chapter shows how dramatherapy creates an "island" of balance in the psychological, interpersonal, and social spheres despite these realities through plays and "silliness". It finds that spirituality is an important factor that must be taken into account when addressing the problem of drinking and when planning successful treatment programs for this group. This observation reaffirms the critique made by Haddad (2004) on why conventional rehabilitation programs for alcohol addiction, which utilize a Western approach, tend to show low success rates (around 3% only).

First, the chapter presents specific issues regarding addiction in Palestinian society in Israel, and creative therapies that have been proposed to address addiction. These subjects were challenging to research as there is almost no literature or materials about them in the Arab world. Second, it presents a case study from fieldwork dramatherapy with addicted men, which I personally facilitated, and through it emphasize the play and playfulness as seen in dramatherapy sessions with addicted Palestinian men. The learning from it was meaningful and the process contributed to the addicted men and their families, as well as spreading the word on dramatherapy in Palestinian society in Israel. The case study has not been published or reported anywhere. All identifying details (names etc.) have been left out for anonymity, and permission was obtained to use the materials for this chapter. Third, it discusses how these men reacted to the different elements of play presented in the case study, and how they contributed to bringing change to their lives.

Addicted Palestinian men in Israel

Palestinian society in Israel

Palestinians are part of Arab people, and share the collective experiences of other Arabs. They lost their homeland to Israel in 1948; while most became refugees and exiles, about 150,000 Palestinians remained inside the new state's borders, and were eventually given Israeli citizenship (Dwairy, 1998). According to the Israel Central Bureau of Statistics (www.cbs.gov.il/he/pages/default.aspx), Palestinian Arab citizens of Israel now number nearly two million people, or 20% of the state's citizenry, with a majority of Muslims (82%), and two smaller groups of Christians and Druze (9% each).

In the hyper-nationalistic Israeli-Palestinian scene, violence is a central mode of behavior. Yet within this complex place, the Palestinians inside Israel are left with no defined role. Despite their citizenship, they are not drafted into Israel's armed forces; and at the same time, they cannot join organized Palestinian armed resistance either. They are at once potential traitors to their national group, and a potential fifth column to their state (Sa'ar and Yahia-Younis, 2008).

Tribalism, Islam and exposure to the West have shaped the collective culture of Arabs. Social norms and values predominantly determine people's behavior more so than personal decisions. The choices that individuals face are often between adherence to society's rules in order to gain familial support, or relinquishing their support in order to pursue individual self-fulfillment (Dwairy, 2006, p. 28). According to Abu Asbe and Abu Nasra (2015), Palestinian society is undergoing a process of change from a conservative society to a modern one. Sa'ar and Yahia-Younis (2008), on the other hand, argue that the Palestinian social field is informed by several meta-narratives, all distinctly modern but which in their institutional forms all reinforce male domination. These narratives yield several models of masculinity, all of which assert hegemony and which are translated into scripts of conduct.

A well-known Arabic motto, "*Al Ard Wal A'ard*" – land and honor – reveals an important layer of the Palestinian man. Massad (1995) points out a very interesting concept regarding paternity and nationality: after the Palestinians were stripped of their land, the essence of being a Palestinian completely changed. Instead of identity being linked to one's birth in the mother country, "territory was replaced by paternity". Here, according to Walters (2010), Massad uses the imagery of rape as having robbed the motherland of her ability to produce Palestinians. An implication of this imagery is that Palestinian men were unable to protect their women, and thus are unable to correctly do their duty. This failure is the epitome of effeminization, and is further aggravated by the notion that the rape was committed by a colonizing force. Sa'ar and Yahia-Younis (2008) suggest that one is not a "real man" in the eyes of Palestinians if they lack the ability to protect their family and homeland.

Palestinian society in Israel has several core cultural features:

1 **Harmony with nature and the supernatural.** There exists a strong belief that Allah/God controls the destiny of every individual,[1] and that prayer heals. Dwairy (2006) claims that listening to authority and understanding the holy texts (*Qur'an and Hadith*) are the main epistemological approaches to gaining knowledge and finding truth. Arab clients frequently refer to traditional healers to seek help and integrate modern treatment methods with traditional healing methods, which has become more recommended in order to achieve greater success (Rosca, Budwski and Haddad, 2015).
2 **Individualism and collectivism.** According to Dwairy (2008), because the Israeli state does not provide all the basic needs of its Palestinian citizens, the family is still the main

institution that ensures these needs. The majority of Arabs live with a dominant collectivism and thus tend to behave more according to social norms than personal opinions. Therefore, the "self" is not differentiated from the familial collective identity (Dwairy, 2006).

3 **Gender roles.** The social structure is patriarchal with male domination, and women's social statues is contingent on being married and on motherhood.

4 **Self-control and responsibility.** Individuals are expected to show good temperament and patience, to avoid showing distress or despair even in critical periods, and to bravely cope with painful moments in life.[2]

5 **Somatization and use of metaphors and proverbs.** Arab patients' communication is impersonal and restrained, and makes use of idioms of distress relying on a complex system of metaphors and proverbs. When explaining issues of health and disease, some tend to place emphasis on an external focus of control, including the intervention of the supernatural, through angels or *jins*, or other people's involvement with the supernatural by means of the evil eye or sorcery.[3]

Addiction within Palestinian society in Israel

Substance addiction results from cultural, social, psychological, behavioral, and biological factors (IADI, 2006).[4] Palestinian society in Israel has experienced a gradual rise in the use of psychoactive substances, including alcohol, especially among adolescents. This trend is mainly attributed to the society's transformation since the 1970s from a traditional conservative culture to a more modern secular one (Abu Asbe and Abu Nasra, 2015; Rosca, Budwski and Haddad, 2015). Another possible reason for the increased use of alcohol and drugs is Arab's lower socio-economic conditions compared to Jews.[5]

Drinking alcohol, which is prohibited in Islam, a religion followed by the majority of Arab society, is considered a social deviance and a taboo. As such, alcohol is not consumed in public but in secret – in hidden places such as orchards or olive fields, often together with drugs; in cars, quiet street alleys and "under the table" at weddings (Rosca, Budwski and Haddad, 2015). In recent years, alcohol consumption has become more frequent among Muslim adolescent boys and young men, who are developing their own character and see drinking as a *rite de passage* from childhood to adulthood – as a measure of assertiveness, independence and power (Rosca, Budwski and Haddad, 2015). Once the taboo border is crossed, they can also move to drugs.

All the preceding factors can put the Palestinian man in a very challenging position, in which he is supposed to take control, stay balanced and behave, protect the woman and the land, and in many cases suppress early childhood experiences, all of which can increase the conditions for substance addiction.

Creative therapies and theatre of the oppressed for addiction

Addiction is both mental and physical; once the stimulus is removed, withdrawal symptoms can be both physically (shaking, sweating, etc.) and mentally (memory, lapses, paranoia, depression, etc.) challenging.[6] The use, abuse and finally addiction to a chemical causes the addict to become "dis-spirited." The role of the arts therapist, therefore, is to help patients regain their lost spirits through creative self-expression (Johnson, 1990).

Zogafrou in Gordon et al. (2018)[7] finds that there is a "beneficial effect of dramatization and dramatic play in improving addicts' spontaneity and ability to self-disclose and self-regulate" (p. 118). A similar conclusion was reached by Bailey in Gordan et al. (2018) who posited that addicts tend to be rebels, and thus fun theatre games might capture their interest in a way that bypasses psychological resistances. In addition, dramatherapy helps addicts connect to, and come to trust, their bodies.[8] According to Newman, T. (2017), during active addiction, the addict intoxicates themselves to reach a sense of disconnection from overwhelming feelings and intrusive thoughts. Dramatherapy offers a safe place where this re-connection can be explored and different emotions can be contained in a healthy manner. Working with addiction in Palestinian society requires special adaptations. The Theatre of the Oppressed was a natural path for Palestinians addicts in Israel. Theatre of the Oppressed creates spaces of liberty where people can free their memories, emotions, imaginations, thoughts of their past and present, and can invent their future instead of waiting for it (Boal, 2002). According to Johnson (1990), the creative-shaman-therapist possesses the tools needed for their liberation: the play space and the stage. These tools give us the needed esthetic distance and the chance to bring some physical form to our experiences.

Creativity is an antidote to shame, connecting us with our Creator-Higher Power and our true selves, allowing us to turn liabilities into assets, and wounds into pearls. Dramatherapists can navigate this exploration through the use of various creative methods, such as storytelling, role-playing, mask-making, playmaking, and developmental transformations, to name a few. Through character roles, letter writing, sculpting and imagination exercises, participants can unlock repressed memories that create unhealthy core beliefs. For addicts, this allows them to empower themselves in relation to their addiction and, most importantly, feel they have been seen, accepted, and regained a healthy group identity (Newman, 2017).

A helpful strategy for normalizing paradox and tolerating contradiction, which can be a therapeutic intervention in the treatment of addiction, is clown therapy, as presented by Gordon, Shenar and Pendzik (2018). Internal contradiction causes the clown to be in a state of perpetual movement between opposites, making inner paradox a form of logic: the logic of flexibility and incessant process. As a role that does not seek resolution to paradox, the clown establishes a dynamic equilibrium that supports self-regulation, allowing individuals to free themselves from deviant behavioral patterns, and to hold the future possibility of individuation. Applying it with addicts, for whom their body is not to be trusted, this kind of work was extremely healing.

The dramatherapist helps to create balance in the client's situation by engaging in a process of collaboration, using creative activities which let them find the full potential to deal with the challenge of personal change.[9] The role of the creative arts therapist in the treatment of addictions is to call, assist, guide, model, encourage, and inspire creativity and self-expression. "The drama therapist encourages patients struggling with their addictions to share themselves through their creativity" (Johnson, 1990, p. 299) and creates a safe play space for exploration. "Our patients stand, paralyzed, at the door of their shame, afraid of what dragons will emerge if they dare to open it" (Johnson, 1990, p. 299).

With collective societies in particular, this work can involve, as Daccache (2016) puts it, "self advocacy", as these marginalized groups hold up a mirror to society that is implicated in the oppression that foments to crime, chaos and madness. A key goal of treatment is thus to help the client establish "a new relationship to one's self; a more compassionate understanding of [others] facing similar struggles; and a new sense of being a valuable contributor to the greater community" with the ultimate goal being recovery (NADTA, 2019, p. 5).

Case study

In this chapter, I will present a case study and through it emphasize the play and playfulness as seen in dramatherapy sessions with addicted Palestinian men. The case took place nine years ago, in 2011. All identifying details (names etc.) have been left out for anonymity, and permission was obtained to use the materials for this chapter.

The group

The group consisted of eight participants between the ages of 20 and 60 years. All of them were male, six Muslims and two Christians. All had undergone physical withdrawal from substance abuse, and were monitored to ensure there had been no relapse. In addition to a range of personality disorders, there was often a history of criminal behavior due to the addictive habits. Some voluntarily chose to enter therapy; others were required by court to undergo treatment to avoid jail sentences.

The sessions took place in a regional day-treatment rehabilitation center for Arabs in northern Israel, run by the municipality. The center was located in a central area of town, on the ground floor of a rundown building, which is physically located underneath the street, resembling a parking area or garage. All the rooms had open windows from the top, and sounds could go through all different spaces of the center. The staff members included a director of the unit, a professional social worker, and a mentor/guide who was a "clean" addict himself. The rehabilitation process went on for nine months – a striking resemblance to pregnancy and birth. The daily program was strict and intensive, and most participants did not have time for additional work. The group members cooked their own meals and cleaned the center themselves.

The process of play and drama

The process consisted of 30 weekly sessions, two hours each. Before we started the therapeutic process, I was on active observations for six weeks to get acquainted with the group and its members; hear "their" language and see their behaviors; also identify group dynamics, roles and other patterns. I remember my mixed feelings about entering this process: insecure, tense, suspicious, enthusiastic, foreign, lonely. My biggest feeling was fear, since I was the only female at the center, and was young, which placed me in a vulnerable position.

During the first five sessions, I introduced icebreaker games. However, each attempt to move/play in the room was responded to with laughter. For example, an invitation to touch a color or a material, or even to take a deep breath and exhale, was taken as a joke by some members. It took them about ten minutes to understand that this was a serious request. After these exercises, when we went back to talking in small groups, all of them were engaged, and even showed creativity in putting the issues we talked about on stage. At this point, the group members were still suspicious and tended to fail every action and initiative I requested of them before even trying. At the end of the second session, one of the participants, in his sixties, said it was a waste of time and that he would not come back to the meetings. Yet he came back; and as we got to know each other better, we managed as a group to build a safe place in which they allowed themselves to explore their issues.

We began the process with a ritual. I said: "Imagine there is a magic string pulling you up; I invite you to enter the session today by tuning your personal strings,[10] and then the group's strings can get ready to work". The playfulness of the ritual was observed in the way in which they used their imagination; gradually, with slight physical movements, moving their hands and heads, and later adding sounds. The purposes of the process were stated at this stage: a) to raise the group members' awareness concerning their vulnerability on account of their past, and how it relates to their daily lives in the present; and b) to widen their current personal choices in creative ways.

In the fifth session, they began allowing themselves to play and to move more, after explaining the purpose of each activity. I suggested a game called "the Colombian hypnosis" by Augusto Boal (2002):

> One actor holds his hand palm forward, fingers upright, anything between 20 and 40 centimeters away from the face of another, who is then as if hypnotized and must keep his face constantly the same distance from the hand of the hypnotizer, hairline level with her fingertips, chin more or less level with the base of his palm. The hypnotizer starts a series of movements with his hand, up and down, right and left, backwards and forwards, his hand vertical in relation to the ground, then horizontal, then diagonal, etc. – the partner must contort his body in every way possible to maintain the same distance between face and hand, so that face and hand remain parallel. If necessary, the hypnotic hand can be swapped; for instance, to force the hypnotized to go between the legs of the hypnotizer. The hand must never do movements too rapid to be followed, nor must it ever come to a complete halt. The hypnotizer must force his partner into all sorts of ridiculous, grotesque, uncomfortable positions. His partner will thus put in motion a series of muscle structures which are never, or only rarely, activated. He will use certain "forgotten" muscles in his body. After a few minutes, the two actors change, the follower and the leader. After some more time, both can extend a hypnotizing right hand, becoming leaders and followers at one and the same time.
>
> *(p. 51)*

The group members started to move after many hesitations. After a few minutes, they appeared to have fun, and laughs could be heard around the room. A few minutes later, we extended the game to have followers and leaders at the same time. The game raised a lot of anger for one member who was a follower and a leader: he stopped and screamed "I'm not anyone's dog, this is bullshit!" His stopping of the game disconnected the chain of the others. Later that day, we learned from the director of the unit that playing the "game" evoked in him a forbidden act he did, involving collaborating with people from the jail. Prisoners had used him to move processes out of the jail, and he felt trapped in this game. A playful group icebreaker woke up his unconscious demons, which may not have been revealed in another way. This incident was hard for other members of the group, with some raising questions about loyalty to the group.

In the sixth session, a new member joined the group and the roles became emphasized: new vs. experienced, one vs. group, past vs. present, and more. At this session they realized the importance of the dramatherapy group in their lives. This indicated that the dramatherapy sessions were different from their daily program at the center, as it gave them the opportunity to be spontaneous and bring out their fears; their resistance especially against the staff; and their unconscious self-limits and interpersonal conflicts. This experience made the weekly sessions "an Isle" in their weekly program.

In the tenth session, the group had a small party before the Christmas season. I suggested the game "*Al H'ilo lal Hilo*",[11] in which each member dedicates a chocolate to a good or "sweet" trait of another member. This marked a step forward in the group members' relationships: they saw it as a sincere and true expression from people who knew them. Some had tears in their eyes. Some said that they had forgotten the last time someone had given them a gift or said something nice and honest to them. This "game" taught them a new or forgotten language they had not heard in years. The words "dedication", "sweet" and "good" were frequently used, and it was clear that these words were felt by everyone in the group.

The main theme explored during sessions 11 to 18 was "polarity". This theme was revealed in the group's icebreakers, and showed how hard it was for them to stay only on one "pole". In the twelfth session, I introduced the story "Hassan and the Ghoul"[12] as a projective tool. The story is about a young boy named Hassan who lives in a small village near a big mountain. The villagers believe a Ghoul lives on the mountain, although none of them have ever seen, heard or smelled it; yet they are scared of it. Hassan grew tired of this and decided to go to the mountain. As he reaches the mountain top, he meets the Ghoul, who is exactly as the villagers described him: three eyes, long sharp claws, bad smell, and hair covering its whole body. Just as Hassan starts regretting his adventure, he sees the Ghoul run and hide behind a rock; the Ghoul was afraid of Hassan. The two talk about their fears and differences, and then befriend and play together.

In this session, I stopped the story at the stage where Hassan reaches the mountain top and turns his head to see the Ghoul. The group was then divided into smaller groups, and participants were instructed to continue the story as they imagined it, and then perform it front of the whole group. The first subgroup suggested that the Ghoul eats Hassan, which was what the villagers believed; the second subgroup proposed that there was no actual Ghoul, and that it is a pity that the villagers are not enjoying the beautiful view from the top; and the third subgroup suggested that Hassan faces and confronts the Ghoul directly.

A new member joined the group after this session. In the ensuing sessions, the members raised many conflicts of their inner selves and with the other members. The polarities derived in part from their personality disorders, from "unfinished business" from their early childhood, and also the 'need to be needed' and loved. It took two months between sessions 19 and 27 for them to face and analyze their childhood memories and how it affects their daily lives today. The last sessions 28 to 30 served as a closure, reflecting on the personal and group processes and saying goodbyes. In evaluating the processes, they explained their feelings at the beginning of the dramatherapy: silly, foolish, time-wasting, and meaningless. Then, over the year, the sessions helped them explore new roles, reveal hidden opportunities and better "confront" difficult issues in their real lives outside the center, especially regarding childhood "ghouls". The dramatherapy thus became a "corrective experience" to address unresolved issues from their pasts.

Conclusion

The nine-month-process in the rehabilitation center resembled rebirth for many of the participants. Like children, the playspace created within the dramatherapy sessions allowed the adult members to explore their own emotional vocabulary; learn the skills necessary for emotional regulation; and develop healthy coping strategies (NADTA, 2019). This rebirthing implies the death of the false addicted self, the "ego" that Bill W.[13] refers to, and the discovery of the true self. Whereas the ego-self is associated with defensiveness and attachment, the Self is associated with creativity and love (Johnson, 1990).

Over the course of the sessions, members were able to explore their younger self, their present addict self, and their future self through role playing. The expressions of vulnerability, sensitivity and reassurance also empowered them by feeling they were heard. The process helped them improving resilience and self-worth. One of the participants said that he became the monster himself. By playing a monster in the dramatic processes, it helped him "re-create" his post-addiction identity. Dramatherapy served to peel off the mask and reveal the silliness and childhood playfulness. As stated by Johnson (1990), miracles can happen when addicts open themselves and discover the beauty of their inner nature. In this case, miracles were possible through playfulness.

Playfulness in dramatherapy sessions with addicted adult men provided the opportunity to explore different roles, speak differently, and move in space and time differently. Such exploration was possible in the case study in three main ways:

1. A story as a classic game of depth in which the members collaborate and work in esthetic distance. According to Landy (1996), acting out a role or situation can achieve a balance between "overdistancing" or denial of painful emotions, and "underdistancing" or becoming overwhelmed with shame and guilt. Through the story of Hassan and the Ghoul, the participants let themselves become Hassan, hence becoming children again, playing, exploring, and laughing. It brought back childhood memories that also brought out the members' own childhood monsters. One subgroup contradicted the story of the Ghoul as a monster; another verified it; and another laughed at it – thus all facing their own "Ghouls" and monsters in different ways.
2. The metaphoric language they use. According to the biopsychosocial model of metaphors by Dwairy (2006),[14] psychological experiences such as anxiety, anger, or happiness are translated through an unconscious process of symbolization. They are influenced by the personal as well as the collective experience, so that many cultural-specific proverbs, idioms and myths become incorporated in the personal images and metaphors. The Arabic language is very metaphoric and the Qur'anic verses include many metaphoric directives (Hourani, in Dwairy, 2006, p. 138). According to psychoanalysis, metaphors are symbols that can be interpreted to gain insights and self-awareness. Alternatively, one may consider metaphorical language to be another legitimate mode of communication which influences the mind and body. As such, metaphors may be utilized in therapy to achieve changes in clients' experiences, without directly bringing unconscious forbidden contents to the consciousness. Playing with the language metaphorically was very helpful for the participants to face their own inner language and dare to bring taboo issues outside their bodies and minds.
3. Icebreakers, taken mostly from Boal's, and psychodrama techniques. The use of the body was threatening for the participants at the beginning, as it was not trusted. Moving in space, adding sounds, and "playing" was also seen as childish and silly. The word "play" in Arabic also has sexual connotations, which is seen as taboo in Arab society and should be dealt with sensitively, especially since some of the participants were sexually abused in their childhood. These games helped them gain a lot in terms of developing a sense of empathy, by putting themselves in the shoes of another, and also in re-trusting their own bodies.

Few of the participants had ever done theatre or any kind of therapy before, but were excited about the chance to express themselves. Because their weekly rehabilitation program was strict and intensive, having an alternate dramatic reality once a week gave them a safe place – an Isle – which enabled them to raise authentic feelings, behaviors and thoughts without being judged. Gordon, Shenarb and Pendzik (2018) claim that while children process and explore life through dramatic play, adults require a theatrical, social or dramatherapy to do so.

Although the story of Hassan and the Ghoul wasn't a specific game from Boal, different techniques and connections from the Theatre of the Oppressed were made at the first stage of the process (e.g. Colombian hypnosis), which prepared the ground for revealing new ways of thinking, feeling, and moving in the space; to liberate themselves from the past; investigate all their polarities; and face the different Ghouls in this "Isle of Silliness". These spaces of liberty enabled them to find the balance they were seeking. Hassan (and other roles) reminded them of happier possibilities and helped them restore their inner and outer balance, enabling them to function better as individuals. At the personal and group levels, the dramatherapy strengthened their resilience, self-esteem and self-advocacy outside of the sessions during the week and later outside the rehabilitation center.

Notes

1 Kanaa, in Rosca, Budwski and Haddad, 2015.
2 Baraka, in Rosca, Budwski and Haddad, 2015.
3 Al-Krenawi and Graham, in Rosca, Budwski and Haddad, 2015.
4 Israeli Anti-Drugs Institution, 2006 (https://www.gov.il/he/departments/antidrugs).
5 Tal-Sapiro, in Rosca, Budwski and Haddad, 2015.
6 Addiction and recovery, in Newman, 2017.
7 In Gordon, Shenar and Pendzik, 2018.
8 In Gordon, Shenar and Pendzik, 2018.
9 http://website.telhai.ac.il/?cmd=programs.419.
10 A term resembles the Eastern musical tool "Oud" and tuning its strings in order to get better playing experience and sounds.
11 The sweet to the sweet.
12 Ghoul – A male Monster, by Taghreed Al-Najjar (1987), *Dar Elfata Ala'rabi*, Beirut, Lebanon.
13 Whitfield, in Johnson, 1990.
14 In Dwairy (2006).

References

Abu Asbeh, K. & Abu Nasra, M. (2015). *Perceptions and attitudes of parents and youths regarding the use of psychoactive substances among Arab society in Israel.* Research Report. Massar Institute for Research, Planning and Social Consulting, Jatt.

Boal, A. (2002). *Games for actors and non-actors*, 2nd edition. New York and London: Routledge.

Daccache, Z. (2016). The unheard stories of those forgotten behind bars in Lebanon. In: Pendzik, S., Emunah, R. & Jonson, D. (Eds.), *The self in performance: Autobiographical, self-revelatory and auto-ethnographical forms of therapeutic theatre*, pp. 227–239. New York: Palgrave Macmillan.

Dwairy, M. (1998). *Cross cultural counseling: The Arab Palestinian case.* New York: Haworth.

Dwairy, M. (2006). *Counseling and psychotherapy with Arabs and Muslims: A culturally sensitive approach.* New York: Teachers College Press, Columbia University Press.

Dwairy, M. (2008). *Culture analysis and metaphor psychotherapy with Arab-Muslim clients.* Israel: Oranim Academic College.

Gordon, J., Shenarb, Y. & Pendzik, S. (2018). Clown therapy: A drama therapy approach to addiction and beyond. *The Arts in Psychotherapy*, 57, 88–94.

Haddad, W.(2004). *Use if alcohol and drugs among the youth in Israel and in the Arab sector.* Personal Communication, National Anti-Drug and Alcohol Authority (IADA). Jerusalem, Israel.

Johnson, L. (1990). Creative therapies in the treatment of addictions: The art of transforming shame. *The Arts in Psychotherapy*, 17(4), 299–308.

Landy, R. (1996). The use of distancing in drama therapy. In: *Essays in drama therapy: The double life*, pp. 13–27. London: Jessica Kingsley Publishers.

Massad, J. (1995). Conceiving the masculine: Gender and Palestinian nationalism. *The Middle East Journal,* 467.

NADTA (North American Drama Therapy Association). (2019). *What is dramatherapy.* London: NADTA Publication.

Newman, T. (2017). Creating the role: How dramatherapy can assist in re/creating an identity with recovering addicts. *Dramatherapy*, 38(2–3), 106–123.

Rosca, P., Budowski, D. & Haddad, W. (2015). Culture-specific characteristics of alcohol related disorders and culture-sensitive approach in Ethiopian immigrants and ethnic Arabic groups in Israel. *Journal of Socialomics*, 4(2).

Sa'ar, A. & Yahia-Younis, T. (2008). Masculinity in crisis: The case of Palestinians in Israel. *British Journal of Middle Eastern Studies*, 35(3), 305–323.

Walters, K. (2010). The masculinity of occupation: The study of Israeli and Palestinian men. Berkeley: University of California Press.

18

GRANNIES ON THE RUN

Into the wild spaces of autistic play

Jonas Torrance

I am a dance movement psychotherapist who specialises in working with autistic children. My clients tend to be children with difficult and entrenched problems and it is common for school to be an exacerbating factor. Sometimes the children in question are hanging on to their educational placement by a thread, or they have already voted with their feet and have joined the unhappy list of 'school refusers' (Garfi 2018: 17). It felt refreshing then to receive a referral from the teachers and parents of Vincent. He was no school refuser, in fact he positively threw himself into all aspects of school. He did his homework, he had friends in the playground, the teachers loved him and he was a long-serving representative for his class on the school council.

So why the referral? Vincent's teachers had concerns that although he was very engaged in school life, he was also becoming increasingly stressed. The school he attended was split onto two sites and at 14 years old, he was due to move to a large new site. He would then be entering his final two years before exams and leaving school and he was feeling the pressure of internal and external change in his life. I met Vincent's parents and they confirmed that he showed clear signs of hyperactivity at home. He had a diagnosis of both autism and attention deficit/hyperactivity disorder (ADHD) (Brown 2017: 11), and at times his liveliness and confusion about situations led him to be controlling and abusive at home. Sometimes Vincent's younger sister was at the sharp end of this behaviour, and his parents were very keen to help him through this transition at school without the stress spilling out more at home. I agreed, in the first instance, to make an assessment of Vincent and his needs and so began seeing him, in an empty and private lounge area in the school. It was early May 2017, and he was due to move school site in September.

Unlocking the story

When I met with Vincent he almost immediately pulled from his pocket a large padlock. In the other pocket he had a jangling bunch of assorted keys. One of the keys fitted and unlocked the padlock. The other keys were a miscellany, some old and antique looking, others appeared to be cheap plastic and metal keys to filing cabinets. Vincent knew what some of the keys unlocked, but others he had collected individually from markets and so on.

Apart from presenting me with a conundrum (the lock) and various solutions (the keys), the thing that most struck me was Vincent's lack of clear cognitive understanding as to why he collected keys and locks in the first place. He could say little more than 'I like them', but I noticed that he had a strong physical connection to the keys. He ran them through his hands and clipped them on a carabiner to his belt as though he needed their heft and movement at his waist. It would have been easy for me to get carried away with such a strong symbolic idea, as therapists tend to love presentations that point to potential subconscious 'action'. But I reckoned that if Vincent was not yet able to think very deeply about the keys, that I would stay with him in the cloud of not knowing, for now.

When we encounter therapy, all of us have uncertainties and vulnerabilities. Autistic children and young people are, in my experience, especially sensitive to observation and investigation (Torrance 2018: 50). Vincent demonstrated this to me by showing me his keys. As in art therapy, he was presenting a separate object (Malchiodi 2012: 70) for us both to examine in parallel, rather than enduring the spotlight of my gaze directly upon him. It's possible that Vincent's relationship with the keys could be described as an obsession. This term has sometimes been used clumsily with autistic children and adults. A true obsession develops a life of its own within its host. It self-perpetuates. The keys seemed to me to be more like companions for Vincent, a steadfast and reliable bunch of miniature, jangling friends who reminded him of who he was as he moved around the unpredictable environment of the school. Also, I noted, he was not alone. A friendly PE teacher who had noticed Vincent's keys got his own bunch out when they met around the school and they happily compared their collections. These encounters helped me recognise another part of the story for Vincent. Keys are strongly associated with power and knowledge. This young man wanted to be connected with the teacher and all he represented.

Attending to the body and movement

As Vincent talked with me he strode around the room. He walked rather like a medical consultant entering a ward. He moved swiftly and with purpose, looking from side to side, his voice was similarly purposeful and direct with an upper middle class accent. He gave the impression of absolute confidence, but there was also a slightly uncertain attention to weight as his foot hit the ground (Bartenief & Lewis 2002: 26), and he walked slightly too fast, with small steps. He was not clumsy, but there was a slight lack of control as he careered around the room. I stood up with him, but I moved slowly from position to position in the room, pausing in each place for some time in relation to his steady, constant movement. I had the sense that if I matched his pacing that we would both end up whirling around in a quickening walking talk. In the first session I suggested that we finish with some simple chi gung movements. We stood facing each other and breathed slowly in and out as we raised and lowered our arms in time with our breath. Vincent was completely compliant in the activity, but didn't seem very interested or curious about it. It also seemed to make very little difference to his pacey movements and speech.

In the second session I tried a different approach. Firstly I was not so directive, I simply produced a prop for us to play with. Secondly, I gave him a prop which stretched. It was a large thick band of lycra of about 3 metres (or a lot more if you stretched it) in circumference. This prop immediately had an effect on Vincent's quick staccato movements – it stretched out time for him as he pulled and pushed the lycra that he wrapped around his body. Fairly quickly he learned that it could be stretched and suddenly released, like a huge rubber band, shocking and unbalancing the partner (me) holding on to the other end. Like any young person he delighted in the power that he had over me and the feeling of risk introduced from the sharp twang of the lycra. Mainly however he became more interested in the band around his own body. He wrapped it tightly around his torso and his pacing reduced as he felt himself being held together by the material.

After a while Vincent started to run at the lycra as I held it up and swished it out of the way. When it was my turn, I made the horns of a bull with my hands and rushed in towards him. Suddenly Vincent was transformed, he became the brave bullfighter, pirouetting around the room as he escaped my attacks. The following week he knew what he wanted. He needed a prop, but the scene was set – for play.

The wild spaces of play

For the next year and a half my sessions with Vincent continued during term time. In that first autumn of 2017, we moved spaces to his new school and were fortunate to have a dance/drama studio to use which was quite large and dark with a black vinyl floor. At first I saw Vincent every week for an assessment period, but he was busy with classes, so we reduced the frequency of sessions to bi-weekly once he was settled into his new school. I always tell my clients that the assessment period is for both of us. As much as I may be checking them out, they are also doing the same with me. For both it's an opportunity to develop a way to be together. Vincent wanted to see me every week, but he also recognised that might create significant difficulties with his school education, so he compromised by seeing me less often.

Although the assessment period is usually around four sessions we both knew earlier than that what we would be doing in our sessions as we went forward. The bullfighting game quickly developed into a host of other scenarios and enactments. When we hit on a play theme which seemed significant, Vincent chose to repeat it in the next session. But the themes did not stay static, sometimes characters stayed the same whilst scenarios changed or vice versa and then at other times an entirely new idea developed from Vincent's inner world. Over the period of two years a great many play scenarios came and went but they are far too numerous to document here. To give a flavour of how things progressed between us, here are two examples.

Hockey grannies

Early on in our sessions in his first school placement, Vincent introduced the idea of an old woman. Originally she was trying to buy things in a shop but was hopeless – she couldn't hear properly and misunderstood what was going on. When this was pointed out to her she became immediately cantankerous and at times comically violent. We both played the old lady in turn, she was stereotypical, like many of Vincent's ideas she seemed to come straight out of a 70s television sitcom. After a while we both simultaneously became two old ladies. This provided a dynamic which was both friendly and antagonistic. His character, Edna, tended to correct my character, Doreen, who was hard of hearing and at times hopelessly stupid, however they were both able to agree about cake, especially their favourite – Battenberg.

When we moved to our second larger space, I introduced two sticks to our sessions. They were plastic, about 2 metres long and covered in a slightly spongy thin foam casing. They are used by martial artists to practise sparring with sticks in relative safety. These sticks began to be featured in nearly all the play scenarios which we developed. Vincent liked to have a prop to manipulate while he moved around the space 'decompressing' from the complex and busy school environment. At first Vincent and I talked about his week and issues at home as we simultaneously explored the drama studio, slicing through the air with the sticks. At times we threw and caught them in the air or slid them along the ground. At some point, Vincent started to go 'into character' as Edna and I reciprocated as Doreen. We often played a kind of hockey game with an improvised puck with Edna at one end of the room and Doreen at the other.

As the play developed, the two women fussed about who was better at hockey and took frequent breaks for imaginary tea and Battenberg cake. Quite often Doreen would have wardrobe malfunctions and would need to adjust her corsets by using increasingly unlikely implements, such as a wrench or a blow torch. The general atmosphere of the scenario was light-hearted, but there was also a steady undercurrent. Doreen never won at hockey and was the subject of constant jokey put downs from Edna. At times Edna would produce from her pocket the controller of Doreen's hearing aid and she would torture her by suddenly turning the volume up or down, sending Doreen into alternating states of confusion or shock. Edna's Battenberg cake was delicious, whereas Doreen's was made by her daughter and tasted so foul that the two of them would vomit.

The two grannies tended to get along better when there was an adversary. Usually these were imaginary young people who were trying to learn hockey. At around the time of the Summer holidays the grannies became air stewardesses, but I certainly wouldn't recommend the airline. The passengers were under the complete control of the grannies, and they operated a very strict like/dislike system of those that boarded the plane. If passengers were not approved of they were routinely thrown off the plane (sometimes in flight) or given terrible food. It didn't take long before the old enmity reappeared and my character, Doreen was ostracised and abandoned mid-flight. I decided that Doreen needed to fight back, so she found various ways to sneak onto the plane or disrupt Ednas tyrannical rule of law. But Vincent (as Edna) always found ways to eject Doreen and keep himself in the position of power.

So what's going on?

The contract that Vincent and I had developed was that of a shared, private space, where play and enactment were enabled alongside the more standard counselling activities of talking about daily life and it's difficulties. I did not express a preference for either activity, but it might be fair to say that our destination in the sessions was the shared play space. Vincent needed some time at the beginning of the session to settle in to this radical change from his school day. So talking about real life issues became our currency until he was ready to make the shift. At the end of the session, similarly, Vincent needed a little time to re-emerge. As we walked back to his class, we chatted more about everyday life. He needed to know that I was no longer Doreen and that he too could leave Edna behind. For a teenage boy this was very important. When working in this way with younger children, they quite often want to play and play, and to remain in the role of their chosen characters. At his age, Vincent was looking for both a way into the play space and a way out. To see me, as a trusted male adult, moving in and out of character gave him permission to do the same. Play and drama are by their nature mysterious. Their power lies in the mystery, so to analyse it too deeply with Vincent would have been unhelpful. He needed the place of power (Castaneda 1998: 17) that I provided and my trust in keeping it private. When I asked him once why we played, he simply replied: 'To get my emotions, my anger out'. At the time that seemed sufficient.

Doreen and Edna seemed to be the manifestation of a conversation that Vincent had been having inside his mind for a long time. Typically for an autistic person, he tended to see things in black and white (Paxton & Estay 2007: 74). Black and white thinking can be a very useful skill in decoding, making quick decisions, seeing through deception and so on. The downside is that inside a predominantly black and white orientated mind, things very often fall into the categories of good or bad, with little chance of being in-between. Vincent's Edna was in control of both the play and my character. Edna wasn't ever shown to be especially superior to Doreen, in fact they were very much a pair and always defined as friends, but in Vincent's mind one of them had to be on top, because equality was just too nuanced. To illustrate this, once Vincent told me of a new birth in his wider family. He was very happy about it and enjoyed meeting the baby, but he nevertheless felt that it was wrong that family members, on arrival enquired about the new baby immediately, rather than say hello to him first. Vincent simply felt that this was not the correct age order of things. If there were a group of children of varying ages, they should all take their turn, with, inevitably the baby coming last. Similarly, I once knew an autistic boy who always made his family enter into the house from the car in strict height order. The important point here is that the boy was not first or even second in his queue and Vincent would happily take his place behind others older than him. Autistic children are hierarchical to try and maintain order in a disordered world, rather than necessarily out of a desire to be in charge. But if, as in the case of the baby, someone disrupts the order, then an autistic child may fight to maintain it. In our play, once Edna was in charge, she inevitably began to oppress Doreen, because Vincent was trying to maintain what felt to him as *the natural order of things*. I realised that Doreen and Edna were playing directly with the relationship between Vincent and his younger sister as well as his own internal conflicts and feelings of confusion, inadequacy and 'disability'.

The prisoners escape

As time progressed through the weeks and months, other definable characters materialised with various attached scenarios. Many of the settings were militaristic, sometimes about battles, but also to do with drilling and parade ground politics. As before, there tended to be a hierarchy, Vincent played a Captain and I was a Sergeant Major. Quite often, soldiers would get thrown into prison for various misdemeanours and then after a while, a new scenario appeared which was set entirely in a prison. Initially, Vincent was a prison guard, whilst I was a hapless prisoner. This time I made my character less comical and more reactive, in order to see what would happen.

The main scenario was that of the dinner queue. Vincent rounded me up into the imaginary queue and then dictated what was being put on my plate as I went down the line. At first the food was standard and pleasant; meat, vegetables and so on, but at some point I was asked if I wanted something unpleasant, usually either custard on a savoury meal or gravy on the sweet. I tried politely refusing, but Vincent insisted. If I vigorously refused I was taken out of line and either shot on the spot with one of the plastic sticks, or sent to solitary confinement. Whichever option, I was either reborn or released in order to face the dinner queue once more. I tried changing the outcomes by accepting and eating the adulterated food but if I pretended to enjoy it, Vincent poured more horrible things on it. If I hated it, he cackled and summoned me once more to the back of the queue. It was a truly Kafkaesque scenario, I felt trapped in a never ending routine of torture by disgust. I rebelled again and was sent back to the cell, which was a pile of gym mats in a corner.

I was wondering how to shift this dilemma when Vincent once again unlocked the imaginary door to the cell. I immediately feigned resistance, but Vincent said in a conspiratorial whisper, 'No, it's *me*. I'm gonna get you out of here!' Suddenly, we were fellow prisoners both intent on escape. The advantage that Vincent had now was one he was willing to share: his keys!

We waited until night time, which was brought on by Vincent turning out all the lights. It was quite dark in the drama space, with just a little light coming in through the door in the opposite corner. The darkness added to our conspiracy, we talked in whispers as we tiptoed down the labyrinthine corridors of the prison. From time to time we had to pretend to be guards in order to get by unnoticed. Whenever we came upon a door, Vincent pulled out his keys and we stole through into the next imaginary space.

We replayed this scenario many times together over several months. At first we would often be caught and put back in our cell, only to try again when night fell. But after a while we started to succeed. One time we managed to escape the prison by finding a boat and navigating the marshes outside the prison at night. Using our sticks as oars we rowed into the darkness of the room, sculling our way to freedom and a new drama.

So what's going on?

When playing in such a dynamic way with a child, it's easy for a therapist to get lost. This can happen either in terms of the therapist's sense of self, or in terms of what's actually happening in the session. The primary concern here for me was one of commitment. Vincent needed both a strong adversary and a strong companion. I knew that I had to commit to the intensity of the play in order to maintain its importance for Vincent. In our sessions I decided to preference the authentic experience of play over analysis or cognitive styled 'problem solving'. However, in this laudable commitment to put play first, what would happen to me? Vincent was literally not taking any prisoners: I was shot, killed, tortured, imprisoned and worst of all consigned to Sisyphus-like tasks. It's hard to describe exactly what it feels like to be an adult in this kind of freefall play with an autistic child. Personally I often felt exhausted at the end of the sessions. Emotionally, the general 'light-heartedness' with which we approached the play helped me, humour meant that I could keep some sense of ease whilst at the same time committing to the drama of the activity. But, Vincent's hyperactivity rubbed off on me. He was a slim, strong teenage boy and he glided through an hour, whereas at the end of each session, I needed time to physically rest.

When ordinarily engaging in play, a father figure tends to thwart or amend the process of the child, to bring them towards a generalised way of understanding a situation through play. In this case with Vincent, I was deliberately sticking with his narrative and direction. I have had this experience with autistic children many times – at first they start to play 'reasonably', metaphorically as though near, but not over a cliff edge. But then they will usually, through play, fling themselves (and you) off the edge, often time and again in order to repetitively experience the visceral feeling of falling. When a 'neurotypical' (non-autistic) (Attwood 2007: 77) child plays, he/she too delights in the falling, but then will always reach towards a possibility of redemption, usually through imaginative and social problem solving.

With Vincent's prison play there were two things going on. Firstly he introduced me to a nonsensical and persecutory experience because it was a reflection of his own experience of being autistic in a neurotypical world. At school especially, he was being constantly bombarded by extreme sensory experiences which he was unable to understand. Stretch this experience over years and it becomes a constant, repetitive disruption in any autistic person's deepest sense of self. Every day Vincent was expected to mix sweet with sour, when he could only cope with one taste experience at a time. When we recognise this autistic sensory phenomenon, we also need to mix in the emotional response and the attendant sense of self. As previously demonstrated through Doreen and Edna, I also provided Vincent the possibility of seeing alternate versions of himself in this prison scenario. While working through being persecutory, he was also able to see the persecuted part of himself in me. It was hard, but sticking with it gave Vincent the opportunity to do what a neurotypical child would have done sooner. He reached out to me to help me, and in doing so he was saving himself from his own prison.

Bringing the outside inside

At some point in therapy with autistic children there needs to be a shift towards directly addressing the problem. In play we address the feelings around the problem. This has the effect of 'loosening up' the child enough to try and think and speak (if possible) about their everyday experience. Because I used this as a way of starting and ending our sessions, it was quite easy for Vincent to make that shift. At times when he had a pressing dilemma at school or home, he began speaking as we moved around the space for longer periods. At times, we talked and moved with the sticks for the whole session. His issues were often social. He struggled massively trying to 'decode' his peers around school. In many ways lunchtime, when he socialised with others, was for him the hardest lesson of the day. I wondered to myself if he felt as tired at the end of 'playtime' as I did at the end of our sessions together. As with most autistic children, he had ups and downs, but significantly his relationship with his sister improved and as he approached his last year of school he found an ideal placement to go to after his exams, where he would get close attention in small groups doing a practical subject that he loved.

When I first met Vincent he was, on the face of it, doing well. This was because he had socially and contextually worked out enough to be able to say and do the right thing at school. Luckily he had some very observant teachers and caring parents who noticed that something was amiss. They were right, Vincent had a blueprint of how to operate, but it didn't fit neatly on top of the real model that he was in. Once we spent some time together he was able to show me the subsequent tension he was experiencing through everyday life. For Vincent to be able to share that with another human being was enough. I liaised with his teachers and parents about some practical ideas and observations, but inside the drama studio he and I had found the tools to reveal and transform complex emotional feelings and ideas. Once Vincent knew that this space was there for him, he knew that he could make it through the confusing and distracted days ahead until the next session. He knew that there was somewhere where someone could help his vulnerable and crazy grannies and where he would be able to set his prisoners free.

References

Attwood, T. (2007) *The Complete Guide to Aspergers Syndrome*, London: Jessica Kingsley Publishers.

Bartenief, I. & Lewis, D. (2002) *Body Movement, Coping with the Environment*, Abingdon: Routledge.

Brown, T. (2017) *Thinking Outside the Box: Rethinking ADHD/ADD in Children and Adults*, Arlington, VA: American Psychiatric Association Publishing.

Castaneda, C. (1998) *The Teachings of Don Juan, a Yaqui Way of Knowledge*, London: University of California Press.

Garfi, J. (2018) *Overcoming School Refusal, a Practical Guide for Teachers, Counsellors, Caseworkers and Parents*, Samford Valley: Australian Academic Press.

Malchiodi, C.A. (ed) (2012) *Handbook of Art Therapy*, New York: Guilford Press.

Paxton, K. & Estay, I. (2007) *Counselling People on the Autistic Spectrum, a Practical Manual*, London: Jessica Kingsley Publishers.

Torrance, J. (2018) *Therapeutic Adventures with Autistic Children: Connecting Through Movement Play and Creativity*. London: JKP.

19

"SOME BODY ELSE"

Harlequin's journey in forensic Sandplay and gender dysphoria

Emma Allen

Distressed by a misalliance of gender identity and sex assigned at birth, Alex (not their real name) struggled with gender dysphoria and suicidal ideation, whilst serving a life sentence for the murder of men, and receiving inpatient treatment for personality disorder in a high-security forensic hospital. Due to being "a fairly mild-mannered, long-suffering individual who buries" their "resentment under rigid but brittle controls", but "under certain circumstances", "may lash out and release all" their "aggression in one, often disastrous, act", before returning back to "over-controlled" defenses (Megargee, 1966, p. 2), Alex engaged in RO-DBT (radically open dialectical behavioural therapy, Lynch, 2018), for over-controlled violent offending in personality disorders (Hamilton et al., 2018), before being directly referred for forensic Sandplay therapy (Allen, 2018); both of which, were first of their kind, innovative approaches in the high-secure setting.

Forensic Sandplay (a psychotherapeutic approach, developed from Kalffian-Jungian Sandplay training) requires the creation of three-dimensional images in a wet or dry sand tray, using symbolic miniature figurines, and sculpting or moulding the sand to express, explore, and embody the inner world (and internal psychic stories) of secure patients and prisoners that are considered a risk to themselves or others. This takes place in a "securely protected" space in which to safely explore "offending behaviour; resilience; control and boundaries; engagement; trust; communication and expressive deficits; self-identity and image; mental health or personality disorder; traumatic experiences and emotional states" (Allen, 2018, p. 61).

This case vignette is an example of how forensic Sandplay assists the over-controlled offender through promoting playfulness (replacing impulsivity with spontaneity) and explores the fool/trickster archetype of Harlequin (in relation to expressions of physical and emotional experience, and gender violence).

Through spur-of-the-moment, symbolic creations in the sand, and descending into the unconscious, "the conscious mind relaxes its control", and "purposeful qualities of the conscious mind are silenced" (Kalff, 1980, pp. xi–xii), where imagery shifts (over time) towards conveying new psychological potential; even with those serving life sentences, and who feel hopeless. Sandplay activates and accesses the limbic system (the brain's right hemisphere), where feelings, awareness, and authentic Self are situated, and where emotional regulation is possible; alleviating emotional inhibition, hyper-vigilant-physiological threat sensitivity, rigidity, over-thinking and indirect communication, and decreasing internal destructive impulses, and preventing escalating emotional tension (Allen, 2018).

Figure 19.1 The fall and breakage of Harlequin (detail, Sandplay 3)

Alex's Sandplay process begins by telling a tragic, sacrificial tale of difference and deception ("playing along") with the false Self and trickster archetype; of hiding, pretending to be, or feeling like somebody else. The journey ends, however, in a recovery from the death of the personal father, and in a recovery of the archetypal feminine. It is a story that uncovers abandonment-trauma and physical abuse before overcoming accompanying feelings of de-personalisation and de-realisation in the search for true Self. Alex's gender, body, identity, sexuality, personality, and ego incongruence, experienced since childhood (potentially associated with Asperger's Syndrome (e.g. Parkinson, 2014) (who "see, hear and feel the world differently to others", (*The National Autistic Society*, 2018)), or a fearful-avoidant attachment and "over-separation" (Colgan, 1987), and long-term consequences of neglect, trauma, and abuse; experiencing chronic distrust and "everything as unreal" (Bowlby, 2008, p. 103), was witnessed through the emergence, development, and movement of Harlequin; a manifestation of the collective shadow (a sacrificed expression in life and repressed "dark side" of the psyche). As an androgynous, sexually ambiguous, gender-fluid, shape-shifter, Harlequin is associated with a discrepancy between mind and body, and reality and representation, and as an archetype, offers a transgression and transition through symbolic alterations. Harlequin's breakage and burial is not only thought to illustrate violent contempt for the body, and a symbolic murder of the masculine, but is also considered a significant phase of the individuation process ("where the ego and the unconscious meet and where the personal and collective unconscious brings self-realisation" (Allen, 2018, p. 51)). As an actor, and performer of magic, in search of love, Harlequin necessitates a need for self-compassion, where a process of initiation ("a meeting point between containment and liberation"; unites internal opposing forces, achieving equilibrium (Jung et al., 1964, p. 156)), and witnessed in Alex's final Sandplay (Figures 19.6 and 19.7).

Alex

Serving a life sentence for murder, Alex (who'd be elderly upon release) struggled with thoughts of death, dying, and suicide, whilst receiving treatment for dissocial and emotionally unstable personality disorder. During psychiatric care, Alex had planned to kill another patient, and whilst imprisoned for attempted murder, had committed homicide, alongside another prisoner. Alex received treatment in DSPD services (for "dangerous and severe personality disorders"), in a high secure prison and the forensic hospital. Like many of those diagnosed with dissocial personality traits, Alex had grown up with parental conflict and punitive, inconsistent parenting (NICE, 2009). The eldest of several siblings, with "distant" parents who divorced when aged seven, Alex became the "father-figure" in the family who always wanted to be "the boss", and in charge, often truanting from school. Self-referred as "the class clown", Alex "craved attention" at school but left without qualifications. After Alex's father remarried, Alex lived with him during the week, and with their mother at weekends. Alex's mother had a series of physically abusive relationships, where one partner disciplined Alex harshly with a belt. Although isolative, and lacking friends, Alex had some brief, "unfulfilling" experiences of heterosexual relationships (attempting to conform, and "fit in"), and had told various professionals that from the age of seven, had felt "different", should have been born a woman, and had "pretended to be female". Alex kept this secret, never telling anyone, until in prison, and struggling with anorexia nervosa. Self-reportedly, "depressive, quiet and shy", Alex was suicidal due to fears that gender dysphoria would never be addressed, believed or taken seriously. Although attracted to men, Alex abstained from relationships due to a desire to be a woman. Now residing in the high secure men's personality disorder service, Alex was one of the first patients to be specifically referred for forensic Sandplay.

Forensic Sandplay

Forensic Sandplay has been pioneered in a high secure setting for the first time, in the world, and is adapted from traditional methods and training (originally developed by Dora Kalff during the 1950s as a non-verbal, psychotherapeutic, Jungian approach) in order to assist in the reduction of risk of harm and violence (Allen, 2018). Sessions involve three-dimensional, image-making, through the use of either one or both (wet and dry) wooden sand trays and a collection of miniature objects. One tray is described as "wet" so that water can be added (for moulding sand), and the other remains "dry" (for meandering), so that different textures provide a different "feel", expression, and level of engagement. Some Sandplay therapists hypothesise that the wet tray requires more physical interaction, and thereby contains more unconscious, masculine content than the dry, and using two trays at a time, can represent processes of progression and regression, integration and deintegration (Grishko, 2014). Sandplay trays are of Kalffian measurements and have a blue base to represent water or the sea, and the collection of miniature figures, symbolise real, fantasy, cultural, and spiritual life.

Jungian Sandplay theorises that deep within the unconscious, there is an instinctual, autonomous drive (given the right conditions) for the psyche to rejuvenate over a long-term process, where "dual (core) processes" of a regression, and descent, into unconscious, internal content, takes place in order to advance towards increased consciousness and self-awareness (Weinrib, 2004; Bradway, 2006). Symbolic stories portrayed in the sand often correspond to inner states that make up a "concentrated expression of a total psychic situation" (Zoja, 2004, p. 151), where archetypal figures, or the sand itself, takes central stage, in supporting the journey towards wholeness and balance. Sand "plays", or scenes, that provide "a visual means for the psyche to reveal itself" encourage "self-expression and reflection and allowing thoughts, feelings and experiences to be shared" (Allen, 2018, p. 51) with therapist and patient as both witness and audience. Creating a series of Sandplays aims towards resolving inner conflict that may exist within the relationship between consciousness and unconsciousness, and where shifts in which symbolic imagery (e.g. constellating the opposites) offers new self-insights and internal restructuring in the aftermath of trauma, neglect, or loss of self-image and identity (Allen, 2018).

A (first) referral

Alex was referred for individual Sandplay (following RO-DBT, (Lynch, 2018)) in order to practise the surrender of self-control by increasing spontaneity, vulnerable self-expressions, openness to feedback from others, self-understanding, and emotional connectedness. Sandplay was also appropriate for trauma, loss, identity, and body-image related work due to feeling "emotionally dead inside and living in a shell" after Alex's parents' divorce. Alex also wanted to be more playful in a self-discovery towards removing "the mask" (a persona that obscured emotions and the inner authentic nature of the Self). Twenty-four sessions took place with Alex (creating 44 Sandplays, and using both trays, 20 times over a 20-month period).

"Killing my self", "never feeling loved"

Alex's first two Sandplays (Figure 19.2) depicted emotional death and restrictive-concrete thinking, as part of a projected, experiential worldview (Dale & Wagner, 2003), that felt conflictual between an "unrealistic, unachievable" fantasy of receiving love verses a reality of "never feeling loved"; separately depicting masculine aspects in the wet tray, and the feminine in the dry. Feminine and masculine does not describe gender here, but to (Jungian) contra sexual psychological characteristics, which manifest in different, opposite modes of perceiving, thinking, acting, reacting, relating, and functioning (Jung, 1960). Gender violence is therefore also considered as a symbolic, unconscious attack upon contra sexual, psychological characteristics.

The first wet tray appeared to embody the masculine, conscious-side of self-identity; containing two men; one, in chains, in a coffin, and carried off by skeletons (to the left), and another man being chased by a ghost (to the right). It suggested the masculine as a terrifying

Figure 19.2 "Never feeling loved" (wet tray, left, and dry tray, right), Sandplays 1 and 2)

conflict in both Alex's internal and external world and seemed to highlight gender dissatisfaction and body suffering. A large bridge led to a brick wall, proposing that a transition would be a dangerous, suicidal, dead-end journey. It alluded to Alex's psychological and physiological incarceration, confinement, and despair, and potential suicide if returned to prison. Ruminations of suicide or damage to the Self, could have been impulsive attempts at control, or a means to escape, and Alex admitted to having asked patients to be murdered by them.

The sand, usually moulded down to the blue bottom after Sandplay creations, and during our joint, verbal explorations, Alex revealed what looked like a severed, separated head (mind, psyche) from its body (matter); showing a mind-body split and denying a physical bodily existence. The man in chains exhibited an experience of the ego wanting to separate, and sacrifice itself, where Alex felt enslaved, and "bound up" by the body, attempting to differentiate from physicality (and with an anorectic history of restrictive eating), perceived the body as existing "in parts or fragments" (Levens, 1995, p. 32). Sand trays can reveal "divided", "split off" aspects of a transgendered self in quest of the feminine (Ballantyne, 2017, p. 209), and Alex's fragmented scenes (with figures distanced from one another) expressed alienation, solitude, and a split defense between devaluation and idealisation, seen in Sandplay with personality disordered client groups (La Spina, 2004). Imprisoned in a distressing paradox, Alex feared rejection from a gender transition, but also feared living a lie. Adamant this predicament could never change (whilst incarcerated in a hyper-masculine setting that could condemn the feminine), Alex felt they would never be loved or be able to love others. The wet tray (*Sandplay 1*) challenged this with a "question mark in the sand", and in the opposite, dry tray, a family scene (parents and son), and an embracing couple, united the internalised parental split, and the feminine and masculine. Although Alex felt "haunted" and "chased" by the past, Alex said this form of visual self-expression felt hopeful in finding a way forward.

"Doing away" with Harlequin

> I can't imagine how to be myself. To be free, to be me – Do away with him!
>
> *(Alex, 2018)*

Harlequin snapped into two during Alex's second Sandplay; reminiscent of being undifferentiated in both body and mind (Figure 19.1). Harlequin, a comic-jester, clown, romantic-trickster-hero (from the Italian comedy play, *Commedia dell'arte*) is an actor and performer of magic, in pursuit of love. Harlequin has attributes of a mindless fool and practical joker, but is also a clever, acrobatic lead in the pantomime, who, through a chequered-costumed-disguise (that covers the body) and black mask (persona, and potentially the DSPD label) (that covers the mind), tells stories through humour, dance, and mime, and is talented at fluctuating between revealing and distancing, or remaining hidden or distant from others, with a twofold identity as Fool and Devil, protector and persecutor (Kalsched, 1996). Alex would need to eradicate this archetype in order to Self-actualise, and for vulnerabilities to be seen.

Harlequin holds a magic wand that brings wisdom from the collective unconscious and is able to transform the play in a carefree, perceptive manner; often disobeying authority (rebelling against the master's orders); representing the surrender of self-control. Harlequin suggests an absence of connection with the outer world and a disregard for full potential, "where such a rigid fixation to the persona, the ego's drive to Self-actualize will create a pressure upon the fixed orientation, resulting in eruptions of compensatory unconscious material or energy" (Turner, 2005, p. 21). In Alex's case, it could be hypothesised that the murder of men, along with a sudden outburst of compensatory unconscious material, could be a projection of a desire to obliterate the self in suicide, or the masculine within the self, where a projection of the relationship with the unconscious, could have been an attack on the shadow, and repressed sexuality. Harlequin's fall, and breakage is sacrificial, and a significant phase of the individuation, and initiation process, where Alex is no longer able to hide a traumatic relationship with the body. The identity that once served Alex, is dismembered, where an alteration of self requires a conscious act of violence, and where processes of change entails a chaotic disorder. Alterations such as this, in an archetype, infer that alterations can also take place in reality (Weinrib, 2004). The broken legs illustrated an internalised body-fragmentation that would require a more integrated entity and wholeness, and the series of frequent empty, and (a state of) "void" spaces in Alex's Sandplays portrayed emotional isolation, disconnection with the world and other, and inner turbulence. Although Alex was sorry, I suggested this rupture, and loss, felt symbolic of internal imbalance.

Harlequin speaks truth, and more importantly, for Alex, represents resilience and standing up to authority. Comfortable with chaos and disruption, Harlequin represents the component in the psyche that is able to revolt, rebel, and trial the irrational and dissociated parts of ourselves. Harlequin overthrow's systems, and is an embodiment of the new, spontaneous and creative ability for new patterns of growth. In order for renewal, Harlequin symbolises fragmentation, and is therefore both tragic and generative; representing a union of opposites in a transcendent function. Harlequin therefore may symbolise Alex's protected relationship with control, ego, personality, and feelings about gender, identity, and sexuality; all undeveloped aspects of Self. Harlequin alerts Alex to self-deception, hypocrisy, deprecation, and contradiction, and as a "shape-shifting quixotic figure who crosses all boundaries" (Kalsched, 1996, p. 51) invites an exciting possibility of transformation, leading Alex to a new identity, or a change in direction or purpose. Alex didn't give Harlequin a gender, whose mysterious characteristics illustrated a "masquerade of appearing clever, emotionally stable and acting all my life" (*'Alex'*, 2019a). The gender-fluidity and ambiguity of Harlequin is similar to that of the devil (a "daimonic" inner figure (Kalsched, 1996, p. 12)), whose enigmatic identity is disembodied, and a shadow of bodily imposter syndrome (and not "belonging" to the body). Both are paradoxical punishers, but are also punished; resembling self-suffering, unconformity, and contemplation towards sin and crime. In "liberating suppressed anxieties" turning "order on its head" and amending "the rigidities of conscious life" (Kalsched, 1996, p. 51), Harlequin reminds us that we can be tricked by our relationship to the unconscious and offers therapeutic guidance in relinquishing control towards the body.

Murder of the masculine

Alex's father (and thereby, the masculine) was described as "never being big on emotion". Alex had learnt that the masculine was avoidant or devoid of emotional connection. Alex's father (cared for in a home for dementia) no longer knew who Alex was; exacerbating original, identity-insecurities. Mistrustful of the world, and intimate relationships, a fearful-avoidant-insecure-attachment style attributed towards feeling alienated from the body; experiencing "over-separation" due to a significant absence of positive male role models (Colgan, 1987). Extreme body-differentiation can be caused by an early disturbance in symbol formation, and an absent or domineering father figure (Limentani, 1979, 1991).

Alex noticed doing things without thinking, or realising in Sandplay, that paralleled past offending behavior. In being spontaneous, and in search of emotional insight, Alex appeared to visually reveal an unconscious disposal of the masculine; throwing a man-shaped cookie-cutter, head-first into *Sandplay 2*, before replacing this with a witch. *Sandplay 4* also had a high volume of trickster imagery that included *The Joker* (playing card), *The Riddler* (Batman), and *Sleepy* (Snow White's seven dwarfs). A skull, placed in some hands (*Sandplay 4*), suggestive of taking matters into the hands through murder, and suicide attempts, was also expressive of consciousness and self-realisation, and of the handling of a new world of being (De Vries, 2004). Rejection and devaluation (and process of deintegration) of the masculine could represent rejection of the father-object that had become pathological. "Becoming someone else instead" can escape trauma (e.g. Heyer, 2018), but Alex did not think it possible to live as a woman (as a "prisoner doing life"), where a surgeon holds their hands up in resignation (*Sandplay 18*). Self-preservation had contributed towards Alex's risk of obliteration and annihilation of self and other, and the murder of men and the masculine, and feeling at threat of engulfment from emotions. Laxative abuse had emptied Alex's body of feelings, perhaps also experienced in the act of murder of men. Alex disclosed details of the index offence (the main, most serious offence, of homicide), describing "playing along" with the accomplice due to fears of being killed. Alex had also been assumed to be "playing along" with therapies too, but in Sandplay, this had felt different this time. Alex's father passed away during therapy, which also evoked an authentic bereavement of the personal, absent, father, alongside an internal separation, and abandonment from a father complex (expressing unconscious imagery pertaining to the image or archetype of the father) took place in Sandplay, in order to strengthen ego development and liberate repressed emotions, and over-controlling impulses. Body trickster archetypes appeared in the use of puppets, namely Pinocchio (associated with telling lies, exaggerating truth, and a desire to become a real boy), who rebelled against his parents, and needed to obey authority, along with being warned to let conscience be his guide, and conscious of the self's regulative conscience (Herrmann, 2008). Pinocchio's wooden body (and "wounded self" (Ballantyne, 2017, p. 215)) alluded to Alex's restricted self-expression and sense of agency. Pinocchio's quest is to "transform a man from within" heal the "divided self, and restore" them "to a state of primordial wholeness" (Heisig, 1974, p. 30). Pinocchio saves the father (releasing the ego in self-realisation) before becoming human and tells Alex's tragedy, and heroic expedition to overcome de-personalisation and de-realisation in the search for true Self;

> I was born in the wrong life, body, not just gender but wrong century and country – a life lost. I'm trapped between the headlights, if released, I'd struggle. . . . I'm just existing, I'll die inside.
>
> *(Alex, 2018)*

In the periphery: emotional separation

Alex's identity-formation existed in the periphery between male and female, masculine (conscious) and feminine (unconscious) energies in emotional detachment, de-personalisation and over-separation. The two trays seemed to divide the (devaluated) masculine and (idealised) feminine that required integration, and the space in-between the two trays was often used as a holding space for miniatures, Sandplay tools, and "detour" signs on each edge, indicating an undifferentiated state or internal divide between a separation-anxiety and fear of abandonment (differentiation) and attachment-desire (Figure 19.2). Placed in this "in-between" were also traffic lights, torches, steering wheels, armed officers, guards, and soldiers that were delaying, protecting, guarding, or maintaining a separation. Alex's need for separateness was experienced in the countertransference of feeling shut down, where I often felt drowsy (similar to in eating disorder work), and was explored as a form of emotional detachment that passively avoided deep repressed rage or sadness and feeling misunderstood, or "invisible". Homicide had induced a state of separateness from Alex's own body and from that of other men in sexual repression; escaping an incarcerated body may have also desired fusion.

Alex would often place miniatures, silently and carefully, alternating, and switching from both trays. Alex described the dry as being "a pleasant place" (perhaps feminine forces in the internal world) and the wet as "more realistic", "needing to be myself" (where perhaps more masculine forces worked to resolve a father complex). Although Alex began to question needing an Asperger's diagnosis for feeling alienated, and overwhelmingly "different", the Sandplays strongly suggested that the masculine and feminine were in opposition. Alex often used a "crying-boy" and a "split-man" ("getting defensive of yourself rather than being one person" (Alex, 2018)). Alex's repeating use of two trays spoke to the separated gender identities, and to the split and divide of parental attachment that is often seen in cases of divorce (Turner, 2005). It also showed a psychic-divorce where initiation would be to reconcile conflicting aspects of personality where Alex could master the Self, whilst breaking free from restrictive patterns of existence (Jung et al., 1964). Differentiation poses risks of abandonment, and the absence of belonging to either parent, but being (psychologically) torn between the two, and in feeling unsafe. Alex often felt that staff only saw the "scary masculine side" as opposed to a positive, feminine side, where Alex was sensitive and insecure to having feelings, desires, and needs. We explored how important it would be to understand that having a healthy relationship with emotions is not just a feminine activity (that Alex didn't need to be female in order to engage with feelings), and that a healthy balance of masculine and feminine parts to self could be beneficial. A man protecting a baby by *Sandplay 8* is at threat of abandonment through homicide (where a sniper and tank aim fire at the (and Alex's internalised) protecting 'father figure', until in *Sandplay 20*, an opened gate invites both trays to relate, or reunite. A ladder leads away from a fragmented body and volcanic fury of the masculine wet, into the feminine dry (*Sandplay 21*). Alex had been split between both parents and conflict. A frequent use of monsters and aliens evoked Alex's disclosure from "thinking I'd died and that an alien had taken over my body".

Condemnation of the feminine

Alex's mother was described as angry, violent, and abusive, and used to hit Alex. Alex had not been believed and had to "exaggerate and lie in the end to be taken seriously"; perhaps witnessed in the use of Pinnochio and Harlequin. Alex said that home, and a sense of self, had never been safe and spoke about feeling deeply rejected from mother, and seemed to be unconsciously, actively, rejecting the feminine in return. Alex deeply craved a loving relationship but could not be self-loving; the feminine was unobtainable. Sandplays contained broken body parts, that expressed feeling numb or cut off and a repeatedly used naked female, threatened by wild animals (e.g. Figure 19.3) or surrounded by three 'wise' (See, Hear, Speak No Evil) monkeys, revealed the feminine emotionality as threat and possibly confirmed body-deception that required a sense of agency. Alex's unconscious instinct had been to threaten the feminine, and vulnerable emotional expressions undifferentiating of the masculine and feminine; an internal tragedy of paternal deprivation. Restrictive eating is also a rejection of the mother, and maternal nourishment. Alex had learned to avoid expressing emotions in case they triggered rejection by others. Re-establishing access to feminine elements of the psyche, that has otherwise been repressed, Alex often asserted the view that to connect with emotions was a female attribute, and that due to not being female was the reason behind Alex's difficulties in connecting to emotional pain. Alex disclosed severe neglect and physical abuse as a child from *both* parents, which had stolen a sense of play, self, and trust in others.

Feminine receptivity to self

Belle (from Disney's *Beauty and the Beast*) makes an appearance (*Sandplays 14 and 22*, Figure 19.4); offering a mirror of self-insight, feminine consciousness of "healing, wisdom and

Figure 19.3 Naked Lady and Lion (detail, Sandplay 19)

Figure 19.4 "Always wishing and never getting" (detail, Sandplay 22)

truth" (Weinrib, 2004, p. 169); persuading Shadow integration. A feminine equivalent to Harlequin, Belle is an archetypal, sacrificial, and compensatory heroine who courageously, and compassionately, gives up her freedom by living with the Beast forever, so that her father can be released. By caring for the father in this way, she is the caretaker of consciousness, clarity of Self and persona. Perhaps Belle can help Alex with their relationship to masculine violence. Alex placed Belle in front of a mirror at a dressing table, for seeing into the unconscious (along with a die (dice) rolled to its sixth face) truth (and true gender-identity), consciousness, and self-knowledge, and, more crucially, the individuation process. Alex described her as "always wishing and never getting"; highlighting an idealisation of the feminine. Belle's double-reflection could be another attempt to differentiate masculine and feminine consciousness or could represent obsessive or paranoid ideations due to seeing the shadow, or looking over the shoulder (Weinrib, 2004). Her reflection could alternatively, humbly, offer a window of hope (and progression), into Alex's "transformational undergoing" that "was, and always had been, rooted in the Self" (Ballantyne, 2017, pp. 214–215). Mediation with the feminine visually increased and harmonised in Sandplay, and gender fluidity is perhaps witnessed in the touching of the dry, gentle sand, which requires more delicate movement and gesture. The mirror reflects the "true face" (Jung, 3:95, Jacobi, 1971, p. 239) that was hidden behind Harlequin's mask, and from the world; reversing depersonalisation in an act of self-realisation. Sat beside Belle, Pinocchio shows a maturing union, and emotional intelligence between the masculine and emerging development of feminine energies. They also visually captured our therapeutic relationship, and perhaps the impact of my gender. The (increased) number archetype of six indicated a new acceptance of Self, in uniting with others and bringing the opposites together (Eastwood, 2002, p. 120).

The ego's development in its relationship to the Self required integrating the feminine elements of the psyche that had previously not been mirrored or met in the maternal relationship. Self, and body-regulation would not be just a male or masculine task, but a feminine one too.

The trickster can do serious bodily harm, emotional death, and murder. Elsa from *Frozen* emerges into *Sandplay 20*. Alex says that "she's had trauma, was cast aside, rejected – she has to find a way to be safe around people too". Her open hand gesture is in direct opposition to the covering, shameful hands of the naked female who covers her body in humiliation, and Alex begins to form social relationships on the ward. A ladder leads out of the wet into the dry, leaving behind the broken legs, of a naked male (*Sandplay 21*), and a man waits at a station, anticipating a change of direction. Sandplay offered a space to safely express emotion without fear of rejection, in forming internal boundaries that were authentic and no longer serving a "chameleon-like" existence.

Harlequin's burial and the fall

The Fool's "clowning often climaxed in a symbolic death and resurrection" (Kalsched, 1996, p. 51) where Harlequin's metaphorical castration ends with a body-burial in the sand (Figure 19.5). Moving towards (shadow) integration, Alex had unconsciously buried resentment towards the masculine body, and the masquerade of hiding tragic emotions. Sandplay may have provided a safe and secure sense of containment within the body (in an act of self-preservation) through Harlequin's self-sacrifice. Harlequin's burial relinquishes the ambiguous serving role

Figure 19.5 Harlequin's burial (detail, Sandplays 24)

of playing along and is no longer required to transcend shame from others; bringing hope for Alex to be true to self, without fear of violence. Burials not only respected the disconnected, cut off, part of Alex, but suggested a fear of engulfment, or in feeling devoured by a "return to the primal parent" (De Vries, 2004, p. 90). Alex no longer needed assistance with letting go of over-control that Harlequin embodied; suggesting initiation into a new phase in life, body, and self. As a master of initiation, Harlequin intuitively holds power in leaving the body, symbolising "the liberation of transcendence" (Jung et al., 1964, p. 149); mediating between consciousness and unconsciousness (seen in Alex's final two trays with a burnt-out house (Figure 19.6), and angelic hands holding a baby (Figure 19.7)). Harlequin's downfall (and process of regression) provides an opportunity for Alex's self-discovery, without needing to perform or please others. Harlequin's dual representation of masculine and feminine forces in the internal world, provides the opportunity of uniting these opposites in a descent into the unconscious and unprocessed

Figure 19.6 Final Sandplay (Sandplays 43 and 44)

Figure 19.7 Baby in hands (Sandplays 43 and 44)

trauma (Kalsched, 1996). Alex wanted our last session to "fall" on the first day of autumn; symbolic of a descent transition, change and alteration, and surrendering control. Alex later commented on Harlequin's journey, and potential recovery from the risk of symbolic homicide;

> The mask I've had to wear is one thing I don't want to do. It's good he's been buried, but not completely. I'm able to show a bit more of myself.
>
> *('Alex', 2019a)*

Initiating new beginnings – containment and liberation

Alex described the baby in the last two Sandplays as "new beginnings". Contact with the primordial, archetypal child in Sandplay shows a need for autonomous separation from 'mother' (archetype, and mediator of the inner world), and in integrating the Father archetype in order to individuate, mature, and expand (masculine) consciousness. A baby crawls out of open, cupped hands, displaying a renewed vulnerable expression; no longer masking Alex's needs through distancing and frightening others away. Romeo and Juliet holding one another in a physical embrace, appeared to embody a marriage and union, and harmonisation of the (masculine and feminine (conscious and unconscious, personal and collective, psyche and soma)) opposites; enabling emotional self-regulation, no longer showing a split. Alex announces having positive feelings saying "I've felt a lot of love recently"; feeling cared for as a result of a memorial service taking place in the hospital, and in acknowledging this loss in Sandplay and symbolic imagery. Alex refers to feminine miniatures (a woman feeding her pigs) as showing feeling nurtured and felt indicative of an effective therapeutic relationship. *Sandplay 23* included a man, with helicopter blades, that can ascend or descend, and in the final Sandplays *(43–44)*, the rolled die, now on its fourth side, shows increased receptivity, balance, emotion, and differentiation, and, as both an addition and a multiplication of two, is the "first even", "female" number (Eastwood, 2002, p. 83); showing progression into feminine qualities. A child-like spirit of play (Signell, 1981) is born and a newly found feminine acceptance, where the hands and baby suggest new psychic qualities, and symbolic offering of embodiment in a new body.

> Sandplay is freeing; relaxing. The hands and the baby – I just instantly felt it fitted.
>
> *('Alex', 2019b)*

Alex's hand-prints, and moulding of the sand, was described as "self-soothing"; potentially offering the relaxation of control and hyper-vigilance, as well as definite self-mark-making. As imprints of early bodily experience, Alex could return to an intuitive body healing. Hand-prints looked to be attempting to catch the void and destructive emptiness that remained after forced body evacuation in family trauma and abuse, where the body now speaks. The miniature hands (Figure 19.7), openly express and symbolise divine intervention, offering a pledge, and gesture, of faith, protection, peace, solidarity, love, and the individuation process itself (De Vries, 2004). As a symbol of both containment (the hands) and liberation (the crawling, moving, baby), this Sandplay illustrated "a journey of release, renunciation, and atonement, presided over and fostered by some spirit of compassion" (Jung et al., 1964, p. 150). Sandplay seemed to support Alex in reestablishing a desire to improve quality of life and living, (in the long-term), as opposed to dying, and in building trust with me over time, helped to drop guarded defenses, and reach out for loving care. Alex's Sandplay process finished with an opening, and offering of a peaceful, compassionate resignation of control, as opposed to closed, defensive, rigid-imagery, exampling how this approach helps to connect mind with body, as when the hands touch sand, there is an immediate connection between the body, thoughts, and feelings (Badenoch, 2008).

Transgression into the body

Alex asked for my help in locating the burnt-out house, as the last miniature to be placed, in the very last moment (Figure 19.6). It felt like a genuinely cathartic act of expelling rage; "accepting" the "shadow into consciousness and no longer projecting it outward through acts of physical violence" (Herrmann, 2011, p. 125). The house, situated in the bottom right, conscious area of the tray (Ammann, 1991), is symbolically illustrative of abandonment-trauma, neglect and deprivation, "the wounded body and mind", Alex's "destroyed sense of self and annihilated ego", and "internalised emotional devastation left behind after psychological trauma and suffering" (Allen, 2019). In the reality of Alex's incarceration, and physicality with restricted exit, is an authentic grieving. Alex's Sandplays progress towards emotional intelligence (shifting into the lower right corner) (Ammann, 1991) and a transgression into the body and to physicality. Eventually, Alex's ego found balance and preservation through masculine and feminine archetypal images overcoming negative aspects of the collective.

Alex needed to relate more to the vulnerable inner child, as opposed to a specific sex. Sandplay promotes the transcendent function, and new perspectives; "bringing the conscious and unconscious together", assisting significant, transformed attitudes towards the unconscious, whilst also giving it a "voice in times of disturbance", bringing clarity to otherwise unknown feelings of rage or destructive impulses, that "fosters the discovery of previously unknown courage and strengths" (Kalff, 1980, pp. xi–xii). Innocent, holding, and containing hands offer redemption, compassion, and hope for a way forward in life, and in living. A parental split, could unconsciously symbolise the separation of masculine to feminine, not only through gender-identity change but in embracing the feminine by engaging with emotions, truly, for the first time, and where the actual death of the father, may also come to represent a death and sacrifice of the masculine. Alex's experiences of both physiological and emotional trauma had disconnected the body and soul through protective and persecutory drives and threat responses. Harlequin is heroic in never falling victim to judgement or shame. Self-preservation was no longer in concealing vulnerabilities, but in showing them. Just as sand pictures change shape and role, Alex is able to unconsciously shift and journey into a new role, a new beginning, and a new identity; no longer running away, but in being held in a renewed sense of self:

> (Sandplay) helped exploring bits of me I'd forgotten and discovering parts of me I didn't know about. It gave me a chance to express myself without over-thinking and not having to worry. It helped me see more of the emotional side of myself; meaning that I'm finding it easier to be myself.
>
> *('Alex', 2019b)*

Conclusion

Forensic Sandplay tells tragically ambivalent tales from deep within the unconscious, and body and mind, of devastating physiological and psychological impact of a parental split, abandonment-trauma, loss, and bereavement, from the masculine (and death of the father), and emotional and physical abuse from the feminine (mother). Using both sand trays at a time, Alex's identity formation existed in the periphery between male and female, and masculine (conscious) and feminine (unconscious) energies in emotional detachment, de-personalisation and over-separation. Sacrificial archetypes, such as Harlequin, surrender the ego to the Self in order for previously *masked* emotional states, and aspects of Self, to become known, in a process of initiation; containing, liberating, and uniting internal opposing forces, and achieving balance (Jung et al., 1964). As a pretending Fool, yet paradoxically, intelligent Trickster, Harlequin transcends resistance and emancipates suppressed anxieties and threat sensitivity. Through uncovering buried, past resentment, and safely exploring the risk of gender violence, control and rigidity are abandoned in a dual representation of both masculine and feminine forces, in order for self-preservation, adaptation, authenticity, acceptance, and self-realisation. Alex's separation of body and mind took authority, and desperately sought a sense of agency. Alex's prohibited sexual and emotional life (perhaps witnessed in hand-printing in the sand) required bodily acknowledgement, and in being seen.

Through Sandplay, and a stable, trusting, and genuine therapeutic relationship, Alex could begin to develop a more secure sense of identity; no longer hiding, playing, or pretending to be somebody else, and learning to show others, true feelings. A search and discovery of self-love and compassion was found; more important than any answer to gender identity. A more harmonious relationship with the body, and self, and between the conscious and the unconscious can develop, where the ego can be restructured and strengthened. Moving towards a more balanced body and mind connection, and where inner fragmentations, and a vulnerability expression of Self, offering, and in opening up to a new beginning and internalised experience of compassion that is symbolically released, held, and "securely protected" (Allen, 2018), (seen in Figure 19.7), the baby in the hands, was representative of the individuation process itself. Instead of a suicidal departure, a baby arrives. The sand embodies body-suffering, shifts away from dysfunctional ego defenses, allowing integration and self-regulation. Perhaps Sandplay offered a fantasy of a relationship with the feminine in both the internal and external world, where Alex could feel more informed in a future decision. Alex can unearth, through the sand, a true role and identity with emotion, where coherence and integrated entities emerged from an inner fragmentation and defensive split between devaluation and idealisation.

Alex's conscious decision to reserve action upon transitioning (to female), within a high secure environment, was explored in unconscious expressions of self (and body)-sacrifice and self-preservation that also contributed towards a risk of obliteration and annihilation of both Self and other, and the murder of men and the masculine. A rejection of, and attack upon the archetypal feminine, and emotionality (unconsciousness, and anima development) (seen in naked, exposed, female figurines, and broken body parts) is perceived as a threat initially in Alex's Sandplay process, before Belle (from Disney's *Beauty and the Beast*), (an additional, and essential, sacrificial and compensatory figure) makes an appearance; offering a mirror of self-insight and persuading Shadow integration. The final burial of Harlequin re-enacts a symbolic murder of the masculine mask where Alex may begin a new journey into true Self, and where a baby carries with it symbolic potential for a rebirth of a newly found ego-attitude/acceptance, self-image, and gender-identity.

Acknowledgements

A unisex pseudonym has been applied to this case vignette and some information has been changed or omitted to protect, and respect, individual's identities. Gender-specific pronouns are also omitted, using plural "they/their". I am grateful to Alex for allowing me to share a part of their journey, and for their consent in writing about their experience of forensic Sandplay.

References

'Alex' (pseudonym). (2018). *Progress Notes*, unpublished.

'Alex' (pseudonym). (2019a). *Transcribed Interview*, unpublished.

'Alex' (pseudonym). (2019b). *Arts Therapies Patient Satisfaction Questionnaire* (S. Compton & C. Beeley in 2012, developed from 2001, H. Odell-Miller & P. Hughes), unpublished.

Allen, E. (2018). Introducing Forensic Sandplay, *British Journal of Play Therapy*, 13, pp. 50–64.

Allen, E. (2019). *"Some People Never Live": Loss of Innocence, Violence of Separation & Self-Realization*, unpublished paper.

Ammann, R. (1991). *Healing and Transformation in Sandplay: Creative Processes Become Visible*, La Salle, IL: Open Court Publishing Company.

Badenoch, B. (2008). *Being a Brain-Wise Therapist: A Practical Guide to Interpersonal Neurobiology*, New York: W. W. Norton.

Ballantyne, D. (2017). Rikki's Awakening: A Journey of Discovery of Transgender, in Turner, B. A. (ed.) *The Routledge International Handbook of Sandplay Therapy*, London and New York: Routledge, pp. 207–217.

Bradway, K. (2006). What Is Sandplay? *Journal of Sandplay Therapy*, 15 (2), pp. 1–3.

Bowlby, J. A. (2008). *Secure Base: Parent-Child Attachment and Healthy Human Development*, New York: Basic Books.

Colgan, P. (1987). Treatment of Identity and Intimacy Issues in Gay Males, *Journal of Homosexuality*, 14 (1–2), pp. 103–123, doi:10. 1300/J082v14n01_09.

Dale, M., & Wagner, W. (2003). Sandplay: An Investigation into a Child's Meaning System via the Self-Confrontation Method for Children, *Journal of Constructivist Psychology*, 16, pp. 17–36.

De Vries, A. (2004). *Elsevier's Dictionary of Symbols and Imagery, Second, Enlarged Edition*, Elsevier Science, London: Brill.

Eastwood, P. S. (2002). *Nine Windows to Wholeness, Exploring Numbers in Sandplay Therapy*, Honolulu: Sanity Press.

Grishko, V. (2014). The Use of Wet and Dry Sand in Sandplay Therapy, *Journal of Sandplay Therapy*, 23 (1), pp. 67–84.

Hamilton, L., Bacon, L., Longfellow, E., & Tennant, A. (2018). Not Everything Is as Its Seems: RO DBT and Overcontrolled Disorders in Forensic Settings, *The Behavior Therapist, Special Issue: Radically Open Dialectical Behavior Therapy*, 41 (3), pp. 109–179, ISSN 0278-8403.

Heisig, J. W. (1974). Pinocchio: Archetype of the Motherless Child, *Children's Literature, Project MUSE*, 3, pp. 23–35. https://nirc.nanzan-u.ac.jp/en/files/2013/07/JWH-Pinocchio.pdf; doi:10.1353/chl.0.0336.

Herrmann, S. B. (2008). Treatment of an Abandonment Trauma, *Journal of Sandplay Therapy*, 17 (2), pp. 51–73.

Herrmann, S. B. (2011). Transformation of Violence, the Case of Jacob, *Journal of Sandplay Therapy*, 20 (1), pp. 117–132.

Heyer, W. (2018). *Childhood Sexual Abuse, Gender Dysphoria, and Transition Regret: Billy's Story*, www.thepublicdiscourse.com/2018/03/21178/.

Jacobi, J. (ed.) (1971). *Psychological Reflections, an Anthology of Jung's Writings 1905–1961, C.G. Jung*, London: Routledge.

Jung, C. G. (1960). *The Structure and Dynamics of the Psyche*, New York: Pantheon Books.

Jung, C. G., Von Franz, M. L., Henderson, J. L., Jacobi, J., & Jaffé, A., (eds.) (1964). *Man and His Symbols*, New York: Dell Publishing, Random House.

Kalff, D. M. (2003/1980). *Sandplay, a Psychotherapeutic Approach to the Psyche*, Cloverdale: Temenos Press.

Kalsched, D. (1996). *The Inner World of Trauma, Archetypal Defenses of the Personal Spirit*, London: Routledge.

La Spina, V. (2004). Sandplay Therapy in the Treatment of Borderline Personality Disturbance, in Zoja, E. P. (ed.) *Sandplay Therapy, Treatment of Psychopathologies*, Einsiedeln: Daimon Books.

Levens, M. (1995). *Eating Disorders and Magical Control of the Body, Treatment Through Art Therapy*, East Sussex and New York: Routledge.

Limentani, A. (1979/1989). The Significance of Transsexualism in Relation to Some Basic Psychoanalytic Concepts, in *Between Freud and Klein*, London: Free Association Books.

Limentani, A. (1991). Neglected Fathers in the Aetiology and Treatment of Sexual Deviations, *International Journal of Psycho-Analysis*, 72 (4), pp. 573–584.

Lynch, T. R. (2018). *The Skills Training Manual for Radically Open Dialectical Behavior Therapy: A Clinician's Guide for Treating Disorders of Overcontrol*, Oakland, CA: New Harbinger.

Megargee, E. I. (1966). Undercontrolled and Overcontrolled Personality Types in Extreme Anti-Social Aggression, *Psychological Monographs*, 80 (3), p. 611.

The National Autistic Society. (2018). *Asperger Syndrome*, www.autism.org.uk/about/what-is/asperger.aspx.

Pickersgill, M. D. (2009). NICE Clinical Guideline: Antisocial Personality Disorder, *The Journal of Medical Ethics*, 35 (11).

Parkinson, J. (2014). Gender Dysphoria in Asperger's Syndrome: A Caution, *Australasian Psychiatry*, 22 (1), pp. 84–85.

Signell, K. A. (1981). The Use of Sandplay with Men, in Bradway, K., Signell, K., Spare, G. H., Stewart, C. T., Stewart, L. H., & Thompson, C. (eds.) *Sandplay Studies, Origins, Theory and Practice*, San Francisco: C. G. Jung Institute, pp. 101–131.

Turner, B. (2005). *The Handbook of Sandplay Therapy*, Cloverdale: Temenos Press.

Weinrib, E. L. (2004). *Images of the Self, the Sandplay Therapy Process*, Cloverdale: Temenos Press.

Zoja, E. P. (ed.) (2004). *Sandplay Therapy, Treatment of Psychopathologies*, Einsiedeln: Daimon Books.

20
PLAYING TOGETHER
The use of directed joint engagement activities in Dyadic Art Psychotherapy

Bridget Rees

Introduction

This chapter will describe some directed activities that have been used with a child and their caregiver in art psychotherapy in order to help foster attachment. Although the work described here is in the context of art psychotherapy, the integrative approach means that the activities could be employed in a variety of therapeutic contexts, including play therapy. Reflections on some material from clinical work are included in order to draw out some of the issues that can arise from this way of working. In the case illustrated, the work was undertaken in independent practice, referred through a Local Authority Children's Services Kinship Team in the UK.

Use of directed activities in art psychotherapy

Art psychotherapy trainings in the UK usually adhere to analytical frameworks, drawing on psychodynamic and attachment theories. Integrative arts training goes beyond the use of visual arts to include all the creative arts – drama, music, art, dance movement and play therapy. As the psychodynamic approach will usually advocate a non-directive way of working, the trainings may not include instruction on therapist-directed activities. Indeed, they may even be frowned upon and not considered 'proper art therapy' (Springham, 2013). However, there is some literature to help the art psychotherapist who is considering using some directed activities, but mainly in a group context (Liebmann, 2004; Silverstone 1997, 2009). Art psychotherapists wishing to incorporate directed activities between children and their caregivers are more likely to find help from the area of Theraplay (Booth, 2010; Norris and Lender, 2019). However, Proulx (2003) devotes a chapter to 'interventions' that can be used with dyads or groups of dyads, mainly with younger children and their caregivers. Directives and symbolic meanings are given with each intervention.

The activities I will describe in this chapter are based on some techniques used in various modalities as well as those devised in-the-moment to respond to the presenting needs of the dyad.

Dyadic Art Psychotherapy

Dyadic Art Psychotherapy (DAP) has been developed as a model of art psychotherapy in UK practice where children and their caregivers engage in the therapy sessions together at least some of the time (Hendry and Taylor Buck, 2017; Taylor Buck, Dent-Brown and Parry, 2012; Taylor Buck et al., 2014). The three main DAP approaches as identified by Hendry and Taylor Buck (2017) are:

- child led with caregiver as helper or witness
- joint engagement
- co-construction of a coherent narrative

Child led with caregiver as helper or witness

When the caregiver is helper or witness, the aim is to focus on the child's self-directed art making. This would have similarities to a traditional non-directive approach, which allows the child to find the expressions they need. The therapist can model ways of being curious without directing, and the therapist, caregiver and child can reflect on the art making together in order to gain understanding of the child.

Joint engagement

In joint engagement, both the child and the caregiver are involved in art making in the sessions, whether on the same, or separate pieces, or a mixture of both. Reflecting on their art making together, they can develop their understanding of each other and their motives.

Co-construction of a coherent narrative

When a narrative needs explaining or exploring, the therapist can support the caregiver to co-construct something coherent with the child, so that meaning is made in a way that the child can understand. The therapist and caregiver plan carefully to decide what information is appropriate to share, and how it will be communicated, which includes rehearsing together.

The main approach that is described in this chapter is 'joint engagement', although at other times in the therapy, the other two approaches were also employed. It is important to point out that the joint engagement approach would not normally be considered to be directive, per se, but in this case, it was felt that having some themes and materials chosen by the therapist could be a helpful way to work with the dyad for a limited number of sessions. One of the key factors in the use of any of the DAP models is the importance of preparing caregivers for their role in the sessions. An example of this, with Sonny and Claire, is detailed in 'Preparing for the Sessions'.

Sonny and Claire

Sonny was referred to art therapy when he was nine years old. He lived with his aunt Claire, together with his sister, who was older by ten months. Sonny and his sister had lived with Claire (whom they called 'mum') since Sonny was ten months old, and prior to that, in foster care since birth. There had been inconsistent contact with birth parents, who were substance users. It had been identified that both children were displaying controlling behaviours in different ways. The sister was appearing very grown up; taking on a bossy parental role with regards to Sonny, beyond what might be expected at this developmental phase. Sonny had difficulties with peer relationships, self-esteem and angry outbursts which would sometimes result in some violent acts towards his peers or family members, especially if he didn't get his way or was confronted. He was described as being domineering, reactive and powerful by school staff. Claire had concerns that Sonny had ADHD, but at this time a formal diagnosis had not been made. Sonny had withdrawn into a world of digital gaming, and was also consuming large quantities of unhealthy drinks and snacks.

Identifying the aims

Together with the social worker and school staff, Claire and I identified these aims and objectives for work with Sonny using a multi-disciplinary approach:

- Sonny and Claire to engage in Dyadic Art Psychotherapy, in order to build on their connection and understanding of each other.
- Claire to have regular 1:1 caregiver sessions with the therapist. This is to reflect on how the therapy is going; to explore her own attachment issues and to support her with strategies to feel confident in being in charge at home.
- Claire to continue to enforce boundaries at home (especially around food and use of electronics) and make this a priority for the time being.
- Family to have home visits from the social worker.
- Social worker to refer Sonny to a mentoring scheme so that there is a male role model in Sonny's life.

Preparing for the sessions

The first time I met Claire was in a meeting at Sonny's school, with other school professionals, and also the social worker from the Local Authority Kinship Team. Claire was very honest about the difficulties the family was facing; she took responsibility for not always enforcing boundaries and was asking for help. It was vital that Claire was to develop a sense of agency in order to help her and the children feel safe. Claire received great encouragement from everybody present at the meeting, with some helpful strategies to put in to place immediately. With support and affirmation from the professional team, Claire was able to see that what she might request from Sonny, in terms of limits, was not unreasonable. It was felt that if she could weather the storm initially, she would see a difference. She managed to do this, and within weeks there were positive changes in Sonny's gaming and eating behaviours.

It was evident that Claire, at times, felt overwhelmed by lone parenting, and the challenges that came with it – both fiscally and mentally. Claire's own feelings of insecurity and lack of confidence were impacting on the children. Claire was very willing to think about making changes, but found it difficult to enforce boundaries. Part of the reason for this was her feeling like she was not in control, and that she becomes 'terrified' of Sonny and therefore doesn't take charge. It became clear in the assessment sessions with Claire that in her own childhood she had experienced fear and control at the hands of an abusive parent, and that addiction was evident in her family history. I met with Claire four times before meeting with Sonny. We decided that Sonny and Claire would attend sessions together, building in 1:1 sessions with Claire at regular intervals. At this point, we had discussed using the approach of caregiver as helper or witness.

As part of the preparatory 1:1 sessions with Claire, I assured her that it was my role to hold the boundaries in the sessions, and I explained the difference between art making in therapy and in other contexts. I modelled some phrases that might help Claire to reflect on Sonny's art making. These included making encouraging remarks, but avoiding comments such as 'that's good', as we wanted to be exploratory without judgment. Eye contact, facial expressions and affirmations were to be welcomed.

Initial sessions with Sonny and Claire

At this stage, Claire was to be helper or witness to Sonny's self-directed art making. In the first session together, Sonny greeted with me a big smile, and immediately sat himself down in the comfy chair. I let him know that I would tell him about what to expect from art therapy, and that he could use any of the materials in the room to make what he liked. He soon got up and started to investigate what was in the room. He brought some paper and pens to the table, and started on a drawing of a character from a digital game. I asked him what he would like mum to do, and he said 'finish the picture'. He didn't give any further instructions, and Claire didn't know what to do. Sonny said Claire had 'failed'. I pointed out that no instructions had been given, so it must be difficult for mum to know what to do. The drawing was then abandoned and Sonny went on to find different materials.

The next piece that Sonny made in that first session showed a self-imposed limit – choosing to leave some materials for another time. I was encouraged by this, and shared this with Claire in our first caregiver session. In his other art making, Sonny displayed desires to use large quantities of materials such as paint, glue and glitter, often using them until they had run out. I was mindful not to replace these materials straight away. In one art piece, Sonny expressed himself as being 'A God', in the context of him being in total control. In the same session, he made a wrestler's belt, which could indicate the power struggles, both externally, and internally.

When I saw Sonny and Claire at their third session together, Claire was very concerned over Sonny's increasingly aggressive behaviour. I observed that Sonny was dominating Claire: he would speak over her; pick up her phone without permission, ignore her and deride her suggestions. I felt it was important to foster a more respectful interaction, and for them to develop a capacity to work together. I proposed that the next few sessions would be much more directive and I would set tasks for them both to do. This was with the aim of providing tighter boundaries in order for both Sonny and Claire to feel more held; and the setting of limits regarding materials, which would hopefully, in time, help Sonny to set and maintain his own limits.

The directed sessions

At the next session, I had two identical sets of materials laid out on the table ready for them. Sonny went to start immediately, but I explained that they would both need to listen to the rules. The instructions for making would change depending on the task I had set, but the basic rules for any task were:

1. Listen to the instructions set by the therapist.
2. Use only the materials put on the table.
3. Listen to each other when talking about their art.
4. If discrepancies occurred, both parties should agree before moving on.
5. Finish when instructed to do so.

Reluctantly, Sonny listened to and followed instructions. There were several times during this session that he reminded both me and Claire that they (Sonny and Claire) needed to agree before moving on. This gave him the chance to explain himself if a request was made. It also gave Claire a chance to explain why she might say yes or no to something, which I guessed was missing a lot of the time at home.

I noticed that Sonny was always moving, and found it impossible to be still. When it was his turn to listen to Claire talk about her art, I gave him some blue tac to play with, which helped him to be busy whilst listening to her. I suggested that perhaps Sonny could also use blue tack in class to help him concentrate.

The activities that were used are set out as follows, in the order that they were employed over seven sessions. Some of them could also be used as directed warm-up exercises at the start of sessions, before the non-directive phase begins, where the themes and materials are often client-led. Therapist comments on each session are added in italics.

1 *The squiggle game*

One person makes a squiggle. The other person then finishes the squiggle into something recognisable and says what it is. This is repeated several times and each participant takes a turn at starting. This is done on one piece of paper, and means that one person is active (drawing) and the other is passive (watching) for a very small amount of time. Winnicott (1971, p. 3) used this game to help him 'get in contact' with child clients at the start of therapy. *Sonny and Claire enjoyed this activity and both said they had 'fun'.*

2 *Building up a joint picture, taking turns*

Each participant starts a picture, and has one minute to do this. After one minute, participants swap pictures and the other person continues the picture the other has started, for one minute. This continues for five minutes. After five minutes, there are two pictures, and participants can take it in turns to discuss the pictures (if one person starts discussing one picture, the other starts the discussion about the next picture). *Sonny and Claire enjoyed the time-limited aspect of this activity and a count-down to 'time up' increased their excitement and sense of fun.*

3 *Extending the picture outwards*

The therapist has prepared two pieces of paper – one small (about 10cm squared) and the other large (A2 size) with a window cut in the centre just slightly smaller than the first piece. Each participant has five minutes to make an abstract picture on the small piece of paper, filling it with coloured lines and shapes. After five minutes, they stick the window sheet over their picture, so it shows through in the centre. They swap pictures. Then the other participant extends the picture outwards, starting from where the other picture left off at the centre of the page. The purpose is to tune in to the mark-making of the first picture; to 'listen' to the other person's style, which can facilitate thinking about someone outside of oneself and aid co-operation. It can also help to develop an appreciation of another person's perspectives and motives. *Claire responded in a way that replicated Sonny's mark-making exactly, and she focused on this rather than filling the page. Sonny used Claire's style of mark-making to completely fill the page, and Claire was surprised that Sonny had paid attention to her visual style.*

4 Making something for the other person

In this activity, I chose pipe-cleaners as they are straightforward to use but quite malleable, but the materials can be varied. Five minutes for making. This is to facilitate thinking about the other person, and the act of giving and receiving a gift, which can invoke warm feelings. I would then suggest each participant repeat the process, but makes something for themselves. *Sonny made Claire a heart, which Claire was surprised by. She was visibly moved by the tenderness of this symbol.*

5 A picture describing a particular event

This can be used when there are different perspectives about a situation that could do with being explored. Ask the participants to represent in a picture, what happened. Set a time limit. Their outcomes can then be discussed, and questions asked about what is happening in the picture, and about any feelings that were felt at the time of the event, or making the picture, or looking now at the picture. *Sonny found this particularly difficult. He denied that a particular event that Claire identified as happening, so I asked him to represent what he thought was happening at that time and place. At first he said he wasn't going to draw anything, but eventually chose to represent a character from a YouTube video. It highlighted how Sonny had a tendency to disassociate when in challenging situations.*

6 A favourite place

Each person represents their favourite place, with the materials the therapist has chosen. Bringing to mind such a place can help a person to regulate and can invoke memories of happy times. This can bring about a sense of safety, so is a useful one if someone is feeling a bit anxious, or if something challenging has arisen, or at the start of therapy. *Claire drew a beach scene, which Sonny instantly recognised and recalled memories. Sonny drew the American flag, stating that America was his favourite place, despite never having been there. In the next caregiver session, I suggested to Claire that perhaps Sonny has not yet internalised a safe place, and so working on a sense of safety would be paramount in our work together.*

7 Making something with a specific set of materials and objects

This activity is designed to encourage creativity, but also to encourage negotiation and co-operation between the participants. Set out on the table exactly the same amount of materials and objects each – I had chosen a cardboard tube, two cotton reels, some coloured string, two paper straws, two craft nests/baskets of different colours, ten coloured matchsticks, two googly eyes, four coloured beads, two wooden pegs and some colored tape. They also had clear sellotape to share. Participants don't have to use all the materials; they can be selective. They could ask the other person to swap colours or items if they wished, without obligation to swap, and they must respect the others' decision. Thirty minutes for this activity. *Straight away, Sonny asked if he could swap colours of the string, straws and nests. Claire obliged immediately. I reminded both of them that they didn't have to swap if they didn't want to. A lot of the time Sonny was trying to tell Claire what to do with her materials, and I encouraged them both to focus on their own piece. Near the end, Claire asked Sonny if she could use his beads, if he wasn't going to use them. Sonny said yes. They made very different outcomes – Sonny made a 3D 'blue monster with baskets as ears' and Claire made a collage picture of a house that was 'homely'.*

8 Making the other person's favourite animal using modelling material

For this activity, we used play-doh. The theme means that the participants are thinking about each other's likes and preferences. Set a time limit, perhaps five or ten minutes, or more if using clay. *Sonny made a giraffe as Claire's favourite animal. At first he tried to make it 3D but found this difficult. He saw that Claire was making her piece in relief and tried that instead, with more success. This was a good opportunity for Sonny to choose to follow Claire's way of doing things without being instructed. Claire made a dog for Sonny's favourite animal. This led on to a discussion about a dog they used to own, which they couldn't keep due to Claire's work commitments. Sonny was clearly very upset about this, which came across initially as being angry, and gave him the opportunity to express this and mourn the loss of his favourite animal.*

9 Guessing objects, blindfolded

One person is blindfolded, the other chooses five objects, one object at a time to be put in the other person's hand for them to guess what it is. This requires a degree of trust between the participants, or to develop trust if needed. *I asked Sonny and Claire who would like to be blindfolded first. By this stage, Claire was brave enough to state that she would, as Sonny usually went first. Sonny said he was going to be blindfolded first. Claire suggested that she go first in this activity and Sonny could go first in the next one. Sonny was adamant he was going first, but Claire stood her ground also. I intervened, and pointed out that Sonny had always gone first in the other activities, which included talking about their art, and that it was perhaps Claire's turn now. Sonny reacted to this by saying he wasn't going to play. I said that was OK, and perhaps Claire and I could play for a while and Sonny could join in when he felt ready. Claire was blindfolded, and I went to pick up a first object. I took my time thinking about this, and as I went to pick something up, I looked over at Sonny and raised my eyebrows as if to say 'what do you think about this one?' He immediately came over and picked it up, and with a smile on his face, placed it gently in Claire's hands, and he was back in the game. They both enjoyed this game, and thought it might be something they could play together at home.*

10 Story telling with toys chosen by the other participant

This is another activity designed to foster a collaborative approach when co-operation is tricky. Each participant chooses five toys/miniatures, and they are instructed beforehand that the other person will be telling a story with them. The other person will tell a story that is different from the one the chooser might imagine when choosing, so there are two potential stories; only one will be heard. The person listening is then asked to repeat the story they have just heard, using the same objects in the same way the storyteller has used them. This helps with listening skills and the capacity to see something from another person's point of view. *Claire appeared a little distracted when listening to Sonny's story (I wondered if she was planning her own story in her head), but she repeated back exactly what had happened. When Claire was telling her story, Sonny was humming and handling one of his objects chosen by Claire for his story – a giraffe (Claire's favourite animal from a previous session). Occasionally this interfered with Claire's storytelling. She told him to stop, explaining that she couldn't concentrate, and he stopped. Sonny re-told Claire's story, almost exactly, but with the addition of an event that neither Claire nor I could recall in her story. Sonny was adamant it had happened, but I suspect it was a strategy to take over the story in his way.*

11 Cartoon strip/graphic novel

This is a good one to do after activity 10, in case anyone has any untold aspects of a story they still wish to tell. Give each participant a large sheet of paper, and ask them to draw some boxes/cells (six to eight) in which the story is shown. This part might be a bit tricky for some, so you could have some drawn up ready. The boxes/cells can be different shapes and sizes, or regular. Each person will draw their story in the boxes/cells. Set a time limit, perhaps 20 minutes. When time is up, they can then 'read' their own story, or the other person can have a go at 'reading' it, depending on the dyad. *As I had suggested six to eight boxes, this wasn't a boundaried enough instruction. Sonny wanted to do six, Claire eight. Even though I said they could do different amounts, Sonny insisted that Claire do six boxes, like he wanted to do. Claire stuck to her guns, and drew eight boxes. Sonny refused to play. After seeing that Claire was continuing, he began drawing again, but insisted on not using any boxes at all. The individual images he drew seemed unconnected and uncontained. There was a dot-to-dot rocket; an alarm clock; a figure saying 'let-go' and he drew round his hand several times. Sonny's story was dis-jointed. Claire drew a story of a girl getting up, having breakfast, tacking up a horse, going for a ride and coming home again.*

12 Messy play

This activity was done just before a break for the summer holidays, and is a good one to build up to. Sonny had requested to work with these materials in the previous session, and we agreed that would happen. Each participant has their own bowl, spoon (if they want one), and ingredients for messy play. I used a jug of water, cornflour, PVA glue, washing up liquid and paint. Set a time limit, perhaps 20 or 30 minutes, depending how the participants usually react to these materials, and leave enough time for cleaning up. *Sonny and Claire both wanted to make something like slime. Sonny asked me for 'borax', which I don't stock, and he accepted its unavailability. Claire's slime managed to hold; Sonny's was very runny and he asked to start again, asking Claire how she had made hers. Both were interacting playfully with each other; feeling each other's mixtures and sharing their experiences of it. It was lovely to witness this warm encounter.*

Reflections

Throughout these activities, the therapist set out which materials Sonny and Claire were to use, and gave clear instructions. As they were both being directed by the therapist, it was a welcome relief for Claire that the boundaries were held by someone other than herself. It meant that she was not in such an anxious state and could relax more. It then gave her the freedom to be more playful, something she said she'd almost forgotten how to be.

Winnicott (2005, p. 72) states that 'if a patient cannot play, then something needs to be done to enable the patient to become able to play'. He goes on to suggest that 'only in playing is communication possible' (ibid., p. 73). Claire and Sonny needed to develop their ways of communicating with each other. When they were working alongside each other with direction, Sonny was not constantly butting up against Claire, which allowed them to develop other ways of communicating and being together. Having the therapist as a third perspective; to witness and reflect their interactions allowed for space between them to develop, and in that 'potential space' (Winnicott, 2005, p. 55), creativity could begin to occur.

These joint engagement tasks produced 'moments of attunement when warmth and playfulness [could] emerge' (Hendry and Taylor Buck, 2017, p. 124). This is especially needed if these qualities are seemingly sparse, and the relationship is at risk of breaking down. The reflective conversations can help the participants to develop a sense that other person has their own thoughts, feelings, perspectives and motivations, and they can begin to understand these. This ability to mentalise, or to imagine what someone else is thinking or feeling, through interactions with them, helps an individual to develop their own psychological mindedness, and is crucial in order to understand the behaviour of others (Fonagy et al., 2018).

By caregivers becoming more attuned, they can develop an appropriate shared emotional language and develop or re-establish a connection with their children. These are key factors of healthy attachments as outlined by Bowlby (1969, 1988) and Stern (1985).

The use of directed activities, as well as the 1:1 sessions with the Claire, were also intended to empower her with knowledge about the development of parent-child relationships; trauma-responses and attachment styles. A psycho-educational approach can help to strengthen the therapeutic alliance, and can be a valid brief intervention its own right (Springham, 2019, p. 27).

Of course, it didn't all go smoothly, as seen in activities 5, 8, 9 and 11. However, it would have been foolhardy to expect it to, and it all produced plenty of therapeutic material. I think it may also have been wise for me to put some limited materials on the table in the first session. But then again, I may not have witnessed the extent to which Claire was frozen and Sonny was displaying the controlling behaviours. After these directed sessions, there was a break for the summer holidays, and Claire reported they had a 'brilliant' time and there was a 'vast improvement'.

The joint engagement activities outlined in this chapter formed only a part of the therapy process with Sonny and Claire. It was devised at a time that they needed to be held very securely in order for them both to develop a sense of safety and containment, and trust in each other. Claire would go on to hold more of a witnessing and reflective role. It was also proposed that when the time was right, space would be given for co-construction of a coherent narrative in joint sessions; for Sonny to explore his very early life, with help from Claire, in order to help him with his sense of identity. This, however, was some way off.

Conclusion

In this chapter, some directed methods for working creatively with a child and their caregiver have been described. It has been suggested that whilst traditional Art Psychotherapy might generally be considered a non-directive intervention, there may be times when a more directive approach could be employed. The activities described here could be used in different therapeutic contexts, where attachments could be strengthened. It may be that listening and collaborating are sometimes problematic, and a different way of communicating and being with each other can be experienced through engaging in these activities. Although the activities are directed, they allow for creative responses and a sense of playfulness to emerge, because neither participants are 'in charge'; they are co-creating. It has also been pointed out that these methods used may not form the entirety of the therapy, but could be useful in the beginning stages to help with engagement and building connections before more exploratory work begins.

It has also been highlighted that, as with any dyadic work, preparing the caregiver before sessions, and offering space for reflection is an important part of the work. In this particular case, the reflective sessions also gave Claire an opportunity to receive encouragement from the therapist, and thus increase her confidence.

Part of the play that happened in this particular case, was the development of a joint engagement approach that employed directed activities. As therapists, we encourage our clients to be playful, with an element of risk-taking to help them try new ways of being and relating. I am grateful to Sonny and Claire for giving me the opportunity to be creative in the approaches that I use as a therapist.

References

Booth, P. (2010). *Theraplay*. Third Edition. San Francisco: Jossey-Bass.

Bowlby, J. (1969). *Attachment and Loss: Vol. 1*. New York: Basic Books.

Bowlby, J. (1988). *A Secure Base: Parent-Child Attachment and Healthy Human Development*. London: Routledge.

Fonagy, P., Gergely, G., Jurist, E., and Target, M. (2018). *Affect Regulation, Mentalization, and the Development of the Self*. Third Edition. Abingdon: Routledge.

Hendry, A., and Taylor Buck, E. (2017). Dyadic Parent-Child Art Psychotherapy with Children Who Have Been Exposed to Complex Trauma. In A. Hendry and J. Hasler (eds.) *Creative Therapies for Complex Trauma: Helping Children and Families in Foster Care, Kinship Care or Adoption*. London: Jessica Kingsley Publishers.

Liebmann, M. (2004). *Art Therapy for Groups: A Handbook of Themes, Games and Exercises*. Hove: Routledge.

Norris, V., and Lender, D. (2019). *The Practitioner's Guide to Theraplay*. London: Jessica Kingsley Publishers.

Proulx, L. (2003). *Strengthening Emotional Ties Through Parent-Child-Dyad Art Therapy: Interventions with Infants and Preschoolers*. London: Jessica Kingsley Publishers.

Silverstone, L. (1997). *Art Therapy – the Person-Centred Way: Art and the Development of the Person*. Second Edition. London: Jessica Kingsley Publishers.

Silverstone, L. (2009). *Art Therapy Exercises: Inspirational and Practical Ideas to Stimulate the Imagination*. London: Jessica Kingsley Publishers.

Springham, N. (2013). Calling an Art Therapy Amnesty. *BAAT Newsbriefing*, pp. 11–15, December.

Springham, N. (2019). BAAT CPD Courses; Art Therapy Based Psychoeducation. *BAAT Newsbriefing*, pp. 27–28, Summer.

Stern, D. (1985). *The Interpersonal World of the Infant: A View from Psychoanalysis and Developmental Psychology*. New York: Basic Books.

Taylor Buck, E., Dent-Brown, K., and Parry, G. (2012). Exploring a Dyadic Approach to Art Psychotherapy with Children and Young People: A Survey of British Art Psychotherapists. *International Journal of Art Therapy*, 18(1), pp. 20–28.

Taylor Buck, E., Dent-Brown, K., Parry, G., and Boote, J. (2014). Dyadic Art Psychotherapy: Key Principles, Practices and Competences. *Arts in Psychotherapy*, 41(2), pp. 163–173.

Winnicott, D. (1971). *Therapeutic Consultations in Child Psychiatry*. London: Hogarth Press, Institute of Psycho-Analysis.

Winnicott, D. (2005). *Playing and Reality*. Classics Edition. London: Routledge.

21
THE "TERROR OF THE SCHOOL" LEARNS TO PLAY

Galila Oren and Susana Pendzik

Introduction and theoretical overview

This chapter describes the process of one year of work by dramatherapist Galila Oren with a second-grade child with severe behavioural problems, who was described by the school staff as "the terror of school". The process is analysed using the Six-Key Model for assessment and intervention, integrating notions of Jennings' EPR paradigm and Landy's role theory. While the authors refer to psychological theories of child-development, the chapter strongly emphasizes that the use of dramatherapy models for assessment, analysis and intervention furthers dramatherapy thinking and strengthens the therapists' sense of professional identity.

Galila Oren, dramatherapist and supervisor, was first trained in NYU by Robert Landy, and later by Sue Jennings in Israel. She integrates dramatherapy with psychoanalytic training in her work with children and families.

The Six-Key Model is founded on *dramatic reality* – a core dramatherapy notion that can be defined as "the manifestation of dramatic imagination in the here and now" (Pendzik, 2018, p. 189). Dramatic reality is an island of imagination that becomes materialized through dramatic as well as other artistic means. In contrast to fantastic reality (Lahad & Leykin, 2012), where "the locus of the *as if* is in the imagination" (Pendzik, 2018, p. 189), dramatic reality gives the *as if* an actual form, so that it is perceived also by other people.

Although there are various genres of dramatic reality, including games, sport, ritual, the aesthetic genres (theatre, dance, and music) (Schechner, 2005), the plastic arts and literature, play is perhaps the most common form of dramatic reality employed by children to "download" the world of imagination. Play is the main tool through which children make sense of the world around them and understand their place in it (Cattanach, 2008). Any dramatherapy or play therapy session involves a roundtrip ticket into dramatic reality and back. This journey makes it possible for a therapeutic experience to occur, in which the contents poured into dramatic reality are expressed, explored, witnessed and validated. In fact, through contact with the potential realm of dramatic reality the contents are elaborated and become transformed (Pendzik, 2006).

People's handling of dramatic reality can provide information about their psychological makeup, social patterns, cognitive abilities, inner world, resilience strategies, etc. Therefore, the Six-Key Model considers all angles pertaining to dramatic reality and its connection with actual life. Dramatherapy processes are conceptualized through six categories (keys) that inform the therapeutic experience. Each key illuminates a particular area, while simultaneously shedding light on the therapeutic process as a whole (Pendzik, 2012).

Table 21.1 describes the domain of each key. The First Key (Passage) refers to the person's ability to transition back and forth from normal to dramatic reality. The Second Key (Quality and Style) assesses the person's capacity to engage with dramatic reality in a way that is "good enough" (Winnicott, 2005) in order to meaningfully explore contents. The Second Key also reflects the individual's personal style – the kind of dramatic reality that a client tends to engage with (for example, play, games, embodied activities, cards, miniatures, role playing, etc.), as well as its unique style. The First and Second Keys are vital to dramatherapy processes, for without them there will be no dramatherapy nor play therapy. Furthermore, they reveal basic aspects of a person's psychological profile, such as ego-functioning (cognitive and emotional), interpersonal and attachment issues, trauma experiences, and more. These Keys also invite thoughts about the EPR paradigm (Jennings, 2012), and the BASIC PH coping and resilience strategies (Lahad, 2017).

The Third Key (Roles and Characters) maps out a client's role-system, providing information about role repertoire (Moreno, 1987), as well as other details pertaining to the role-system (Landy, 2009; Landy & Butler, 2012). The Fourth Key comprises the content patterns that emerge throughout the therapeutic process – themes, plot, conflicts, symbols, etc. As an integrative tool, the Six-Key Model intersects with the therapeutic approach embraced by the therapist. Thus, a dramatherapist with a Jungian orientation may perceive the contents in a different light than one coming from a narrative perspective; and each may act on the contents according to their own theoretical framework (Pendzik, 2008).

The Fifth Key (Response) involves the clients' response to dramatic reality, whether through ascribing meaning to it, criticizing it, or reflecting about it. It provides information about

Table 21.1 The Six-Key Model

1st key Passage	Transitioning between realities: entering and exiting dramatic reality	✓basic ego-functioning (cognitive and emotional) ✓interpersonal issues (intimacy, trust, collaboration) ✓attachment issues
2nd key Quality & Style	Creating a good enough *as if* and defining its unique style	✓traumatic experiences ✓EPR ✓BASIC PH
3rd key Roles & Characters	Role-system structure	✓role-repertoire ✓role-system plasticity (flexibility/ rigidity) ✓role-development ✓role/counterrole/guide system ✓ability to move from role to character
4th key Content Patterns	Themes, plot, conflicts, symbols	✓main/secondary issues ✓unfinished business ✓problem saturated stories and plots ✓recurring conflicts ✓imagery world
5th key Response	Self-reflecting of the client about the experience in dramatic reality	✓self-reflection capacity ✓self-esteem ✓self-criticism ✓integration capacity (life-drama connection)
6th key Meta-level	Subjective, unconscious, and unexpressed contents	✓transference/countertransference ✓unconscious process ✓repressed memories ✓unprocessed trauma

people's capacity for self-reflection, possible issues connected to self-esteem and self-criticism, and hint at their ability to integrate the life-drama connection (Jones, 2007).

The Sixth Key (Meta-level) is the locus of subjectivity, the depository of unexpressed, unconscious contents elicited during the therapeutic process. This Key accounts for transference-countertransference phenomena, repressed memories or emotions that may be on the verge of surfacing, as well as holding collective processes (national, global), the numinous, and other ineffable contents.

The Six-Key Model is a qualitative, non-interpretative method based on the subjective observation of the dramatherapist of single sessions or across processes. The keys are not hierarchical and are organically connected to each other, creating together a gestalt of the process. Therapists may use them as a "anchor points" for reflection (Smitskamp, 1996), as they may help to make sense of the complexity of the sessions, and point at particular keys from which interventions will be more effective.

In the following case study, we will show how the Six-Key Model can be used to create a map that provides direction in dramatherapy work, as well as help the therapist to integrate other theories and concepts, while keeping the focus on dramatherapy-based thinking.

Case study: the "terror of the school" learns to play

> Ah, how oft we read or hear of
> Boys we almost stand in fear of!
> For example, take these stories
> Of two youths, named Max and Maurice
>
> Who, instead of early turning
> Their young minds to useful learning,
> Often leered with horrid features
> At their lessons and their teachers.
>
> *(Wilhelm Busch, 2009/1902)*

Nir (pseudonym) is a child in second grade who had been portrayed as "the terror of the school". He usually skipped classes, hit other kids, and got in trouble with everyone. His parents' divorce had been quite violent, and his father had a history of psychiatric hospitalizations. When Nir started therapy his father was functioning normally at work and was mentally well. The parents lived in the same neighbourhood. At the beginning of the treatment, Nir saw his father very often, although he didn't have visitation permissions. Nir was treated at a public mental health institution. His mother was in therapy with another therapist from the team.

First period: connecting with the wild and scared kitty (from Galila's diary)

In the first couple of sessions, I was astonished to find a cute, handsome, quiet, and shy boy. But quickly enough the tables turned around, giving place to a turbulent period in our relationship: Following my suggestion to play with hand puppets, Nir started hitting them hard and throwing them all over the room. He also began throwing them at me on purpose; and when he managed to hit me, he ran away from the room. Out of helplessness, feeling worried and hurt, as I looked for him everywhere, I decided to turn my search into a game. I said out loud (as if playing a game of hide-and-seek with a little baby) "Where is my little kitty. . .? Maybe he is behind the door. . .? Maybe under the table. . .?" Defining the search as a game helped to calm Nir down: He looked out and fled, he hid and was found; and thus, we played until he gradually returned to the room.

As he kept on hitting the puppets and throwing things at me in the following sessions, I suggested teaching him to hit them as real actors do in the movies: "as if". We rehearsed for hours beating up the puppets "theatrically" in different styles; we spilled red gouache on them, the way that ketchup is poured on actors in films. Often his play became wild, and I had to hold up the rules again, as a theatre teacher would do. To cope with his running away from the room, we played tag and hide-and-seek, working a lot with the body. Slowly, the mood calmed down, the quality of the play became a bit more stable, and we moved to games that comprise elements of projection (in addition to playing hide-and-seek, "hot-cold", and so on).

At some point I suggested playing calmer games and Nir chose "war cards". Eventually, this game also became wild. Nir began to cheat openly: He secretly stole powerful cards and changed the rules every time he was about to lose. I experienced painful emotions: I felt helpless, faced with a wild and domineering rival. I felt that I was living in a world where the rules were arbitrary, where I had no chance of succeeding. I was filled with despair. The quality of the game became rough, and the borders between ordinary and dramatic reality were often disrupted.

In an attempt to reinforce the boundaries of the game, and to pour my own feelings inside dramatic reality, I said at some point: "No wonder that you want all the powerful cards: Every warrior, before going to the battlefield, should ensure that he has the appropriate weapons." I suggested that before we play the "war game" Nir should deal the cards in a way that he knows exactly what weapons he has and which ones I have. We started playing cards like two armies: a powerful one and a weak and helpless one; the strong army could change the rules if it stood in danger of loss. This intervention calmed Nir down. The quality of the game improved, and we revisited together the stages of dramatic play development (EPR).

Towards the end of this period one could say that Nir was able to enter and exit dramatic reality reasonably well. The boundaries between game and reality were stabilized, and a bond of trust had been established. A truce was felt, and occasionally Nir even asked to play "normal" war (i.e. playing by the rules) – without fear and without losing his temper.

Analysis of the first period

During the first months the work focused on the First and Second Keys. The therapy dealt mainly with developing the ability to enter and exit dramatic reality. Reinforcing the boundaries between the two realities is essential to develop the ability to stay in dramatic reality without being overwhelmed with destructive aggression. Winnicott (2005) views the instincts as a threat to play and to the ego. He states that in order for play to contain an experience in a way that facilitates development and enables therapy, it is necessary for the client to learn to play in a healthy manner. Following the child's failed attempts to play, the therapist's task is to structure the spontaneous play, so that it will enable a distinction between realities.

In this case, the Sixth Key – meta-reality – contained experiences associated with "projective identification". According to Klein (1946), this concept is an initial defence mechanism. Bion (1962) and Ogden (1979) see it more as an unconscious and primal way to communicate emotions that cannot be accepted by the psyche: Clients project onto other people emotions and feelings that they cannot tolerate. Others experience these emotions "as if" they were their own, and sometimes even act accordingly. This primary form of communication is present during early infancy, when the baby cannot yet speak or mentally integrate experiences and is overwhelmed with feelings and sensations. The role of the environment during infancy (and of the therapist during adulthood) is to acknowledge these feelings, process them, and return them to the client in a way that they can be contained. Bion (1967) calls this process "reverie". Turning emotional subtext into structured play is a way to induce reverie while developing the client's ability to play in the dramatic reality.

Galila experienced intense emotions as a therapist: She felt attacked, helpless, and in despair. These feelings threatened her ability to play, and it seemed as if the space was filled with hidden contents – feelings and emotions that Nir apparently could not contain. Such intense feelings usually indicate an activation of the Sixth Key. When the Sixth Key is turbulent and charged, it is necessary to turn the inexpressible subtext into overt content (Pendzik, 2012). In verbal psychotherapy this is the place of interpretation. In dramatherapy, however, the initial therapeutic goal (particularly with children with communication difficulties) is to help the client to develop an ability for healthy play. Hence, instead of interpretation, Galila structured the play in a way that would enable the expression of the subtext, while skipping questions such as "why does he feel this way?" or "what's the source of the helplessness?" – which would involve making interventions in the Fourth Key (content patterns). The intervention made in the Sixth Key involved channelling the elicited contents into dramatic reality (First Key) – which in turn improved the quality of play (Second Key). Introducing the hidden content into dramatic reality helped to stabilize the quality of play. In terms of Jennings's (2012) EPR Paradigm, play progressed from wild and destructive embodiment to projective play (the cards were the weapons), eventually evolving towards role (armies).

Second period: playing in hiding – the knowledge of the second key

One day, while we were playing hot-cold, I hid a miniature figure in the sandtray. Nir really enjoyed looking for the toy inside the sand and went on to hide things there. He started playing with the figures in the sand and asked me to sit with my back towards him, so I wouldn't see what he was doing. At first, he was afraid that I would peek, and kept checking on me; but over time he relaxed and immersed himself in dramatic play on the sandtray.

At his request, I sat with my back towards him for several weeks. Every so often I would hear sounds of war or pain in his play; but otherwise, I was disconnected from the contents of the play. Nir played quietly, with concentration and enjoyment.

When I presented Nir's therapeutic process at a staff meeting, a heated discussion ensued. Some of the staff believed that we may be replaying a traumatic experience that occurs in the family, in which the mother is present, albeit unaware of what is happening to the child. They believed that it was my role to disclose the secret. Other members thought that respecting the boundaries that Nir requested and allowing him to play without inquiries was part of an important process of building a relationship and restoring Nir's trust in adults.

For several weeks I let Nir play while my back was turned. Every time, after playing alone, he would calmly ask me to play hide and seek or a card game with him. We played calmly and enjoyably.

One day, after the usual sandtray round, he took two lion figurines and started playing with them on the rug next to me: The lions were fighting each other in plain sight! When I asked about them, he said: "I'm training the lions, just like you're training me". "How am I training you?", I asked surprised. "You teach me to pretend to fight, so that I don't fight for real", he answered.

This statement deeply moved me. I was amazed at his ability to give meaning to dramatic reality. At this stage there was a certain improvement at school, and the behavioural problems had slightly decreased.

Analysis of the second period

The atmosphere quietened as Nir's ability to enter and exit dramatic reality improved. Playing became enjoyable and meaningful. The sandtray created a clear boundary between the two realities, allowing Nir to enjoy the sensory sand and move back and forth from Embodiment to Projection. The game of "hide and seek" expanded from being a bridge in and out of dramatic realty, to developing projective play (hiding the figures in the sand), as well as giving Galila the role of the "blind witness."

The question raised by the staff regarding playing behind Galila's back is diagnostically crucial: Is this the re-enactment of a traumatic experience that Nir is going through, in which the adults don't see what's happening under their noses? Or is this about building trust with a boy who is afraid to tell his story?

Looking at Second Key can help with this question: When play is mostly a repetition of trauma, it does not heal. The quality of the dramatic reality generated in these cases may be boring, turbulent or fragmented, but by all means, not "good enough" (Winnicott, 2005); leaving dramatic reality may be sudden and accompanied by frustration, rage, or other discontents. However, when play contains an experience, there is a sense of deep concentration and aesthetic enjoyment even if its contents are difficult (Pendzik, 2008); the boundaries of play are maintained, and exiting is accompanied by a sense of fulfilment.

Nir was deeply invested in the sandtray. After playing there, he was calm, relaxed and fulfilled. In his article "the capacity to be alone", Winnicott (1958) emphasizes the importance of playing in the object's presence for the purpose of building trust in the object and in oneself. This kind of play helps in itself to develop ego strength and enables working through difficult emotions. Galila's diagnostic decision was based on the Second Key: The quality of the play reassured her that it was important not to disrupt Nir's play behind her back.

Third period: the scary man and the innocent child – Third and Fourth Keys

During this period, Nir created a play in which I played a boy and he played all the other roles. The boy meets a man who seemed to be good, but suddenly changes and demands from him to commit crimes and steal for him. At first the boy refuses, but the man threatens him and becomes very cruel. Scared, the boy commits the crimes, and while stealing, a policeman appears, and he is caught. The boy is brought to trial and the judge sentences him to life imprisonment with hard labour. In jail, he suffers from a cruel prison guard. When he is finally released, the man is waiting for him outside, and the story repeats itself.

We played this harsh plot repeatedly for several weeks. While the boundaries of dramatic reality were maintained, Nir was a domineering director and scriptwriter, making every tiny decision regarding the plot. After performing it a few times, the domineering quality softened, and I had more freedom to invent text for my character. In the role of the boy, I began to express feelings: I shook with fear when facing the bad man and begged for my life in front of the judge. In prison, I sang songs that voiced my helplessness and suffering. Although Nir paid careful attention to my texts, the cruel plot was not altered by them.

The staff members raised the possibility that Nir was being abused by his father. It was clear that Nir had a close relationship with his father and was afraid of him. We knew that his father had dragged him into many arguments and the staff debated whether to contact the social services or not.

In one of the sessions, while the following scene developed between the Judge (Nir) and the Boy (Galila):

Judge: Thief, what do you have to say in your defence?

Boy: There's no point in speaking as you won't listen anyway.

Judge: You have to talk! You're on trial!

Boy: But the punishment has already been set. Why should I talk?

Judge: Speak, I said!

Boy: I didn't want to steal but I feared the man. He has a large gun. I don't know if it's a real one or a toy-gun. I am afraid he'll kill me! I hate myself for being a coward. I don't know if the man is really dangerous or not. I am scared of him, and I also feel sorry for him!

Nir was quiet for a moment, and then as the judge he said:

Judge: Don't worry, boy. I'll set you free and make sure that you are far away from the man, and that he'll never catch you again.

Throughout this entire period, Nir's mother had continued her therapy sessions and there was a huge improvement in her parenting. Concurrently, welfare services were contacted, and visitation rights and supervised meetings with the father were organized.

Analysis of the third period

In this period, the Second Key had the style of a *nightmare*: The plot was frightening and cyclically re-enacted, as often happens with children whose play has been hindered by trauma (Oren, 1995; Ramsden, 2011). Although there were several characters, there were in fact only two roles: The boy was the "victim" and the adult figures played the role of "aggressor". In Landy's (2009) role theory, this reflects the dynamics of a role and counter-role, without a guide.

At this stage, when the dramatic reality was well established (First Key), and stable enough (Second Key), the Third and Fourth Keys (both associated with contents) became meaningful: The characters and roles that clients bring into dramatic reality are parts of themselves, just as the characters that appear in our dreams constitute parts of our psyche (Perls, 1992). People may often cast the therapist in a role that represents their vulnerability – a role that may scare or otherwise overwhelm them. Casting the therapist in a sensitive role is the client's natural way of creating aesthetic distance (Landy, 1996). Clients may project onto the therapist parts that they are currently incapable of bearing, so that the therapist plays them out from them in dramatic reality. The roles in every drama simultaneously represent "interpersonal" and "intrapersonal" relationships (Landy, 1993). We may presume that the "bad man" in Nir's play is a parental figure; yet it appears that all the authority figures – policeman, prison guard and judge – who seemingly represent educators, are perceived by Nir as penalizing, merciless and cruel. However, their cruelty also reflects "intra-personal" relations, embodying Nir's merciless aggression toward his own helplessness and fears. Through casting of Galila in the boy's role, Nir could explore and elaborate these aspects.

The therapeutic goal at this stage was to soften his aggression toward himself, to help Nir develop empathy toward what he believed to be his "weakness." Galila's intervention in the Third Key allowed a meaningful dialog between role and counter-role, as well as an expansion of Nir's character and role repertoire.

Fourth period: the child looks for a home – interaction between the Keys

After the session referred at the end of the third period, a change occurred. To my surprise Nir asked to play the role of the "the boy," who wondered around, alone in the world. I played sad music to accompany his lonely journey. From time to time, he knocked on an imaginary door, and I became all sorts of people. He asked them for shelter, and the people kicked him out. During this period, the quality of Nir's dramatic reality changed. There were almost no words: I played music and Nir danced as the sad boy. There was a great deal of tenderness and harmony; it was aesthetic and intimate. Nir would occasionally exit dramatic reality and comment: "I'm lucky I have a home" or . . . "this is such a sad story". Then he'd stepped back and continued to play.

After several sessions, the boy finally found a home at a shawarma shop. The shawarma seller adopted him and taught him to work. The boy trusted him and told him all about his past. At this stage there was a great improvement in Nir's behaviour and mood. Nir and I continued to work together on various issues for another year, and his mother continued therapy for even longer.

Analysis of the fourth period

In this period, all the keys connect and interact with each other. As soon as dramatic reality became a safe and stable place that allowed for appropriate aesthetic expression (First and Second Keys), difficult content could be explored, complex feelings projected and processed. Moving onto the third and fourth, Nir developed the ability to pour his themes in dramatic reality. He acknowledged that the boy in the play was both "me and not-me" (Landy, 2009), and his responses from the audience did not violate the boundaries of dramatic reality nor interfere with the flow of the play. They were more like "asides," which reflected the development of his internal audience. In theatre, the aside technique serves as a bridge between the character on stage and the audience. It is as if the character is speaking to himself, behind other characters' backs, thus creating a sense of complicity with the audience as well as indicating an ability for self-reflection

(Pavis, 1998). This is the essence of the Fifth Key, which assesses the client's capacity to observe the therapeutic process. During therapy, Nir acquired all these skills – the ability to enter and exit dramatic reality, to play in a way that allowed him to elaborate experiences, to communicate in a meaningful way through role and metaphor and to observe what is being done.

Discussion

In this chapter we wanted to illustrate the use of the Six-Key Model in an ongoing process of assessment and intervention in dramatherapy.

In the first period, Nir had significant difficulty playing. He was overwhelmed with emotions that he could not contain, and they manifested in his tempestuous behaviour. The Sixth Key (Meta-reality) was filed with turbulent emotions. Acknowledging and reframing them as simple games with set rules (hide and seek, play-fighting, cards, etc.), reinforced the boundaries of dramatic reality, providing a measure of aesthetic distance. Galila's awareness of her feelings in the Sixth Key illuminated unconscious aspects of Nir's behaviour. However, the intervention was done in the First Key: No attempts were made to intervene at the level of content patterns or subtext during this period. The work focused on defining the boundaries of the play and channelling the inexpressible feelings within it. In this period Galila and Nir only held two roles (and there were barely any characters): Galila was "the helpless confused one" while Nir acted out the "almighty demanding aggressor". As these roles found their way into dramatic reality only in the third period, it would have been futile to attempt any role-related intervention at this stage.

The second period was a sort of incubation stage. The boundaries of the play stabilized; the quality calmed down. Contents, stories and characters quietly began to emerge under the sand in the sandtray. During this period the main work was done in the Second Key (quality) – an understanding that was crucial in defining the clinical debate of the staff.

In the third period, Nir created a play. Although characters and plot were defined, the quality of dramatic reality was still repetitive and rigid: Nir was a domineering playwright and director, Galila an obedient actor. Nir's need for control found its way into dramatic reality through the characters of the adults in the play (man, judge, etc.). In this instance, no specific intervention was made, except for allowing the creative process itself to take course enabling the transformation. The more Nir played the "domineering" characters in dramatic reality, his need for control gradually faded away, and he could engage in a meaningful dialog between the characters.

Just as the establishment of the First and Second Keys was a continuing process, the development of "good enough" roles and characters took time; for despite the large number of characters at the beginning of this period, there were only two roles: a helpless boy, and aggressive and exploitative adults – role and counterrole without a guide. This is a dead-end situation. The guide is meant to contain both sides and to allow conflict and paradox to exist (Landy, 2009). When there is a role and counter-role without a guide, there is a constant struggle lacking empathy and compassion. During this period, the Sixth Key (meta-reality) was very active again, raising the question of whether Nir was being abused. Since the active key at this time was the Third Key (roles and characters), Galila chose to bring the subtext to the stage through the boy's voice. She symbolically asked if the man had a *real* gun, or if he was mostly just frightening. The question implied an acknowledgement of the possibility that actual danger was at stake. The boy also recognized being scared and also caring for the "bad man". Through her character, Galila introduced the idea of holding paradoxes, which ultimately resonated in the "judge" played by Nir. Voicing contradictory feelings produced the sense of empathy and consolation that enabled Nir to transition to the fourth period – in which the "guide" was born.

At first the guide was the music, which held the feelings of the boy's sadness as well as of those who kicked him out. When the guide was born, Nir could finally embody the role of the boy – the vulnerable one. The work was done mainly through dancing and music, with hardly any words. Its quality and style (Second Key) echoed the silent period of the quiet sandtray, only that this time the content was open, and Nir and Galila played together without words. A healthy Fifth Key developed in the form of an observing ego, concurrent with Nir's comment: "I'm so lucky that I have a home!", signalling an awareness of the life-drama connection (Jones, 2007). The appearance of the shawarma seller helped to elaborate and support the role of the guide as a responsible empathic adult.

The Six-Key Model enables the integration of artistic and therapeutic worlds, making it possible to understand dramatherapy process using dramatherapy's unique language. The model emphasizes the importance of helping the client develop the ability to meaningfully engage with dramatic reality. Without this ability dramatherapy cannot be conducted. Moreover, the ability to enter, exit, and maintain a "good enough" dramatic reality helps clients build within themselves ego strength to bear life's hardships and develop healthy coping mechanisms (Winnicott, 2005; Lahad, 2017).

Play develops naturally in childhood. In normal development it is part of normal maturity processes, like language development (Jennings, 2010). Numerus psychologists have discussed the importance of play in emotional development. In *Childhood and Society* (1950), Erik Erikson emphasized the importance of play in building the child's ego. He viewed the ability to play as an indicator of mental health and part of the developmental process. Winnicott (2005) took this understanding a few steps forward when placing the ability to be in the "transitional space" at the centre of his theory, also stressing the deep connection between normal development, ability to play and mental health.

In dramatherapy processes, when we work on stabilizing the quality of the dramatic reality, we are actually working on several levels at the same time. We are strengthening the client's ego, developing the therapeutic alliance, and creating the space in which therapy can occur. The work is done simultaneously, since a "good enough" dramatic reality cannot be created in the absence of trust, or if the ego is too weak or vulnerable. Winnicott (2005) defines this well when he states that being able to play is a basic condition for meaningful therapy to occur – and not only with children. Although dramatherapy with adults differs from dramatherapy with children, the differences manifest in the means or the method, not in the essence: The ability to create a dramatic reality is a necessary component in every arts-based therapy; and therefore, the Six-Key Model also guides us when working with adults (Pendzik, 2003, 2008, 2012).

When the client finds it too difficult to enter and maintain dramatic reality, we believe that the therapist must seek the gateway to the client's playfulness. For some clients this involves finding what modality of EPR helps them to experience dramatic reality as meaningful. Often therapist and client will search for an artistic medium in which the client feels free to create "an island of imagination" that can hold the content.

Although many dramatherapists and play therapists may prefer certain forms of dramatic reality over others, every genre and every modality of dramatic reality has its own features and benefits. In working with Nir, Galila used a variety of mediums – including games, sandtray, drama, music, and dance. Her therapeutic interventions began by acknowledging the emotions that emerged in the Sixth Key, but rather than acting upon them, she structured them as simple games to reinforce the boundaries of dramatic reality and provide aesthetic distance. The Six-Key Model enables us to expand our understanding of how to find a person's creative opening into dramatic reality. When there is a gateway to creativity and its quality is "good enough" it is experienced as a place of depth, intimacy and transformation.

References

Bion, W. R. (1962). *Learning from experience*. London: Heinemann.

Bion, W. R. (1967). *Second thoughts: Selected papers on psychoanalysis*. New York: Basic Books.

Busch, W. (2009/1902). *Max and Maurice: A juvenile history in seven tricks*. Project Gutenberg ebook. Downloaded 9–6–2019, www.gutenberg.org/files/28847/28847-h/28847-h.htm

Cattanach, A. (2008). *Play therapy with abused children* (2nd ed.). London: Jessica Kingsley Publishers.

Erikson, E. (1950). *Childhood and society*. New York and London: W. W. Norton.

Jennings, S. (2010). *Healthy attachment and neuro-dramatic-play*. London: Jessica Kingsley Publishers.

Jennings, S. (2012). Drama therapy assessment through embodiment-projection-role (EPR). In D. Johnson, S. Pendzik, and S. Snow (Eds.), *Assessment in drama therapy* (pp. 177–196). Springfield, IL: Charles Thomas Publisher.

Klein, M. (1946). Notes on some schizoid mechanisms. In M. Klein (Ed.), *Development in psychoanalysis*. London: Hogarth Press, Institute of Psycho-Analysis.

Jones, P. (2007). *Drama as therapy: Theory, practice, research* (2nd ed.). London: Routledge.

Lahad, M. (2017). From victim to victor: The development of the BASIC PH model of coping and resiliency. *Traumatology*, 23(1), 27–34. http://dx.doi.org/10.1037/trm0000105.

Lahad, M., and Leykin, D. (2012). The healing potential of imagination in the treatment of psychotrauma: An alternative explanation for the effectiveness of the treatment of PTSD using fantastic reality. In S. A. Lee and D. M. Edge (Eds.), *Cognitive behavioral therapy: Application, methods and outcomes* (pp. 71–93). New York: Nova.

Landy, R. (1993). *Persona and performance*. New York: Guilford Press.

Landy, R. (1996). The use of distancing in drama therapy. In *Essays in drama therapy: The double life* (chapter 2, pp. 13–27). London: Jessica Kingsley Publishers.

Landy, R. (2009). Role theory and the role method of drama therapy. In D. Johnson and R. Emunah (Eds.), *Current approaches in drama therapy* (pp. 65–88). Springfield, IL: Charles Thomas Publisher.

Landy, R., and Butler, J. (2012). Assessment through role theory. In D. Johnson, S. Pendzik, and S. Snow (Eds.), *Assessment in drama therapy* (pp. 148–176). Springfield, IL: Charles Thomas Publisher.

Moreno, J. L. (1987). The role concept: A bridge between psychiatry and sociology. In J. Fox (Ed.), *The essential Moreno: Writings on psychodrama, group method, and spontaneity* (pp. 60–65). New York: Springer.

Ogden, T. (1979). On projective identification. *International Journal of Psychoanalysis*, 60, 357–373.

Oren, G. (1995). Dramatherapy with children who are trapped in their spontaneous play. In S. Jennings (Ed.), *Dramatherapy with children and adolescents* (pp. 40–60). London: Routledge.

Pavis, P. (1998). *Dictionary of theatre terms, concepts, and analysis*. Toronto: University of Toronto Press.

Pendzik, S. (2003). Six keys for assessment in drama therapy. *The Arts in Psychotherapy*, 30, 91–99.

Pendzik, S. (2006). On dramatic reality and its therapeutic function in drama therapy. *The Arts in Psychotherapy*, 33, 271–280.

Pendzik, S. (2008). Using the 6-key model as an intervention tool in drama therapy. *The Arts in Psychotherapy*, 35, 349–354.

Pendzik, S. (2012). The 6-key model: An integrative assessment approach. In D. Johnson, S. Pendzik, and S. Snow (Eds.), *Assessment in drama therapy* (pp 197–222). Springfield, IL: Charles Thomas Publisher.

Pendzik, S. (2018). Drama therapy and the invisible realm. *Drama Therapy Review*, 4(2), 183–197.

Perls, F. (1992/1969). *Gestalt therapy verbatim*. Gouldsboro, ME: The Gestalt Journal Press.

Ramsden, E. (2011). Joshua and the expression of make-belief violence: Dramatherapy in a primary school setting. In D. Dokter, P. Holloway, and H. Seebohm (Eds.), *Dramatherapy and destructiveness* (pp. 53–64). London and New York: Routledge.

Schechner, R. (2005). *Performance theory*. London and New York: Routledge.

Smitskamp, H. (1996). Brief dramatherapy: The need for professional diagnosis. In A. Gersie (Ed.), *Dramatic approaches to brief therapy* (pp. 98–110). London: Jessica Kingsley Publishers.

Winnicott, D. W. (1965/1958). The capacity to be alone. In *Maturational process and the facilitating environment: Studies in the theory of emotional development* (pp. 28–34). London: Hogarth Press, Institute of Psycho-Analysis.

Winnicott, D. W. (2005). *Playing and reality*. New York: Routledge.

22

THE INTERNATIONAL APPEAL OF FILIAL THERAPY

Values, methods and its use in Turkey to empower children and families and the therapists who assist them

Filiz Çetin and Risë VanFleet

Few family interventions cross international and cultural lines as easily as filial therapy. Co-created in the late 1950s by Drs. Bernard and Louise Guerney in the United States, filial therapy is a form of family therapy that uses play as its primary modality. It is a theoretically integrative and psychoeducational model of therapy that empowers and strengthens families to enhance the functioning and well-being of all family members, including children. With therapist instruction and supervision, parents learn to conduct special nondirective play sessions with their own children, offering them a microcosm within which to develop new skills, permit their children to work through various problems, and to realign their attachment relationships for the long term.

Because of the universal values underpinning filial therapy (FT), its collaborative, strengths-based, and empowering approach, and the flexibility with which it can be applied under different cultural and family circumstances, it has been steadily gaining international interest over the past 20 years. This contribution provides an overview of this modality, a summary of the research on it, and describes how it is applicable to a multicultural international audience. It includes training programs that have been conducted in many different countries, the train-the-trainer model implemented to ensure fidelity to model by professionals within each country and culture. To illustrate, the first author's FT program in Turkey is described in depth, including the training, supervision, and certification program that has been active there for over a decade. The impact of this program on Turkish families is illustrated with case material. Identifying information in the case material has been changed, and some composite information is used, to protect the privacy of the families. In all cases, the material represents the realistic use and outcomes of FT.

Overview of filial therapy

Filial Therapy (FT), developed by Bernard and Louise Guerney (Guerney, 1964; Guerney, 1983, 2003; Guerney and Ryan, 2013; VanFleet, 2014), is a theoretically integrative and evidence-based approach that combines play therapy and family therapy. The therapist trains and supervises parents as they learn to conduct nondirective play sessions with their own children. When properly trained and supported, parents eventually are able to conduct these special play sessions at home without direct supervision, although the therapist continues to monitor the process with regular meetings with the parents and, on occasion, watching video of the home sessions.

FT is a truly integrative model, drawing from a variety of theoretical orientations, including humanistic, psychodynamic, behavioral, interpersonal, cognitive, developmental/attachment, and family systems. The ways these various theories and practices are incorporated are detailed in VanFleet (2014) and Ginsberg (2003).

FT helps meet therapeutic goals of all family members, including all of the children, the parents, and the family as a whole. For children, FT has been used with a wide range of presenting problems. The parents learn to create a safe and accepting environment in which the children can express their feelings fully through their play, gain a better understanding of their world, solve problems, and gain confidence in themselves as well as in their parents. FT helps parents become more responsive and empathic with their children, better attuned to their needs, and better at solving child and family problems. It is a goal for parents to develop better coping skills as well as to strengthen their family relationships to help prevent future problems. Parents develop better communication and problem resolution skills while also increasing their confidence and competence as parents. In essence, FT helps the family build a network of healthy, secure attachment relationships. Despite its ability to address intensely serious problems, the process is often enjoyable for the entire family, and many families incorporate the play sessions as well as the daily interaction skills into their lives long after formal therapy has ended (VanFleet, 2014).

Because of its focus on building strengths and teaching psychosocial family skills simultaneously with addressing child distress, FT has always been a time-limited approach. Families presenting with mild to moderate issues typically require 15 to 20 one-hour sessions. Families with more serious problems take longer, but the model allows for that to happen. FT has also been used with families without specific complaints but who simply wish to enhance their relationships with each other.

FT was originally developed as group family therapy and is still conducted that way today when feasible. The original Guerney group model typically runs from 16 to 24 weeks (Guerney & Ryan, 2013). VanFleet (2014) discusses a variety of shorter-term group models that are drawn from or are similar to FT that are very useful for a variety of purposes (e.g., Bratton & Landreth's CPRT, 2019).

Filial play session skills

FT therapists teach parents four specific skills that they use to conduct their parent-child play sessions. Parents are asked to conduct a half-hour play session with each child each week if at all possible. After parents master these skills during the play sessions, therapists help them generalize their use to daily parenting situations. The four skills are detailed in VanFleet (2012, 2014) and are briefly described here:

1 The *structuring skill* helps children understand how the play sessions work. Parents learn how to explain the sessions to their children, what to say when starting the session, and how to handle the end of the session, including how to deal with resistance from the child when it occurs.

2 The *empathic listening skill* builds parents' attunement to the child, helping them show greater understanding and acceptance of the child's feelings and needs. Parents learn to rephrase the child's main behaviors and emotions expressed in the play session, and also to convey interest in their child nonverbally. Parents also learn to refrain from leading, teaching, questioning, or directing the children's play. The play sessions remain nondirective throughout, with the child leading the way and the parent following that lead except in cases where limits are needed for safety.
3 The *child-centered imaginary play skill* offers another form of attunement between parent and child. The parents engage in imaginary play when invited by the child, and they do so in a way that the child wishes. They learn to act out different roles or characters the child requests. Parents follow the child's ideas for the direction of the play so that themes of importance to the child can emerge and be worked through.
4 The *limit-setting skill* creates safety within the play sessions. Parents set and enforce the rules needed for safety. The number of limits is minimized, but they are important nonetheless to help children understand boundaries, how to redirect their energies when their desired behaviors are risky, and to respect their parents' authority when it must be used. The therapist teaches parents a three-step limit-setting process: (a) state the limit clearly and specifically when the child is imminently going to do something harmful or dangerous, after which the parent redirects the child in a general way (e.g., "You may not throw the blocks at me, but you may do just about anything else."); (b) give a warning on the second infraction of the same limit in the same session, informing the child that the play session will end if they try it a third time, and (c) enforcing the consequence for the third infraction of the same limit by ending the play session. It is rare that children ever push the limits to this third level, and this process quickly reestablishes parental authority when it is needed.

In addition to the skills, therapists help parents learn to identify and interpret the play themes in the play sessions. Each session is followed by a discussion of what the parents and the therapist observed, and using the context of the child's and family's life, some possible meanings of the play are discussed (this is done only with the therapist and the parents). Ideas of play themes are considered "hypotheses" which may be later confirmed or disconfirmed by further play sessions and patterns.

Sequence of FT

The initial stages of FT include an intake, a family play observation, and a play session demonstration. This means that the therapist first watches the family members play together without any training, and then discusses what happens with the parents, including how typical or atypical the interactions were. If FT looks like the best route, the therapist demonstrates a nondirective play session with the child while the parents watch. The child is then excused and the therapist answers questions and points out aspects of the play and how it might relate to the recommendation of FT.

Going forward, the therapist holds two to three one-hour training sessions with the parents during which the skills are presented and practiced. Mock play sessions are used in which the therapist pretends to be a child and plays in ways to help the parents practice the four skills. The therapist offers some feedback as they go along, mostly commenting on what the parent is doing well, and making just a couple suggestions for improvements.

After this, the parents start the play sessions with their own children. These are half-hour sessions with one parent and one child at a time. The therapist (and sometimes the other parent) watches. The child is excused and the therapist has a discussion with the parent. They go over the skills the parent was using and also discuss the play themes or what the play might have meant. Each time the therapist elicits the parent's impressions first. The therapist provides feedback, concentrating on what the parent did well and giving only a couple suggestions to work on for next time. This continues until the therapist has observed each parent four to six times, or until the parent has become quite skilled. The focus eventually begins to emphasize the play themes as patterns emerge.

When parents meet certain criteria, they begin holding the play sessions at home. They take notes or make videos of the sessions and meet with the therapist without their children to go over the home sessions and to talk more about play themes and other concerns about life in general. Usually by this point there is considerable resolution of the presenting problems and parents are feeling much more confident in their ability to handle them constructively. In the final stages of FT, the parents learn how to generalize the play session skills to daily life, and there is often some discussion of whole-family play that might be useful as well.

Values

Core values of FT include honesty, openness, respect, genuineness, empathy and acceptance, relationship, empowerment and self-efficacy, humility, collaboration, playfulness and humor, emotional expression, family strength, and balance. These are described in more detail in VanFleet (2014), and are lived by therapists throughout the FT process.

Essential features of FT

In recent years, some inaccurate information has been circulating about FT, and other interventions have arisen that claim to be similar, when in fact they are not. So how does one determine if an intervention actually is FT? Drawn from her training and work with the Guerneys for many years, VanFleet (2011, 2014) has outlined these required elements. They are briefly described here.

The client is the relationship, not the individual

Too often, systems of care require that a single family member be identified as "the problem." In FT, the focus is always on strengthening the relationships among family members, with the empirically supported belief that change most readily occurs within relationships.

Empathy is essential for growth and change

Empathy plays a vital role in FT. The therapist shows deep empathy to the parents while they learn to be empathic to their children's distress, wishes, and motivations. FT therapists must first become proficient in nondirective play therapy which uses empathy as one of the core values and skills. Parents are never villified, but are treated as collaborators in the therapeutic process as they learn better ways of interacting with their children, and often with each other.

The entire family is involved whenever possible

FT was conceptualized as family therapy. Every member of the family has an impact on other members. Because of that, both parents or carers are included in the process, and all children are involved as well. The play sessions are conducted with one parent and one child at a time, but when everyone has play sessions it can impact the family dynamic and system. Furthermore, as relationships within the family improve, the family becomes more able to balance its attentions and can engage in more whole-family activities that are enjoyable for everyone. It has the power to reintroduce the family to the many positive features of each other and the family system as a whole.

A psychoeducational training model is used with parents

FT is founded on the principles that much human distress comes from a lack of knowledge and skill about how to do things better. FT is designed to provide that knowledge and skill. While FT is most often used to address clinical problems of children and/or their parents, it is based on a skills-based educational model rather than the medical model. There is a belief that if people know better ways of behaving, they will engage in them. The therapist is there to teach, and later, to support this process.

Therapists provide live supervision of parents' early play sessions

A key element of FT is to create the circumstances through which parents are successful. To ensure this success, therapists first teach the play session skills and then directly observe several of the parent-child play sessions for each parent and child involved. The therapist then gives feedback to the parents so they can maintain behaviors they are doing well and adjust those that are not working for them. In the early stages, parents are asked to use the new skills only in the play sessions and not in daily life, but a generalization phase of the approach is added once they have mastered the skills and are more able to apply them in the complexities of daily life.

The FT process is collaborative in every way

Therapists involve parents as partners in the therapeutic process. It is not true, however, that parents are taught to be therapists. They learn to conduct the play sessions, become better at it, and receive support from the therapist throughout. They are considered the primary change agents for their children, but the therapist remains the therapist and guides the process. On the other hand, therapists conducting FT welcome and encourage parent input at every stage of the process. Parent ideas are taken seriously. Therapists ask parents to reflect on their own play sessions, how they felt during them, and how they see them based on what they know of their child and family life. Therapists might add their own thoughts about play themes, but in a manner to help the parents understand better, never to contradict them. The therapist-parent relationship is decidedly collegial. As is often said to parents, the therapist has some expertise of skills that have been shown to strengthen families, while the parents have expertise on their own family, the dynamics, their children's behavior, and so on. The therapy is stronger when these two facets are combined through frank, open, and friendly conversations characterized by empathy and consideration.

Research

There are 55 years of research conducted on FT and some of its variants. It is considered an evidence-based approach. It has been shown to effectively improve children's behaviors and symptoms for a wide range of difficulties, parental empathy and attunement, parenting skills, parent confidence, and family cohesiveness, to name a few. Controlled research on FT is reviewed in VanFleet, Ryan & Smith (2005). A meta-analysis of play therapy research (Bratton et al., 2005) showed that parents in FT performed 1.06 to 1.15 standard deviations better than nontreatment groups, and it clearly showed that parent involvement (as in FT) had large advantages over play therapy conducted individually (which were also shown to be effective). Topham et al. (2011) explored the predictors of success in FT. They found that higher levels of parent distress and poorer child regulation of emotion at pre-test were predictive of significant reductions in child behavior problems. Poorer emotion regulation in parents at pre-test was predictive of significant increases in parental acceptance across treatment. In sum, this study showed that parents and children with poorer emotion regulation were likely to make positive gains in FT. The research continues in a number of different countries.

Multicultural applicability

FT has been used with many families representing a wide variety of ethnic and cultural backgrounds throughout the world. The values of FT resonate with many cultures, and most cultures value strong family systems. While the structure and practices of families varies culturally, FT has the flexibility to adapt to those differences. Several factors are believed to play a role in this adaptability.

First, children throughout the world play. If given the chance, children play, and they play within the cultural environment in which they live. The nondirective nature of the play sessions allows children to play out their concerns within their own cultural context. Second, when therapists work with children who are culturally different, there's an increased chance that the therapist will misunderstand or misinterpret the children's play. In FT, the parents are partners, and they often have the knowledge of culture and the children to provide more accurate input to the process of understanding. Furthermore, FT emphasizes the need to interpret play themes through a lens of family and cultural context, and parents provide that lens in most cases. Third, in FT the respect for the family is vital. Therapists develop open and empathic relationships with parents or carers, and they encourage parents to speak freely, even if they disagree with the therapist. Parents are encouraged to educate the therapist about their own cultural practices, beliefs, symbols, and interpretations. This relationship-focused form of cultural sensitivity usually results in true partnership between parents and therapist, all with the common goal of helping the child.

International filial therapy training

As interest in FT has increasingly grown around the world, the need for training has increased. FT is deceptively simple and requires considerable training to do well. The Family Enhancement & Play Therapy Center (www.play-therapy.com) has long offered a certification program in FT, and since 2002, this has become an international venture. A train-the-trainer model is used in which local professionals first learn to conduct FT with the families in their communities. As they develop proficiency, they are eligible to become FT supervisors and instructors, and after further training and demonstration of competencies, they can be certified FT instructors who can offer trainings themselves. The entire process parallels the FT process, with much empathy and support for the candidates as they develop their understanding of the subtleties and complexities of the approach and become better able to teach it to others. There are now over 15 FT instructors running their own programs in as many countries, in addition to those in North America.

Parent materials are valuable for parents who like to read, and they can make the approach more accessible. *A Parent's Handbook of Filial Therapy* (2nd ed.; VanFleet, 2012) has been translated by filial therapists (to ensure accurate translation of concepts) into Arabic, Chinese, Spanish, and Turkish, with several others underway. This, along with the translation of professional books, including *Filial Therapy* (VanFleet, 2014), has helped make the international awareness and use of FT more prominent. A social media group called the Filial Therapy Forum (Facebook) also brings together international professionals and parents interested in the method.

The rest of this chapter is devoted to the active FT program being run by the first author (Filiz Çetin) in Turkey. This provides an example of how this process of multicultural application works.

Filial Therapy in Turkey

Getting started: my journey in Filial Therapy (Filiz Çetin)

This part of the chapter is intended to enable the reader to understand how I have arrived at the clinical practice of Filial Therapy (FT) in my career and to illustrate its effect with a short case study. After I graduated from the Psychological Counseling department and had my master's degree from Bogazici University in Istanbul, I started my career working as a preschool counselor in 1999. In the preschool, it was the first time I strongly realized how play is effective at reaching children and helping them express much about their lives. On this journey of discovery, it has become increasingly clear to me that play opens the doors of vitality and intimate connection.

In the early 2000s, I wanted to dive further into the psychological underpinnings of children's emotional worlds and methods of healing. I started to work in a private counseling center with children who had emotional and behavioral problems. Being a young practitioner in the field, I needed to learn intervention methods that included play in helping children and families. When I explored more, I saw that there were many trainings in play therapy in Europe and the United States. I attended several workshops and seminars in directive and nondirective forms of play therapy. Through this I realized that, beyond any specific intervention, it was the development of a safe and accepting relationship that formed the core of all mental health interventions, not only for children but also for parents.

Parents are the most important agents to create healing in children's emotional lives. It is almost impossible to leave the parents outside of the therapy process of their children if one wants good results. If the parents are not convinced and motivated for therapy, then the child loses the chance to get help. As a therapist, I knew I needed to establish the therapeutic alliance with the parents and maintain it from my first day of working with them throughout the therapy process. Having these realizations in mind and having experienced them in sessions, I looked for a more grounded therapy model which would satisfy my need to integrate parents effectively into the child's therapy. I found out the FT trainings given by Dr. Risë VanFleet, who was well known in this field. Since FT was a psychoeducational and therapeutic method that aims to enhance the relationship between parents and children, I wanted to learn more about it.

After my discovery, I read more, concentrating on Risë VanFleet's books on FT and also watching her videos in her online course on FT to gain more information. I realized it was not an approach mastered only by reading and watching videos. The experiential part of the training was a must.

Through this process I found that FT was very congruent with my clinical posture which strongly believes that parents also need to feel safe and be understood as well as their children in therapy. In my clinical work with children and parents, I had often found that parents felt guilty for their children's difficulties. It was not easy for them to attend therapy and face their roles or behaviors that might have contributed to their children's problems. I could see that it would be painful for them if the therapist failed to show empathy and ignored their anxieties. As I studied more about FT, I learned that not only was it an effective therapeutic method to work with children and families, but it also embodied this empathic attitude toward parents and saw them as a valuable source of help for their families.

In 2007, I invited Risë to come to Istanbul and introduce FT to colleagues. Play therapy in Turkey had been improving since 2000 and there was a huge interest in learning more about FT. After that, I immersed myself more into the theory and rationale behind it and attended the intensive FT training with Risë. The intensive filial therapy training helped me to see the need to expand beyond using FT only to help families, but also to teach it to the professionals who work with families. I continued with the FT trainings, including the advanced FT training where problematic cases were covered in greater detail, and eventually, the FT instructor training. I had the very great fortune to co-lead intensive FT professional training groups with Risë in Turkey and Ireland.

In sum, my discovery and journey with FT started by meeting with Dr. Risë VanFleet, and, since the first contact, she has been my mentor. My way of approaching and working with parents in therapy has changed by having contact and supervision with her over the years. My interest gradually extended from the child to the parent-child relationship. I shifted from a stance of leaving parents outside of the play room to understanding how their inclusion and cooperation enhanced their family relationships. Working with the relationship allowed me to study the influence of parent attitudes toward their child and the child's attitudes toward their parents. Even in cases of biologically mediated developmental problems and special needs, interaction with their parents had a very positive effect. While it is very hard to determine when and how negative family interaction patterns first shifted into a vicious cycle, it was clear that when one of the people in the family gets better, it affects the others' well-being. Helping parents improve helped the relationship, and consequently, the other family members.

Filial Therapy training programs in Istanbul

In addition to practicing FT directly, I also have been conducting training workshops since 2008. Each year I organized a group of trainee therapists and helped them learn how to apply nondirective play therapy and FT in their work. This then evolved into a full-fledged child-centered (nondirective) play therapy and FT certification programme based on Risë VanFleet's (and ultimately, the Guerneys') work.

The idea of offering FT trainings arose from (a) a wish to spread interest in FT and reaching more professionals and families, (b) the need to increase the awareness of the impact of parents as therapeutic agents in their child's emotional well-being, and (c) an interest in exploring parental attitudes and strengthening their skills while remaining in the child-centered conceptual framework.

Before participants ever learn FT, they obtain intensive training in child-centered play therapy (CCPT), and then receive supervision as they practice CCPT with children. This is to ensure that they gain enough confidence and compatibility with CCPT before they start teaching the basic CCPT skills to parents.

The certification program that I offer includes a four-day intensive filial therapy training. The training is limited to six participants so that the trainer can give considerable attention to each participant, providing individualized feedback during role-played practice of each of the skills involved in FT. In the first day, trainees learn the history, rationale, and theoretical background of FT. They are reminded the CCPT skills by engaging in role plays. In the following days, they learn how to teach parents the four special play session skills, as well as how to show empathy to the parents and how to deal with any resistance or concerns from the parents. The whole workshop flows in a process of building up, or scaffolding, the necessary skills to conduct FT.

After this intensive FT training, participants take part in supervision. The six people from the training are supervised each week by the trainer. Usually the group is separated into morning and evening groups formed according to the working hours of the participants. These subgroups of three participants meet for two hours per week for four months. The supervision process is modeled on the way in which FT offers observation and feedback to parents. This further reinforces what they have learned about working with the parents. Group supervision also includes discussions of complex relational dynamics between therapist and parents, therapists and children, and parents and children as described by the therapists or shown on a video. The trainees share their opinions and thoughts related to each case and have the opportunity to learn from each other's experiences. Particular attention is paid to the stages through which therapy is going, while also looking carefully at the interactions mother and child, father and child, child and siblings, and mother and father. In other words, the entire family system is considered through the play sessions and the discussions with parents after. This supervision process helps the professional participants to develop their skills and their depth and breadth in using FT.

When supervision is nearing completion, I offer the Advanced FT Training in which common and complex problems that arise in FT are discussed and role-played. Participants bring many questions about their cases and they can find solutions to the problems at different stages of FT. A respectful, collaborative problem-solving model of working with parents is presented. They are taught to overcome resistance in various forms and maximize their effectiveness in conducting FT with many challenges.

Once all the requirements are completed, the participants apply for the certification which is defined clearly by Family Enhancement & Play Therapy Center. The detailed information can be found at the website, www.play-therapy.com.

Play therapy interventions continue to increase in Turkey. Since FT is a very humanistic and universal model, it can effectively be applied to Turkish culture. A brief explanation of Turkish family structure will be helpful before explaining FT practices here.

Turkish family structure

Geographically, Turkey is a bridge connecting East and West and the family systems are under the influence of both cultural systems. In several parts of Turkey, the family relations show variety, ranging from more traditional to more modern urban characteristics. In general, family ties are very important. Although mothers often seem to be more involved in terms of child rearing and emotional closeness, the number of the fathers engaging in the lives of their children and in therapy has been increasing. Twenty years ago when I started clinical practice, I made first appointments mainly with mothers. They usually came alone to the first interview because the fathers didn't want to participate or they didn't have any time to arrange. Today this scene had changed considerably. The parents are more likely to come together and to act more like partners in FT, supporting each other.

Usually the size and structure of the households differ from a nuclear family of one or two children and the parents to an extended family where more children or grandparents are also very close and included. There is a hierarchy in the family accorded by age, and in some families emotional and economic support can be shared among the members. The grandparents might be the main helpers during the birth and early years of the children if they are available. In some cases grandmothers or grandfathers or other family carers who are important in children's lives are also included in the FT.

In the recent years, parents are more mindful about encouraging the emotional expression of their children. Even so, negative emotions can still be kept suppressed within the family. One of the difficulties I observe in the sessions are the parents' silences when their children express these hard feelings. In that sense, FT training helps parents realize the importance of recognizing and reflecting all emotions expressed by the child. When children's conflicted feelings are understood better by the parents, the children can express them and then find better solutions to their problems. Since there is a generational gap between parents and children in terms of expressing emotions, some parents need guidance in recognizing, describing, and expressing their emotions. In past generations, there was a tendency to show conformity to authority. Due to close ties in the family, it could be difficult to show conflict, or to act independently and autonomously. Parents who were raised in these older practices can have difficulty allowing children to express their disagreement or independence.

One of the emerging relational problems I observe in families involves parents' inability to set effective boundaries due to their feelings of guilt because of the absence of quality time spent together with their children. In the last decade, the number of the working mothers increased in the big cities. As the economic and social transformations happen, the mothers are more engaged in work, so the time they used to spend with their children gets less and they feel guiltier. FT gives every member in the family a voice to express themselves and help each other to show acceptance to each other's emotional world and needs. Without going too far at both ends, the limit setting skill coupled with building empathy creates a good balance between the old and new child rearing practices. The parents can learn how to show empathy by maintaining their emotional closeness with their children. At the same time, they can put effective boundaries in place without having fears of losing their power to maintain structure. It is the holistic combination of skills in FT that allows this balance.

Over the years, hundreds of children and parents have taught me how critical empathic listening and attunement are in family relations. The Guerneys' FT model, conducted and designed as a certification program by Risë VanFleet gave me a tool to tap into every child's potential to attach and grow. Helping the parents to show affective presence, attunement, and patience to the child and to increase their receptiveness to even minimal changes in the child, fosters the child's growth. The following case is shared as a model to explain the effectiveness of FT. The names and ages and other identifying information are changed to ensure confidentiality.

Filial Therapy case study

A five year old boy, Brian, was referred for the therapy due to his father's death. His mother came to the first appointment alone. In the first meeting, she gave Brian's developmental history and information about her family with particular focus on her spouse's death due to traffic accident. She had many questions in her mind. Her main concern was the need to explore and understand her son's innermost feelings and empower him to deal with the loss. Because Brian expressed no affection, but only behavioural difficulties in social settings, she felt anxious and wanted to help him. She wondered why Brian never asked about his father and wanted to understand what he was feeling inside about his dad.

Since play is the natural way to communicate for a child, I explained that I wanted to do a family play observation first. I also mentioned FT and how it could help them. I recommended FT to the mother so that she could better help him express his confusing emotions arising from his father's loss. I thought that engaging the mother and supporting her playing a more active role in her child's therapy would strengthen and help her to deal with her helplessness also. Because feelings of loss can be very scary for children to talk about, in the mother-son play sessions, Brian could get some distance and emotional safety through his symbolic play. Instead of confronting the loss directly, the nondirective play skills could keep him from feeling overwhelmed by his feelings of loss. I explained to her that one of the best ways she could help him was to listen and understand his story through play. It was a healing experience for him to play as he wanted.

Family play observation

I did a family play observation, simply watching the mother and child play together, to better understand the mother-child relationship. During this observation, Brian began exploring the toys in the room and his mother asked him the functions of the toys. As he took the ambulance, she asked what the ambulance is called and what do we do when we see an ambulance in traffic, how does the ambulance sound, and so on. Brian was silent and didn't answer any of her questions. It seemed he disliked being examined by questions; he just needed to play. I made a note to discuss with the mother about the way she tried to communicate with him. She seemed to relate to him by asking questions. Parents usually do this to communicate, to engage or encourage the child to talk during play. As seen in this case, however, children usually feel interrupted and pressured to give an answer. Since they don't feel understood, in the end they might stay silent or stop what they play or even get frustrated.

Later in this same observation session, the mother's attention turned to the other toys and she directed Brian in the room. It was clear, however, that Brian said he still wanted to play with the ambulance. He then took a man figure and another car and crashed it into the wall – an accident happened. Then he immediately took the ambulance and gave the man figure a shot. The mother was silent. I saw the anxious look on her face because of his aggressive play and she suggested that he switch and play a teacher-school game. Brian then expressed his frustration and anger with his mother. She was scared of the Brian's angry feelings, and it appeared she had difficulty handling his emotions so she wanted to change the play.

Brian then stopped his ambulance play and accepted his mother's suggestion. This time a school scene emerged. In his play he wanted his mother to be the teacher and asked her to draw a little star on his hand because he wanted to get rewarded for his performance. Then he created a scene where the little boy figure had a sudden accident and fell off a bicycle. He asked his mother (in her teacher role) to call the boy's father. As Brian played like this, his mother grew silent and seemed puzzled by his mention of "father" in his play. Brian repeated his need for the teacher to call the father by phone in this imaginary play, and she complied. Brian then asked her to play the father role. He asked to be taken to hospital. They both went to the hospital in the play. The mother (still in the father role) asked him if they should call the mother, too, but he refused, "Only father and me." Then after spending some time in the hospital, he said the child was feeling better, so he left the hospital with the father (played by the mother) and returned to school.

The mother was playing both the role of the father and the teacher in the play. Brian seemed to identify himself with the child doll. After getting back to school, he produced a theme of insecurity and expressed his need of protection by creating another scene in his play. This time the school yard was under threat by bad people who had guns. The mother tried to reassure him by saying, "There is no need to be insecure in school, school is a safe place." Brian continued with this play. He hid children figures behind the school yard, and then took the father doll who threw stones at the bad guys. The bad guys went away. The father doll was the one who provided security in his play.

His need of the father was also observed in the following scene where he asked the mother in the teacher role to call the father again during the school time. He said, "Now children stay in the class, you come and tell me 'your father comes.'" The mother acted as he asked, and he wanted her to repeat this sentence again – "Brian, your father comes." His smile could easily be seen in his face as he was looking at the mother and hearing those words. His longing for the father was being displayed in his play, and he expressed his need of being with his father by creating the following scene; where he and his father went to eat ice cream after school.

Discussion of family play observation with mother

After half an hour of mother-child play observations with all of the scenarios just described, Brian was taken to the waiting room. The remaining time was spent with the mother for feedback and discussion.

I asked for the mother's thoughts and impressions and discussed the following ideas with her. In Brian's play, the main theme was to seek safety. The death of a father could shake the child's belief that the world is a safe place. The play themes of security and his trials of exploring his feelings about his father were covered. The mother shared her surprise and confusion about his play, and she recognized she needed help in terms of tuning her responses according to his needs in the play and in real life. In the end I explained FT again and recommended it so the mother could feel empowered in helping her son and allowing them to feel more connected. Creating an environment of acceptance and empathy while setting certain boundaries in FT would help Brian to express his feelings (especially about father's death) and his needs more clearly. I offered a play demonstration session as the next step.

Play session demonstration

In the play session demonstration in which I played with Brian using CCPT, the mother observed attentively and took some notes about the non-directive play. After the demonstration, Brian was led to the waiting room and I did a short debriefing with the mother. The mother shared that she realized how directive and educative she usually behaved in her play with Brian after watching the demonstration. At this point, I empathized with the mother's self-realization, and I explained that if she learned to do nondirective play with him, it would provide an opportunity for her to help him express his inner world in a safe and nonjudgmental environment. She could see how this might work and agreed to move forward with FT.

FT parent training sessions

The mother's FT training sessions went smoothly. Since she was a teacher working with young children, she was eager to learn basic principles of play therapy. She easily understood the rationale of the child-centered play therapy skills. She carefully listened and expressed her concerns. During this training phase of the filial therapy, besides practicing the four child-centered play therapy skills with the mother, her concerns were also held gently.

Supervised parent-child play sessions

When the training phase was completed, the mother was ready to begin the play sessions with her son. In the first play session, she tracked him very well and reflected his emotions. Although she asked a few questions to understand his play better, in general she tried to let him play as he wanted. In the next play sessions, there were some pauses of her reflections during the play, especially when he started to play aggressively and expressed the security needs again. In the supervision feedback to the mother, I emphasized her improving empathic listening skills and discussed with her how she felt when she became silent during some parts of his play. When the mother's fears and anxieties about aggressive play were addressed, she found an opportunity to express herself and this feedback time helped her with her overwhelming emotions. Since she was a single parent, discipline was an important topic to discuss, too. After talking about his play themes, she realized his imaginary aggressive play (which did not actually harm anything) reflected his need for safety in his world and the importance of acknowledgement and acceptance of his feelings like anger and grief. We also discussed her improvements in showing respect for his own way of coping in his play, and how important that was.

In the following play sessions, the mother did much better with her play session skills, and the feedback period was more focused on play themes. As Brian trusted his mother more in the sessions, he took more risks and created more play scenes about car or plane accidents. He was quite silent as he played. In the feedback meetings with me, one of his mother's concerns was about Brian's silence and she wondered why he was not still feeling safe enough to use his voice during play. Brian, however, was expressing and exploring his own vulnerabilities and emotional needs at his own pace. In the beginning sessions, he was not actively playing, but later he started to play without speaking. Then his mother realized that he should not be forced to speak aloud and he needed his own time to confront his issues. She shared with me that he might be trying to explore the details of the accident that his father had. We discussed ways that she could talk more with him at home about his father's death and how she could respond to his reactions, all with the intention of helping him feel safe and able to share with her. As she understood that he was already communicating his feelings through his play with her, she felt more comfortable talking more about his father at home in a natural way.

When a family member dies, the whole family can feel fractured and disorganized. It's not easy to give the right explanation to the child, and it is quite natural for parents to withdraw for a while as they cope with their own grief. During that time, however, the child can feel lonely and disconnected from their grieving parent. Internally, the child might feel helpless and anxious with unspoken questions in his or her mind, such as, "Who will look after me?" "Did I do something that caused this loss?" "Will he come back?" It was important for Brian to find answers and still feel connected to his father's memory and also with his mother. When his mother was ready, she gave him developmentally appropriate information about his father's death. The mother's understanding of his play patterns and themes, with the therapist's help, allowed her to express her own feelings and to gain perspective on her role in helping her son with their shared grief.

Home play sessions

After the supervised FT play sessions in the therapy room, the home play sessions started. The mother continued playing with Brian at home and attended therapy once a week without him to discuss the home sessions and daily concerns. She shared her observation that he was more relaxed and more talkative during the play hours.

Filial Therapy helped this mother and son reconnect and increase the security feelings of the child. His mother's acceptance of his scenarios expressing his feelings created this safety and provided him with the reassurance to feel secure again. She respected his own way of coping by following his lead in the play. Also, reflecting his emotions helped him to deal with his overwhelming emotions in the daily life.

Summary

Filial Therapy has a long history of clinical effectiveness with a wide range of child and family problems, as well as a well-documented research history that provides evidence for its efficacy. Involving parents as the change agents for their own children makes sense because they already have a relationship, and the work of the therapy is to strengthen that bond. Within the parent-child, or carer-child relationship lies the potential for great healing.

Because of its use of the universal child language of play and its emphasis on empowering parents to be very involved in strengthening their own families, it has much international appeal. Inviting parents into the process allows us to show them the power of play in healing many problems while building a safe, comforting network in their relationships for the future. When all family members are included, it can enhance not only the individual parent-child relationships, but also the relationships among siblings and between parents. Wherever they are in the world, if they are permitted to, children play. They play about the things that are important to them. Giving parents a full view into their children's worlds empowers not only the children, but also the parents.

Much of child and family treatment continues to focus on problems, but FT was far ahead of its time in looking for family strengths and then capitalizing on them, in using them to help address the problems and build a stronger safety net for families. Research on strong families around the world has repeatedly identified the key features of strong families: commitment, appreciation and affection, positive communication, time together, spiritual wellbeing (broadly defined), and the ability to cope with stress and crisis (DeFrain, 1999). As DeFrain mentions, "As individuals and as families, we are all unique and different. Yet deep down, as human beings, we are all quite similar." Filial Therapy helps build many of these features of strong families, and it is flexible enough to honor and accept the differences while reaching deeper to our basic needs as individuals and families. It is exciting that this is borne out by its expanding use around the world.

References

Bratton, S.C., & Landreth, G.L. (2019). *Child-parent relationship therapy (CPRT) treatment manual* (2nd ed.). New York: Routledge.

Bratton, S.C., Ray, D., Rhine, T., & Jones, L. (2005). The efficacy of play therapy with children: A meta-analytic review of treatment outcomes. *Professional Psychology: Research and Practice, 36* (4), 376–390.

DeFrain, J. (1999). Strong families. *Family Matters, 53* (Winter), 6–13.

Ginsberg, B.G. (2003). An integrated holistic model of child-centered family therapy. In R. VanFleet & L. Guerney (Eds.), *Casebook of filial therapy* (pp. 21–48). Boiling Springs, PA: Play Therapy Press.

Guerney, B.G. (1964). Filial Therapy: Description and rationale. *Journal of Consulting Psychology, 28*, 303–310.

Guerney, L.F. (1983). Introduction to filial therapy. In P.A. Keller & L.G. Ritt (Eds.), *Innovations in clinical practice: A source book* (vol. 2, pp. 26–39). Sarasota, FL: Professional Resource Exchange.

Guerney, L. (2003). The history, principles, and empirical basis of filial therapy. In R. VanFleet & L. Guerney (Eds.), *Casebook of filial therapy* (pp. 1–20). Boiling Springs, PA: Play Therapy Press.

Guerney, L., & Ryan, V. (2013). *Group filial therapy*. London: Jessica Kingsley Publishers.

Topham, G.L., Wampler, K.S., Titus, G., & Rolling, E. (2011). Predicting parent and child outcomes of a filial therapy program. *International Journal of Play Therapy, 20* (2), 79–93.

VanFleet, R. (2011). Filial Therapy: What every play therapist should know (Part two of a series). *Play Therapy Magazine of the British Association of Play Therapists, 66*, 7–10.

VanFleet, R. (2012). *A parent's handbook of filial therapy* (2nd ed.). Boiling Springs, PA: Play Therapy Press.

VanFleet, R. (2014). *Filial Therapy: Strengthening parent-child relationships through play* (3rd ed.). Sarasota, FL: Professional Resource Press.

VanFleet, R., Ryan, S.D., & Smith, S. (2005). A critical review of filial therapy interventions. In L. Reddy & C.E. Schaefer (Eds.), *Empirically-based play interventions for children*. Washington, DC: American Psychological Association.

23
THE COLOR OF PLAY
Breaking through walls within a child's world

Neal Brodsky

Introduction

This chapter explores possible trajectories for expressive dynamic play therapy (McCarthy 2012) through two anonymized case studies of work with an 11-year-old and an 8-year-old child. The play and interplay with the therapist demonstrate expressive techniques adapted from body-centered Core Energetics (Wilner 1999; Brodsky 2015) and rooted in the somatic psychotherapy lineage (Reich 1945; Lowen 1995). The importance of integrating awareness of the family and larger eco-system housing the child is demonstrated by offering readers a window on conversations with parents who are encouraged to consider how doing their own therapeutic work can contribute positively to their children's development (Siegel 2010).

Case one – Katie

"Tear Down that Wall!" Eleven-year-old Katie is smashing through the wall she's created with soft furnishings in my large New York City office. She had built this painstakingly over the past ten minutes, and now throws her thin body again and again against the wall, until an opening emerges that she squeezes through. She grins at me in triumph. "Now your side is my side!"

Katie had been brought in by her divorcing parents six months ago. She had been yelling at them non-stop for a year before and refusing the entreaties of her teachers to participate in activities with the other kids. I'd spoken with the psychologist at her school about mirroring the activities she'd begun in her therapy, bouncing on a large exercise ball to loosen up some of that anger and fear locked in her musculature which she may have developed when her father came home night after night to their small apartment, smelling of too many beers. Or when her mother began traveling to distant cities for her job. The school had not listened. For a period, we would have to double up on what had been once weekly sessions.

The conflict held within Katie's body was clearly not restricted to the one in her nuclear family. We were meeting in a country whose leader was calling for the building of a wall to keep Latino families on the other side of the border. Katie was Latina Catholic on one side and White Ashkenazy Jewish on the other. She was being taught Spanish in her school while beginning to study elementary Hebrew after school. This girl embodied a complex 21st-century mélange of influences she was trying to integrate through her active, dynamic play in session.

As a family therapist, many of my cases come with a measure of parental conflict and confusion. This one fits the mold. One parent had been adopted at birth. The other had been physically abused as a child. I wondered about the epigenetic influences this young girl held, not something I can diagnose within the framework of the DSM V. Parents want me to stabilize their children so they can thrive within their existing family and school environments. I cannot change the genetic cards these kids have been dealt. And I can help to create conditions where young people become more resilient.

Eva Reich M.D., the daughter of the founder of body-centered psychotherapy Wilhelm Reich, spoke of the prevention of neurosis from birth on.[1] I have taken on the continued exploration of such work. Because my own mother was trained as a teacher in the methods of the Open Classroom where the individual child learns at her or his own pace, I look at each child with the curiosity of what meditation teachers have called "beginner‘s mind," a perspective that honors the potential for what each new young person may present by way of openings into their healing. Wilhelm Reich posited that many children grow into adults who are destructive in their expression of life force. This happens because of shocks received to the developing emotional system in childhood and suppression of a child's natural reactions (including rage and grief) by parents, schools and the cultural eco-system. The result most often is an "armored" human who acts compulsively from defense and whose body is characterologically shaped and limited in its breathing patterns and movements (including intra-psychic, social-emotional and relational movements vis-à-vis others).[2]

Therefore, my goal as a therapist working in the realm of play and play therapy is to "recharge the energy system" of the child, allowing for freedom of expression in my office. We are working with, rather than suppressing processes seen by parents and caregivers as symptomatic behavior that ranges widely from explosive opposition to scattered learning and attention challenges or regressed interactive styles. Once in therapy, the natural "pulsations" of play, revive and re-balance the energy system of the child over time, allowing new and healthier patterns to gain traction.[3]

My attention is riveted by Katie's use of clay. She has taken a large gloppy clay ball from the plastic container where it awaits hands that will shape it in session and is ripping it apart, bits of clay falling onto a flat square of brown cardboard. One of my goals is to give this girl the opportunity to be creatively destructive using this clay.[4] As she tears the clay into smaller and smaller pieces, Katie tells me how she was bullied recently by a boy in her class who kicked her while she was standing in the lunch line. I am noticing how her breathing quickens and her shoulders seem to stiffen. I ask her if a part of her wanted to smash the boy's face. Whenever I say something like this in therapy, I remember how I, as an otherwise very well-behaved seven-year-old had punched and bloodied the nose of another boy and was called down to the principal's office. I still cringe today when I think of this and I prepare parents for the explosive power of play in the therapy room with the words "better here than outside with you or in school."

Katie gradually brings her clay shavings together into a ball and I hand her a smooth wood handled white rubber mallet. I tell her, "What I'm encouraging you to do here is just for this room, do you agree?" She holds up her hand in the age old peace sign. I say "OK, really flatten him. Go for it!" And she does. While her facial expression appears subdued, I am noticing that her breathing deepens as her assault on the clay continues, and her shoulders look more relaxed. The rhythm of her hits has taken on what feels to me in their regularity a sense of insistence and releasing of frustration. I encourage her to make a sound as she hits, inviting her to be as loud as she wants. Katie looks at the door to the room and stops. I ask her if I can make a sound with her. She nods. I say, "The sound comes from deep inside the belly." I demonstrate by opening my mouth wide, bearing my teeth and growling like a bear. She mimics me and as she gets louder, I lower my volume so she can hear herself with my tone in the background. Her mallet strikes continue accompanied by our soundings. The clay is now a pink-brown flattened pancake, merging with fibers of cardboard, pierced by her blows. I ask her how she feels. Katie smiles at me. "Good," she says.

In our next session, the "wall" story expands to a new theme. Using a crazy quilt of influences, Katie moves us into a wider war. On one side of the wall are the Egyptians. On the other, the Romans. She scales the wall of blue and red corduroy foam cubes, screaming at the top of her lungs, her eyes moving up into their sockets like the girl possessed in "The Exorcist" movie. "You . . . die!" she yells, giving me a piercing look. I hand her a red foam covered "bataka encounter bat." I take a blue one. Out of character in this moment, she takes a deep breath, looks me straight in the eye, and says: "Now really fight me. Don't hold back." She charges at me with a force I can hardly believe. "OK Katie, see if you can knock it out of my hands. I dare you!" I angle my bataka so she can hit downwards while I am holding on for dear life. This is my "Core for Kids" version of Core Energetics where we bring energy down from the head and shoulders so clients can feel the full energy of their neural network all the way down to their feet. It's a move to "ground" upwardly displaced energy that contributes to cycles of anxiety, depression, perseverating and imbalance.[5]

Katie continues her focused assault. I am in the fight, conscious that I'm going to allow her to experience winning this today. Aware that my body strength is greater, I feel the pain of what it is like to be in a world where people more powerful than me are in control. I feel exhilarated by the clean power of Katie's battle and I don't want this to be too easy for her. She smashes once, twice, three times. Focused, determined. On the fourth hit my bataka falls. Katie drops hers. In shock she looks at my bataka on the floor. Standing at her full height she jumps up and down on the corduroy-covered foam cube. I'm afraid she will injure her fingers on the ceiling, and I show her how to flatten her palms so her hands can safely make contact with the ceiling. She notices her palm prints on the white ceiling paint. I tell her it's fine, that another therapist I know has a ceiling full of palm prints. Maybe next week we'll get some finger paints and she can put those prints up there in color . . . for good.

Case two – Matt

Seven-year-old Matt is a country boy through and through. The only problem is that even the thought of dirt makes him squirm. Sitting frozen on the office couch with his blue hoody sweatshirt covering his head, he speaks in barely audible word snippets. It is a windy day outside and the trees we can see out the window are genuflecting deeply. Matt doesn't want to move. I ask him whether this is what I have taken to calling a "fast track" session. The kind where he will do something physical, to get his energy moving. Lips pursed tightly, shoulders collapsed into a hunch, he glances at the waiting sandbox, his eyes moving slowly to the clay, to the purple exercise ball. He barely moves his head to the right and left, and whispers "No."

Matt is here because he demands that everyone at home uses hand sanitizer before they touch him or anything he might make contact with including door knobs, plates, spoons, or chairs. He also gets frightened at night and waking from a dream, yells, disturbing everyone in the family. His parents are at their wits end. They want this to stop. They had told me when I first met them in my initial parents-only session that they were looking for a cognitive behavioral therapist. I had said "Well then, I'm probably not your guy."

After a tour of my play spaces they decided to give it a try. Matt lives in a two-bedroom subsidized housing apartment with his mother, father and younger brother in the center of an affluent suburban community that over the last one hundred years had been developed from farmland. Both parents had come from farming families who had lost their farms and been working odd jobs ever since. Matt's brother had been born prematurely a year ago and his mom had been hospitalized along with the baby for several weeks. The fear of dirt had first surfaced then.

Among the constellation of Matt's symptoms was difficulty choosing food to eat at home and at school. In his first session, he looked at the shelves of plastic Sandplay creatures and asked me what I thought he should select. When children come in with energy that is scattered or scared, they often need some period to "settle in" before choices are made. Still, my goal was to support this boy to make an eventual choice on his own even if it would take time. I smiled at Matt and said, "Wow, that's a good question!"

Matt decided after some minutes that he wanted to make a "dinosaur stew" and picked a few of the smaller dinosaurs to put in the basket I give children to put their sand play figures in before they place them in a sand world. The basket then became a cooking "pot" and I would be the guest receiving the largesse of Matt's "cooked" meal. He mixed the stew slowly with a long-handled wooden spoon and "served" it onto a "plate" which was a red frisbee. As I made loud "slurping noises" wolfing down Matt's stew, I noticed how the boy's shoulders seemed to relax. We had shared a meal. The ice had been broken.

My next thought was to get us outside beyond the confines of my office. In the back of my property behind my home office in Connecticut is a one room cottage that had been a writing studio for an earlier owner. It holds a history of stories, and Matt I will come to learn, is a story teller. He imagines the cottage is the possible destination for a treasure hunt. While he imagines this, he creates a world in the sand. The persona in charge of this sandbox world is "Map Boy," a figure with a scarf around his head from my shelves that looks like a cabin boy from a pirate ship. Matt tells me that Map Boy will also cook the food needed for our journey.

Money is clearly a big stressor in Matt's family, and I notice that he has begun to quantify the value of what he is selecting for "food." I have a small basket holding polished flat glass "gems" which Matt tells me are pearls. Vegetables cost one "pearl." Stuffing for dinosaur stew cost half a pearl. "And all the food," Matt sweeps his hand in an arc across his body, pointing to me, "costs three pearls." On the shelves, Matt also discovers the figure of a California goldrush prospector from the 1800s, who carries a shovel over one shoulder. The boy dubs him "The Shoveler" for his quest to find the treasure. Refusing my invitation to go outside and visit the cottage, Matt, for the moment, wants to stay inside and cook some more.

In subsequent sessions he begins selecting larger and larger items to populate his sand worlds including rocks I have brought back from places around the world. He tells me that these rocks which he buries in the sand are really rare where he comes from. Over the next few sessions, these rocks are dug up by "Map Boy." Eventually the rocks migrate to the stews that Matt "cooks" for us after I tell him about the book "Stone Soup" that my parents had read to me as a child and that Matt's parents then bought to read to him after I told them about it.

Meanwhile, reports are coming in from school. Matt has begun to advocate for himself more, asking for help in classes where he needed a teacher to pay more attention to him. While there had never been a great deal of hand sanitizer use there, Matt had occasionally refused to participate in school activities. He was one of the physically smaller children in his grade, though ironically, he was one of the oldest because his parents had held him back a year in pre-school. Matt, his parents told me, was embarrassed about his physical size and had even lied to other kids, telling them he was a year younger than he actually was. One day on the school lunch line, a much larger student had pushed him, and Matt pushed him back hard. The attacker had hit the ground and Matt was sent to the school office. This had shocked Matt and his parents. In session, I re-framed this for the parents as assertion in the face of aggression. I asked Matt's dad to come in with Matt the next time, and we discussed how Matt could advocate for himself in situations where he didn't feel safe. I showed Matt how he could focus his eyes, breathe deeply and send the bully a look that indicated he was not to be "messed with" and would likely have the bully thinking twice.

One night I dreamed about Matt. We had gone to visit the cottage behind my house. Matt creates a sandbox world made up entirely of figures from a hospital, nurses and doctors who watch over a group of resting babies, swaddled in pink and blue blankets. In the middle of making the sand world, Matt stands up ramrod straight, screams and runs out the door. Matt is running so fast I can't catch him. He is running past his mother's waiting car, down my driveway and towards the busy street. I yell out for his mother to follow me and together we run, chasing Matt. I can't keep up and neither can his mom even though she is ahead of me. Suddenly at the last possible moment, at the edge of the street, with cars careening past, Matt stops dead in his tracks. "Thank God," I tell his mom. "He stopped himself. . . . He stopped himself." When I awoke, I knew that while I could help, Matt would need to be the agent of his own healing. He would need more experiences that let him feel his feet more solidly on the ground.

In the next session, I look out the window and ask Matt if he'd come for a walk with me to the stream down the lane from my home office. He nods. As we walk down my driveway, I pick up a few small sticks. Matt sees this. Our eyes meet. He touches nothing. It is Spring in the Connecticut woods and we can hear the stream roaring. Matt's eyes, which had been cast down toward his shuffling feet, are now focused ahead. He walks faster, beginning to jog and then runs fast ahead to the stream. I take my sticks and I throw them one by one into the bubbling current. I pick up a two more sticks and ask Matt if he wants one. He shakes his head "no" and then grabs one of his own. With a small flick of his hand, he tosses his in and watches it being carried away by the water. He picks up two more and throws them in. Mirroring him, I do the same. No words are spoken. He throws three. I throw three. He picks up a pile of stones and branches and with a great heave all fly through the air, the stones plopping into the water and sinking to the bottom, the branches floating away. Matt, his eyes alight with energy has caught sight of a large branch that has fallen in a recent storm on the lane. Together we drag it to the stream. Matt is laughing and we are both panting with the exertion. We tip the branch up and over into the stream and watch as this new co-made dam catches floating, rotten leaves. I think this dam will be a fixture until the wood itself rots in the water. We walk back in silence to the office.

After the session, on my recommendation, Matt's parents, though tentatively, entered into an agreement with him. They would remove the large bottle of hand sanitizer that had occupied a central place in the family kitchen. Over the past weeks, less and less hand sanitizer had been used and even Matt had agreed with me that he didn't need to use it anymore. "You know, it just made me feel stronger," he shrugged. "Almost," I had told his parents, "like a second skin."

Now that Matt was experiencing his skin as thicker, I had another experiment in mind. The summer was approaching, and Matt would be headed off to summer camp, his first summer away. In the Fall he would be celebrating his eighth birthday. I did not know whether Matt would need to come back after the summer and I wanted to offer him a demarcation ritual to mark his progress, envisioning the possibility of an ending where he could feel the power of his emergence into a greater sense of his own strength. He had used sand play figures as symbolic helpers on his path, brewing stews to nourish his growing confidence. Now I would invite him to use the power of his own body as a symbol for what could be completion, for the moment, of his therapeutic work with me.

It is late afternoon and through my office window, the sun is shining deep red-orange through the trees. Matt creates a story where he is the manager of a grand hotel. I am his only guest. From the room with the sandboxes, Matt brings in plastic bowl after bowl of food for me to consume. Each bowl is different. A smorgasbord of stews, made of whole cows, of giraffes and gazelles, of zebras, of elephants. Somewhere we both know this may be Matt's last session. Today, there is something regal in his bearing. He makes me two drawings. One is of his alter ego "Map Boy" who hugs people, and the other is of Neal who hugs "Map Boy." I look at the clock. Five minutes before the end of session. From my office closet I take a large, brightly colored cylinder and unfurl it into its fullness, a six-foot-long "play tunnel" that only children and smaller adults can fit their bodies through. Matt's mother shows up early at the door. She surveys the room of bowls on the floor filled with plastic animals. I place the tunnel so one end faces my office exit and the other where Matt and I stand in the room. I motion to the tunnel. "Matt," I say, "It's time to go." He bends his body toward the tunnel, turns for a moment to look back at me and crawls through on all fours towards his mother. On the other side, he exits, uncurls his body to standing, stepping on his own over the threshold of my office into the outdoors.

Postscript to Matt

Matt's parents told me in their final parents-only session that the waking behaviors at night had decreased markedly. We discussed the need of children for unconditional love and how the perceived removal of that sense of love and connection when parents become reactive to symptoms in a child can really hurt, feeling like a threat to their very survival.[6] As I was now handing Matt back fully to their care, I asked them to monitor future experiences with Matt, noticing where their own need to maintain an idealized self-image as a "good parent" could have them over-controlling Matt. I encouraged them to "scan themselves for shoulds," judgments of themselves or their son. Because Matt's behavior would likely continue to differ from how they remembered their own as children especially with their sometimes unreasonable expectations of him. He would feel their disapproval. It would hurt. I asked them to remember how it felt in their bodies when their caregivers banished them to a room or resorted to corporal punishment when they were children[7] and to have the pain of this memory give them pause before they react to their son. To put contributing to their son's wellbeing ahead of any potential exercise of parental power which might in the end prove counter-productive.

There was one last thing on today's family therapy menu. I asked Matt's parents to consider whether their own childhood wounds might need some care. Neither one of these parents had experienced the benefits of dynamic, movement oriented play therapy to ease the emotional constrictions which begin in childhood. How might their son be taking on an anxious operating style feeding off their own unaddressed anxieties? Would learning how, as their son was learning, how to be more in the present moment, by noticing his own body sensations and tensions, be a skill that would serve both them and their family? Would learning to track their own body reactions, cueing them to where perceived threats push them over the edge into fight-flight-freeze response be helpful? What unresolved triggers created as they grew up could have them feeling threatened and emotionally unsafe as they tried to parent together, to the point of compromising their choice of wise responses with their sons? So, I spoke to them about the guidance provided by psychiatrist Daniel J. Siegel. That our children's development of secure attachment, influencing flexibility in their future lives and relationships will depend in large part on their parents having a mindful and coherent narrative of what happened to them as children.[88] I was gently challenging the parents of my young client to consider doing their own therapeutic work to build a clear picture of their own inner lives as the young people they had been for the benefit of their developing children.

Conclusion

In the case of Katie, an 11-year-old girl, bouncing on an exercise ball to loosen up held-in feelings was one of the first moves in the play room. Productive narrative play could then be more easily initiated through the building up and breaking down of soft pillowed walls, symbolic of the boundaries needed to protect her delicate developing psyche in a world not always perceived as safe.

For children, whose natural expressions of emotion – including anger and grief – have been suppressed within the family, the therapeutic goal is often channeling "creative destruction" in the service of re-balancing and re-building the energetic/emotional system of the child. As the child's perception of safety grows in play with the therapist, young people such as Katie take greater risks in their play, even more so when the therapist is willing to "mirror" the child. This can be done first as the child experiments in the arena of sound, listening to the rise and fall of volume in their own voice. Later, as the safe therapeutic container grows, the child can be encouraged to play in in battles using padded "bataka" encounter bats which help ground upwardly displaced energy in the body of an anxious or angry child. This also contributes to a growing "felt sense" within the child of her own power. In parallel, positive reinforcement and support by the therapist for the child's expressive capabilities, models adult support in real time for what is authentic to that child, a fuller range of natural emotional expression as it modulates and pulsates within the body.

In the case of seven-year-old Matt, we have parents who must initially decide to trust the process of dynamic play to deal with their son's symptomatic behavior which they had hoped to arrest through cognitive behavioral therapy. The endless possibilities of a narrative menu in the course of this play include miniature dinosaur figurines, the "meat" of a "stew" which is "cooked up" by the child to share with his therapist. In a family facing economic challenge, Matt begins to develop a language through his play that quantifies what may be of worth in his world and maps a quest for hidden treasure. He sets out like the characters in a book to make soup, even if the only ingredient you have to start with is stones. Through inventions of his own making, he begins to address fears that had closed doors to the point he had needed to "sanitize" much of his world before he could interact with it. Now as he begins to expand his creative narrative, Matt embarks on a journey of adventure where the next step becomes an excursion outside in nature, accompanied by his therapist. The boy is delighted as he moves outside the therapy room where parents send you to be fixed. I encourage therapists to consider the value of such venue variety.

Where the presenting problem was a child's anxiety around cleanliness, the sphere of safety widens. Cautious hands became courageous ones. The seed planted in the therapist's mind by parents initially seeking cognitive behavioral therapy, flowered in a different way than the family might have imagined where the petri dish of the dynamic play palette allows water, wood and sand to be among the ingredients employed in healing. And in what may be the close of a chapter in therapy, the boy invents his way out of his own symptomatic predicament. Emerging through a colorful play tunnel, he returns, re-energized to his family. While he entered into play therapy as an individual, on another level, his bolstered creativity can serve as a beacon, beckoning others in the family to challenge old narratives and explore new ones. To take the risks required to live and play as fully as possible in a story of one's own design.

Notes

1 Martin, J. (Dir.). (1995) An Interview with Eva Reich, M.D. Gentle Bio-Energetics: Prevention of Neurosis from Birth on [VHS]. Fort Bragg, CA: Flatland Books

2 Reich, W. (1945) *Character Analysis*. New York: The Noonday Press, Farrar, Straus and Giroux, pp. 155–158

3 McCarthy, D. (2007) *"If You Turned into a Monster" Transformation through Play: A Body-Centered Approach to Play Therapy*. London and Philadelphia: Jessica Kingsley Publishers, pp. 49–61
4 McCarthy, D. (2012) *A Manual of Dynamic Play Therapy: Helping Things Fall Apart, the Paradox of Play*. London and Philadelphia: Jessica Kingsley Publishers, pp. 81–122
5 Wilner, K. B. (1999) 'Core Energetics: A Therapy of Bodily Energy and Consciousness'. In D. J. Wiener (Ed.), *Beyond Talk Therapy: Using Movement and Expressive Techniques in Clinical Practice*. Washington, DC: American Psychological Association
6 Lowen, A. (1995) *Joy: The Surrender to the Body and to Life*. New York: ARKANA, Penguin Books, pp. 246–251
7 The Center for Nonviolent Communication. (2000) *Parenting the Nonviolent Communication Way by Marshall B. Rosenberg, Ph.D. [Brochure]*. La Crescenta, CA: The Center for Nonviolent Communication
8 Siegel, D. (2010) *Mindsight: The New Science of Personal Transformation*. New York: Bantam Books, pp. 171–173

References

Brodsky, N. (2015) 'Emergence: A Tale of Two Boys'. In D. McCarthy (Ed.), *Deep Play – Exploring the Use of Depth in Psychotherapy with Children*. London: Jessica Kingsley Publishers.

The Center for Nonviolent Communication. (2000) *Parenting the Nonviolent Communication Way by Marshall B. Rosenberg, Ph.D. [Brochure]*. La Crescenta, CA: The Center for Nonviolent Communication

Lowen, A. (1995) *Joy: The Surrender to the Body and to Life*. New York: ARKANA, Penguin Books.

Martin, J. (Dir.). (1995). *An Interview with Eva Reich, M.D. Gentle Bio-Energetics: Prevention of Neurosis from Birth on*. Fort Bragg, CA: Flatland Books.

McCarthy, D. (2007) *"If You Turned into a Monster" Transformation Through Play: A Body-Centred Approach to Play Therapy*. London and Philadelphia: Jessica Kingsley Publishers.

McCarthy, D. (2012) *A Manual of Dynamic Play Therapy: Helping Things Fall Apart, the Paradox of Play*. London and Philadelphia: Jessica Kingsley Publishers.

Reich, W. (1945) *Character Analysis*. New York: The Noonday Press, Farrar, Straus and Giroux.

Siegel, D. (2010) *Mindsight: The New Science of Personal Transformation*. New York: Bantam Books.

Wilner, K. B. (1999) 'Core Energetics: A Therapy of Bodily Energy and Consciousness'. In D. J. Wiener (Ed.), *Beyond Talk Therapy: Using Movement and Expressive Techniques in Clinical Practice*. Washington, DC: American Psychological Association.

24

LISTENING TO PAN

Helping children who panic

Dennis McCarthy

> "we need to go down, to find the earth in us, the animal root-impulse of all our being, that which spontaneously with all the force of instinct reconnects us to our path in this life, and through it to the heart of being itself".
>
> – *Ann Belford* Ulanov 2004*:283*

For much of my 40 plus years of working with children in crisis, I have found that the symptoms that bring children into treatment often have the resolution of the symptoms, as well as the underlying problem, embedded within them. In recent years I have written and spoken in keynotes at play therapy and dance therapy conferences internationally about using the energy and imagery of symptoms to treat the underlying problem. I have found in my work with children that the urgency of certain symptoms such as panic attacks and phobias makes them easier to treat if viewed in the right light and used dynamically. "Becoming the Storm" has been a common theme in these conferences, based on my experience treating a number of children with storm phobias. These children all had initially been afraid of thunder and lightning, which had morphed into fear of rain and then cloudy days. By the time I saw these children they were housebound even on sunny days because, as several of them told me, " you never know". They all resolved the phobia rather quickly by "storming" in my play therapy space. These storms involved them leaping and falling to the accompaniment of drums and cymbals, repeatedly and very loudly. By embodying the aggression of storms they mastered their fear of them. This seems like an outrageously simplistic treatment plan but it has always worked, not just with storm phobias but with other phobic reactions and with panic disorders as well. This is especially true with latency age children. The panicking or phobic child will need their family to make room for a more spirited child. But the severity of the symptom encourages parents to accept the challenge. Most families prefer a louder more outspoken child to one immobilized by panic or housebound due to a phobia.

A panic attack is the abrupt onset of intense fear or discomfort that reaches a peak within minutes and includes at least four of the following symptoms: palpitations, pounding heart or accelerated heart rate, sweating, trembling or shaking, sensations of shortness of breath or a smothering feeling. The panicking child contracts upwards into their bodies trying to avoid these sensations, which of course doesn't work.

Figure 24.1 Pan

Source: Francisco Rivera.

The idea of panic arises from the myth of the Greek god Pan, who was conceived by the union of the god Hermes and a nymph. Hermes was the messenger of the gods, who attempted to resolve conflict among the gods and was the only god who could go to the underworld. He was also a trickster, and is the god often associated with the art of psychotherapy. Pan has something of the trickster/therapist innately in him. But he is also associated with the instincts and therein lay his powers in the play therapy process. Pan has the legs and the horns of a goat and the upper body of a man. He wreaked havoc among the gods and goddesses on Mount Olympus with his antics. Like a mischievous child, he made them laugh even as he unsettled things. He was also known for his capacity to scream very loudly when annoyed. A character that can unsettle the gods is obviously not an easy one to deal with. The child who is plagued by panic needs to unsettle things in their own psyche. They need to make noise and maybe make trouble. Their defense system needs to be strengthened, their capacity to both tolerate and express negative emotions and aggressive impulses enhanced. It is said that the sound of Pan's flute incited panic in those who would not surrender to it and great joy in those who would, those who danced to the music. Helping a child tolerate the sound of his flute and surrender to the dance is the path we must take.

Pan was the god of unbridled wildness, of chaos, of creativity, of human and animal procreation. Worshipped in groves in the forest or caves rather than temples, he lived on the earth, not Mount Olympus, and was thus closer to being human, despite being part animal. And he made the gods laugh! With children it is important that their therapy treatment include a big dose of levity. Most of the monsters drawn by children that stare down at me from my office walls are humorous even as they are also monstrous. The paradox of play therapy is its capacity to allow the child to express huge emotions in the context of play, protected by the raw intensity of these emotions by the very nature of play.

It is helpful to explore the history of both the child and the symptom to make sure there is no underlying medical issue to be resolved. But often it is pediatricians that refer panicking children for treatment and the potential medical issues have already been considered and ruled out. In my extensive experience, if the child can play with and through the panic rather than have the symptom medicated, then they will emerge a much more self-possessed child. Simply put, in inciting panic, Pan also stirs up the very impulse needed to resolve panic.

An eight-year old boy began to panic right after witnessing a fistfight between two adult men at a soccer meet. Neither man was hurt very badly but it had been a scene of unbridled anger and aggression. This high functioning and popular boy quickly became immobilized by panic.

In our first meeting, I invited him to imagine he had drunk a magic potion that turned him into a monster and then asked him to draw the monster. I often do this when I first begin working with a child, in part to invite all of the child into the play space and therapeutic relationship. The monster is often an ally for the child in crisis. And with panicking children, the monster can be a very important guide for both child and therapist. This boy seemed surprised by my request, as he thought we would simply be talking and that I would teach him relaxation techniques to use to cure his panic. I explained that this drawing would help. He nodded and quickly set to. His monster was somewhat blah and tenuous, and its power was that it helped people. This might seem like a good thing, but this boy had always been a helper. Even when he played soccer, which he was quite good at, he always set up other players to score. Because I thought this tendency was part of the problem, I suggested that his monster drink a magic potion and turn into a wilder monster. He agreed and the second monster was larger and was beginning to seem dangerous. But there was still something missing.

Ordinarily I would not push a child to go beyond their limits so quickly but given the severity of his symptoms and my confidence that together we could help them abate, I pushed him further. So I urged this monster to also drink a potion and turn into an even wilder monster. He smiled, almost as if waiting for me to suggest this, and then drew a huge beast with numerous heads with spikey teeth. His whole body and being were involved in the drawing of this monster. It grew numerous arms with long claws. It reminded me of the Scylla, a terrifying monster in The Odyssey that devours many of Odysseus's crew. The Scylla is a monster that Odysseus must face. He can't avoid it. This monster looked impressive and the child looked enlivened. I asked him what his monster did that made it such a monster and he told me simply that it ate everyone.

My play therapy office has walls that are covered with monsters drawn by children with a wide variety of presenting symptoms. Children find comfort, amusement and inspiration in looking at other children's monsters. They are never frightened of them as they see in them expression of emotions, aggression and bigness that they too feel but cannot verbalize. There are many one eyed monsters as well as many monsters with numerous heads. In addition to the walls of monsters there is a deep sandbox with thousands of miniature figures that can be placed in worlds made by children in the sand. There is a dollhouse, a large basket of blocks and two small guillotines for chopping the heads off of figures made with clay. There are several rubber mallets for pounding clay, and of course drawing and painting materials. This is the space in which we two sat.

What I knew about this child prior to seeing him is that he came from an intact family that loved him. His parents told me that during the so-called terrible twos he was an angel, unlike his older sister. In fact he never did go through a period of saying no but had always been complacent. What I didn't know at first was that during the age of two his father was seriously ill. For over a year it seemed that he would die. The family held their breath. Then when this boy was four his father was cured and healthy. But the child had missed an essential developmental stage.

He and I talked about his panicking in our first meeting and I assured him that I could help. I went directly to his need to learn to say no and assured him he could master this by playing with it in my office. So we explored ways of him asserting himself using his body. He pounded and ripped clay, fought me with foam bats, pounded with a tennis racket, and threw things at me that ranged from large balls to paper airplanes. In all these activities the focus became not only his putting his whole body into the movement but urging him to really let go of the object when it was thrown. He also spent time each session stomping around my office. This began hesitantly, but when I urged him to stomp so hard that the people in the office below would bang on the ceiling, he began to stomp in earnest. There were in fact no people in the empty office below me but he imagined he heard banging in response to his stomping and this made him very happy. He was really putting his foot down at last and feeling more grounded in himself. He began to wear very large boots to his sessions so that his pounding was louder and more disruptive.

Within a few days after my first meeting with him his mother reported that she found him sitting on the floor in the kitchen banging a cupboard door, the way a toddler might do. She asked him what he was doing and he said that he had no idea. But she realized he was playing with an impulse that serves a function in the developing child. She saw it as a good sign, coming in the wake of his first session.

Meanwhile as he pounded and stomped in my meetings with him, he continued to draw himself as a monster each time. These monsters got larger each time until one monster spanned three sheets of paper. It was so big that some small bumps at the bottom of the page were the Himalaya Mountains! I thought that his monsters had become a bit too big, almost ineffectual in their hugeness. I assumed they would/should begin to shrink, which they did. His last monster was short and muscular and very powerful, probably like the men whose battle had triggered his panic. Meanwhile his panic quickly subsided, replaced by episodic nausea. His parents were able to distinguish that the nausea occurred when he did something he really didn't want to, or ate something he really didn't want to eat. In these instances they began to urge him to assert himself rather than be compliant. This took effort, due to his having gotten so much praise for always being agreeable. But the family collaborated to help him master this.

Once I learned of his father's serious illness, which had occurred when this boy was two, I urged the family to talk about this as a family. His father was reluctant to do so, as he felt so relieved that he had survived and was now healthy. That phase of his life had been so terrifying. But with encouragement he was able to see how deeply it had affected his son and so he agreed to talk about it. The family reexamined that period in their lives, and they all wept, both recalling how frightening it was and out of relief that he and they as a family had survived.

His panic and nausea receded after this. He was now playing soccer without holding back. He was also asserting himself on the playground when needed, still seen as a leader by his peers but one that could put down his foot when needed.

In our last visit he was sitting in my play space with his mother and older sister at the end of the session. As his mother chatted with me about how well things were going, I saw his sister reach over and pinch her brother's butt. His body had the immediate reflex of smacking her hand away. His sister smiled in amazement. He had arrived at a place of instinctual assertiveness with no ambivalence. I can't think of a healthier manifestation of autonomy. His reflexes said "no", without any extra drama or remorse. It was a fitting end to our work.

Most of the children I have worked with suffering from panic attacks had for various reasons not gone through the so-called terrible twos or the age of "no". In some cases it was due to parental conflict that took precedence over what should have been an innate developmental stage. At other times it was a parent being seriously ill or the child himself being seriously ill. And sometimes the child was naturally precocious and seemed to leap ahead developmentally not due to anything specific. But the absence of no set the stage for panic, even if it didn't emerge for several years. This stage, also called "rapprochement", is the beginning of individuation, marked by an alternating between clinging and pushing away behavior and is essential to becoming autonomous.

> Knowledge is a function of discrimination. To know what A is, it must be distinguished from all that is not A. Knowledge arises from the recognition of differences. The first difference that an organism can recognize is that between what feels good to the body or pleasurable and what feels painful. In the growing child, the "no" functions as a psychological membrane. It prevents the individual from being overwhelmed by outside pressures and allows them to discriminate, guarding against exaggerated impulsiveness and defining the ego boundaries of an individual. Saying no is an expression of opposition, which is the cornerstone of the feeling of identity.
>
> *(Lowen 1970:155)*

Looked at in light of Lowen's commentary on "no" it is easy to see the panicking child as lacking an adequate filter, unable to discriminate what is them from what is not them. All the subsequent stages of development happen, but without an underlying sense of self-awareness. No amount of intelligence or empathy can compensate for the lack of this filter, and for the boundaries, self-protection and self-awareness it fosters.

An 11-year-old girl had been battling serious stomach ailments for roughly one year. She had gone to numerous specialists and undergone numerous tests. There were various theories about the origin of the symptoms but no clear treatment plan. Then she began having panic attacks that were so debilitating that she was missing a great deal of school. She began treatment with me for the panic attacks. She told me in our first meeting that she was convinced that both the stomach ailments and the panic were rooted in some negative emotions. Her parents had been battling a prolonged divorce when she was young, and her father had attempted suicide. Her mother told me that her child had gone through a long period of what seemed like the stage of no, but in examining this it was really inconsolable crying jags. It seemed that she had been expressing the anger and sadness that the whole family was feeling, and not really discriminating who she was separate from them, which is what this stage is about.

So in our first meeting we leapt right into working with this idea of self-assertion. Her first monster drawing was called "Belly Blaster" and it erupted energy from its belly that knocked over buildings and trees and people. She was surprised at her monster even though she had created it. She also spent time hammering a lump of clay. Although she enjoyed the hammering her body wasn't fully committed to it, and it felt ineffectual. She talked about a series of tidal wave dreams she had been having as well. In these dreams, the huge waves were not a threat to those around her, only to her. I assured her that they were "her waves" which she was surprised but happy to hear.

In our second meeting she drew "Belly Blaster" again, this time it was blowing up the entire world with its blast of belly energy. She was quite thrilled with it. Her pounding had more commitment this time. She was starting to spend the entire day in school whereas when we met she often had to leave after a few hours. She reported that her stomach felt better as well. She mentioned that one of her great fears was going into an elevator. She knew this could be avoided for the most part but she was determined to resolve it. In her third session she reported that she'd had an elevator dream. In the dream she and her mother entered an elevator, going down, carrying armloads of her stuffed animals. The animals began to spill on the elevator floor but then a live polar bear that was also in the elevator bent down and picked them up for her. She was not afraid to ride in the elevator with the polar bear inside as well. This rich dream suggests that the resolution of her fear is embodied in the bear, now aligned with her in what had once been an image that evoked panic. I asked her what her association was with polar bears, and she said "they are ferocious and protective of their children".

The next "Belly Blaster" was a black hole that blasted other black holes. She was very happy with the unfathomable monster she had created. She couldn't quite believe that she had conceived of something so huge. But she herself had become bigger, taking up more space in her family and her school and actually seeming taller. After this she discovered some boxing gloves that I have and used them to hit balls that I would throw at her. I threw different sized balls until she found one that seemed just right. It was a large yoga ball that allowed her to throw her whole body into the movement. She was thrilled by this, and asked to do it again the next time I saw her. We repeated this ball boxing each time and I realized it felt a lot like Belly Blaster. In fact, the way she threw herself into each attack did involve both using and relaxing her stomach muscles. She had become a belly blaster!

Her next monster was not a belly blaster at all but a powerful female deity that had evolved from the black holes. Her figure stood firm, rooted in herself. She reminded me of the Hindu goddess Kali, who embodied Shakti or female power and who was both the protector and destroyer, or the Greek goddess Artemis, the protector of young girls and the destroyer of those who threatened them.

She had also created a series of worlds each week in the deep sandbox in my play room and these were pivotal. In her first two sand worlds, a powerful goddess/sorceress with a pet Pegasus stood on a mountain, controlling a village below. The villagers had been under her spell and unable to think or act for themselves, but were finally revolting. It seemed unlikely that this revolution could succeed given how powerful the goddess was. And yet the goddess seemed to be waiting for them to figure out how to succeed. She was provoking them to revolt even as she was making it seem impossible. The paradox of this was important.

Then she created a world in which a dragon fell from the sky into a deep crater that had been made by a meteor. The dragon fell on top of a group of animals that were living in the crater. But the animals weren't killed. In fact they had been woken up by the dragon's fall and were climbing out of the crater. The goddess incited revolt even as she curbed it, and the dragon fell on the animals and provoked escape from imprisonment. It's easy to see these worlds as speaking metaphorically about her early experience within her family. So much emotion was being expressed and also resolved. She too was attempting to revolt against her own stasis, to free her animal instinct.

After creating the dragon world the panic stopped. But she still suffered from stomach aches at times. Together we figured out that these aches began the day before she was scheduled to go to her father's house. Although she liked seeing him, he was someone she still could not assert herself with. So I met with the two of them to figure out how he might help her do so. Her father hung a punching bag at his house for her to use when she was there. This gesture did the trick. His giving her permission to aggress while with him kept her from regressing into the internalized aggression that had become stomach aches and then panic. Like her panicking, the stomach aches were a symptom that lead to its resolution. Like the goddess and the dragon, they provoked a revolution in her ineffectual defense system.

Aggression, far from being the diabolical, destructive principle that classical psychoanalysis makes it out to be, is really an essential part of the life-preserving organization of instincts (Lorenz 1966:48). Right from the beginning, aggression contributes to discovery of the real world around us. It helps us make things real, external, existing in their own right. Aggression links us to the reality of the environment and opens its riches to us, if we can endure the necessary transformation of aggression (Ulanov 2001:107).

There are children whose panic arises from not having been allowed to assert themselves, prevented from doing so by violence or the threat of violence by a parent or caregiver. For them the healthy aggression of the no turns against the self due to another's will. Pan calls to them via panic so they will finally rebel. These children may need to play with the violence that initially stopped them. I have a small guillotine that I allow children to use to decapitate clay figures. This may seem like an outrageous play activity but it is a pivotal one for many children as they understand the metaphor behind the act. The child thwarted by parental violence uses the symbolic violence of the guillotine's blade and witnessing this action feels like a firm no. This activity rarely happens after the first few sessions. It is a defining act that may cut through an ineffectual defense system and allow a more functional one to develop. It is especially helpful with a child stuck in a state of regression.

Years ago I saw a six-year-old child who had been bitten by a large dog and struck by a car in the same week. Surprisingly neither of these had actually hurt her bodily beyond a few scratches. But she had completely regressed and was clinging to her mother constantly and unable to go to school or be out of her mother's sight. And she was panicking even while in her mother's arms, saying that she wanted to climb back into her mother's belly. She sat in my waiting room with her mother and her older sister. The older sister was rolling her eyes at having to be sitting in a therapist's waiting room. The six year old was refusing to come in, partly because it meant leaving her mother's arms and partly because of her sister's eye rolling rejection of therapy. I stood there stymied by this impasse. So I told her that I had a small guillotine in my play space that was sometimes used to chop the heads off of older sisters. Would she like to try it out? At this she leapt out of her mother's arms and ran into the play room. After we had both entered, she slammed the door shut. Then she sat down to begin the playful assault of her sister.

Although she enjoyed the head chopping it felt like a largely symbolic act, one that helped pull her out of the regression. She laughed at what she was doing and quickly began to shape the clay into a bowl. She was a child who had in fact gone through the stage of no at the appropriate age and was basically a healthy self-possessed child. But the attack on her body by the dog and the car had regressed her and left her panicking. For her, coming in to chop her sister's head off accomplished several things. The act of aggression, even my offering her this and her accepting it, pulled her out of the state of regression and the ensuing panic. You could say the desire to aggress won out. Secondly, it altered her relationship with her sister, who looked at her with new respect when she and I emerged from the playroom at session's end.

She came one more time with her big sister, who was determined to get inside the room that had seen her guillotined. The two drew monsters for my wall together and chatted about this and that. The older girl was an excellent artist and her monster was quite complex. But she made sure to compliment her little sister's monster, something I suspect she would not have done prior to this.

Encouraging destructive play seems like a mistake to many therapists but it is deeply therapeutic for most children, especially those who are in a state of panic. DW Winnicott wrote extensively on the importance of aggressive and destructive play.

> Destructiveness arises as one part of our inherited instinctive aggression. Aggression manifests itself firstly as motility, beginning in the womb and then outside it, in kicking, in biting, in gesturing spontaneously. Aggression at first means impulse – impromptu, unstudied. Aggression starts by showing itself as energy – crude, unhibited, bubbling up from the body – to be housed by each of us. Such dynamic animation leads to movement and then to exploration of the environment.
>
> *(Winnicott 1987:232)*

I myself suffered from childhood trauma related panic attacks when I was a young adult, so I do appreciate how debilitating panic can be. For me, beginning to dance and then working with children and their monsters, plus a great deal of therapy, helped me slowly replace panic with self-assertion. I was finally able to listen to Pan fully. My whole approach to working with children emerged out of my own symptoms and the underlying emotions and stymied aggression. The potency within panic, when listened to, can free us.

I had an experience with Pan during the birth of my second child. I was present with my wife and the midwife, and as the labor amped up I suddenly felt myself begin to panic. I felt a few moments of pure terror, wondering how to endure it. Then I began to growl. This was pure instinct, as it was my body and not my mind that was growling. It was a low belly growl and had the effect of immediately stopping the panic. I felt totally present. The midwife said to me "I don't know why you are doing that but keep it up. It's really helping!" I myself was amazed, not sure where this belly growl had emerged from. But ever since when I have felt panic looming, often before doing a presentation, just sensing the place that a belly growl might arise from helped the panic abate. To reiterate the lines I opened this chapter with, "We need to go down, to find the earth in us, the animal root-impulse of all our being, that which spontaneously with all the force of instinct reconnects us to our path in this life, and through it to the heart of being itself."

It may be Pan's having the legs and horns of a goat that make his flute music so compelling or terrifying, that make him so liberating for the child who is panicking. It may be that his very name, which means "all" that makes room for the whole child in his playful provocation. He is god of all the basic instincts; aggression, hunger, sexuality, creativity and spirituality, albeit a pantheistic form of spirituality. And his lower half still resides in the animal realm. Children are basically animistic, i.e. they perceive all things – animals, plants, rocks, rivers, weather systems, human handiwork and perhaps even words – as animated and alive. In panic, this spirit of aliveness has been thwarted. The absence of "no" also hobbles the "yes". But innately when offered the chance to connect with this instinctual animism, they will do so.

Pan's music is said to have inspired leaping in animals and humans who surrendered to it. This reminds me of the best definition of play, reiterated by Erik Erikson from Plato's *Laws*.

> In Plato's *Laws* he sees the model of true playfulness in the need for all young creatures, animal and human, to leap. To truly leap, you must learn how to use the ground as a springboard, how to land resiliently and safely. It means to test the leeway allowed by given limits; to outdo yet not escape gravity. Thus wherever true playfulness prevails, there is always a surprising element, surpassing mere repetition or habituation, and at its best suggesting some virgin chance conquered, some divine leeway shared.
>
> *(Erikson 1977:17)*

The aforementioned girl, whose sand world depicted a dragon falling on trapped animals and freeing them, speaks for all of children who are stuck. It is the animal impulse freed by the play experience that brings the child more fully into their body, which is the dwelling place of emotions and intellect, the vehicle of all self-expression and all relationships. Like the dragon, Pan provokes us to grow into more fully realized beings. But to do so we must rediscover if need be and/or retain our connection to the earth, to the animal, to the wild.

All case material in this chapter has been thoroughly altered so as to make it unrecognizable. That said, I have worked with thousands of children, many with the same presenting symptoms and similar play processes.

References

Erikson, Erik (1977) *Toys and Reasons*, New York, W. W. Norton.

Lorenz, Konrad (1966) *On Aggression*, New York, Harcourt Brace and World.

Lowen, Alexander (1970) *Pleasure*, New York, ARKANA, Penguin Books.

Ulanov, Ann (2001) *Finding Space – Winnicott, God and Psychic Reality*, Louisville, KY, Westminster John Knox Press.

Ulanov, Ann (2004) *Spiritual Aspects of Clinical Work*, Einsiedeln, Daimon Books.

Winnicott, Donald Woods (1987) *The Child the Family and the Outside World*, Cambridge, MA, Perseus Publishing.

25
THE STORY OF ANGELA

Katerina Couroucli-Robertson

The history of dramatherapy and playtherapy in Greece

The Herma Institute of Dramatherapy and Playtherapy, which I founded in 1992, has adopted Dr. A. Cattanach's definition of playtherapy:

> Playtherapy is a way of helping troubled children cope with their distress, using play as the medium of communication between child and therapist. The method is based on the central assumption that play is the place where children first recognise the separateness of what is 'me' and 'not me' and begin to develop a relationship with the world beyond the self. It is the child's way of making contact with the environment.
>
> *(Cattanach 2003 pg. 24)*

Furthermore, we practise non-directive playtherapy. The child chooses those materials or objects which make most sense to him or her and uses the play experience together with therapeutic relationship in their own way and in their own time. The child is in control of the process and directs the agenda, focus and timing (Axline 1989). The non-directive playtherapist allows the child to lead the process, while at the same time acknowledging the choices the child makes in a non-critical and empathetic manner. A core aim of the non-directive playtherapist is simply to be with the child and to provide the optimum conditions that allow the child to develop their inherent potential Gordon (2015).

The first dramatherapy workshop in Greece was given by Dr. Sue Jennings in 1984. A handful of people were interested and set up a pilot training course under the supervision of Dramatherapy Consultants, which later became the Institute of Dramatherapy. In 1986, The Arts Therapy centre in Greece began the first dramatherapy training, where many renowned professionals taught, including Dr. Sue Jennings, Pr. Robert Landy, Dr. Mooli Lahad, Alida Gersie, Dr. Ditty Dokter, Steve Mitchell, Aisi Minte, Dr. Murry Cox M.D. and others. Around the same time, Dr. Sue Jennings also began giving Playtherapy workshops, which aroused a lot of interest, resulting in a Playtherapy training, which was also supported by Dr. Ann Cattanach, who often came to Greece to give workshops and supervise our trainees. Later, the training became part of the Dramatherapy and Playtherapy Association, where it ran for ten years. Now, dramatherapy and playtherapy are established four-year courses, run by Greek instructors with international visitors and external examiners. I qualified in 1989 and have been co-ordinating both dramatherapy and playtherapy trainings ever since. My background has been in special education, where I have worked for over 20 years, both with physically disabled children as well as with adolescents with learning difficulties, my focus being on theatre. Gradually, I left special education to devote myself to dramatherapy and playtherapy training.

There are now three four-year courses in dramatherapy and playtherapy in Greece:

- 1992 – The Institute of Dramatherapy and Playtherapy "Herma"
- 1994 – The Institute of Dramatherapy "Aeon"
- 2005 – The Greek Institute of Playtherapy and Dramatherapy "Athyrma"

The story of Angela

Angela was eight years old when she came to me for therapy. She had been adopted at the age of four. She had a small scar on her upper lip, having undergone an operation to correct a harelip. Her parents brought her to me because Angela appeared to be hyperactive. She had difficulty concentrating at school and at home was unable to stick to her daily routine.

According to Russell Barkley (2018), the inability to inhibit a behaviour is the underlying problem for individuals with ADD. They simply act without thinking, regardless of the consequences. Angela may have appeared to have ADD, but I believe that other factors were also influencing her behaviour. Her parents were going through a difficult time in their marriage. Her father insisted, however, that Angela's erratic behaviour had started prior to their marital difficulties. I could not dispute this, but in my experience marital problems are present long before the couple fully acknowledge them. So, I suspect that her parents were experiencing difficulties before Angela's father became involved with another woman. Children are often aware of far more than we realise and quite possibly Angela was unconsciously or consciously aware of her parents' precarious relationship, which may have been a cause of possible disturbance.

After meeting her parents, I used drama/playtherapeutic methods to work with Angela for nine sessions. On the tenth, I had a session with Angela and her parents together to talk about the issues that emerged during therapy.

Angela was a lively child, who enjoyed playing, but was not very talkative. During the first sessions, she wanted us to play football and then do a role-play. Whenever we did role play, she was very specific about the role I should take and how to play it. She usually took on the part of a young boy and I was to be the 'bad' person who would chase her and capture her. Stagnittin and Cooper (2009) maintain that play not only affects a child's development and inner world, but contributes to his or her development and facilitates growth, development and individual competence. They go on to say that what children play largely reflects their culture and environment, as I believe was true in the case of Angela. Why she chose to play the part of a boy I could not say, possibly because she felt safer in that role, having seen that boys appear to be stronger and more assertive.

In the last session, her story changed. She was still the boy but this time she was employed by a couple as their servant. I played the part of the housewife who gave her orders. I told her to wash the dishes, bring me coffee, sweep the floor and then slowly, as the story progressed, she told me that I would grow to love her and adopt her as my child. She said that her own mother had not kept her because she had a mark on her mouth but this new couple was going to keep her. It is interesting I believe that Angela gave herself a menial role, indicating perhaps her low self-image and that she felt she was not in a position to change the course of events.

During this last session, she gave the impression that she had begun to accept her adoptive parents and they her. Storr (1972) believes that creativity can reflect a compulsive need to order and control. In Angela's case, I would suggest that Angela was directing our play in a way that she wanted her life to go. She wanted to create what she envisaged as a perfect family, in which she was loved unreservedly, despite her harelip and the fact that she was adopted. The important thing is that by creating the perfect family through play she began to believe that it was possible in real life.

For his research in areas relating to children and young people, inclusion, children's rights, the arts and therapy, Pr. Phil Jones (2008) proposed that he interview me on a specific case study. I chose the story of Angela. Through the questions and answers, we discussed the therapeutic process and how drama/playtherapy is more beneficial to children than verbal therapy. The following paragraphs sum up the outcome of the interview and our discussions together.

I strongly believe that the dramatherapist-playtherapist who follows the principles of non-directive playtherapy accepts the concept that at birth all human beings possess the ability to be aware of their experiences and be creative. Moreover, the playtherapist accepts the principle that all children's reactions stem from their intrinsic need to stay healthy and protect their true nature. 'The body is a primary means of learning (Jennings 1998) and early somatic experiences influence later physical and cognitive growth' (Jennings 2011, pg. 19).

The main process in my work with Angela was to establish a secure relationship with her so that she could trust me. Once this was achieved, I believed she would be able to communicate with me and perhaps share what was going on in her life. At first, she needed to express her anger, which I allowed her to do, as long as she did not damage anything deliberately. She wanted us to play football during our first sessions and once she accidently broke a vase. I saw the fear in her eyes and reassured her that it did not matter, as she had not meant to break it.

She gradually felt safer and suggested that we do a role play where she gave me instructions on how to play. It was very important that she took on the role of the instigator because in this way she felt that she was in charge. When working with children, I find that they tend to treat the therapist as a teacher and often, although not in Angela's case, want to please them. I needed to get it across to her that I was there to play with her under her instructions, not the other way round.

The fact that Angela gave me a role to play in her story meant that she had started to trust me, even though I had the role of a nasty person. She took on the part of a young boy and I was to be the bad person, who would chase her and capture her. Once I had caught her, I would tie her up. Sometimes she asked me to kill her. I think when she used the word 'kill' it was mainly to indicate her fear of this 'nasty' person I was impersonating. Although she wanted me to capture her and 'kill' her, she enjoyed proving to me that I had not succeeded, by coming back to life each time. This made her feel stronger, I believe.

Bowlby (1988) suggests that attachment relationships and attachment behaviours are suffused with emotion. If the security is broken, anxiety and anger appear. I believe that in Angela's case this had taken place at some time prior to her adoption, possibly when she was separated from her birth parents when she put into an institution.

I remember clearly playing the role of her persecutor time and time again and how when she left the premises it took me some time to roll up the twine I tied her with. I did not enjoy the role she gave me, but I tried to maintain it as best I could, which meant putting as little emotion into it as possible. I believe this role play made her stronger, as she proved to herself that she managed to survive the ordeal. At the end of each session, I remember how she hid from her father under the large cushions we had used in the role play, when he came to pick her up. Her father joked that he was her pa-teras (father in Greek). If you take away the 'pa', the word *teras* is left, which means monster. This may be an indication that he felt some guilt, indirectly admitting that, in her eyes at least, he was a 'monster'. I believe that Angela had a great fear of adults and that role play helped her to establish that not all adults are dangerous and that they can even be played with. The dramatherapist-playtherapist acts as an "undistorted mirror" that facilitates the child to restructure her/his sensory experiences respecting her/his pace and time, so she/he can explore her/his abilities, start trusting her/his inner truth, and thus discover and express her/his creativity.

It occurred to me that she had previously either suffered or felt that she had suffered situations similar to those in the role play. From her parents I knew that she had lived in some form of institution, so perhaps her memory was that of an older child bullying her or of some abuse by an adult. Rycroft (1968) talks about 'acting out' as 'the replacement of thought by action, and as a substitute for remembering past events'. He suggests that there are three implications: (a) the impulse acted out has never acquired verbal representation; (b) the patient lacks the capacity for inhibition; (c) the impulse is too intense to be discharged in words. Perhaps in Angela's case all three applied. What Angela may have undergone had never been expressed in words, nor was it possible for her as a child to do so. Play, though not a verbal expression of the cause of disturbance, did allow her to express her feelings indirectly and in some way overcome the 'nasty' being that had caused her disturbance.

My role as a therapist both with children and adults is to help them understand what is happening in their lives and how they wish to respond to these situations. So, when I played with Angela I didn't wish to influence her story but simply to become her helper. My relationship with Angela was to restore her faith in people. I hoped that through our role play she would be able to learn how to trust and how to form a relationship with another person.

The story of Angela may have influenced me too on a personal level, as it brought back memories of my own childhood. When I was her age, my family returned to Greece after having spent about five years in Canada. At school, I felt very different because I could not speak Greek. Aigen, K. (1991), points to the importance of therapists being able to open up vulnerable places and to relate to their patients from that place.

Through her trust in me, Angela was able to indirectly bring her issues out into the open so that both of us could witness them. Geldard and Geldard (2009) stated that the skilled counsellor is one who can take advantage of opportunities that occur through play and intervenes in a purposeful way. Had I been working with Angela using verbal psychotherapy as opposed to drama, she would not have been able to say anything significant about her life, as Angela was not a talkative child. She could not function verbally. Even when she drew a picture she did not like to comment on it. Through the story she made and taking part in the role play, she was communicating in a way that came easily to her and harboured no threats or fear of repercussion.

Her parents and possibly her teachers were afraid that Angela's problems were not only emotional, resulting from being institutionalised, but that she may have had learning difficulties. So, by letting her introduce a story/theme that we could play together, she was, indirectly and unconsciously, able to tell me about herself in a symbolic/detached manner. In this way, she brought out the issue of her adoption spontaneously through the role play. I believe that in the same way that dreams reveal aspects of the subconscious to adults, play reveals what is going on in a child's mind. A child will play out what he or she is going through at that specific time or something she or he has experienced in the past, by re-enacting his or her story. In this way, during the process the child becomes more conscious of his or her life. He or she communicates this to the therapist and this process may give him or her some form of release and allow him/her move on with their life.

Angela was able to play a certain role and slowly get stronger. So, in the final session she was able to try out a new role, one that matched her current circumstances. By trying out different roles, she built up her role repertoire. As she had gained trust in me, she was able to give me a more positive role. I was no longer her captor but a woman she worked for, who eventually became her mother. Landy (1990) defined role as 'the container of all the thoughts and feelings we have about ourselves and others in our social and imaginary worlds' (p. 230).

Holmwood (2014) maintains that the roots of dramatherapy lie in each individual, cultural and societal acknowledgement that drama which exists in all cultures and societies is present in a myriad of approaches. As Holmwood implies, role play and drama seem to be an innate part of human behaviour. We take on different roles all the time, depending on changing circumstances and the fluctuation of those around us. It stands to reason, therefore, that play/dramatherapy should be an ideal therapeutic process, as expression through 'drama/play/acting' comes naturally to human beings, particularly children, who have not yet replaced expression through play and action with expression through words.

The contract that we had agreed on from the beginning was, firstly that Angela would only come if she wanted to and that at the end of the ten sessions we would meet with her parents and tell them what we agreed she wanted them to know. Angela did not speak herself during the final session, maybe because it was hard for her to mention issues that the parents were not talking about openly, issues that had to do with their relationship. So, instead she drew storms, possibly indicating what was going on at home. The parents did not comment on this, however, when I pointed it out to them.

Murray Cox and Alice Theilgaard maintain that:

> metaphor exerts its mutative effect by energizing alternative perspectival aspects of experience. This means that material which the patient has endeavoured to reliquish, avoid, or deny so that it is 'safely' classified, categorized, and 'filed away', appears again in the 'pending action' file.
>
> *(1987 pg. 99)*

In the case of Angela, her metaphor had not been filed away and did appear very clearly in the role play that we acted out together.

The parents had wanted me to see Angela and tell them whether anything could be done about her hyperactivity and lack of concentration. They did not seem aware of the fact that their breaking up could also be upsetting her. They were going to leave their home and go to an area closer to Angela's mother's parents' house and her dad was going to move out as he had another relationship. This he had confided in me at our first meeting, not in front of Angela. At the final meeting, they said that they would not be moving yet, implying that Angela would not be immediately affected. I told them that they must be more open with Angela and not imagine that a young child is not aware of what is happening between her parents. If she might have to change house and was not told about it openly, it was not surprising that she was disturbed and unable to focus. She needed a stable environment, where both parents would support her and be honest with her about what was going on (Bratton et al. 2006).

What brought about change was that Angela had told her story. She believed her biological mother had left her because of her harelip, which was why she was put into an institution and later adopted. Winnicott (1971) referred to therapy as a process that gives back to the client what she/he brings to therapy and argued that a good therapeutic relation acts as a developmental experience that facilitates the individual to identify her/his being, be aware of her/his existence, feel her/his own truth.

At the last meeting with Angela's parents, I believe that they were made more aware of how their behaviour was influencing her. Landreth (2006) maintains that we need to help the parent to understand that the child's play will reveal her experiences and her feelings as well as her reactions to her experiences. I would like to have seen them again but another meeting did not take place.

References

Aigen, K. (1991) The Voice of the Forest: A Conception of Music for Music Therapy, *Music Therapy*, 10(1), 77–98.

Axline, V. (1989) *Play Therapy*, London: Churchill Livingstone Press.

Barkley, R. (Ed.). (2018) *Attention-Deficit Hyperactivity Disorder: A Handbook for Diagnosis and Treatment*, Fourth Edition, London: Guilford Press.

Bowlby, J. (1988) *A Secure Base: Parent-Child Attachment and Healthy Human Development*, London: Routledge.

Bratton, S.C., Landreth, G.L., Kellam, T. and Blackard, S.R. (2006) *Child Parent Relationship Therapy (CPRT), Treatment Manual,* New York and London: Routledge.

Cattanach, A. (2003) *Introduction to Play Therapy*, London: Brunner-Routledge.

Cox, M. and Theukgaard, A. (1987) *Mutative Metaphors in Psychotherapy*, London and New York: Tavistock Publishers.

Geldard, D. and Geldard, K. (2009) *Basic Personal Counselling,* Sixth Edition, Frenchs Forest, NSW: Pearson.

Gordon Clark, L. (2015) Play Therapy, in Sue Pattison, Maggie Robson and Ann Beynon (Eds.), *The Handbook of Counselling Children & Young People*, London: Sage, BACP.

Holmwood, C. (2014) *Drama Education and Dramatherapy*, London: Routledge.

Jennings, S. (1998) *Introduction to Dramatherapy Theatre and Healing*, London and Philadelphia: Jessica Kingsley Publishers.

Jennings, S. (2011) *Healthy Attachments in Neuro Dramatic-Play*, London and Philadelphia: Jessica Kingsley Publishers.

Jones, P. (2008) Research into the Core Processes of Drama Therapy: Vignettes and Conversations, *The Arts in Psychotherapy*, 35(4), 271–279.

Landreth, G. (2006) Child Parent Relationship Therapy, in S.C. Bratton, G.L. Landreth, T. Kellam and S.R. Blackard (Eds.), *Child Parent Relationship Therapy (CPRT), Treatment Manual*, New York and London: Routledge

Landy, R. J. (1990) The Concept of Role in Drama Therapy, *The Arts in Psychotherapy*, 17, 223–230

Rycroft, C. (1968) *A Clinical Dictionary of Psychoanalysis*, London: Nelson.

Stagnittin, K. and Cooper, R. (Eds.). (2009) *Play as Therapy: Assessment and Therapeutic Interventions*, London: Jessica Kingsley Publishers

Storr, A. (1972) *The Dynamics of Creation*, London: Penguin Books.

Winnicott, D. (1971) *Therapeutic Consultation in Child Psychiatry*, London: Hogarth Press, Institute of Psycho-Analysis.

PART IV

Play therapy

Introduction

In this final section of the book we focus specifically on play therapy. We have purposefully structured the book so that the reader goes on a journey or continuum through the multi-faceted aspects of play, from play in the most generic terms to play therapy and all the shades in between. It is important to note that we don't consider this continuum from a hierarchical perspective, each section and area that involves play is as important and essential as all the others. The difference being here is that we focus on therapists who have specifically trained as play therapists and their practice is specifically "play therapy". It is interesting to note that this section focusses mainly on work in the US, Canada the UK and South Africa, where play therapy as a specific profession is more clearly defined due to training and licenses to practise.

We begin the section with Chapter 26, which, similarly to Chapter 25 at the end of the last section, makes connections between dramatherapy and play therapy. McFarlane (a dramatherapist) and Chown (a play therapist) consider the use and benefits of a combined drama and play therapy approach with children who are at risk of being excluded from mainstream education in the UK. Reminding us again of the liminal borders between these professions. This is contrasted by Chapter 27 by Crenshaw et al. who consider from a US perspective a range of play therapy-based approaches to working with young people who have been traumatised. Crenshaw et al. bring a range of practical solutions to their approach but based in in-depth theoretical notions including 'Trauma-Informed Index and Treatment Implications.'

In Chapter 28 Gregan, an experienced play therapist shares her experience working with two children on the autistic spectrum using a specific approach to play therapy, the DIR®/ Floortime™. Murray's Chapter 29, reminds us of the fundamental aspects of play therapy as a way of repairing play deprivation, which has a massively negative impact on overall childhood development and has some connections with Chapter 26.

In the following chapter, Chapter 30, Yasenik and Gardner from Canada share their play therapy dimensions model. They consider the role of the therapist within play itself, with clients, and consider the role of the conscious and unconscious and directive and non-directive approaches to play therapy with some theoretical rigour. As a contrast to this, in Chapter 31, Chown considers the challenge of play therapy outside of the therapy room in green spaces. After helpfully defining the differences between various forms of play she considers what role does nature and the environment bring to play in play therapy?

Chapter 32 also considers play therapy within schools. Drewes' considers key ideas around the notion of how play therapy should occur in schools which has some interesting connections to McFarlane and Chown (Chapter 26).

We then consider animal assisted play therapy within a bullying context in South Africa. VanFleet one of the co-authors has been championing this approach for several decades. Our penultimate chapter returns to the UK where experienced play therapist Galloway considers play therapy from the perspective of working with young people in a secure setting, and the importance of creating a safe place for children to learn, develop and play. The final chapter in this volume contrasts this with an introduction to Ho's pioneering play therapy approaches used at the "Creativity at Heart" creative arts centre in Malaysia. Reminding us of the potential of play to allow children to heal, a fitting end to this volume.

26

SWINGS OR ROUNDABOUTS?

The case for using a combined play therapy and dramatherapy approach as an effective intervention with children at risk of exclusion

Alison Chown and Penny McFarlane

We live in a 'doing' rather than a 'being' society, in an age in which success is measured by how much we have as a result of how busy we have been, how many boxes we have ticked and how many rungs up the ladder we have climbed. It is not surprising therefore that our approaches to containing and supporting children at risk of exclusion in schools should reflect this philosophy. Hitherto we have tried to 'fix' the behaviour of the child as if the child were the problem, by doing some kind of intervention rather than 'being' alongside them and hearing what they need. Moreover, we have tried to do this as quickly as possible, so that he or she can return to that all – important academic progress chart with a minimum number of boxes or rungs missed.

And yet these approaches are not working. We are failing our children and will fail our children's children if we do not change this approach.

All children have a right to be educated at their local school, one which recognises individual needs and does its absolute best to meet those needs through an inclusive and nurturing ethos and flexible curriculum approaches. That is why this chapter advocates an entirely different approach to that of ticking boxes. At its core it is preventative rather than compensatory, holistic rather than specific and qualitative rather than quantitative. Its basic principles can be applied to the 4- and 14-year-old alike. Furthermore, therapeutic support for children who are 'missing the mark', even if it exists, has, up until now, generally been seen as an 'add on', an extra-curricular series of sessions which are, more often than not, viewed with impatience, irritability or downright incomprehension as time away from the curriculum which teachers are under pressure to deliver. Here we suggest that in order to effectively help our children of today, this support has to be put in at a grass-roots level, has to be embedded in the ethos of the school, part of the fabric of every aspect of the school curriculum and have a tangible presence in every classroom across the country.

So what would this support look like? Why are we advocating a combined play- and drama-therapeutic approach and why this vision now? To answer these strategic questions fully we need to go back to the beginning of our professional relationship and explain the common ideas, principles and beliefs, which drew us together, detail our approach and provide some case studies which highlight the ways in which this has worked. First however, and most importantly, we need to explore the 'why now?' Why is our vision so vital at this point in our society?

Why this vision now?

In July 2018 the *Guardian*, based on figures supplied by the Department of Education, reported that 'the rate of pupil exclusions from state schools in England rose sharply last year, with the total of 7,700 (equating) to more than 40 permanent exclusions a day during the 2016–17 school year' (Adams 2018). In August 2018, the *Times* reported that self-harming by girls had doubled in the last 20 years according to NHS figures with blame attributed to school and social media pressures (Coates and Lay 2018). Most recently, on 12 January 2019, the *Guardian* headlined with 'Inner London students placed in excluded pupils' schools almost double national rate. Exclusion is 'correlated with layers of disadvantage' (Weale and McIntyre 2019), quoting research carried out by 'The Difference' education charity. We are, patently, missing something.

With the rise in such social issues as domestic violence, street gangs and substance abuse, it is clear that many home environments are struggling to provide the safe, secure and nurtured start to life that is needed for healthy, holistic development. In addition, we, no longer, on the whole, have the strong sense of community, of belonging to something, fostered historically by the Church, family and local business communities. People are isolated, separation and divorce rates are on the rise, couples or single parent families are forced to move to find employment and live far from their own original family and source of support and companionship. High rise flats and large housing estates, often with no focal centre, accentuate the sense of isolation. Children without a secure and safe start in life and lacking in self -worth, grow up seeking love, attention and a sense of belonging, and may produce children who they in turn are unable to nurture and parent effectively and the cycle starts again. For many children and young people, the only secure and safe community they know is their school, and for some even this does not exist with the increase in those schools operating the idea of 'zero tolerance' particularly prevalent within academies. If the school environment cannot rise to meet the emotional and social needs of these children as well as the learning needs, then they will go without the sense of psychological safety and security needed in order for them to be available for learning.

Many of our most troubled and troubling young people will have had a number of Adverse Childhood Experiences, or ACEs, which are broadly categorized as child maltreatment through physical, sexual or emotional abuse, parental separation or divorce, domestic violence, mental illness, drug or alcohol abuse within the family or a family member in prison. The greatest negative impact comes from experiencing four or more ACE's which may only apply to a small percentage of the school population, but which would apply to a significant number of those who have complex social, emotional and mental health needs (SEMH) or are at risk of or excluded from school.

Many schools are beginning to recognize that to meet the needs of these kids with complex SEMH requires the skills of professionals with training beyond that of school-based staff. Fortunately, and at long last, the government is now in agreement . . . in principle.

In a news story for the Department for Education and of Health and Social Care published as recently as December 2018, Secretary of State for Education Damian Hinds said:

> Children today experience pressures that we as adults often find hard to appreciate, or possibly even understand. We are much more aware of mental health in the education sector now than in decades gone by and rightly so, and teachers are often able to recognise the early warning signs of changes in their pupils' behaviour or mood, but they are not mental health professionals.
>
> *(Dept. for Education and Dept of Health and Social Care 2018)*

Curiously however, this is somewhat at odds with the same governments' previous Green Paper entitled Transforming Children and Young People's Mental Health Provision (2017). In this, in addition to pledging funding of £1.4 billion over five years they stated that:

> every school would be incentivised to have a designated senior mental health lead to provide rapid advice, consultation and signposting with children and young people's mental health services identifying a link for schools and colleges.

So although Damien Hinds may say teachers are not mental health workers, the reality is that there will be an expectation for them to be so and indeed, many already are particularly since 'austerity' cuts have led to the loss of many significant support services providing for children and young people with mental health needs and at risk of exclusion. Additionally, many mental health practitioners feel that the promised additional monies will not even start to recover the ground lost under seven years of underfunding.

In relation to the Green Paper proposals, it needs to be said that in many, many schools, we already have senior leads in mental health – the school's Special Educational Needs or Pastoral Co-ordinators, and additionally, an increasing number of schools already employ trained and experienced play and creative arts therapists to provide support for pupils and further training for staff. What is desperately needed now is full and fair funding for all schools to ensure that where high quality provision is available locally from independent providers this can be quickly accessed and fully funded. This would mean that early intervention is freely available when specific individual SEMH needs indicating a risk of exclusion, are identified that cannot be resolved with universal school provision such as emotion coaching and mindfulness. Additionally, there needs to be a very significant increase in CAMHS provision to allow quick access to specialist services to avoid difficulties in accessing educational opportunities from becoming entrenched.

If there is now agreement that in order to reduce the risk of exclusion, the most complex children need something additional to and different from that available at universal level and from the special needs systems in most schools, we must also recognize, that the crisis in funding has also severely restricted 'in service' training opportunities for teachers, training which would enable them to be more fully aware of mental health needs, respond appropriately and refer on to the Senior Mental Health Lead (SMHL) in their school. This suggests to us therefore, that initial teacher training must include an element of therapeutic training, so that there is a consistent approach to supporting mental health needs, at 'grass-roots level' across all schools. However, if this is to be done, time needs to be made available to all teachers and particularly SMHLs so that this is not just another piecemeal, tick the box approach from government.

There is now clear evidence that the current reliance in education on hard, numerical data has created an 'exam factories' culture (Hutchings 2015) that is causing high levels of stress and anxiety amongst pupils of all ages and those who teach them. Since the current government came into power in 2010, there has been a significant shift away from the idea that **the learning process** which we take children through, and their ability to engage with it, is crucial to the outcomes for all children but particularly for those whose life experiences have had a detrimental impact on their social, emotional and mental health development.

The mental health charity Young Minds and partner organizations have developed a resource for schools called the Academic Resilience Programme aimed at supporting students to 'achieve good educational outcomes despite adversity' (youngminds.org n.d.). The guidance notes for staff accept that much of what is proposed is not new, but is best practice in fostering an ethos conducive to effective learning and with an understanding of the process of learning. The programme echoes earlier well-established approaches such as Nurture UKs National Nurturing Schools programme (n.d.) based on the work of Marjorie Boxall and the THRIVE approach, developed by Fronting the Challenge Projects (n.d.). Since these all show good outcomes, and underline the importance of best practice in inclusive schooling, similar content should be part of initial teacher training, rather than something that has to be commissioned. However, for some children who have experienced more than four ACEs in their early life, resulting in 'developmental trauma' (Van der Kolk 2015), individual therapy is needed in order for long-term changes to be sustainable. A truly inclusive approach, which would include access to additional support, places the child in the middle of the curriculum – as the learner engaged with the learning environment. It will also emphasize the importance of the reciprocal partnership where adults and children jointly construct understanding and knowledge, rather than teaching a body of knowledge or providing 'experiences' to meet a set of predetermined objectives or outcomes. In doing so it will reflect both the child centred nature of play and dramatherapy and the co-construction of the therapeutic relationship as well as promoting the idea of giving children the skills, and disposition for learning, echoing the self-healing process of client-centred therapy (Smith 2011:153).

Why drama and playtherapy?

Ali's story

When I first met Penny on a supervisors training course run by Dr Sue Jennings, we realized we had a 'connection'. Over several lunches we found we had much in common including starting life as teachers with a shared view of the impact of what we saw as a society dominated by 'masculine' principles and agreed there was a need for more compassion, humanity and nurture within the education system, particularly at secondary level and even more so in many of the academy chains.

When we were approached by Sue to join her at the Summer School in Romania to work with students on the combined play and dramatherapy course, we were delighted to have been asked, on both a personal and professional level. On the supervisor's course, Sue had used this combined play and dramatherapy approach and the integration of the two modalities was very powerful for me.

Working with young clients, I was confident with a wide range of play therapy tools such as the sand or water tray, clay, puppets and endless glue, paint, glitter and paper, but realized I lacked confidence in those more dramatic tools which would sit under the R of EPR (see later in the chapter) and might be needed particularly when I worked with older clients. Planning the delivery of sessions on EPR in Romania was a great opportunity for Penny and I to exchange ideas, experience, knowledge and understanding which strengthened and enhanced both our work with the students as we describe later, and our individual practices.

As an experienced play therapist and dramatherapist respectively, our work had hitherto been primarily confined to our own fields of expertise. Working together on the project in Romania crystalized the realization that not only did our disciplines share many commonalities but that we felt then and continue to feel, increasingly strongly, that a society that places a higher value on what is often seen as the more traditionally masculine principles of control, achievement, profit, possession, competition and power rather than the more traditionally seen feminine qualities of acceptance, nurture, compassion, equality, listening and caring is one which is dangerously out of balance. Whilst this might represent a broad-brush approach, within most major institutions and organizations, including the present government, we believe this difference holds true.

Dr Sue Jennings, both a dramatherapist and play therapist, has been a life-long advocate of the importance that nurture and acceptance of the individual and his or her uniqueness plays in effective learning and of the 'reciprocal partnership' between adult and child, as discussed previously. In the mid 1980s she introduced the developmental play paradigm of Embodiment, Projection and Role (EPR) which corresponded roughly to the stages of child emotional development as outlined by J. Piaget (1970) and E. Erikson (1968). This model has been adopted and used extensively by dramatherapists and play therapists in schools ever since. Just as the Nurture Group and Thrive approaches focus on the developmental age of the child, the EPR paradigm guides us to where the traumatic interruption to development has occurred.

As a dramatherapist and play therapist, we have found this particular model very useful as an overall structure when considering our joint working practice, since, as previously stated, the two disciplines, have much in common. In particular these commonalities relate to the principles of coming from the stance of being the non-directive, non-expert, having a sense of curiosity, using symbolic and metaphoric story and play and having a sense of fun. In addition, there is no need for a level of linguistic and cognitive ability that would be required for talking therapy such as counselling, and which is needed for interventions such as emotional coaching. In short, both play therapy and dramatherapy are not a 'done unto' intervention but rather a 'start from where the child or client is' and 'accompany them' on their journey of self-discovery, healing and emotional and psychological growth.

In Romania, trying to explain the EPR model to students, we found that it served our purpose and the project to make a rough subdivision into the following: play therapy as a wonderful exponent of the Embodiment stage and dramatherapy as the Role stage with both disciplines using Projection. In saying this, we are not suggesting that play therapy does not use Role or that dramatherapy does not use Embodiment but rather that the two therapies could be said to span the whole EPR model in greater depth in this way. Since the ideas of Erikson emphasise the issues of emotional conflict which may arise at particular stages and ages in life, they are a useful resource for both drama and play therapists. For example, in the Embodiment stage, which according to Erikson, corresponds roughly in chronological age from birth to two years, the overriding issue here is 'Trust versus Mistrust'. An infant, through secure attachment to the primary caregiver learns to trust another, see herself mirrored in their eyes and actions and thus receives approbation to trust herself.

> Through the physical attachment to our caregiver we are able, later, to let go and explore independently. The body of another person holds and contains and allows us to experience body boundaries. Embodiment is the first stage of our dramatic development.
>
> *(Jennings 2011)*

Since Piaget views this stage as 'sensory-motor' it would be easy to see how mirroring, messy play, rhythm etc. might be useful tools here. For children and young people with complex SEMH needs, there is often little or no or even a distorted sense of the 'body' self and therefore a need to return to this embodiment stage of play in order to process experiences held by the pre-verbal sensory system.

> Within our sessions together, . . . he became more expansive in his use of paint, which he splashed everywhere, and the big sand tray which I used on the floor in an effort to ground him. He poured in all the available water, took off his shoes and socks and splashed and jumped and stamped with all his energy. Then he suddenly took off his shirt, went to take off his shorts and pants and shouted 'Me lie down!'
>
> *(Chown 2014:90)*

Once a child becomes aware of the world outside herself, she is able to progress to the Projection stage and make use of puppets, soft toys, etc., to explain and explore her world. It is this stage, which, Penny, in her work in schools, has found most useful for initially getting to know a child. Some children may progress then quite easily to the Role stage when they are able to see themselves through another's eyes. For some children this stage is around age seven but for many it comes much earlier and, indeed, a child may go backwards and forwards through the stages depending on events in their lives and their environment. Since, according to Erikson, this stage is concerned with establishing identity, it is easy to see how many teenagers in our schools are struggling with these issues, whether to do with image, ability or gender.

Most children will progress healthily and through these stages.

> If, however, a child does not make a natural progression for reasons such as illness, neglect, abuse or trauma, they need the therapist to use emotional and not chronological age-appropriate techniques to allow the child to explore this stage until he or she is ready to move on.
>
> *(McFarlane and Harvey 2012:45)*

For example, a seven-year-old child who cannot sit still, and shows no interest in telling stories through puppets or small figures but prefers to cuddle the soft toys will need to remain in this world of sensory exploration in order to 're-do' and find resolution to negative early experiences in order to move forward on a healthy developmental pathway and be open to learning.

In advocating individual therapy for children and young people, we need to accept that by 'therapeutic' we mean 'healing', and we need to understand that healing can happen on different levels, that of mind, body and soul. While interventions such as the preceding do much to heal the mind, we would put forward the premise that they do not, and cannot, offer effective healing for the body or soul. Children who are suffering from pre-verbal or epigenetic trauma cannot put their pain into words. Consciously they do not understand it although subconsciously they feel it, somatically through their bodies, and existentially through their spirit. Inclusive ethos and mind-based or cognitive therapies are not enough for these children. Play and dramatherapists work by addressing the right side of the brain, that area responsible for intuition, feeling, visualisation, imagination, rhythm, creativity and holistic thinking where unresolved trauma is held. Through play, ritual, metaphor, storytelling and creative arts and by 'being' with the child rather than 'doing', the therapist can hold the space, feel the energy and allow the natural healing process to take place.

The importance of ritual cannot be overestimated. Most indigenous societies have used ritual to mark important events in their lives including the passing of the seasons and the coming of age. In this way they remained connected to nature and natural processes whereas we, in our modern society, through industrialisation and technology have become disconnected.

Although mind-based interventions can go some way towards helping our children understand and deal with their feelings, unfortunately we are beginning to see more and more young people who cannot be reached in this way. They need something more: something that will plumb the depths of their suffering and address it on both a body and a soul level. They need a therapist skilled in working with this level of pain.

We are not advocating an either/or scenario with regards whether it is a dramatherapist or play therapist who works with a child since many play therapists use role play and dramatherapists, messy play. What we *are* suggesting is that the combined approach of play and dramatherapy can more effectively cover the whole spectrum of needs for those children experiencing developmental delay due to Adverse Childhood Experiences.

Working with drama and play therapists provides the opportunity for young people to experience more than just a key relationship with a significant adult. We provide a level of attunement and attentiveness which school staff are not able to do. It is the quality and depth of the therapeutic alliance and the use of these 'emotional and not chronological age-appropriate techniques' which underpin our advocacy for the use of a combined play-and dramatherapy approach in schools and which we believe are desperately needed in order to address the tsunami of mental health issues arising in our schools today.

Case study one – Luke, aged seven

Luke was seven when I started working with him in his primary school. He could not work within the classroom since he had little sense of himself beyond the boy who 'couldn't', no sense of himself as a learner and was on an individualized timetable supported by staff from the nurture group. Much of his day was spent in 'play' activities, appropriate to his developmental stage which we gauged was around 6–12 months with support staff acting as primary attachment figures and creating the safe and secure base he needed in order to begin to develop more healthily.

He spent much of his first 12–14 sessions with me using a large tray of water or sometimes sand and all the small world toys to recreate the chaos and toxicity of his world – one which involved drug abuse, chaotic and volatile relationships and an all-pervading sense of being annihilated. He would hurl the toys in and mutter about poison and toxic sea, all the while accompanying this with various action noises, grunts and groans.

He and I were fighting figures, armed with bows and arrows and clothed in special suits that kept us from being obliterated and enabled us to 're-spawn' if we did get contaminated. Throughout this period of time, his relationships with both adults and peers were also extremely volatile.

Later play changed as he developed a stronger sense of himself as an autonomous being and he fought and killed me with fire, ice, acid and electricity. This went on week after week until I felt we had become stuck again. Discussion with my supervisor about my own feelings of needing to move on highlighted the parallel processing that was going on between myself and Luke and the sheer intensity of the feelings he had towards key family figures and aspects of himself. Although he felt more empowered, he did not know how to use this new-found feeling in a constructive way and needed some direction. As a group of supporting adults, we told him we wanted to stop play that was about hurting someone and think about other ways of dealing with problems and making ourselves feel better. In our sessions we focused on telling the story about what we were going to do and how we could help each other or the characters of the play in order for him to begin to be clearly to look at himself in a new light, as a boy 'who could' be different and find pleasure in both activities and relationships.

When he left us to go to specialist provision some two months later, he had a strong sense of himself as an individual who could succeed, rather than a boy who couldn't. While we were worrying that he might find it distressing to leave his 'secure base' behind, he was happily waving us goodbye! Without those early sessions of wet, messy chaos, he would not have been able to process at a deeply sensory and pre-verbal developmental level and move forward towards healthy integration.

The document State of Children's Rights in Scotland 2016 underlines the potentially serious impact of early traumatic experiences on a child's education

> Trauma in childhood can lead to reduced educational attainment. . . . Children who have experienced abuse and trauma will perform more poorly at school than their peers. A child living with fear and helplessness from experiences outside school cannot focus or learn to the same standard as other children. Issues at home which may affect a child's learning in school include alcohol or substance misuse, poverty, mental or physical ill-health, and children's caring responsibilities.
>
> *(Harris et al 2016)*

Case study two – Lucy, aged six

Lucy was just coming up to six when she started play therapy sessions with me. She was referred because if she was asked to do anything she did not want to do, she was reduced to wailing and tears which could last for several hours. Staff considered her 'older than her years' and it was her way or "No way!" In the playroom she needed to be in control of everything that happened, including when I spoke, usually telling me not to make a sound, when I moved and where I sat, often directing me to face away from her. She seemed to be sending the message that in places in her life, she had been and continued to be overwhelmed by the demands of adults.

For the first four months or so, although acknowledging my presence with various commands, she rarely responded to my reflections or wonderings other than to tell me to stop talking and face the wall! I was merely there to do her bidding. Yet alongside this role she needed to take, she mixed paint and water, and later flour and sugar, sometimes carefully spooning it into various containers and sometimes pouring it into a large tray and immersing her hands into it, squidging it and smearing it, reminiscent of early eating experiments of the developing infant. This part of her was deeply within the embodiment realm.

My task was to bring these two parts together, to enable her to experience the control she lacked whilst enabling her to allow the 'small person' within, whose needs at a deeply sensory level appeared to have been neglected or distorted, to experience pleasure in a safe and contained way. I am fortunate enough to have Sue Jennings as my supervisor and found her dramatherapist's view of the role aspect of Lucy's sessions really helpful to my understanding of what my own role might be in order to support the integration of these two aspects of her being and of how I might enable her to retain control but also not be just a passive observer.

This inability to form relationship and regulate emotions may have drastic consequences if a child reaches secondary school age without proper therapeutic support. According to information issued by the Department for Education (DfE) '14 year olds (also) have the highest rate of fixed period exclusion, and the highest rate of pupils receiving one or more fixed period exclusions' (DfE 2015). At this age, together with hormonal and neurological changes the overload for a child who has unresolved traumatic stress and emotional delay may prove too much.

Case study three – Darren, Katie and Gina, aged 14

Darren, Katie and Gina, all 14, had been on short term exclusions from school over the past year. The next time, I was told, as a dramatherapist brought in to fire fight, would be a permanent arrangement. Would I see them as a group? It was a last resort. I would and did. Although I own that it was with some trepidation that I entered the allocated room where three surly teenagers were eyeing me with suspicion bordering on hostility. We worked with the hostility. There was a great deal of 'acting out' behaviour but gradually, as they began to realize I was not shocked and that I continued to turn up, week after week, to listen, to empathize and just 'to be', the hostility turned into amusement, a 'poking fun', but in a vaguely proprietorial way. My expectations of them experimenting with the functional and dysfunctional roles which might (or might not) equip them for society were very soon dashed. They were not interested in any form of role play whatsoever. Neither were they interested in projecting any of their stories onto paper, small figures, puppets or any of the other copious materials I had struggled to bring in.

For eight weeks, week after week, the only activity they wanted to do was to play 'Changing Rooms'. Based on a television programme, using lengths of brightly coloured material and taking it in turns they decorated the room in different ways according to the wishes of one of the trio. They stroked the material, wrapping themselves up in it at times and always transforming the rather drab looking furniture into something new, bright and exciting.

After eight weeks they were noticeably quieter, calmer and willing to listen and confide in each other. In between the play they exchanged their stories of abuse and neglect, commiserating and offering each other advice and support (along with the usual fallings out of course). School reported progress. They were no angels but at least they were off the 'at risk of exclusion' list.

> Years later, I caught up, through a mutual friend with Darren, now a Dad and in a full – time job to support his family. Gina, always bright, had astonished everyone to go on to take A' Levels and attend university. Katie, after a difficult period had returned home to look after her ailing Mum and was attending a hair and beauty course. Darren wished to be remembered to me, my friend said.
>
> 'She annoyed the hell out of us.' He said, 'But she was all right. It made a difference'.

Whilst interventions such as Emotion Coaching, based on the principle that nurturing and emotionally supportive relationships provide the optimal context for the promotion of children's outcomes and resilience may work at universal level, they demand that young people have a good degree of self-awareness to recognize their emotions, the language and communication skills to articulate their feelings and the cognitive development to engage in problem solving. Our most complex SEMH children who have suffered significant developmental trauma are unlikely to have these skills, as noted earlier. What play and creative arts therapies offer is a space which is free from the linguistic and cognitive demands and instead has an emphasis on being and creating within the context of an accepting, non-judgmental, attuned and affirming relationship with an adult who is 'trauma informed' and understands the need to engage in a relationship which is child directed and not adult controlled. This particularly so for young children and those who may have language or learning needs.

Darren, Katie and Gina did not have the language and communication skills, or the degree of self-awareness even to begin to understand or process their internal emotional landscape. All they knew was that this landscape needed an overhaul. Together they shared this need and together they created, week after week, for each other, something, new, bright and exciting. They 'played' together, as children of five or six might do, sorting, arranging, compartmentalizing and creating fantasy worlds where they could play out their lives.

Whether we are dealing with a child who has experienced violence and abuse at a pre-verbal level, and therefore cannot find the words to express his pain and disempowerment such as Luke, or a child such as Lucy who is struggling to find her own personality, or teenagers like Darren, Gina and Katie whose adverse childhood experiences have stunted their emotional and psychological development, the fact remains that these children need far more support and intervention than is provided within our current educational system.

What would an alternative approach look like?

Our suggestion is a 'four -pronged approach'

1 At grass-roots or universal level for initial teacher training courses to include training to recognize red flags for mental health problems and offer broadly therapeutic strategies to deal with them within day to day relationships, curriculum planning and interactions in the classroom.

If teachers are to be expected to have an understanding of mental health issues, be able to recognise them and work in ways which support good emotional and mental wellbeing then training for this needs to be embedded in teacher training programmes to avoid schools having to spend money on high cost schemes which require regular revalidation or become 'history' in the face of next year's new ideas. Initial teacher training needs to ensure that all teachers:

- Have a full understanding of the impact of poor attachment and ACEs on pupil's ability to make and sustain meaningful relationships, to be available for learning and access the curriculum.
- Have knowledge and understanding of child development, and how the preceding adverse experiences impact on brain development and learning.
- Are confident in their ability to both recognize the early signs of potential mental health issues in all ages of pupil, and use relevant, therapeutic and supportive strategies to meet identified needs.

However, we also need teachers to recognize that emotional and mental wellbeing issues that do not respond to school based strategies within a reasonable time scale are likely to become more serious and therefore require signposting to relevant specialist practitioners whether bought in or through outside agencies. Lead teachers must be given the additional time that is needed to do this in a way that is more meaningful and appropriate than merely another tick box exercise on the road to exclusion and the financial recognition that would underline that this is a serious issue to which the government is fully committed. Additionally, all teachers need increased time away from the classroom to reflect on pupils learning needs and discuss with colleagues how they might meet these.

2 A truly inclusive ethos in all schools to raise awareness of mental health needs and make creative and appropriate provision to meet these needs.

Any attempts to understand and provide for young people's mental and emotional well-being will only be effective within the context of an inclusive and nurturing environment which places pupils at the heart of all practice. It will not be successful if there is an ethos of 'zero tolerance' and a use of exclusion to 'weed out' those pupils whose needs require additional and different provision.

3 Full and fair funding to enable schools to 'buy in' from independent providers.

Since education support services in the majority of councils have been drastically reduced as a result of swinging cuts to budgets, funding to schools needs to be radically increased to allow for more time for staff to reflect on pupils needs and the provision required to meet these and to allow schools to 'buy in' relevant support from qualified and trusted local providers rather than large organizations offering training at high prices.

4 An increase in CAMHs provision.

As discussed earlier in the chapter, CAMHS has also suffered significant underfunding over the last decade or so. The current plans from government will not go far enough in restoring provision and there needs to be a significant increase in mental health funding to allow for a real reduction of waiting lists, a breadth of earlier interventions to avoid young people reaching crisis point and immediate access to specialist services when they do.

Perhaps the time has come to stop swinging backwards and forwards or going around and around in circles but have the nerve to take on an entirely different vehicle for movement and change. Having climbed slowly and painfully to the top of the slide from where we can have a good view of what is happening around us, perhaps now is the time to take that next step and propel ourselves forwards in a positive, dynamic and rewarding direction to give children and young people with mental health issues the appropriate support to allow them to achieve *their* potential and offer the possibility of change for later life outcomes.

References

Adams R. (2018) 'Sharp Rise in Pupil Expulsions from English State Schools.' *The Guardian*, 19th July [Online]. Available at www.theguardian.com/education/2018/jul/19/sharp-rise-in-pupil-exclusions-from-english-state-schools [Accessed 7th January 2019]

Chown A. (2014) *Play Therapy in the Outdoors – Taking Play Therapy Out of the Playroom and into Natural Environments*. London: Jessica Kingsley Publishers.

Coates S. and Lay K. (2018) 'Self-Harming by Girls Doubles in 20 Years.' *The Times*, 6th August, [Online]. Available at www.thetimes.co.uk/article/self-harming-by-teenage-girls-doubles-in-20-years-x2vbzm87m [Accessed 23rd April 2019]

Department for Education. (2015) *Permanent and Fixed Period Exclusions*. Available at www.gov.uk/government/collections/statistics-exclusions [Accessed 23rd June 2019]

Department of Health and Social Care and Department for Education. (2017) *Transforming Children and Young People's Mental Health Provision*. Available at www.gov.uk/government/consultations/transforming-children-and-young-peoples-mental-health-provision-a-green-paper [Accessed 3rd March 2019]

Erikson E.H. (1968) *Identity: Youth and Crisis*. New York: W. W. Norton.

Harris J., Millership E., Moodie N. and Shepherd J. (2016) *State of Children's Rights in Scotland 2016*. Edinburgh: Together Scottish Alliance for Children's Rights.

Hinds D. (2018) *NHS and Schools in England Will Provide Expert Mental Health Care, Department for Education and Department of Health and Social Care*, 20th December [Online]. www.gov.uk/government/news/nhs-and-schools-in-england-will-provide-expert-mental-health-support [Accessed 7th January 2019]

Hutchings M. (2015) *Exam Factories: The Impact of Accountability Measures on Children and Young People*. London: National Union of Teachers.

Jennings S. (2011) *Healthy Attachments and Neuro-Dramatic-Play*. London: Jessica Kingsley Publishers.

McFarlane P. and Harvey J. (2012) *Dramatherapy and Family Therapy in Education*. London: Jessica Kingsley Publishers.

The National Nurturing Schools Programme. (n.d.) Available at www.nurtureuk.org/news/national-nurturing-schools-programme [Accessed 23rd June 2019]

Piaget J. (1970) 'Piaget's Theory.' In P.H. Mussen (ed.), *Carmichael's Manual of Child Psychology*, volume 1. New York: John Wiley Publishers.

Smith A.B. (2011) 'Relationships with People, Places and Things: Te Whariki.' In L. Miller and L. Pound (eds.), *Theories and Approaches to Learning in the Early Years*. London: Sage.

The Thrive Approach. (n.d.) Available at www.thriveapproach.com [Accessed 9th June 2019]

Van der Kolk B. (2015) *The Body Keeps the Score: Mind, Brain and Body in the Transformation of Trauma*. London: Penguin Books.

Weale S. and McIntyre N. (2019) 'Inner London Students Placed in Excluded Pupil's Schools Almost Double National Rate.' *The Guardian*, 12th January, p. 5.

Youngminds.org (n.d.) *Building Pupil Resilience in Schools*. Available at https://youngminds.org.uk/youngminds-professionals/our-projects/academic-resilience/ [Accessed 29th June 2019]

27

CREATIVE INTERVENTIONS WITH CHILDREN AND ADOLESCENTS WITH COMPLEX TRAUMA

David A. Crenshaw, Maura Collins, Amanda Krülle and Shyla Dauria

Introduction

Some children, despite the assault on their developmental process by complex trauma, are particularly resilient and their self-reparative resources emerge in child-directed therapeutic play. These children have their own internal compasses and know what they need to do in order to heal. Other children and teens need direction and support through this process. The trauma-related stories may be too difficult for them to develop, even in symbolized form, and they may require a therapist-directed approach to work through some of the traumatic material. A wider range of therapeutic tools enables the clinician to tailor each intervention more specifically to the child and the traumatic events that he or she experienced (DeNigris, 2008).

The nature of complex trauma often requires clinicians to use more creative and out-of-the-box interventions to initiate and facilitate the therapeutic process. Recent work in the field of play therapy has yielded new interventions that have shown to be particularly useful in the population of children with complex trauma histories, including animal-assisted play therapy and a range of play-based and expressive arts interventions. This chapter will introduce some new and creative interventions, including letter writing exchanges within the context of animal-assisted play therapy, letters of growth strategy, the pizza strategy, the seeds and flowers intervention, and sand art feelings jars (Jenkins, 2019).

Overview of complex trauma in children and adolescents

As children are in the process of development, they are particularly vulnerable to the devastating effects that trauma can produce. Complex trauma in children refers to multiple exposures and/or sources of traumatic events in childhood, typically those of interpersonal and invasive nature that tend to have a wide variety of long-term consequences (National Child Traumatic Stress Network, 2007). The effects of trauma can be widespread and may impact the cognitive, affective, behavioral, or physical realities of the exposed individual (DeNigris, 2008). Some children have a greater risk of developing mental health problems due to a variety of risk factors such as their genetic vulnerability, temperament, family dysfunction, poverty, caregiver separation, and abuse histories (Newsome & Gladding, 2014).

It is imperative that clinicians be skilled at executing prevention and intervention strategies that address each level of the child's system as described by Bronfenbrenner's ecological model (1994; Newsome & Gladding, 2014). Research has indicated that adolescents that have experienced multiple traumas often develop more mental and physical complications compared to their counterparts who have experienced a single trauma (Finkelhor, Ormrod, & Turner, 2007). As complex trauma is associated with individualized and distinct effects, it requires effective trauma-informed treatment interventions by the play therapist (Green & Myrick, 2014). Clinicians need a larger repertoire available in their efforts to provide paramount support to the youth (DeNigris, 2008), and to engage them in ways that establish safety and trust. Safety and trust are necessary cornerstones of trauma-informed work but the impact of complex trauma poses formidable challenges to achieving this critical foundation to the therapeutic process.

Since the extent of the trauma and the multiple sources of adversity are cardinal features of complex trauma, acting-out behavior, sometimes extreme and frightening to caregivers, is not surprising. In the 50-year career of the first author working with children of trauma, I was inspired to create the following trauma-informed index to assist clinicians and caregivers to maintain perspective in dealing with the acting-out behavior of youth suffering complex trauma.

Crenshaw trauma-informed index and treatment implications

1. The degree of acting-out behavior will be proportional to the extent of the experienced internal hurt of the child/youth;
2. When the hurt is validated and acknowledged by a trusted helper, the need to act-out will be lessened;
3. A child/youth will only share their internal pain (hurt, woundedness) with a helper who has gradually earned their trust, makes them feel safe, and convinces them that they truly want to see and hear their hurt;
4. Disclosures under other conditions especially coercive, such as court proceedings, will not only be unhelpful, they can be harmful;
5. An often unrecognized contributor to the healing process, is the active declaration of trust when youth courageously take the risk of sharing their trauma story which leaves them exposed and vulnerable;
6. Surrounding youth with a caring and supportive community at their most vulnerable times contributes significantly to the healing process.

The compelling advantage of play-based and expressive arts interventions

Symbolic play provides children with a way of safely working through traumatic memories in a non-threatening manner that does not overwhelm their emotional capacities. Specifically, play is the natural language of children ages seven and below. Play therapy allows children to create their trauma narratives, which is an essential component of trauma-informed therapy in the language (play) they most comfortably and naturally speak. As children get older, particularly during the adolescent years, children may resist symbolic play as they may perceive it as being more appropriate for younger children. However, these older children may still struggle in their abilities to engage in verbal expression of the traumatic events that they have experienced.

Adults also are known to go "mute with terror," as neurobiological findings have shown that often trauma is encoded within non-language areas of the brain (van der Kolk, 2014). Thus, the question arises of what child therapists should do when the pre-adolescent or adolescent is confronted with the "unspeakable."

Our colleagues in the creative arts therapies, in art, music, drama, dance, movement, and sand therapy, have offered a wide range of alternatives that allow children and adults who no longer comfortably engage in symbolic play to express their trauma narratives safely and creatively without relying solely on verbal expression. For example, play therapists frequently use sand therapy as a non-verbal expressive strategy, which provides children the freedom to express their inner psyche without direction or instruction. Sand therapy acts as an alternative modality to engage a child in the therapeutic environment without requiring verbal exchanges of communication that may seem challenging or intimidating to a young child (Allan, 1988). Sand therapy involves an alternative medium of expression using a sandtray as a vehicle to get children to their destination of what will result as a sand world, or the product of their expression (Knoetze, 2013). Sand therapy is one example of a repertoire of alternative methods and interventions used when working with traumatized youth that yields an opportunity to create a story narrating the events of their lives. Though sand therapy has proven to be highly successful when working with children, not all children may be receptive to engaging in this technique. As a result, clinicians should have a variety of methods at their disposal when working with traumatized youth. In this section of the chapter, we will share work with children suffering complex trauma utilizing letter writing in combination with animal-assisted therapy, sand art feelings jars, the pizza technique, the treasure chest, and the seeds and flower technique. These techniques are all out-of-the-box and thoughtful, carefully considered interventions to safely work with complex trauma in children and adolescents.

Letter writing in play therapy

It is human nature to tell stories. Our stories are told universally, across time, location, context, and inevitably are told through therapeutic conversation (Gersie, 1997). We, as a population, construct and continuously reconstruct life events and narratives in order to both understand and assign meaning to our lives and identities (Knoetze, 2013). The act of telling stories can actually give one's self a sense of rootedness in time (Knoetze, 2013); an instance that cannot escape us, even when everything else has dissipated or has been seemingly ripped away from our young hands. Telling a story gives one an opportunity to be heard, a concept that may be few and far between in the life of a young child who has little say in the direction of his or her life. The act of telling a story gives empowerment and worth to the storyteller. With telling a story comes listening to what the storyteller has to say. This teller-listener relationship that is developed gives the teller a chance to capture the listener and their agreement to attend to the relevance of the story at hand. This, in turn, creates a consensual engagement between teller and listener that without consent, the story could not be told (Knoetze, 2013). New meanings do not have a chance to begin to formulate without the communication of narratives. Where verbal communication is not always an option for a young child or adolescent, non-verbal communication, such as letter writing, can pose powerful benefits to the creation of a narrative and the psychological well-being of a child.

Therapeutic letter writing has been affirmed by psychoanalytic therapists since the earlier part of the twentieth century, while the publication *The Use of Written Communications in Psychotherapy* (Pearson, 1965) has more recently provided context for the conceptualization of the sort for the first time. Murray Bowen (1993), a pioneer of family systems therapy, suggested clients use letter writing in order to develop relationships outside of the therapeutic relationship (Jolly, 2011). Even with its proven therapeutic benefit, with the rise of new technology, traditional letter writing is at risk of falling to the wayside. New interventions are being created for use with youth, particularly young children with trauma histories that include the exchanging of letters such as animal-assisted letter writing exchanges, or "Letters to Ace." With new interventions comes the hope of rejuvenating the art of the paper and pencil mode of communication, the form of coping, and the benefits that result.

Benefits of letter writing

Research conducted over more than two decades ago suggests that emotional disclosure through written means enhances one's physical well-being (Pennebaker, 1997). Extensive research has been conducted to find links between the emotional expression of adverse life experiences through written means and the improvement of physical and psychological health (Mosher & Danoff-Burg, 2006). In a study conducted by Lepore and Greenberg (2002), the results indicated that the act of expressive writing may be used as a buffer for individuals from the possible

Figure 27.1 Meet Ace

negative health outcomes that can arise and be associated with interpersonal stressors. Forgiveness is a concept of letter writing that is defined as "a prosocial change in a victim's thoughts, emotions, and/or behaviors toward a blameworthy transgressor" (McCullough & Witvliet, 2002, p. 447) and is related to a decrease in depression and anxiety, as well as a higher self-rated report of physical health and lower cardiovascular stress response (Mosher & Danoff-Burg, 2006). It is through letter writing that an individual can reach a place of resolution of troubling emotions conceived through interpersonal offenses that may catalyze forgiveness and promote positive health outcomes (Mosher & Danoff-Burg 2006).

Ace is currently a nine-year-old Golden Retriever certified and trained as a facility dog, working full time at The Children's Home of Poughkeepsie in Poughkeepsie, New York. Ace appears to provide Carl Rogers'(1957) basic premises of empathy, genuineness, and unconditional positive regard in the presence of traumatized youth, as he has the natural ability to come forth as a blank slate, a concept we as humans have long strived to be. Ace is a symbol of safety to a traumatized child that can discover no words to formulate for the ears of a human recipient. He is a canine that will form no judgments, call no authorities, and will never turn a child away for any story they may tell. He gives children with complex trauma a chance to love something unconditionally and feel that same love in return, a concept that may otherwise seem unfathomable to such a population. His non-judgmental, non-evaluative, and unconditionally accepting response, coupled with his uncommon calmness is soothing to a traumatized child, as they frequently experience abnormally elevated cardiovascular functioning due to the need to be constantly vigilant while scanning their environment for signs of danger (Perry, 2009). Ace provides a great sense of relief to children in crisis, however, he is not always on campus during a crisis or time of need. Using the "Letters to Ace" mailbox system, children are able to reach out to Ace 24/7, making him virtually always available to provide comfort to a child in need.

Letters to Ace

Using animal-assisted letter writing exchanges, youth on The Children's Home campus are able to readily write and exchange letters to Ace whenever they want or need to (Gil & Crenshaw, 2015). Mailboxes were created and placed in every cottage at The Children's Home, enabling children to place letters inside of their designated mailbox at any time. Each mailbox is placed in an area of the cottage that is in eyesight and easily accessible to the children at any time. Mail is collected from each mailbox every week and is responded to by a Mental Health Counseling Graduate Intern acting as Ace, himself. Each response is written to be positive and empowering, signed "Love, Ace," stamped with a paw print and placed in an envelope. Each envelope is also stamped with a bone, addressed to its recipient and marked with a return address, "Ace from The Children's Home of Poughkeepsie." The return letter is then brought to the cottage, accordingly, for the child to read at their own leisure.

Letters to Ace has a psychoeducational component, as it creates a creative outlet for children to vent things, events, or ideas that are troubling them as well as teaching basic letter writing and mailing skills. In the process of providing a fun and therapeutic activity, children learn the proper way to mail an envelope, and all included components in letter writing. They learn where to put a stamp, the location in which their name is filled out on an envelope as well as the destination of the contents inside. They learn a sense of patience, as they may have to wait several days from the placement of their letter inside the mailbox, to the pickup of the mail, and the receipt of their return letter. Patience is a trait that may be of utmost importance for a traumatized youth to learn 'Meet Ace' and acquire, as immediacy is a concept quite common among children with trauma histories. In this way, youth are forced to wait for the mailing process to proceed through all steps before they are able to receive the end result of a return letter.

Example letters and responses

Letters to Ace collected from the children and responded to by the Mental Health Counseling Graduate Interns range in length, content, and intention. All letter responses are written empathetically to the child's individual situation and content of the letter. Letters may be informative as some contain psychoeducation on a given topic brought up by the child. Letters to Ace responses show interest, love, and pride in the children they correspond to. No matter how a child is feeling at the time of writing the letter, response letters attempt to make the child feel good about themselves while providing an expressive outlet for their feelings.

> Dear Ace,
>
> You are the coolest dog I've ever met. I love spending time with you and taking pictures with you! Ace you are adorable and sweet! Thank you for making CHP a good environment for kids/us.
>
> – Faith* @ North Cottage

> Dear Ace,
>
> I miss you and hope to see you soon. Lately I've been so anxious but a little depressed about the holidays and not being with Jen* yet. I am also doing great in school for someone who started late. I hope you feel okay. I will always love you Ace. XOXOXO. Thank you for always being such a great dog.
>
> – Ariel*

> Dear Ace,
>
> I feel sad because I miss my mom and dad. Can you help me please feel better. I have one more week to stay here at The Children's Home. I cried in my bed.
>
> From Jeremy* for Ace

> Dear Ace,
>
> I don't know much about you but I would really like to meet you and learn more about you.
> Sincerely, Daniel*

Response:

Dear Daniel,

You know, I don't know a whole lot about you either! I would love to learn more about you, too! I would love it if you wrote me a letter telling me about you and your life. I hope to meet you very soon, I love making new friends. In the meantime, here is a little bit you might like to know about me: I am a Golden Retriever, I am three years old, I have a brother named Hansen, and I love making kids feel safe and loved! I hope to see you soon! Write again!

Love, Ace

Response:

Dear Ace,

It started out as a fun life than I got put in foster care. I've been in it for two years. It's been boring, I've just thought not to think about it. Now I'm here in The Children's Home in Poughkeepsie. So thank you for the last letter.

Sincerely, Daniel*

Dear Ace,

I am having a hard time right now. I need your help to stay strong and keep my head up. I need your help to help me move on in life and don't look so depressed all the time. I need your help to make me feel like I am here and I am in the light. Just say you have faith in me. Just say you hope I do okay as I grow up.

Love, Angel*

Response:

Dear Angel,

I'm sorry to hear you've been having a hard time lately. I want you to know that I think you are a wonderful girl and I have and always will believe in you. I enclosed my picture in this letter. Keep it with you and whenever you feel sad, look at it and know that I love you and have faith that you will grow up into an incredible woman.

Love, Ace

Dear Ace,

Hi it's Marybeth*. How are you? I am sorry you miss Hansen [*Ace's Brother*]. I want to say that you make me happy that you can take time to see children to help them. Depression is a really bad sin of problems. You help people with their problems and that's cool. I want to thank you for that Ace <3

Response:

Dear Marybeth,

Thanks for your letter! I'm doing great, how are you? You make me really happy too. Thank you for coming to spend time with me every week, I always look forward to seeing you. Sometimes we all go through difficult times and it can make us feel sad or depressed, but me and the other staff are here to help all of the kids with their problems. I'm always here for you and I love you!

Love, Ace

*names have been altered to protect confidentiality.

Letters of growth

Writing letters to express emotions can be a safe and valuable outlet for children. Similar to the Letters to Ace technique, Letters of Growth provide this outlet with a therapeutic purpose. This activity involves writing a letter, burying it and planting a seed to add a positive ritual to the experience. The child is provided with a general prompt, although if the child requests more guidance, examples can be provided. The general prompt allows the child to take control and decide what they want to write about; what they feel is pressing at the time. Children who have experienced trauma often feel powerless, and allowing them to take control and make their own decisions can be beneficial and empowering. Therefore, giving children the freedom to write about their past, current, or future feelings provides enough guidance and enables the child to choose the details.

Topics that could be included vary, as children may choose to write about their feelings, memories, fears, hopes and dreams, or even a letter to a specific person. The children choose what to write and the type of flower that they would like to plant. The flower choice can be meaningful and provide insight as well. The child may pick a flower that reminds them of the person, or the meaning behind a flower supports what they are trying to accomplish. The pots utilized in the case studies are biodegradable. This allows the child to place the letter in the bottom of the pot, then plant the pot, letter, and seeds in the ground.

This technique is an alternative to writing a message in a bottle and throwing the bottle into a body of water. Letters of Growth allow the child to express their feelings, bury them out of sight, with the added growth aspect. This can provide a concrete and externalized example of positive growth from a negative memory, emotion, or experience. Providing hope and beauty can help the child sort through their emotions; helping them to accept that these emotions are normal and do not have to last forever. This technique also requires that the child tend to and take care of the flower. Having something to take care of can help provide the child with purpose and a sense of responsibility. The child can watch the flower grow at each stage, which can be done outside in the ground or inside in a pot (depending on the flower the child uses). The tending to the flower provides length to this technique as well as responsibility for the child.

An important consideration is to pick flowers that will survive where the child lives as well as explaining to the child that sometimes seeds do not germinate. Having extra seeds is useful in case the first attempt does not germinate. Some children may want to bury the letter in the biodegradable pot without the flower seeds. The diversity of this technique allows the children to have autonomy over how they handle the experience. Autonomy and control over the exercise is important in supporting the children therapeutically.

Figure 27.2 Seeds

Case examples

*Tracey**

Tracey, a 14-year-old female, wanted to write a memorial letter for her deceased father. She picked forget-me-nots as her flower of choice. She liked the way the flowers looked, as well as the name. She never wanted to forget about her father, but she also knew in order to move on with her life, it was important to find peace. This activity helped Tracey find peace. She was worried the flower would die, and therefore decided to plant it in the community garden on campus. Tracey made a cross, decorated it, and placed it where she planted her growing pot. She utilized this intervention as a memorial to her father. Tracey was able to go to the garden and visit the cross she made knowing the letter she wrote to her father was in the ground. She did not tend to the flower herself, as it was in the communal garden, and she was afraid to kill the flower.

Allowing her to adjust the technique to support her needs was important. Adding the home-made cross was her idea and a way to further symbolize her feelings and thoughts around the letter she wrote to her father. Tracey shared the letter with the practitioner and she read it out loud before she folded it up and put it in the bottom on the pot. Providing the space for the child to share the letter can support healing as well as support the therapeutic relationship.

*Anna**

Anna, a 16-year-old female, wanted to write letters expressing her feelings about a past trauma. She chose the tulip as her flower, as it was her favorite flower and she wanted to attempt to associate one of her most distasteful memories with something she favored. This technique can have many layers of healing. Anna had an extensive trauma history; as a young child, she and her siblings endured years of abuse. She took to this technique quickly and wrote a few letters. One was a general letter about her trauma, one was to her brother who was also living at the Children's Home, and the last letter was to her abusers. Anna did not wish to share the letters, though she did discuss the general letter and the one to her brother with the practitioner. The general letter had hope and sorrow, and the letter to her brother was supportive and deep. Anna expressed how the abuse affected her and her relationship with her brother. The letter to her abusers was the longest. She did inform the practitioner that she had been wanting to express herself to her abusers, and this letter gave her that ability.

The beauty of this technique is the control that the child has, as they can alter the process to best support their journey. Anna decided to plant her tulips in the memorial garden that was created for Ivy (the previous therapy dog at the Children's Home of Poughkeepsie). Anna had a strong bond with Ivy and wanted the flowers to be near her memorial, adding another layer to this intervention.

*Amelia**

Amelia, a 16-year-old female, decided to write two letters: one about her hopes for the future, and one about her negative past trauma. She chose pansies for her future and forget-me-nots for her past. Amelia was very insightful. She loved the flower pansies and decided that they would represent her future goals. She chose forget-me-nots for her past trauma, which was interesting. Amelia felt that her past traumas were important since they helped shape her. Although they were difficult, during her therapeutic work at the Children's Home of Poughkeepsie she was able to accept her past and utilize it to help fuel her future. Amelia's hopes for her future were based off her trauma, as she planned to attend college and work in the criminal justice field.

Amelia did not want to push her past trauma away; she did not want to bury it away and forget. She chose the forget-me-nots to help herself never forget. Amelia decided to utilize her past as support for her future, and hoped that her story could help others heal. Amelia shared the hopes for her future letter by reading it out loud before placing it in the pot. Amelia did not want to read her past trauma letter out loud, though she did allow the practitioner to read the letter. Amelia planted the pots in the service dog memorial spot, as she and Ace had grown very close. Amelia wrote a letter to Ace weekly and wanted her flowers to be in the garden where Ace and the other residents and staff could enjoy them.

*names have been altered to protect confidentiality.

Sand art feeling jars

Description of the technique

In this technique, the practitioner prepares eight large jars, each labeled with a different emotion. The emotions include four positive emotions, including excitement, happiness, hope, and love; and four negative emotions, including anger, confusion, fear, and sadness. Each labeled jar should have its own corresponding color of sand: orange (excitement), green (happiness), yellow (hope), purple (love), red (anger), brown (confusion), black (fear), and blue (sadness) (see Figure 27.3). The practitioner should also have an assortment of smaller jars and a funnel available for the activity.

In facilitating this intervention, the practitioner should have all eight jars displayed in a line for the child. The child can choose his or her own small jar. The child will be instructed to read the labels (or have labels read to them) and choose which emotions they are currently feeling. If a child does not know one of the emotions, the practitioner can provide psychoeducation on that feeling. The practitioner should ensure that the child is not choosing the jar based off of sand color, but rather the emotions that the child is truly feeling at the time. The child will then be asked to use the funnel and pour in different layers of those identified emotions. The proportion of each layer should correspond to the extent in which the child is experiencing the given emotion.

Once the child fills up the jar with the emotions that he or she is feeling, the practitioner should give the child the opportunity to discuss the jar that he or she has made. The child will be asked to describe each emotion that was chosen and explain what life events may be contributing to each identified emotion. The practitioner should ask the child's permission to take a picture of the jar, and the child is to keep the jar that he or she has made. This is an activity that can be repeated at any time throughout the therapeutic process, though it has been found to be particularly helpful during an initial session and then again during the termination process.

Research support

This technique was studied in an urban population of children that had high rates of trauma exposure in order to assess its effectiveness in improving children's overall confidence in their emotional state (Jenkins, 2019). Specifically, children were asked during pretest to identify their current emotions and then rate on a confidence scale how confident they were in their self-identified emotional states. After the intervention was employed, the children were asked posttest to re-rate their confidence level on the same scale used during pretest. Results of the pre-post analysis showed that children's confidence in their emotional states significantly improved as a result of the intervention (Jenkins, 2019). Thus, this intervention's effectiveness is an important contribution to the play therapy world, as many child therapies require emotional identification and expression in order to engage in effective change (i.e., TF-CBT) (Jenkins, 2019).

Discussion of the technique

This activity is a creative way for children to visualize and learn how to get in touch with the emotions that they are currently experiencing. A core component of empirically supported trauma-focused protocols is a psychoeducational portion that focuses on the identification and labeling of feelings (Cohen, Mannarino, & Deblinger, 2006). Not only does this technique help children to identify and label their feelings, but it also allows them to distinguish between a wide array of feelings that they may be experiencing simultaneously. It is common for children with complex trauma to experience many different and confusing emotions at the same time, and this activity can be a psychoeducational experience for children to begin to identify and differentiate those complex emotions.

This technique can also be used to promote and facilitate conversation about how the child is feeling in a creative, non-threatening, and indirect manner. In addition, the sensorimotor aspect of this activity is therapeutic, as experiential and sensorimotor expressive activities have often been found by research to enable survivors to access trauma memories and facilitate their trauma narratives (Cohen, Mannarino, & Deblinger, 2006). This activity, mainly used in trauma work with children, has been a useful technique to facilitate the expression of emotions that might not otherwise be expressed in standard play or talk therapy.

Case examples

To give an example of how this activity benefitted a client, a jar made by Alex*, an eight-year-old male, will be described. Alex had been removed from his home by Child Protective Services three months prior to making the jar. Alex was a child who internalized all of his feelings and rarely discussed his emotions. In the jar he made, Alex put happiness (green), hope (yellow), excitement (orange), confusion (brown), and anger (red) (see Figure 27.4). He explained that he was happy and excited because he was having a visit with his parents the next day. He reported being hopeful that he would end up going home with his mother soon, but was confused because he was unsure about whether or not that would actually happen. He also reported being angry because he was not currently home with his parents, and had been away from them for three months. This activity was one of the first instances in the three months that the therapist had been working with him in which he was able to express the emotions that he had been feeling. This was a breakthrough moment in the therapeutic process, and led to very valuable conversation in which the therapist was able to validate his feelings, reinforce his expression of emotions, discuss with him what the removal from his home was like, and prepare him for the possibilities of not being able to return home.

Another valuable example of the utilization of this technique was with Jordan*, a seven-year-old male. Jordan had just recently come to the Children's Home of Poughkeepsie (CHP) when he did the sand art feelings jar together with his therapist. He and his siblings had been removed from his home by Child Protective Services because his sister had been exposed to

Figure 27.3 Sand art feeling jars

Figure 27.4 Alex's sand jar

sexual abuse, which Jordan was unaware of. In his jar, he put in every emotion (happy, excited, love, hope, anger, sadness, confusion, and fear) (see Figure 27.5). Jordan reported being happy and excited because he was having a visit with his parents the next day, and put in love because he had so much love for both his mother and father. He said that he was hopeful that he would go home with them soon, and was sad because he was not already home with them. He reported being angry because he felt as if he was one of the only children in the cottage (his place of living at CHP) who behaved, and he was angry that the other children chose not to behave. He described his confusion about why he had been brought to CHP and did not understand why he could not go home yet. Lastly, he described his fear that he would not be able to go home with his parents and that he would never be able to see them again.

Figure 27.5 Jordan's sand jar

This was a breakthrough session for Jordan, as he was able to describe his feelings towards his sister, as he believed that the removal was her fault, in addition to the intense worry and fear that he had been experiencing since the removal. He also discussed his anger towards his father, who had been promising that he would take the kids home soon, but then was not able to follow through with those promises. Jordan was almost at the brink of tears when he looked up at the therapist and said, "moms and dads don't lie, but he broke this promise to me." Jordan was also able to describe his familial role to the therapist, in which he was supposed to look out for and take care of his two younger siblings, and the heavy burden that that responsibility had placed on him.

This session was highly valuable, as the counselor was able to validate his feelings and experiences that he had been through. The counselor and Jordan were able to address his confusion about the removal, and the practitioner was able to explain to him that the removal was not his sister's fault, and that the purpose of the removal was to ensure his and his siblings' safety. They were also able to discuss some of the trauma associated with the removal from his home, and talked about the possibilities of not being able to return home. The counselor was amazed at Jordan's receptiveness and his ability to verbalize all that he did at seven years of age, especially since the counselor had only met with him once prior to this session. This activity really facilitated his openness and was a non-threatening modality to allow him to discuss what was going on for him emotionally. As they left the session, he smiled at his counselor, held up the jar, and said, "Maybe I'll give this to my sister". *Names have been altered to protect confidentiality.

Pizza strategy

Description of technique

The pizza strategy is another intervention that was designed to create a non-threatening and creative way for children to express the feelings and emotions they have for significant individuals or places in their lives. Using paper, cardboard, or any flat surface that will allow you to cut and write on, create a circle and draw four lines within the circle to create eight slices of pizza. Coloring the circle to look like pizza, create eight or more individual pieces of crust that can be added to each pizza slice. Each crust should be labeled with important people and places that are relevant to the child's life. Examples of crusts used in this activity have included the counselor, mother, father, brother, sister, grandmother, grandfather, foster mother, foster father, friends, home, school, therapy, doctor, etc. It can also be helpful to have blank crusts, and have the child come up with their own labels that might be more personal for them. Pepperoni and mushrooms (or any pizza topping) are then cut out and labeled with an emotion prior to working with a child (e.g. anger, frustration, happiness, joy, confused). Blank toppings are also helpful for providing children the opportunity to identify their own emotions that may not be listed on the pre-made toppings.

The child is instructed to choose eight labeled crusts that are the most applicable or important to them in their lives. They are then asked to add as many feelings as they wish to describe their emotions toward each of the significant people or places in their lives. This is a very helpful way to aid children in getting in touch with their emotions and facilitate valuable conversation that can be used to further inform the therapeutic process.

Discussion of technique

Like the sand jars, this activity is another creative and non-threatening intervention for children to visualize and label the feelings that they have towards significant individuals and places in their lives. This is particularly important in traumatized populations, as a major component of empirically supported trauma-focused protocols is a psychoeducational aspect that targets identifying and labeling feelings (Cohen, Mannarino, & Deblinger, 2006). Children with complex trauma histories are likely to experience varying and confusing emotions simultaneously. This technique not only provides a way for children to identify and label their feelings, but it also helps them to differentiate between a multitude of emotions that they may be experiencing concurrently. Furthermore, like the sand jar intervention, this technique also contains a sensorimotor aspect, which is valuable in facilitating trauma narration and/or processing (Cohen, Mannarino, & Deblinger, 2006).

Case examples

Harold*, a 13-year-old male, created a pizza using the Pizza Strategy in which his crusts consisted of himself, his grandfather, foster mother, foster father, biological mother, biological father, brother, and sister. The majority of his pizza toppings were on the slice of pizza that represented himself. These included the emotions "safe," "worried," "love," "rescued," "selfish," and "thankful." Harold purposefully laid "thankful" and "grumpy" across his slice and the slice representing his grandfather, of whom he expressed he was very thankful to have in his life. He went on to explain that they both shared the trait of being "grumpy people." The slices representing his brother and sister share the pizza topping of "hurt," as he explained they were both hurt, and they had hurt Harold before his removal from his home. Harold's biological mother and father shared the pizza topping "jealous," though he was unwilling to disclose the meaning behind this word. Harold's foster mother and foster father shared the pizza topping "hate," as Harold was experiencing animosity towards them at the time for bringing him to a non-secure facility. He expressed feeling abandoned by them and couldn't understand why he was not at home living with them, causing him to feel what he expressed as hatred towards them. Harold proved to be a very reserved individual and was generally cautious about disclosing information about himself and his family through means of traditional talk therapy. The Pizza Strategy

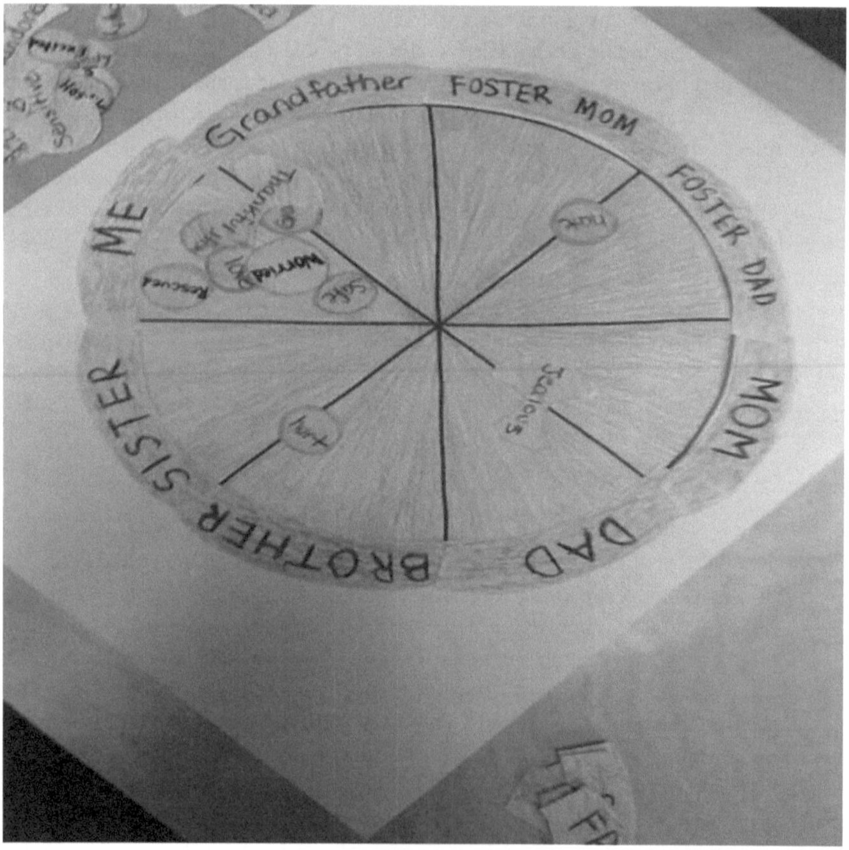

Figure 27.6 Daniel's pizza

served as an indirect and safe means for Harold to open up about the different people that serve meaning to the life he is living presently.

Daniel* was a 12-year-old boy, who the counselor had been working with for six months at the point where the Pizza Strategy was tried for the first time. Daniel had been very reserved and reluctant to talk for most of their work together, so the counselor had used many different indirect and creative interventions with him. Up to this point in the work with Daniel, he had not spoken very much about his family, and this activity proved to be quite valuable, as it facilitated conversation about the way Daniel felt about and viewed each of his family members in a way that was non-threatening to him. When Daniel and the counselor did this activity together, he chose the following crust categories: mom, grandfather, staff, school, the Children's Home, brother, dad, and sister. For each slice, he added on toppings that had the feeling words on it. For his mother, he placed the toppings of "love" and "trusting" to describe her and commented that he had a strong connection with her. For his grandfather, he used the topping of "sorry," and mentioned that his grandfather, who he had been close with, had passed away. For the staff, he put "angry" because he did not like it when they told him what to do. For school, Daniel used the toppings of "confused" and "worthless," because he did not like school, did not see the point of going, and struggled a lot with certain subjects. For the Children's Home, he placed the topping of "safe," and reported that he felt safe living there. For his brother, he used the toppings of "kind" and "disappointed," and reported that his brother is kind, but is sometimes considered a disappointment because he does not do well in school. For his dad, Daniel used the topping of "accepted," and reported that he feels accepted by his father, although he does not get to see him as much as he would like. Lastly, for his sister, Daniel used the topping of "selfish," and commented that she is sometimes selfish when she talks back to their mother.

Daniel had been engaged in the activity and was able to verbalize how he felt about his family members, school, the Children's Home, and the staff. This activity facilitated Daniel's openness about these important people and aspects in his life and gave him an opportunity to express his feelings in a way that would not likely have been possible with the use of standard talk therapy. *Names have been altered to protect confidentiality.

Conclusion

In order to engage and make meaningful connections to the disconnected children and youth who suffer complex trauma, therapists and counselors may need to reach deeply into their toolbox, or call on their creativity and imagination to find novel ways to excite the interest of their wary clients. The creative interventions described in this chapter were developed with careful consideration of keeping trauma-informed therapy safe for those who suffer complex trauma since it is well-known that if the youth don't feel safe, they will shutdown, or even worse, be retraumatized. Except for the sand art jars strategy (Jenkins, 2019), no empirical support can be offered for the efficacy of these strategies, although the premises of these techniques are highly compatible with the psychoeducational and coping components of the empirically supported trauma-focused protocols in wide use today. Furthermore, the strategies presented in this chapter ensure safety by allowing youthful clients to self-pace on their therapeutic journey to confront their multiple traumatic events, and in some cases an early life replete with interpersonal trauma.

References

Allan, J. (1988). *Inscapes of the child's world: Jungian counseling in schools and clinics*. Dallas, TX: Spring Publications.

Bowen, M. (1993). *Family therapy in clinical practice.* New York: Jason Aronson.
Bronfenbrenner, U. (1994). Ecological models of human development. In *International encyclopedia of education* (vol. 3, 2nd ed.). Oxford: Elsevier.
Cohen, J., Mannarino, A., & Deblinger, E. (2006). *Trauma-focused CBT with children and adolescents.* New York: Guilford Press.
DeNigris, P.N. (2008). Trauma in youth: Reactions and interventions. *Journal of Psychiatry and Law, 36*(2), 211–244.
Finkelhor, D., Ormrod, R.K., & Turner, H.A. (2007). Poly-victimization: A neglected component in child victimization. *Child Abuse & Neglect, 31,* 7–26.
Gersie, A. (1997). *Reflections on therapeutic storymaking: The use of stories in groups.* London: Jessica Kingsley Publishers.
Gil, E., & Crenshaw, D.A. (2015). *Termination challenges in child psychotherapy.* New York: Guilford Press.
Green, E.J., & Myrick, A.C. (2014). Treating complex trauma in adolescents: A phase-based, integrative approach for play therapists. *International Journal of Play Therapy, 23*(3), 131–145.
Jenkins, M.L. (2019). *Alexithymia in minority children: Assessment, intervention, and implications* (Publication No. 13807242) (Doctoral dissertation). Regent University, Proquest Dissertations Publishing.
Jolly, M. (2011). What I never wanted to tell you: Therapeutic letter writing in cultural context. *Journal of Medical Humanities, 32*(1), 47–59. doi:10.1007/s10912-010-9127-z
Knoetze, J. (2013). Sandworlds, storymaking, and letter writing: The therapeutic sandstory method. *South African Journal of Psychology, 43*(4), 454–469. doi:10.1177/0081246313506663
Lepore, S.J., & Greenberg, M.A. (2002). Mending broken hearts: Effects of expressive writing on mood, cognitive processing, social adjustment and health following a relationship breakup. *Psychology and Health, 17,* 547–560.
McCullough, M.E., & Witvliet, C.V. (2002). The psychology of forgiveness. In S.J. Lopez & C.R. Snyder (Eds.), *Handbook of positive psychology* (pp. 446–458). London: Oxford University Press.
Mosher, C.E., & Danoff-Burg, S. (2006). Health effects of expressive letter writing. *Journal of Social and Clinical Psychology, 25*(10), 1122–1139. doi:10.1521/jscp.2006.25.10.112
National Child Traumatic Stress Network. (2007). *Complex trauma.* Retrieved from www.nctsn.org/trauma-types/complex-trauma
Newsome, D.W., & Gladding, S.T. (2014). Ethical and legal aspects of counseling. In *Clinical mental health counseling in community and agency settings* (4th ed., pp. 45–71). Upper Saddle River, NJ: Pearson Education.
Pearson, L. (1965). *The use of written communication in psychotherapy.* Springfield, IL: Charles Thomas Publisher.
Pennebaker, J.W. (1997). Writing about emotional experiences as a therapeutic process. *Psychological Science, 8,* 162–166.
Perry, B.D. (2009). Examining child maltreatment through a neurodevelopmental lens: Clinical application of the neurosequential model of therapeutics. *Journal of Loss and Trauma, 14,* 240–255.
Rogers, C.R. (1957). The necessary and sufficient conditions of therapeutic personality change. *Journal of Consulting and Clinical Psychology, 21,* 95–103.
van der Kolk, B. (2014). *The body keeps the score: Brain, mind, and body in the healing of trauma.* New York: Viking Press.

28

PLAY THERAPY/ THERAPEUTIC PLAY WITH CHILDREN WITH AUTISM

The journey: wait, watch and wonder

Audrey Gregan

Introduction

This chapter is about sharing my experience of going on a journey with two children on the Autism and presenting with multiple symptoms. My first contact with the parents and children was meeting them at a team assessment for entry to the hospital and special school, where I was in full-time employment as a member of the multidisciplinary team.

The assessment of the children was based in my therapy room with the SLT psychologist, principal and myself. It was a play-based observation of the child's reactions relating to us and toys in the room. Other team members, a social worker and a psychiatrist reviewed the family history with the parents as was presented with files that came with the application. It was agreed by the team that these children should have individual play therapy sessions as part of their intervention. As with any therapeutic approach it was very important for me to choose a suitable approach which will provide for their needs.

Autism

Autism spectrum disorder is a Diagnostic and Statistical Manual diagnosis that is usually given after a thorough psychological evaluation wherein the evaluator measures the child or adolescent's behaviour across a myriad of tests, assessments and observations. The disorder is a spectrum disorder meaning the symptoms vary in intensity from severe to very mild. Common terms used to describe the variance include high and low functioning, or severe or mild impairment. Children and adolescents with an autism spectrum disorder will likely have similar problem areas, but the severity of their difficulty and the presence or absence of other features (fine motor clumsiness, normal intelligence, increased or decreased verbal output) varies (Coplan, 2010).

Children and adolescents with autism spectrum disorder can and usually do exhibit multiple symptoms. The most common symptoms that children and adolescents with autism seem to display at some level include impairments in social skills and functioning; impairments in communication; impairments in the ability to regulate and modulate emotions; difficulties establishing and creating meaningful relationships and connection with others; restricted, repetitive and stereotyped patterns of behaviour, interests and activities; sensory processing issues; and difficulties handling transitions and change.

In regard to play skills, adolescents with autism spectrum disorders usually struggle in the areas of pretend or imaginary play and peer or group play. In regard to peer or group play, children and adolescents with autism spectrum disorders typically do desire to have friendships and interact with others peers but simply lack the social ability and skills to interact successfully. Thus, most attempts at some type of interaction usually are met with rejection and anxiety for the child with autism. Directive play interventions can help children and adolescents learn social skills to interact more successfully with peers and participate more fully in group and peer play. Pretend and imaginary play skills can also be increased, but present a greater challenge. Due to the deficit in pretend and imaginary play, practitioners should avoid play-based interventions that rely heavily on abstract concepts, pretend play and metaphor (unless specifically working on improving this skill). Play-based interventions should be concrete and clearly communicated to the child and literal in addressing the child's issues.

A summary of the DIR/Floortime approach

The DIR (Development, Individual-Difference, Relationship-Based) Floortime approach is a comprehensive framework for understanding and treating children challenged by autism spectrum and related disorders. It focuses on helping children master the building blocks of relating, communicating and thinking, rather than on symptom alone (Greenspan & Weider 2006).

- "D" is for Developmental. Understanding where your child is in his emotional development is critical to planning a treatment program. The six developmental milestones are the building blocks that every child must master as a foundation for healthy emotional and intellectual growth.
- "I" is for Individual-Difference. Each child has a unique way of taking in the world – sights, sounds, touch, etc. – and responding to it. Difficulties in processing and/or responding to sensory information may be part of your child's individual differences and can interfere with his or her ability to relate, communicate and think.
- "R" is Relationship-Based. Building relationships with primary caregivers is a critical element in helping a child return to a healthy developmental path. Floortime, a centrepiece of the DIR approach, encourages parents and others important in the child's life to tailor their interactions with him, based on where he is currently in the development and in tune with his individual differences, in order to support his development.

Six developmental milestones

Milestone 1: Shared Attention and Regulation
Milestone 2: Engagement and Relating
Milestone 3: Purposeful Emotional Interaction
Milestone 4: Social Problem Solving
Milestone 5: Creating Ideas
Milestone 6: Connecting Ideas Together, Thinking Logically

Six basic developmental skills, or milestones, lay a foundation for all our learning and development. Children without special needs often master these skills relatively easily. Children with challenges often don't, not necessarily because they can't, but because their biological challenges often make the mastery more difficult.

Floortime

Floortime, a systematic way of working with a child to help him climb the developmental ladder, is the heart of the developmental approach to therapy. It takes a child back to the very first milestone he may have missed and begins the developmental process anew. By working intensively with parents and therapists, the child can climb the ladder of milestones, one rung at a time to begin to acquire the skills he is missing.

The developmental approach to therapy consists of three parts.

- Parents do Floortime with their child, creating the kinds of experiences that promote mastery of the milestones.
- Speech, occupational and physical therapists; educators; and/or psychotherapists work with the child using specialised techniques informed by Floortime principles to deal with the child's specific challenges and facilitate development.
- Parents work on their own responses and styles of relating with regard to the different milestones in order to maximise their interactions with their child and create a family pattern that supports emotional and intellectual growth in all family members.

Floortime is precisely that: a 20- to 30-minute period when you get down on the floor with your child and interact and play. How can playful interactions help your child master the milestones? The answer has to do with the nature of the interactions. Certain types of interactions with other people promote a child's growth.

Sensory play (messy play)

My interest in sensory play was influenced by Maria Montessori (1870–1952), who believed that by using their senses, children are able to build a picture of the concept of the reality of the world and that every child carries within them the desire to explore their world. It is through their senses they learn to discover and make sense of the world around them.

Sue Jennings Embodiment Projection and Role played a very big part in my work with the two clients I am giving case history of.

The body and its relationship with other bodies – through touch and other senses – form the basis for development of identity in all human beings. However it is important to remember that our identity depends on the relationship between bodies not only touch and senses, therefore there needs to be a balance between touch and separation, between the we and the I (Sue Jennings, 1993.

Many children coming for therapy spontaneously enter into a therapeutic regression knowing to which stage they need to return. The therapist helps by accepting this, but for children with autism some have not moved, they are still at the early stage and need support to progress through their developmental stages. In this case the therapist has to meet the child where they are developmentally and create an environment that provides the child with experiences of satisfaction for the child to move forward.

The therapist has to take the role of nurturing parent who can regulate stress and meet the therapeutic needs of the child.

I chose to start the therapy sessions with these children by using these two approaches because I felt they were presenting with a need for early nurturing work.

Highlighting areas of need

I spent some time in the classroom observing Sean and learning about his difficulties that brought him to a diagnosis with autism. I did some with Paul, whom I observed in his residential care home. Then decided as to what was the priority for intervention for play therapy which would help the staff and the children to develop a meaningful relation and give the children some skills to connect with the people close to them.

Case one – Sean

Sean was referred to our service at the age of seven. He was extremely withdrawn, tense, tearful and presented with multiple symptoms of autism. Our school was suitable to provide for his needs. He had been to several clinicians for intervention for autism but he finally was accepted in our special school. He had no speech, walked on his tippy toes, was very fearful of unfamiliar environments, was on a bottle with formula milk and did not eat solids but powdered baby food. He found touch frightening, his parents were very stressed and he was the eldest child with two younger siblings who were both in mainstream school. It was agreed by the team meeting that Sean should have individual therapeutic play sessions.

Setting the play therapy room

It was very important to have the room in a way that would not be overwhelming to Sean and material suitable for his very sensitive nature as he was presenting as a child who needed to be nurtured, being at a very early stage of development, very infantile. It is very important for a practitioner working with children with autism to have a great understanding of child development so that one can start at the level that the child is presenting.

Stage 1

I used the soft room to make a connection with Sean. It was simply a room with cushions and soft padding on the walls, big windows that let in light and different sizes of soft balls. It was quiet; he was distant but willing to come along into the room. This went on for four weeks: just going in and playing with the balls which were colourful but not connecting with me. I called his name but he found it difficult to make eye contact, but every now and then he would raise his head to try and connect; he needed gentle cohesion. It was difficult for him to engage with people. Gradually he accepted my presence in the room and the sound of my voice. Every now and again he would raise his head and give slight eye contact; I reflected by calling his name and rolling the ball towards him. At first he did not roll the ball back but after a couple of sessions be began to pick up the ball and roll it but not directly toward me. I followed the ball and rolled it to him. He smiled and we began to play ball together. Slowly he started to explore the few objects in the room and I mirrored him.

He found it overwhelming to be in a classroom but soon accepted the presence of his peers even though he did not engage in group work. He sat and watched what the others did. I walked with him around the building to familiarise him with the environment and meeting staff, looking into different rooms to reduce his fear of places. I also encouraged him as we were walking to let his feet down instead of tip toeing; he liked this and turned it into a game. He would tip toe and turn towards me so that I could say "feet down." He did so with a smile and that really was a beginning of our connection and relationship development. The level he was functioning at there was no need for a therapy room with too much equipment and non-directive work. It had to be directive and nurturing; it was going to be a long-term intervention as he was in our special school. There were a lot of people involved in his care and they all needed to be part of play therapy sessions too – teacher, parents, SNAs. The way I was working with him had to be taught to all those involved, so that we all engaged with him at the level he was functioning at.

Stage 2

This stage the intervention was to introduce solid food to him, so I introduced, "messy play" as the activities, again being careful that he did not feel threatened by the activities. We played with pureed food, tasting it using our hands. I let him feed me and I just encouraged him to feel it on his lips and gradually leaving some lumps in it; it was very slow and sometimes very upsetting to him, he wanted to but was afraid. He really needed encouragement as a baby, so I fed him and he would put his hand on mine to show that he wanted to try and eat. I had to model chewing by using dry foods, such as crackers or crisps; it was a very difficult time for both of us.

Parents came to observe some of the sessions and found them to be helpful, with the support of the social worker we were able to bring to the home and support the parents to do the same as I had done in the school. Most of the work with Sean was at a very early stage of development even though he was seven years old, emotionally he was a baby, needing nurturing with developing attachment and relationship. He developed trust and his feeding improved and was able to make choices of food by texture and smell.

Stage 3

Change of therapy room

We progressed to a more integrated play therapy room with a variety of toys and Sean began to make choices of toys to play with, he was more relaxed and giving some eye contact. I began to introduce a non-directive and followed his lead, engaged with him at his choice of activity, but he was still not talking, but communicated by using PEC's to make his needs met. He was much happier and was more interested in exploring new places and making contact with people around him. His play improved, it became more symbolic and interactive including me most of the time it was fun.

Approach used

The DIR/Floortime had a big influence in my relating and engaging with Sean, following the Axline Principals gave the theory I needed to accept Sean as he was. In order to support and help develop his emotional process at his development level I feel it's important to discuss how! I see my role within the autism setting and therapeutic play environment; I see myself as providing missed experiences that encase the use of all the senses that have a synergistic effect on the developing brain. The first three years of seeking and effectively engaging the world are crucial for the success of every boy and girl.

As some of the children I work with are at a very early stage of development their interaction and relationships with people around them. They need therapeutic play that provides engagement at their level of development.

In Gerhardt's book, "Why love matters," a nursery school study was carried out which showed that it was not the absence of an adult figure who would be able to respond effectively to their states moment by moment. She goes on to say that what a young child needs is an adult whom is emotionally available and tuned in to help regulate his state. This would suggest that the role of the therapist is to be a nurturing parent who can regulate stress and meet the therapeutic needs of the child to enable and facilitate the growth of self. It is summed up perfectly by Edith Sullwood when, describing the mystique of the client–therapist relationship providing a surrogate lap (O'Connor and Schaefer, 1994).

I feel by being sensitive and providing a play therapy room and materials that promote exploratory play gives the child the opportunity to satisfy their intrinsic need to explore emotionally and physically through sensory play.

Sean's progress was very slow and individual sessions were on a daily basis lasting three-quarters of an hour. This was to settle him in school and also to give him an attachment figure at his developmental level. I found that the therapeutic environment allowed him to be himself; he got autonomy in the presence of an adult who was non-judgemental and was able to enter his world as a follower and be fully present with him in the moment. As we shared the space and being present and his feelings were validated not dismissed, he began to relax every time he was in the therapy room. The crying reduced when he realised his feeling's where understood I reflected back that I felt he was crying because he was afraid and not sure of what to do, he stopped crying and started to engage in play as I mentioned earlier.

We began to engage in a lot of movement, ball play in the soft room and chasing, he enjoyed this activity and remained relaxed during the play session but this needed to be extended to the school environment. I invited his classroom assistant to join the sessions too. The parents came to observe Sean at play and to learn how to follow his lead at play at home. His confidence developed he was beginning to take part in the group activities with his class with the support of his SNA and his siblings at home. He is a very happy young man now and will be graduating from his school to go to an adult service, 20.06.2014. Children with autism have difficulty to express themselves but can do so through bodily stimulus and senses. It is for us to be very observant when working with these children. He made great progress his speech has developed and was able to develop friendships with peers by using music and sing in a group with his peers. Sean is very independent and will do very well in the adult services.

Case two – Eoin

Eoin was four years old when I met him. He was referred to our service after getting diagnoses of ASD. It was a very traumatic time for the family. The family were going abroad due to mum's illness, she was wheel chair bound and had lost her speech. This happened during the time when Eoin was going through assessment for ASD, by the time the diagnoses was made mum was not able to physically engage with Eoin, so he had to be taken into care to help the family, but was able to go home two nights a week. It was agreed by the team in the service that I should give him time for Play Therapy. He was a very active child just could not sit quietly, he climbed on furniture squeezed himself into tight corners and spend a lot of time inside boxes and containers or duvet covers. He had no speech, he liked clutching in his hand bits of paper, holding them tight or liked to run out and had no sense of danger, running off was a game to him having somebody chasing him. He wanted to make contact but just didn't know how. It was important to create the environment to gratify his needs at his level. He was a happy child but needed a place where he could use his energy safe and relieve a part of the primary experience, to help restore his missed sensory experiences.

Stage 1

I brought him into the soft room with cushions, boxes to climb into, and blankets to wrap him in as he liked deep pressure and box of small bits of paper from book binding machines as I mentioned early he liked these bits of paper to hold. This seemed to regulate him when he got tense and worried. For the first few weeks he explored all the materials in the room as we developed our relationship of trust and having fun together. He slowly was engaging with me and spending less time in the boxes and asking to be wrapped in the blanket making more connection with me, looking for nurturing from his caregivers.

Stage 2

Connection following the child's lead

During the following sessions these took place in his day room so that his cares could get involved in his play and he liked this very much.

1. We used blankets as a hammock and gave him a swing.
2. Rolled him to and from between two people on a mat.
3. He initiated jumping on cushions and a small trampoline, we mirrored his movements.
4. We used a large full length mirror to interact and point to parts of the body for self-awareness. He loved the mirror and helped with dressing, tooth brushing and also giving eye contact.
5. We provided soft music and did some rocking movements.

Communication

He was no longer running to try and get somebody to chase him. The climbing reduced because he had the activities to do. He would always come and take a member of staff by the hand to play with at a chosen activity. His concentration improved and listened more to adults around him, enjoyed music and movement, took part in group activities with his class. He still needed the support of his SNA in the class room. He was given speech and language sessions and this improved his communication skills. He still has no speech but can make his needs met by using his ipod. Eoin is still having a lot of difficulties he needs supervision through the day. Play therapy improved his social skills but he will continue to need residential care. He is a lovable young man full of energy which needs direction.

Conclusion

Play is a wonderful tool for helping children with autism to engage with others and move beyond autism's self-absorption into real, shared interaction. Used well play can allow children to explore their feelings, their environment and develop their relationship with parents, siblings and peers. Both approaches I used gave me the skill to understand and relate to children with autism. They also contributed greatly to our on-going interactions to the point that children actively wanted to share more and more experiences with me.

This was a journey with two very different young people who taught me about the individuality of people on the autistic spectrum. Working with Sean and Eoin was not about changing their personalities but accepting who they are and how our everyday way of leaving affected them. They see the world differently and need time to see it our way and us to see from their perspective. It is not an easy journey for the children, their parents and all those involved. There is a lot to be done in developing relationships, trust and respect for children diagnosed with autism and neurological disorders. We have a lot to learn from them, the children are amazing they bring so much experience to the therapist. Once the relationship is formed the journey opens the door to tears, laughter, pain and joy. I have shared my time and space with wonderful children and their parents. Worked with great professionals who were supportive when the work was difficult and we learned from each other. I am grateful to the parents and children let me into their world. It is so true when it is said follow the child's lead they bring you to places and thoughts beyond your imagination. I call it discovering Ephesus a place of wonder. What it takes is to "WAIT" to be accepted by the child, "WATCH" when the relationship develops and "WONDER" at the experience of learning.

References

Coplan, J. (2010) *Making sense of autistic spectrum disorder: Create the brightest future for your child with the best treatment options*. New York: Random House.

Greenspan, S. Weider, S. (2006) *Engaging Autism*. Cambridge, MA: Da Capo Press.

O'Connor, K.J., & Schaefer, C.E. (1994) *Handbook of play therapy. Vo. II: Advances and innovations*. New York: Wiley.

29

PLAY THERAPY

The ideal environment for play development and the repair of play deprivation

Sonia Murray

Play is essential for humans but is often not recognised as significant. Play is a developmental process and requires a number of elements to facilitate its growth. Unfortunately, without these elements, children's play skills are underdeveloped and play deprivation occurs. By exploring a few key play developmental models, plus providing a brief overview of play deprivation and its potential consequences, this chapter will explore how play therapy provides the ideal environment to readdress the impact of play deprivation and help the children gain the necessary play skills needed for child development. Furthermore, it will explore the play developmental stages in the context of brain growth.

Play is fundamental to all aspects of human development and is essential to gain the necessary play skills required for adulthood. However, it is often minimised, disregarded and undervalued in modern society as it is considered frivolous and a 'waste of time'. There is a greater emphasis on academic achievement, limiting opportunities to play. This is impacting on the development of children's skills and there is an increase in play deprivation. The impact of play deprivation has potentially life-long physical and psychological implications (Gray, 2011).

The necessity of play

Play is vital for healthy development. As stated by Sheridan (1977) 'play which provides suitable opportunities to strengthen the body, improve the mind, develop the personality and acquire social competence, it is as necessary as food, warmth and protective care' (p. 14). Through play children develop their individual self, whilst learning to be part of society.

Play is challenging to define as it is multifaceted. However, Schaefer and Peabody (2016) define play as 'an activity with the following key attributes: freedom from constraints of reality, positive affect, flexibility, intrinsic motivation, inner control, and a focus on the process of the activity rather than the outcome' (p. 23). Howard (2017) determines 'play not just as a behaviour, but also a process and an approach to a task' (p. 4). It is perceived by McLean and Hurd (2014) as an activity that connects us socially with others, the world and our internal self.

Play enhances all aspects of children's holistic development. From a physical perspective, it develops their gross and fine motor movements and advances their hand and eye co-ordination. They learn where their body begins and ends. Play paves the way for cognitive development, by providing opportunities to develop problem solving skills, cognitive sequencing, concentration capacity, and enhance their working memory. Play's cognitive and social elements provide the ideal environment for language and communication skills to develop. As noted by Jenkinson (2001) 'Play prepares children to enter the game of life without undue fear' (p. 19).

As stated by Brown, (2008) 'nothing lights up the brain like play' (10.23). Play is one of the brain's natural ways to learn, interpret the world, create new ideas, problem solve and create social cohesion. Panksepp identified a number of systems in the brain and play is one of them (Panksepp, 1998, 2009; Panksepp & Biven, 2012). It is through play that relational connections are formed and maintained. It provides a safe container to explore, experiment and develop new ways of being. Within the parameters of playing, there is the opportunity to practice and master new skills, thus reducing the fear of failure. Additionally, play provides a repetitive, consistent and predictable environment to create neural networks to develop children's social, emotional, cognitive and regulatory skills (Perry, Hogan, & Marlin, 2000).

Play enables children to experience, understand and learn to manage their emotions. As Panksepp (1998) notes, play is 'perhaps one of the major brain sources of joy' (p. 281). It provides opportunities to experience and learn to regulate a variety of emotions. Play can create immense excitement, happiness and/or joy, but also at times generates fear when involved in play that involves an element of risk. It provides a distance to experience emotions safely. For example, playing Hide & Seek evokes excitement about hiding and being found, even joy at being found, but by also outwitting the seeker. However, disappointment can be elicited by being found too early and fear evoked from the risk of not being found. Play can take children to the extremes, but within the safety of play. It can push and extend the person's capabilities whilst keeping them within their window of tolerance. Play and its therapeutic powers provide the foundation for all learning (Schaefer & Drewes, 2014). As Gordon and Esbjörn-Hargens (2007) remark:

> The implicit or explicit limits that bind play in space and time make it safe for the player to surrender to the playful urge, take chances, try on new roles, and attempt tasks that, under normal circumstances, might be avoided as too difficult or unpleasant. It is a place where the novelty and risk of a new situation or experience only add to the intensity and pleasure of play. The player is able to be in control of being out of control and so enjoy a sense both of risk and of mastery simultaneously.
>
> *(p. 216)*

The developmental process of play

It is a crucial for play therapists to have a deep understanding of typical child development and the developmental process of play. The developmental process of play is the series of changes in children's play associated with the growth and development of children's physical, cognitive, language, emotional and social skills during childhood. Play begins in the womb with the baby playing with the amniotic fluid, flexing their limbs and clapping their hands. Then in the first few months of life, play starts with very simple play actions of exploring their body and manipulating an object in the first few months of life. However, by the age of four years, children are undertaking complex role play activities with others. Children need play to develop, but they also need child development to advance their play. As Howard (2017) notes that 'play both influences and reflects development' (p. 17).

Play is intrinsic to human development and there is an innate drive to play (Panksepp, 1998). However, play is a developmental process and requires children to be in a safe, calm, alert state (Perry, 2006) to be able to learn to play and then play. Additionally, Polyvagal theory collaborates with this thinking (Porges, 2011). When children have an internalised perception of safety, then they are able to access their cerebral cortex. They are accessing their social engagement system; thus, their ventral vagal parasympathetic system is engaged. When children are feeling safe, they can risk exploring and being curious (Panksepp, 1998; Porges, 2011). Not only do children need to be in a calm, alert state, they initially need a regulated consistent other to support the play. The other person needs to stimulate, extend and regulate the child's play and the emotional responses evoked by play. Children then have the capacity to learn to play and develop the joy in play. 'True play is not only about being able to play, but it is the enjoyment of the play' (Stagnitti, 2017, p. 2).

There are many theories that may be useful for play therapists to explore, such as Piaget's (1962) three stages of play linked to cognitive development; Erikson's (1963) play types, which were focused on the emotional benefits; and Smilansky's (1968) play schemes of functional play, constructive play, dramatic play and games with rules. A greater discussion into these is beyond the scope of this chapter. However, for the focus of this chapter, Parten's (1932) six social stages of play, Stagnitti's (2017) Pretend Play Enjoyment Developmental Checklist model and Jennings' (2005) Embodiment, Projective and Role play paradigm will be explored in more depth.

Firstly, Parten's (1932) developmental play model provides a basis to understand children's emerging social capabilities. From her study of preschool children, she determined the children's social capabilities and then linked to them to the differing stages of play. This mirrors children's increasing ability for social interaction and is a hierarchal model that is linked with children's age. She identified six developmental stages of play as shown in Table 29.1. Unoccupied play generally occurs between birth and three months and is descriptive of children playing with their own body or watching anything with fleeting interest. Solitary play is where children play alone, whilst onlooker play is where they observe others playing. In Parten's view, parallel play progresses from onlooker play, to where children play alongside others, but with little or no play interaction. This proceeds to associative play, which involves children beginning to play with others, but they are playing to their own play agenda and with no cooperation. The final stage is cooperative play, which requires children being able to negotiate, compromise and accommodate others' needs within the play. Howes and Matheson (1992) expanded and adapted Parten's theory of parallel play, by adding an additional category of parallel aware play. This is like parallel play, but the children have some social contact. Additionally, they separated cooperative play into cooperative social pretend play and complex social pretend play. Cooperative social pretend play is where the play with others incorporates pretend play without any planning or conversing about how the play will be. Whereas, complex social pretend play is deemed to be the most multifaceted play stage as it requires the children to 'both coordinate their roles with their partner and be sufficiently verbally fluent, coherent, and articulate to coordinate the

Table 29.1 Parten's (1932) six social stages of play with Howes and Matheson's (1992) adaptations

Ages	*0–3 months*	*3–9 months*	*12–19 months*	*19–23 months*	*24–29 months*	*30–35 months*	*36–41 months*	*42–47 months*	*4 years*	*5 years*
Stages	Unoccupied play	Solitary play	Onlooker play		Parallel play		Associative play		Cooperative play, cooperative social pretend play – complex, social pretend play	

planning and maintenance of the play' (Howes and Matheson, 1992, p. 962). Parten's model with its expansions is particularly useful when ascertaining children's play interactions both in the assessment process and with the play therapist in play therapy.

The preceding theory links with Stagnitti's (2017) Pretend Play Enjoyment Developmental Checklist model. Stagnitti's model not only considers children's social skills but also the children's increasing physical, imaginative, language and cognitive ability. Stagnitti (2017) breaks down the development in areas that are useful when exploring and understanding children's play in play therapy, as shown in Table 29.2. The Pretend Play Enjoyment Developmental Checklist model defines play areas as 'play scripts, sequence of actions, object substitution, social interaction, doll/teddy play and role play' (pp. 1–2). Play scripts are the stories that are developed in the play and are linked with children's narrative development. The greater the narrative skill, the more complex the play scripts will be. Sequences of play actions is the formation of different play actions in to a logical and sequenced format. According to Stagnitti (2017) 'sequencing play actions is a thinking play skill' (p. 2) and incorporates the development of executive functioning skills. The next category is object substitution which develops as children's cognitive capacity and ability to pretend increases. It is where children pretend that an object is something different to what it actually is (Stagnitti, 2017). Doll/teddy play enables children to project meaning on to another that is separate to themselves. Role play requires children to be able step into another's shoes metaphorically and enacts their role. Like Parten's model, Stagnitti's social interaction section describes the developing ability of social interaction within play.

In conjunction with these two theories, the Embodiment, Projective, and Role paradigm provides an additional overview of children's play developmental process (Jennings, 2005). This incorporates three development stages of play – embodiment, projective and role – as shown in Table 29.3. Embodiment play begins in the womb and develops through the first year of life. It is where children explore themselves and their world through their body, together with their senses. Projective play is the second stage and develops as children's cognitive pretend and symbolic capacity increases. Children explore their world using objects and toys separate to themselves, plus can project meaning, thoughts, feelings and concepts on to the object. The final stage role play is like Stagnitti's role play section and the pretend play elements of Howes and Matheson (1992) model. This model like the others is a trajectory and Jennings' posits that for typical healthy holistic development, children need to experience and incorporate all the different play stages.

These models can be tracked alongside the developmental path of the brain. The different developmental stages of play reflect the development of the different hierarchal structures of the brain (Perry, Hogan, & Marlin, 2000). Using Perry's model, the embodied play enables the brain stem to help the child to regulate their arousal state. Whilst the embodied gross motor and fine motor sensory play helps the midbrain to manage the child's motor skills and learn to incorporate the multifaceted sensory information. The beginnings of symbolic and dramatic play help to develop the limbic structure that develops children's emotional regulation, empathy and social interactions. As the play begins to involve others and with greater cognition, it engages and enhances the neocortex area by developing creativity, imagination, language skills and executive functioning capabilities. As noted by Tonkin and Whitaker (2019) 'the development of play and playfulness mirrors the evolution of the brain in terms of its size and complexity' (p. 27), as indicated in Table 29.4. It has been suggested that the greater the play capacity, the brain is physically larger (Gauntlett et al., 2010).

The play skills from the Pretend Play Enjoyment Developmental Checklist model (Stagnitti, 2017) are used with permission of the author.

Table 29.2 The pretend play enjoyment developmental checklist model

Ages	*0–3 months*	*3–9 months*	*12–19 months*	*19–23 months*	*24–29 months*	*30–35 months*	*36–41 months*	*42–47 months*	*4 years*	*5 years*
Play scripts	None	None	Scripts related to body	scripts – own life experiences in the home	own life experiences inside & outside the home	own life experiences and less frequent life experiences	own life experiences and fantasy scripts	own life experiences and fantasy scripts	play scripts that expand over a few days, incorporating greater problems	play scripts – never personally experienced
Sequence of play actions	Imaginative play absent or random – manipulates and explores the object		One simple imaginative action – fleeting	two or three similar play actions – are illogical and repetitive	Simple, sequential, logical – Actions are intentional	spontaneous, logical, detailed & logical- 6–7 actions, no planned story line	Multiple actions in a logical sequence	A play strategy is present	pre-planned story line with complex sequence and sub-plots	Pre-planned, organised with complex sequences and sub-plots – able to embed problems into the play actions
Object substitutions	Manipulates and explores objects		Relates to objects functionally – for example a spoon is placed in a cup	Uses similar looking objects for needed object – for example paper for a blanket	Inanimate objects – for example a box as a table or a car – may use imaginary objects – pretend a piece of cake on a plate	Inanimate objects used for two or more functions	Inanimate objects used for many functions	Body parts used as objects – imaginary objects referred to in the play	Objects with a distinct function used in substitution – for example a hat for a boat	Any object used – language used to describe an object and its function
Social interactions	Imitates adults	Actions – plays peek a boo	Imitates a pretend play action e.g. giving a drink	Imitates an adult using object	Asks adult for objects needed in play, plays alongside other children	Imitates another child, plays alongside – interest in what other children are doing	Plays in association with others, i.e. play the same activity beside another child, but separately	Plays in association with others, i.e. play the same activity beside another child, but separately	Co-operation/ negotiation occurs	Co-operation, negotiation occurs – play is organised and a story evident – solutions are found to conflicts
Doll/teddy play	Manipulates objects	Able to place the doll upright	Imitate feeding	Doll play begins – child sees doll as separate to themselves. Separate single actions	Sees as alive – more actions undertaken	Identifies with the doll	Doll play active – may have emotions	Doll play active – one or two emotions	Doll has its own life – includes feelings & thoughts	Doll has its own life – includes feelings & thoughts – can plan events for doll even when not present
Role play	Watches people		Copies an action of an adult	Carries out actions they have seen others perform	Copies a characteristic of another person	Indicates they are someone else – brief – changes role quickly	Indicates a role is being played – time spent is short	Indicates a role is being played – time spent is short	May play several roles	May play one role for whole play scene – understands the roles of others – able to play roles in a group

Source: Stagnitti, 2017.

Table 29.3 Embodiment, projection and role play paradigm

Ages	*0–3 months*	*3–9 months*	*12–19 months*	*19–23 months*	*24–29 months*	*30–35 months*	*36–41 months*	*42–47 months*	*4 years*	*5 years*
Stage		Embodiment			Projective				Role	

Source: Jennings, 2005.

These three theories provide a developmental basis for children's play skills to be analysed in play therapy. This enables the play therapist to identify children's play ages and modify the therapist's approaches to meet the child's needs and any play deficits. While it is useful to have a theoretical model to provide a baseline, it should not be considered in isolation. It is imperative to gain information of children's play from a variety of sources such as caregivers and school staff, observing their play in their own settings, plus repeated observations in play therapy. This will provide a holistic view of children's play skills.

Play deprivation and play bias

Many children referred for play therapy may be play deprived for a number of reasons. It is essential that play therapists have an understanding of play deprivation and its impact. Play deprivation is defined by Hughes (2000) as 'a chronic lack of sensory interaction with the world; a form of sensory deprivation' (p. 21). Whereas his definition of play bias is 'the loading of play in one area of experience or another, having the effect of excluding the child from some parts of the total play experience' (Hughes, 2003, p. 68). Modern society has reduced the opportunities for play and free play with others (not directed by adults) has decreased (Gray, 2011; Brown, 2013). There has been an increase in play deprivation and a potential increase in play bias (for example sports directed play), and this is impacting on children's holistic wellbeing, including resiliency. Resilience is not being developed as children are not playing as frequently with others. In playing with others, they learn to fall in and out of friendships, find ways to repair relationships, resolve conflicts, understand situations from other's perspectives and bounce back from setbacks. Additionally, due to the limited free play opportunities, children's imagination and creativity may not fully develop and these are needed for the ability to problem solve in everyday life.

Play deprivation according to 'Play Wales' (Brown, 2013) is on a spectrum. One end being children experiencing play deprivation through extreme neglect and abuse; to the other end where children are not experiencing free play because society has become risk adverse (Gill, 2007). Unfortunately, the media bombardment about antisocial behaviour and the potential risk of abuse from a limited number of incidents, has skewed society's view into believing that these incidents are occurring on a higher frequency than in reality. This leads to a belief that there is an exacerbated risk to children if they play outside. While, the increase use of technology has reduced the opportunities for outside free play with others. However, a significant contributory factor for play deprivation is the greater emphasis on academic achievements. However, play is the foundation for all academic learning.

There is little rigorous statistical research as it would be deemed unethical to purposely deprive children of play and it is unclear at the present time what the long-term implications of play deprivation are going to be. However, already Gray (2011) believes there is a contributory connection between the decline in play and the increase in psychopathology in children and

Table 29.4 Linking play developmental models to the sequential neurodevelopment of the brain

	Six Social Stages of Play Parten, M. B.(1932) Howes, C., & Matheson, C., 1992	*EPR Play Paradigm Jennings, S. (2005)*	*Pretend Play Enjoyment Developmental Checklist Stagnitti, K. (2017)*						*Neurodevelopment Gooey Brains, 2015*
Ages	*Stages*	*Stages*	*Play scripts*	*Sequence of play actions*	*Object substitutions*	*Social interactions*	*Doll/teddy play*	*Role play*	*Stages*
0–3 months	Unoccupied play	Embodiment	none	Imaginative play absent or random – manipulates and explores the object	Manipulates and explores objects	Imitates adults Actions – plays peek a boo	Manipulates objects Able to place the doll upright	Watches people	Brain stem – regulates level of arousal and keeps the body functioning
3–9 months	Solitary play		none						
12–19 months	Onlooker play		Scripts related to body	One simple imaginative action – fleeting	Relates to objects functionally – for example a spoon is placed in a cup	Imitates a pretend play action, e.g. giving a drink	Imitates feeding	Copies an action of an adult	
19–23 months		Projective	Scripts – own life experiences in the home	Two or three similar play actions – are illogical and repetitive	Uses similar looking objects for needed object – for example paper for a blanket	Imitates an adult using object	Doll play begins – child sees doll as separate to themselves. Separate single actions	Carries out actions they have seen others perform	Midbrain – regulates motor skills (fine and gross). Allows integration of complex sensory input
24–29 months	Parallel play		Own life experiences inside and outside the home	Simple, sequential, logical – actions are intentional	Inanimate objects – for example a box as a table or a car – may use imaginary objects – pretend a piece of cake on a plate	Asks adult for objects needed in play, plays alongside other children	Sees as alive – more actions undertaken	Copies a characteristic of another person	Limbic area – emotional reactions, tolerance, empathy, affiliation. Increasingly understand social relationships
30–35 months			Own life experiences and less frequent life experiences	Spontaneous, logical, detailed & logical – 6–7 actions, no planned story line	Inanimate objects used for two or more functions	Imitates another child, plays alongside – interest in what other children are doing	Identifies with the doll	Indicate they are someone else – brief – changes role quickly	Neocortex – controls concrete and abstracts thoughts. As area grows, gain skills, such as creativity, complex use of language, morality and respect

(*Continued*)

Table 29.4 (Continued)

	Six Social Stages of Play Parten, M. B.(1932) Howes, C., & Matheson, C., 1992	*EPR Play Paradigm Jennings, S. (2005)*	*Pretend Play Enjoyment Developmental Checklist Stagnitti, K. (2017)*						*Neurodevelopment Gooey Brains, 2015*
Ages	*Stages*	*Stages*	*Play scripts*	*Sequence of play actions*	*Object substitutions*	*Social interactions*	*Doll/teddy play*	*Role play*	*Stages*
36–41 months	Associative play	Role	Own life experiences and fantasy scripts	Multiple actions in a logical sequence	Inanimate objects used for many functions	Plays in association with others i.e. play the same activity beside another child, but separately	Doll play active – may have emotions	Indicates a role is being played – time spent is short	
42–47 months			Own life experiences and fantasy scripts	A play strategy is present	Body parts used as objects – imaginary objects referred to in the play	plays in association with others i.e. play the same activity beside another child, but separately	doll play active – one or two emotions	indicates a role is being played – time spent is short	
4 years	Cooperative play, cooperative social pretend play – complex, social pretend play		Play scripts that expand over a few days, incorporating greater problems	Pre-planned story line with complex sequence and sub-plots	Objects with a distinct function used in substitution – for example a hat for a boat	Co-operation/ negotiation occurs	Doll has its own life – includes feelings and thoughts	May play several roles	

Source: Adapted from Table 2.1 (Tonkin, 2019, p. 27).

young people. Brown (2014) suggests from his 45-year study of thousands of play histories that children who experience moderate to severe play deprivation in their first ten years are more likely to have a greater incidence of emotional dysregulation, including depression, difficulties with social, emotional and behavioural regulation, impulsivity, interpersonal challenges, struggle to adapt to situations and appear to present as less resilient to the complexities of life.

If children perceive their environment as unsafe and/or dangerous, they will be unable to access their play system and therefore experience play deprivation. Play and play development can only occur when children are in a calm, alert state (Perry, Hogan, & Marlin, 2000). However, if children's sense of safety is being challenged and they do not feel connected, they will revert to survival strategies to ensure their safety. When using their survival strategies, children are functioning from their lower brain stem and their nervous system moves into a hyperaroused or hypoaroused state. At this point, their energies and behaviours are focused on survival and not play. As Kestly (2014) states 'these behaviours dampen our cognitive functioning, narrowing the amount of new information we take in so that we can focus exclusively on the source of threat' (p. 15). Therefore, if children experience repeated instances where they need to be in survival mode, their play skills are likely to be deprived.

As mentioned earlier, children need an attuned adult to initiate and facilitate playful interactions. However, when the caregiver is underskilled, neglectful, abusive or physically and emotionally unavailable, children are left in a toxic stress state and are not able to access a play state. As suggested by Kielhofner et al. (1983) 'play is extremely sensitive to environmental conditions' (p. 305), therefore living repeatedly in a toxic stress environment will adversely affect children's play development. In Cooper's (2001) study of maltreated preschool children's play, he undertook a comparison of children with a maltreatment history and non-maltreated children. His findings suggest that the maltreated children showed less play enjoyment, limited play imagination, reduced play concentration, and the play behaviours were more cognitively disorganised. It also indicated that aggressive play was more prevalent in maltreated boys, which would corroborate as often they are in a fight survival state (2001).

Play therapy

The play, the therapeutic environment and the therapeutic relationship that occurs in play therapy provides the ideal space for play deprivation to be repaired and the development of play skills. Sometimes, play therapy assumes that children who are referred can play. However, children may be referred to play therapy for many reasons and these reasons may have impacted on their play development. These include the impact of trauma, early holistic neglect, attachment disruptions, interpersonal developmental trauma (Van Der Kolk, 2005) inconsistent serve and return interactions (Center on the Developing Child at Harvard University, 2017). For other children, they may have experienced medical trauma and were just not well enough to learn play.

Some children have limited play skills. Their social, emotional and play skills are younger than their chronological age and typical stage of development. This has been the author's experiences of working with children in a specialist education provision for children experiencing severe social, emotional, mental health and behaviour challenges. Many of the children who were referred to play therapy in this setting had experienced sustained chronic interpersonal abuse, trauma and neglect in their early years and in some cases it was ongoing. Their play was significantly underdeveloped. For some they had never the opportunity to learn to play and were 'stuck' at a younger development stage. From assessments undertaken, many had not developed to the developmental stage of the play scripts (Stagnitti, 2017) an aspect of projective play (Jennings, 2005), let alone role play (Stagnitti, 2017; Jennings, 2005) and complex social pretend play (Howes & Matheson, 1992). This would provide an understanding of their limited social and emotional skills as the skill of sociodramatic play is a clear indicator of children's ability to function socially (Smilansky, 1968). This linked with Brown's (2014) findings about play deprivation, a large proportion of the children were unable to regulate their emotions and behavioural responses and were unable to sustain peer relationships or function productively within the group settings required for educational environments. However, this is not only significant in specialist educational provisions, but is common in mainstream schools. Unfortunately, with opportunities for play under pressure in the education system, more and more children are becoming play deprived and are not learning the skills needed to manage the everyday requirements of life. Play therapy provides a perfect environment for children's play to evolve and mature. The play therapy orientation discussed in this chapter incorporates Norcross' (2005) assimilation integration with Axline's (1947) non-directive values, whilst being grounded in a humanistic approach (Murray, 2019). It is further informed by holistic child development, including neurobiology, the impact of trauma and attachment theory. The predictability, repetition and constancy of the play therapy and the play therapist creates an environment of not only physiological safety, but also psychological safety. The therapeutic relationship is a criterion to facilitate change (Murray, 2019). The therapeutic relationship replicates the compassionate attuned caregiver, who facilitates playful interactions with their child. Within the safety of the therapeutic environment and the therapeutic relationship, children can experience and re-experience early play developmental stages. By meeting the child where they are emotionally and socially in a permissive 'free' play environment, enables them to develop at their pace. It is a time where they can be free of adult direction, similar to play of previous generations (Gray, 2011). Children who are play deprived through abuse and neglect will need to learn to trust the therapy environment and the play therapist, before they can even begin to learn to play. Then they will need to catch up on earlier developmental stages before being able to play at a typical age appropriate level. A consideration that needs to be given with children who have experienced play deprivation and underdeveloped play skills is that they will need time. To repair and rebuild takes time and is highly unlikely to be resolved in a short period.

Incorporating Jennings' (2005) paradigm to the play therapy resources enables children to explore at their play level. For the embodiment stage resources consist of cushions, blankets, sand trays, clay, cornflour, slime and other sensory toys. Whilst for projective play, miniatures, environments, figures and puppets are essential to meet this developmental stage. Finally, at the role play stage, equipment such as dressing up clothes, telephones, magic wands, handcuffs, play money, play crockery and shopping tills will facilitate this development.

Whilst it is necessary to provide a play environment, with children who are play deprived, play therapists will need to draw upon their own playfulness and ability to play, even more so than with children who have developed play skills. The play can be repetitive to the play therapist and sometimes boring, but the pure nature of the repetition is what the child needs. It is important for play therapists to cultivate their playfulness (McInnes, 2019) as children will know when the play therapist's playfulness is not congruent and is 'constructed playfulness' (Chown, 2015).

By using their own play ability, play therapists can model different play skills at points where the child is stuck. For example, one child and the therapist were undertaking parallel play with painting (under the child's direction) for several weeks, but the child's painting was restricted and joyless. She repeatedly filled the paper with painted boxes, there was little spontaneity or imagination. The therapist had been directed to paint whatever she wanted, so she decided to be playful by splashing paint on her paper. The child looked up in shock and amazement. After a few moments of watching, she tentatively replicated the same play. This was a turning point in her development as she noticeably became freer in her play actions, to the point where she was standing up and using large gross motor movements to splash her paint. The laughter and joy emanating from her was an honour to watch.

Other ways of modelling imagination, projection and playful interactions are by providing playful voices to characters. For example, with a board game, the play therapist talked as the character in a playful, squeaky voice reflecting how the character felt every time he got sent back to his base (which was every throw of the dice).

Children who have been play deprived and their play skills developmentally delayed can progress through the developmental stages in play therapy, but it takes time, persistence and ideally a multisystemic approach. Part of the approach is understanding that children live within a context of a family, a social network and a school community. Therefore, it is helpful to work alongside the caregivers and in the author's opinion, the child's school. By understanding play development, play therapists can educate and support the adults around the child to incorporate play and playful interactions into the child's everyday world. This will expediate the child's developmental progress and more importantly enhance their relationships. This can be done through by providing psychoeducational information about the necessity and benefits of play, training caregivers in filial therapy (VanFleet, 2013), Child Parent Relationship Therapy (Landreth & Bratton, 2015), Theraplay® techniques (Booth & Jernberg, 2010) and Play Development Programme for Schools (Murray, 2017).

Case illustration

The following brief case study illustrate a child's developmental progress in play therapy. Neil, aged seven, was initially referred to play therapy as his behaviour regulation was presenting significant challenges both at home and in school. (Factors have been changed for anonymity purposes.) He was at high risk of exclusion. Everyone around him were at a loss about how to help him. Following an extensive assessment involving all parties, it was concluded that a multisystemic approach was needed to support this child. Neil in his first year of life had been exposed to abuse, neglect, drug and alcohol misuse and domestic violence. The environment and care were deemed detrimental for him and he was looked after by the local authority until he was adopted shortly after his second birthday. Due to his life experiences, he was significantly delayed in physical development, speech, social interactions and his play. Furthermore, he was constantly in a state of hyperarousal. The reports from his first year would suggest that he was in such toxic distress that he was in a hyperfreeze state (Rothschild, 2017) and therefore unable to access any learning. He was unable to use a fight or flight response due to his age, so his only survival option was to freeze.

The approach incorporated long-term individual play therapy, therapeutic parenting support incorporating child–parent relationship therapy, and attachment informed behaviour support for school staff including a play development programme. When Neil first came to the playroom, he crept up the stairs behind his mother's legs and was highly anxious about attending. The play therapist had to have a gentle, calm, confident presence about her to enable him to risk entering the playroom. For many months, he needed to check where his mother was, and his play was just manipulating the toy soldiers he found the first week in the sand (embodiment play). He would repeatedly pour sand over the soldiers and bury them. He would want the therapist to be right beside him, but was unable to engage in the relationship. His play was at the embodiment solitary stage and with no play script or object substitution or sequence of events. His play was representative of a child under the age of one year. He was hypervigilant to every sound or movement. The play therapist worked from a non-directive stance in Quadrant III of the Play Therapy Dimensions Model, where the child takes the lead and the therapist is a non-intrusive follower (Yasenik & Gardner, 2012).

Gradually, the play developed to hiding the soldiers in the sand and needing to find them. Initially he could not tolerate the wait for the therapist to find them so he would show her. This play continued for approximately 12 sessions and only when he was ready, was he able to move forward. Alongside this stage, in play therapy, the interventions were being implemented with both family and school therefore things were becoming more settled for him.

His play was still solitary, and he controlled every aspect of the play, but the play was chaotic with limited sequential order and was hard to track. At times, it was difficult for the therapist to stay connected to him. His sessions, the play and the disconnection would evoke significant tiredness in the therapist. This was evident in other relationships too. Over time, his soldier play began to incorporate a play script, that involved other soldiers fighting the other team. He would give the therapist a team, but directed every aspect of the play. This was understandable due to having felt so out of control in his first few years of life he had learnt to be controlling to feel safe.

The play developed to more complex projective play, where he was able to create play scripts with a play strategy for their associative play (Stagnitti, 2017). The play scripts were factually based and the different teams fighting and tricking each other. Then cooperative projective play emerged, where Neil and the therapist's teams worked together. The therapist was given the occasional opportunity to co-create. Initially, the play was concrete with limited symbolic representation or object substitution, but increasingly dragons appeared in the play, the chair became a hospital and shiny stones became portals to move between bases. Sometimes he would arrive for his session with a pre-planned story.

It took three years of consistent, persistent support, but Neil's play developed to a significantly more age appropriate level. He was able to undertake complex social pretend play with the therapist and they co-created role-plays. The role-plays incorporated police officers and robbers (beyond personal experience, Stagnitti, 2017), who would become invisible, move stealthy around the room not being heard (physical regulation) and escape from places with pretend keys. Often, he would want the therapist to lead the play and he would improvise from the therapist's play script. The play would start with a plan, but evolved cooperatively with subplots and often symbolically involved issues related to interactions with his peers. Within the play, he was able to learn to regulate his emotional responses. His play became flexible, fun and connected. He had learnt through the play to accommodate others, negotiated and compromise. He had found his self. He now presented as more confident and the behaviour challenges at home and school had significantly reduced.

The play therapist had been able to track his development using the models described earlier. She was able to understand the play not only from the thematic representation, but also from a developmental perspective. Alongside this, he used the safety of the therapeutic powers of play (Schaefer & Drewes, 2014) and the therapeutic relationship to process his early attachment disruptions and the emotions evoked by trauma. The therapist was able to see the consistent progression, even though the process was long. However, it is the author's belief that this process was necessary and would have taken significant longer without the complete commitment of the adults around him to undertake and incorporate the identified interventions.

Play is an essential skill for humans to enable them to negotiate the complexities of life. However, it is a skill that needs to be learnt and can only be achieved when children are feeling safe. By drawing on the play development theories and understanding the impact of play deprivation, this chapter has shown, using a brief case illustration, how play therapy can provide the ideal environment for some children to address their play developmental deficits as well as the impact of interpersonal developmental trauma. The understanding of play development and play deprivation are essential prerequisites for play therapists, but I would suggest a knowledge and understanding of this is also critical to parents, educators, child-care practitioners and policy writers.

References

Axline, V., 1947. *Play Therapy*. New York: Ballantine Books.

Booth, P. B., & Jernberg, A. M., 2010. *Theraplay: Helping Parents and Children Build Bettter Relationships Through Attachment Based Play*, 3rd ed. San Francisco: Jossey Bass.

Brown, F., 2013. *Play Deprivation; Impact, Consequences and the Potential of Playwork* [Online]. Available at: www.playwales.org.uk/login/uploaded/documen . . .

Brown, S., 2008. *Play Is More That Just Fun* [Online]. Available at: www.ted.com/talks/stuart_brown_says_play_is_more_than_fun_it_s_vital [Accessed 27 May 2019].

Brown, S. L., 2014. Consequnces of Play Deprivation. *Scholarpedia*, 9(5).

Center on the Developing Child at Harvard University, 2017. *Serve and Return* [Online]. Available at: https://developingchild.harvard.edu/science/key-concepts/serve-and-return/ [Accessed 1 June 2019].

Chown, A., 2015. *Play Therapy in the Outdoors; Taking Play Therapy Out of the Playroom and into Natural Environments*. London: Jessica Kingsley Publishers.

Cooper, R. J., 2001. *The Impact of Intra-Familial Child Abuse and Neglect on Preschool Children's Play*. Brisbane: Univsersity of Queensland.

Erikson, E., 1963. *Childhood and Society*. New York: W. W. Norton.

Gauntlett, D., Ackermann, E., Whitebread, D., Wolbers, T., & Weckstrom, C., 2010. *The Furture of Play: Defining the Role and Value of Play in the 21st Century*. Billund: LEGO Learning Institute.

Gill, T., 2007. *No Fear: Growing Up in a Risk Adverse Society*. London: Gulbenkian Foundation.

Gooey Brains (2015) Using Play to Build the Brain, retiried from https://gooeybrains.com/2016/04/10/using-play-to-build-the-brain (accessed 22.8.20).

Gordon, G. A., & Esbjörn-Hargens, S., 2007. Are We Having Fun Yet? An Exploration of the Transformative Power of Play. *Journal of Humanistic Psychology*, 47, pp. 198–222.

Gray, P., 2011. The Decline in Play and the Rise of Psychopathology in Children and Adolescents. *American Journal of Play*, 3(4), pp. 443–463.

Howard, J., 2017. *Mary Sheidan's Play in Early Childhood,* 4th ed. Abingdon: Routledge.

Howes, C., & Matheson, C., 1992. Sequences in the Development of Competent Play with Peers: Social and Social Pretend Play. *Developmental Psychology*, 28(5), pp. 961–974.

Hughes, B., 2000. *A Dark and Evil Cul-de-Sac: Has Children's Play in Urban Belfast Been Adulterated by the Troubles?* s.l.: Anglia Polytechnic University.

Hughes, B., 2003. Play Deprivation, Play Bias and Playwork Practice. In: F. Brown, ed. *Playwork – Theory and Practice*. Buckingham: Open University Press, pp. 66–80.

Jenkinson, S., 2001. *The Genius of Play – Celebrating the Spirit of Childhood*. Stoud: Hawthorn Press.

Jennings, S., 2005. *Creative Play with Children at Risk*. Milton Keynes: Speechmark.

Kestly, T. A., 2014. *The Interpersonal Neurobiology of Play: Brain-Building Interventions for Emotional Well-Being*. New York: W. W. Norton.

Kielhofner, G., Barris, R., Bauer, D., Shoestock, B., & Walker, L., 1983. A Comparison of Play Behavior in Non·Hopitalized and Hospitalized Children. *American Journal of Occupational Therapy*, 37(5), pp. 305–312.

Landreth, G., & Bratton, S., 2015. *Child Parent Relationship Therapy (CPRT): A 10-Session Filial Therapy Model*. New York: Routledge.

McInnes, K., 2019. Being a Playful Therapist. In: P. Ayling, H. Armstrong & L. Gordon Clark, eds. *Becoming and Being a Play Therapist*. Abingdon: Routledge, pp. 98–109.

McLean, D., & Hurd, A., 2014. *Kraus' Recreation and Leisure in Modern Society*. Burlington, MA: Jones & Bartlett.

Murray, S., 2017. *Play Development Programme for Schools*. s.l.: Unpublished.

Murray, S., 2019. Symbolic Solutions: Establishin a Sense of Safety for Relationship Repair. In: L. & G. K. Yasenik, eds. *Turning Points in Play Therapy and the Emergence of Self – Applications of the Play Therapy Dimensions Model*. London: Jessica Kingsley Publishers, pp. 237–254.

Norcross, J. C., 2005. A Primer on Psychotherapy Integration. In: J. C. &. G. M. R. Norcross, eds. *Handbook of Psychotherapy Integration*. New York: Basic Books, pp. 3–45.

Panksepp, J., 1998. *Affective Neuroscience: The Foundations of Human and Animal Emotions*. New York: Oxford University Press.

Panksepp, J., 2009. Brain Emotional Systems and Qualities of Mental Life: From Animal Models of Affect to Implications for Psychotherapeutics. In: D. Fosha, D. Siegel, & M. Soloman, eds. *The Healing Power of Emotion: Affective Neuroscience, Development, and Clinical Practice*. New York: W. W. Norton, pp. 1–26.

Panksepp, J., & Biven, L., 2012. *The Archaeology of Mind: Neuroevoluntionary Origins of Human Emotions*. New York: W. W. Norton.

Parten, M. B., 1932. Social Participation Among Preschool Children. *Journal of Abnormal and Social Psychology*, 27(3), pp. 243–269.

Perry, B., 2006. Applying Principles of Neurodevelopment to Clinical Work with Maltreated and Traumatized Children: The Neurosequential Model of Therapeutics. In: N. Webb, ed. *Working with Traumatized Youth in Child Welfare*. New York: Guildford Press, pp. 27–52.

Perry, B., Hogan, L., & Marlin, S., 2000. Curiosity, Pleasure and Play: A Neurodevelopmental Perspective. *Haaeyc Advocate*, August, pp. 9–12.

Piaget, J., 1962. *Play, Dreams and Imitation in Childhood*. New York: W. W. Norton.

Porges, S., 2011. *The Polyvagal Theory*. New York: W. W. Norton.

Rothschild, B., 2017. *The Body Remembers, Volume 2; Revolutionizing Trauma Treatment,* 2nd ed. New York: W. W. Norton.

Schaefer, C. E., & Drewes, A. A., 2014. *The Therapeutic Powers of Play: 20 Core Agents of Change,* 2nd ed. Hoboken, NJ: John Wiley & Sons.

Schaefer, C. E., & Peabody, M. A., 2016. Glossary of Play Therapy Terms. *Playtherapy*, June, pp. 20–24.

Sheridan, M., 1977. *Spontaneous Play in Early Childhood: From Birth to Six Years,* 1st ed. Windsor: NFER.

Smilansky, S., 1968. *The Effects of Sociodramatic Play on Disadvantaged Preschool Children*. New York: John Wiley & Sons.

Stagnitti, K., 2017. *The Pretend Play Enjoyment Developmental Checklist*. Melbourne: Learn to Play.

Tonkin, A., 2019. Playing for a Healthy Brain. In: A. Tonkin & W. J. Tonkin, eds. *Play and Playfulness for Public Health and Wellbeing*. Abingdon: Routledge, pp. 20–33.

Tonkin, A., & Whitaker, J., 2019. *Play and Playfulness for Public Health and Wellbeing*. Abingdon: Routledge.

Van Der Kolk, B., 2005. Developmental Trauma Disorder. *Psychiatric Annals Thorofare*, 35(5), May, pp. 401–408.

VanFleet, R., 2013. *Filial Therapy: Strengthening Parent-Child Relationships Through Play*. Sarasota: Professional Resource Press.

Yasenik, L., & Gardner, K., 2012. *Play Therapy Dimensions Model: A Decision Making Guide for Intergrative Play Therapists*. London: Jessica Kingsley Publishers.

30

THERAPEUTIC USE OF SELF AND THE PLAY THERAPY DIMENSIONS MODEL

Lorri Yasenik and Ken Gardner

> Five-year-old Robin walked tentatively behind me towards the playroom door. I am in jeans and stocking feet. There is one step just before entering the room and I hop off the step and skip into the room. I look behind as Robin hops off the same step and follows me into the playroom. I know Robin knows I am no ordinary adult. I feel Robin's uncertainty and I feel it dissipating. My voice tone indicates a moderate level of excitement as I introduce the special play space. My body does not move in an abrupt way. I move to one side as Robin claims his space in the room. As I quickly drop down to a kneeling position, I am now at his height, and together we visually take in the play materials available. Robin sees a play sword he likes and picks it up and I follow his lead. He says: 'I am a knight and you are a bad guy'. I provide space between us, accept the sword he provides, and promptly change my voice tone and timbre to accommodate his instruction.

Many decisions are made when considering the use of self in the play and expressive therapy settings. But first, let's visit the idea of 'self'. Erwin (1997) notes that many authors of psychotherapy have referenced to the self as a person, the ego, set of self-representations, an inner agent etc. Alternatively, constructionists view the self as constructed through interaction with others (Andrews, 1991; Cashdan, 1988). Cashdan (1988) states that the child's self is incrementally evolving through interacting and engaging with others. Interaction with others begins the formulation of how we begin to develop a self-concept. Things are communicated to us and about us through early relational experiences and initially we are told who we are. This sense of self is, however, viewed as fluid and shifting in response to the environment (Andrews, 1991; Cashdan, 1988; Gergen, 1996, 2009).

Wosket (1999) differentiated the person of the therapist and the therapist use of self. The person of the therapist is revealed inadvertently through such things as "dress, accent, age, voice intonation, skin color, involuntary changes in movement or facial expression, mannerisms, the furnishings and state of orderliness of the counselling room and so on. Yet inadvertent self-disclosure is not the same as intentional use of self" (p. 11). Intentional use of self in the therapy requires self-reflection and self-awareness. It is not enough to have academic knowledge to be helpful and effective, rather you must be aware of who you are as a person. It is the integration of the personal, professional and technical self. Edwards and Bess (1998) identified three areas of consideration: (1) inventory of self; (2) development of self-knowledge; and (3) acceptance of risks to self. Wosket (1999) emphasizes the importance of acknowledging ones' own unique helping attributes over therapy models as well as the value of therapist failure and breaking the rules of counselling.

How and what do we reveal of ourselves to our child clients? Let us return to Robin. What occurred from the time of entering the play therapy space to the fantasy sword play scenario? Purposeful use of self begins before entering the therapy space. My attire is informal and relaxed sending an initial message to Robin that this person is ready to play. Sensing Robin's apprehension, I indicate playfulness by entering the space in a fun manner. I think about my physical self, my emotionality (pitch, excitement, intensity and tone of my voice) as I introduce the room. I am aware of our proximity to one another and I continue to shift my body position to provide a safe distance and my gaze follows Robin's; all the while remaining in proximity to Robin. What do I notice in my body? What is my self-system telling me? Is this a comfortable and safe feeling for me and for Robin? I remain present in the space and fluidly enter the fantasy play and shift my voice tone and timbre to become another character. I feel relaxed in my body and I take Robin's direction. Robin appears engaged, in the lead and pleased.

What makes play and expressive therapies 'work' and look like 'magic'? Certainly not the prescribed activity, the words used in the therapeutic reflections, the therapist's model or theory or the quality of the playroom or therapy space. It is the purposeful use of the qualities of the play therapist. It is timing and understanding, it is being present in a variety of ways that are mostly purposefully woven throughout the session with the child. Even if the therapist experiences a therapeutic 'miss,' risks are better taken than consistently avoided. The dance between the therapist and client can always be reviewed in order to explore what happened to either keep with the rhythm or take the dance out of step.

There are few places to receive training in the area of therapist use of self, and there are even fewer systematic guides to use of self (Wosket, 1999). This chapter attempts to raise the play and expressive therapist's awareness to the various ways one makes use of the self in a play or expressive therapy session. The thing that differentiates play and expressive therapists from those primarily using talk therapies is that play and expressive therapists use their whole body when engaging in therapy. Additionally, entering into fantasy play or a drama requires the therapist to be both in character *and* maintain an overview and direction of the overall process. This extension of use of self is complex and requires multiple uses of self in order to remain helpful and not get lost in the process or interactions. During the 1990s models of multiple 'selves' as related to self-development were discussed in the literature. The focus was on how the presentation of 'self' varied across situations, relationships and through multiple roles people adopt (Gergen, 1991; Kihlstrom, 1993; Markus & Cross, 1990; Rosenberg, 1988). Considering the multiple 'self' theories, the play and expressive therapist not only operates differently within relationships, across situations through different roles, but they also add dramatic roles as *assigned* or requested by the client. It is a challenge to remain actively aware of all of these moving parts.

Nalavany et al. (2004) added to what others have found to be important factors in their study of therapist qualities, competencies and skills for play therapists in particular. The research identified seven clusters of qualities: being attuned to and reflecting the child's verbal and nonverbal behavior and feelings; being sensitive to the child; being warm, empathic, genuine and accepting of the child; being open to personal awareness and growth; having the skills for working with parents and families; having a theoretical understanding of the process of child therapy; and having a structured, intentional approach to the therapeutic process. Certainly all of these factors will be helpful when evaluating the use of self, but again, these qualities alone do not reflect *how* the play therapist will actually *be* in relation to their client in the therapy setting.

As a way to assist all play therapists in tracking and evaluating therapeutic use of self, the *Degree of Immersion: Therapist Use of Self Scale* (Appendix A) (Yasenik & Gardner, 2004, 2012) was developed. The scale is useful for purposeful reflection after play therapy sessions as practitioners can decide if the ways in which they made use of themselves was facilitative or interruptive to the process. By taking a closer look at the potential domains of therapist use of self, the play therapist can raise their awareness of what to pay attention to. The literature has reference to play therapy skills (Kottman, 2011; Ray, 2004; O'Connor, 2000) and techniques (Goodyear-Brown, 2010; Kaduson & Schaefer, 2003; Malchiodi, 2008; Rubin 2008; Schaefer & Cangelosi, 2002; Lowenstein, 2006) but (with the exception of Ray's (2004) contribution for child-centered play therapists) there is little reference to *how* play therapists reflect on the various ways they could engage in these activities. To further explore *how* the play therapist makes use of the self in sessions, the following six areas outline what the therapist could begin to identify and evaluate in a given session and sessions over time: verbalizations, reflections, emotionality, physical, embodiment, and interpretations.

What is the Play Therapy Dimensions Model?

The Play Therapy Dimensions Model (Yasenik & Gardner, 2012) is a way to conceptualize the play therapy process according to two primary dimensions: directiveness and consciousness. These dimensions help define the therapeutic space from a variety of different philosophical and theoretical schools of play therapy. The model is a decision-making and treatment-planning tool. It guides the play and expressive therapist to identify the critical elements of the play therapy process. It assists therapists to think about their interventions as related to their client readiness to receive or make use of the intervention. A four-year-old, for instance does not often have the cognitive or emotional capacity to make use of activities that require higher order thinking and high levels of consciousness about a presenting issue.

When thinking about making use of the Play Therapy Dimensions Model there are three fundamental overriding assumptions: first, each child is unique regarding his/her skills and abilities; second, all children follow a common developmental pathway; and third, the play therapist has a central role in facilitating change and optimizing growth. Specifically, the ultimate goal of the model is to aid play therapists to answer the *who*, *what*, *when*, *why* and *how* of the play therapy process.

The *consciousness* dimension reflects the child's representation of consciousness in play, while the *directiveness* dimension relates to the degree of direction provided by the play therapist. Play therapists working in more direct ways with children tend to be more immersed in the play, use more interpretative comments and often provide structured activities for the child. The **consciousness dimension** (Figure 30.1) is represented by the child's play activities and verbalizations. For many children there is a need for emotional distance from the issues they are attempting to reorganize. There is often a weaving process, representing movement up and down this dimension, moving from greater levels of consciousness to lesser levels or vice versa. The child's play could be very direct and literal, accompanied by verbalizations, indicating that the child is working with a certain level of conscious awareness. At other times, the child needs distance and protection from troublesome thoughts or feelings, and utilizes play scenarios and objects in a less conscious and more symbolic manner.

Wilson, Kendrick and Ryan (1992) refer to the dimension of consciousness in relation to symbolic play and its role in play therapy. They refer to Piaget's (1977) developmental theories of adaptation, assimilation and accommodation in relation to symbolic play. When children experience disruption through an outside experience, their mental schemas may be distorted and conflicted in relation to the way they see themselves and/or others. Through symbolic play, dissociated thoughts and feelings can be made conscious. Symbols used in play can assist in the organization of cognitive schemas, and the child may begin to assimilate new possibilities into a past representation, which in turn, helps the child grow and change. The capacity for conscious 'awareness' for children must be viewed from a developmental perspective. The younger the child, the less likely it is that the child will represent awareness in a direct manner due to the fact that language and cognitive schemas are still developing. Decisions about facilitating greater degrees of conscious representation of dissociated thoughts and feelings in the play are critical.

The second dimension, **directiveness dimension** refers to the degree of immersion and level of interpretation of the play therapist. Immersion refers to the therapist use of self during a given session such as use of verbalizations, physical and emotional use of self, the level of structure of the interventions and so on (see Appendix A, *Degree of Immersion: Therapist Use of Self Scale*). At the lowest level of directiveness, the therapist is tracking the play through observation and reflection and is not actually involved in interactive play with the child. At the high end, the therapist has entered the play as a co-facilitator, and is actively taking part in elaborating and extending the play. Very directive play therapists may stay with the symbolic representation of the play, but begin to name possible feelings, behaviors and future actions during the play session. Others may structure play activities for the child as related to the interrupting events. Some children enter therapy and alert the therapist to the fact that they need to openly explore and directly talk about their worries or concerns and therefore the play therapist becomes very directive (and conscious) and intermittently leads the verbal discussion.

As represented in Figure 30.1, in the play therapy dimensions diagram (Yasenik & Gardner, 2012) the directiveness and consciousness dimensions intersect, forming four quadrants. Depending on the case conceptualization, and the theoretical approach of the therapist, a therapist might choose to focus therapy activities primarily in one quadrant. Alternatively, there may be a number of indicators that suggest movement is required amongst the quadrants. Furthermore, movement may occur within a session, or across sessions, as the therapy process evolves. This conceptualization assists therapists in navigating the complex client–therapist–treatment interactions in order to tailor treatment approaches and optimize effectiveness. This integrative approach also offers a process-oriented framework, providing guidance for tracking important change mechanisms.

The four quadrants

The following briefly describes the four quadrants of the Play Therapy Dimensions Model displayed in Figure 30.1. The quadrants provide an organizing structure for play therapists. No matter what theoretical orientation therapists primarily work within, they will be able to identify the levels of their directiveness, immersion, and the degree to which they facilitate the child's conscious awareness of emerging play themes and activities. At first glance one may interpret the four quadrants as a means to identify different ways to practise as a play therapist. Although this may be possible, the four quadrants also provide a window into the use of many play therapy models and interventions during a session or series of sessions. The same therapist may work in all four quadrants depending on the style of the child, the presenting need, capacity for play, and the child's developmental stage. The four quadrants signify a model of possible movement – both in theoretical choice and in therapist use of self within a play therapy theory. The therapy process could begin in any quadrant depending on the decision-making variables. There is no prescribed order.

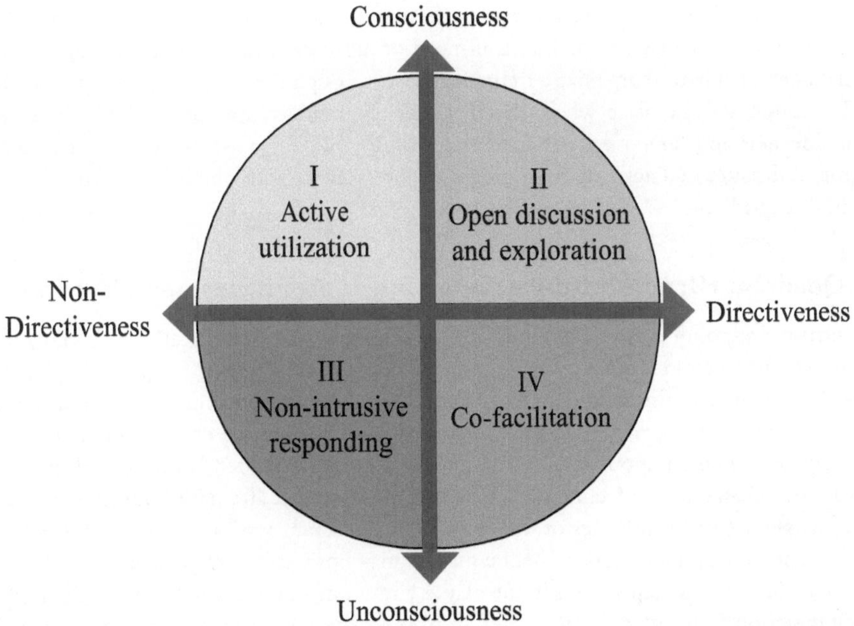

Figure 30.1 Play therapy dimensions diagram. Reproduced with the permission of Jessica Kingsley Publisher through PLSclear. From Yasenik & Gardner (2012, p.34), Play Therapy Dimensions Model: A Decision-Making Guide for Integrative Play Therapists.

Quadrant I: active utilization (conscious/non-directive)

In the upper left corner of Figure 30.1, Quadrant I, Active Utilization is identified. In this quadrant, the child initiates the play using his/her own metaphors, symbols and/or concrete verbalizations. This quadrant is placed in a more non-directive, yet conscious position on the diagram. It is different to working in Quadrant III (which is also on the left side of the diagram) due to the intermittent interpretive comments initiated by the therapist that trigger conscious responses from the child. The therapist at various points enters the play with the child and expands the play into the realm of conscious awareness. Active utilization is typically entered into in a brief, time-sensitive way. It is in this quadrant that a model of interpretation is valuable to have. Those therapists who value conscious awareness believe in the importance of interpretive comments, and think they have a purposeful role in helping children to reorganize dissociated affect, behaviors and thoughts will find themselves periodically working in this quadrant.

Quadrant II: open discussion and exploration (conscious/directive)

Quadrant II, located in the upper right corner of Figure 30.1, is Open Discussion and Exploration. A therapist working in this quadrant would be observed as initiating and structuring a play activity relative to a child's presenting problem. A child may have been referred for a particular issue such as self-control, anxiety, depression, sexual abuse or aggression, and the therapist will introduce the child to concrete, highly conscious interventions. When working in this quadrant, the therapist is primarily utilizing a developmentally sensitive, cognitive play therapy approach, and will engage in conscious processing of the child's presenting issue. Structured play-based activities may include therapeutic board games, drawing exercises, externalizing activities, role-playing, Sandplay reenactment, feeling card games, etc. The therapist may choose to work in this quadrant when a child is in need of more structure, feeling language, or if the child has been unable to reorganize a traumatic event through the normal course of play. Quadrant II is considered the place where therapists are observed as introducing the highest degree of consciousness and directiveness to the child. Some play therapists were trained to provide direction and structured activities as a means of intervening with children and may primarily work in this quadrant.

Quadrant III: non-intrusive responding (unconscious/non-directive)

Non-intrusive responding is situated in Quadrant III, Figure 30.1. A therapist is working in this quadrant when the child is observed initiating and completely directing the play. If a play therapist works primarily in this quadrant, the play is viewed as having intrinsic value and is process oriented. The therapist generally facilitates the play by following the child's lead. The therapist may engage in tracking responses, such as "now the baby is getting fed," or make reflective statements about a character or a feeling state. The therapist does not attempt to interpret, talk about what is happening or bring issues or themes up in a conscious way with the child. Depending on the orientation of the therapist, she/he may or may not join the play when requested to do so by the child. If the therapist joins in the play as a result of an invitation from the child, she/he takes all direction from the child as to how to play her/his character, including what the character should say. Much emphasis is placed on the inner abilities of the child to process traumatic or stressful matters through the play without the structured intervention of the therapist. The therapist remains fully present and observant and tracks the sequences and themes in the child's play for reflection and analysis outside of the session.

It would appear that roughly equal numbers of therapists have been primarily trained in directive play therapy approaches as those who have been trained in non-directive play therapy approaches. The Play Therapy Dimensions Model views both directive and non-directive therapy approaches as important and useful and, depending on the child and therapist, sometimes utilized by the same therapist in a given session.

Quadrant IV: co-facilitation (unconscious/directive)

Quadrant IV, Co-facilitation, can be viewed in the lower right corner of Figure 30.1. Initially the child is observed as directing the play. Therapist activity in this quadrant differentiates this quadrant from all of the others. It is here that the therapist has entered the play at the invitation of the child. The therapist has observed and tracked a number of themes and patterns and makes the decision to test a hypothesis or elaborate the play by inserting comments, actions and soft interpretations in the context of the play. Staying in the play and elaborating the play to aid the child to interrupt looping (circular, incomplete segments of play) or compulsive repetition is what differentiates this quadrant from the others. Therapist activities are more directive than Quadrant III, but the therapist never attempts to directly discuss or interpret the play with the child in a conscious manner. Therapists may test their hypotheses about what they believe a child may be trying to portray or potentially disclose through his/her play characters or actions. The therapist may introduce new characters (not directed for use by the child) as helper figures, helpless figures, or he/she may use more than one character to demonstrate a child's inner conflict. The goal of working in this quadrant is to become a co-facilitator of the play in order to open new avenues for the child to express, process and internally differentiate emotions and experiences.

Applications of the play therapy dimensions model to supervision: therapist's awareness of use of self

When applied to the context of supervision, the Play Therapy Dimensions Model supports the supervision of those trained in one or more approaches to play and expressive therapy. Importantly, the model emphasizes the purposeful use of self and provides tools for reflection on key areas in use of self – areas the practitioner may choose to increase, decrease or, perhaps, access for the first time.

Although numerous models of supervision exist, there are few that specifically apply to play/expressive therapists. As noted by Hewson (2001), supervision models generally provide a process to deliver theoretical knowledge, reliable frameworks, insecurity management, techniques and intervention strategies. These components are all necessary and fundamental to supervision. However, using a decision-making framework, in addition to a core supervision model, helps expand supervisee conceptualization skills.

The model encourages the therapist to look at both client and therapist moderating factors. In terms of therapist moderating factors, there is a range of questions to explore in supervision. At the outset questions might include, "How confident are you working with the presenting concerns? What is your experience level? What is your current connection to the client?" As the therapy process unfolds, deeper self-reflection is required. Questions might include, "What did you notice or understand about your use of self and . . . the client's response?" "What parts of your self are you aware of holding back or containing?" (Fonagy & Target, 1997). Questions such as these guide practice and therapeutic activities.

The *Degree of Immersion: Therapist Use of Self Scale* (Appendix A) (Yasenik & Gardner, 2012, 2019) was developed to identify the supervisee's process in the understanding and responding to their client. Because this is a difficult construct to pinpoint during discussion with supervisees, specific therapist behavior and interaction is described as 'immersion.' So instead of asking, "To what degree did you immerse yourself in the play?" we initially divided how one *could* become immersed in a play therapy session into three categories: verbal, emotional and physical use of self. This tool was elaborated in 2019 when a fourth category – self-system – was added (Yasenik & Gardner, 2019). The latter category, which emphasizes embodiment, helps supervisees examine their body/energy and internal mental states. When body-mind awareness is online we begin to monitor our internal experiences in relation to the child, and potentially make better use of our self in service of the child's needs. Each category and sub-category is exemplified in the supervision case study that follows.

The level a therapist engages in any of these activities indicates the degree of 'immersion.' Recognizing that it would be pointless to ask therapists to score themselves without providing a rating of the child's response to their action or non-action, therapists also rate the child's response to their level of immersion. An analogy would be adjusting the water temperature in a shower. We tend to make small adjustments while taking note of our feelings about the water temperature, gradually tweaking it to the 'just right' temperature. In discussing the Trauma-Play^{Tm} model, Goodyear-Brown (2019) stresses that within each component or stage of trauma work there is a need for 'nuanced titration' of the clinician's therapeutic use of self. Questions such as "how much?" and "in what dose?" are fundamental to the titration process. Accordingly, after each rating on the degree of immersion scale, therapists are asked to provide three clinical indications as to the effectiveness of their use of self in that category. This is part of a recursive/reflective process that helps therapists find the right 'dose.'

Once play therapists have scored themselves on all categories, a total score is derived. Higher scores are not better! The fundamental question is whether there was sufficient immersion in each category to support safety, self-expression, re-organization or movement for the client. This reflection can be illuminating to play therapists. Some therapists are surprised with this score because they did not view themselves as immersed at all; others believed they were *fully* immersed. After reflecting on each category and sub-category (including the child's corresponding response), the supervisor and supervisee can more thoughtfully discuss how to vary use of self to better meet the child's needs.

In our experience careful reflection of what happens in the play therapy space, interpersonally and intrapersonally, helps supervisees identify why the process is either moving forward, stuck in neutral or even regressing. Kottman and Meany-Walen (2018) note that it is important to get 'underneath' the presenting problem by exploring these two dynamics. Kottman and Meany-Walen (2018) further emphasize that one's theoretical orientation influences the way intrapersonal and interpersonal dynamics are conceptualized. The degree of immersion scale, when used repetitively, increases supervisee's awareness of these dynamics through exploration of the many levels of immersion. Recursive self-reflection fosters attunement to the client-therapist relationship. Self-reflection also fosters awareness of markers of transference and countertransference, and the potential impact on the play therapy process (Crouch, 1997; Yasenik & Gardner, 2019).

The *Degree of Immersion: Therapist Use of Scale* breaks down each of the four categories of immersion and links each sub-category to a quadrant(s). During supervision a critical question is: Which therapist or client factors contributed to the decision-making for remaining in, or moving between quadrants? The immersion scale keeps therapists on track, asking them to continually reflect on the interaction between their use of self, the client and movement in the therapy process. This is what informs the attuned play therapist (Yasenik & Gardner, 2012).

Brown (2009) notes play is a cornerstone of all personal relationships, and a 'state of play' occurs when there is spontaneous and mutual delight between a child and a parent. Just as a parent who demonstrates attuned, sensitive and responsive interactions with their child strengthens the relational bond, a play therapists who demonstrates interest and understanding of the child's emotional world, on a moment-to-moment basis, strengthens the therapeutic relationship. Through attunement the therapist and parent each tap into an evolutionary, wired-in aspect of our social brain and move forward with the lifelong tasks of building, maintaining and repairing relationships (Badenoch, 2008; Siegel, 2007).

By carefully reflecting on the 'state of play' the therapist strengthens the relationship. At the core of this 'state' is the therapist's ability to access their playful self. McInnes (2019) indicates that playfulness is the heart of the relationship, rather than the act of play. In the literature, there are two views of playfulness, each having a multitude of related articles and research. The first view is that playfulness is a personality trait that resides in the individual. For instance, Leiberman (1977) identified five core traits of playfulness, leading her to view playfulness alongside a cognitive style involving creativity, imagination and divergent thinking (McInnes, 2019). The second view, playfulness as an approach or attitude, links it to internal affective qualities. For example, playfulness might include the affective qualities of enthusiasm, motivation, willingness to engage, and freedom to choose or try out new ideas (Moyles, 1989). Chown (2015) asserts that where there is playfulness there is attunement, affection and affirmation.

Next, we examine a client-therapist relationship that became 'stuck' to the degree the therapist struggled to find and make use of his playful self.

Supervision case study: Brad and Sammy

Brad was a play therapist with three years of clinical experience. He was generally confident in his ability to make connections with children but in supervision he readily acknowledged feeling disconnected in his relationship with Sammy, an eight-year-old boy. In speaking with his supervisor, Katherine, Brad wondered if he was a 'good fit' for Sammy and whether Sammy was ready for play therapy. Hearing this, Katherine invited Brad's exploration of a range of interpersonal and intrapersonal dynamics to get 'underneath' the presenting problems of this case.

By history, Sammy was a young person who witnessed domestic violence over the course of his first four years of life. Sammy's mother struggled to care for him, particularly after his father was removed from the home. At age seven, Sammy was placed in care and then experienced two placement breakdowns. At the point of referral Sammy had just entered a new foster home. Brad noted he felt pressure to somehow make this placement work for Sammy.

Katherine first asked Brad to reflect on specific factors that characterized Sammy's presentation in the play space and his interactions with Sammy. Brad noted Sammy appeared to have low play skills for a child his age. Brad elaborated that Sammy seldom explored the playroom; upon entry to this space Sammy often ran to the swords and began swordplay without any verbal narration or context. Brad further commented that Sammy's play quickly 'ramped up' to the point where Brad needed to set a therapeutic limit. At other times, Sammy, suddenly pivoted from one play activity to another. For example, Sammy would abruptly move from swordplay to repetitively burying items (including his hands) in sand. These observations led Brad to emphasize that others had viewed Sammy's self-regulation skills as a concern. Brad further commented that Sammy exhibited a range of relational difficulties, inside the play space and in his school and home/community settings. For instance, Sammy frequently got into fights with peers; in the play space Sammy often tested limits. When Katherine asked about Sammy's worldview, Brad commented it appeared negative – a negative sense of self and a lack of trust in others. At the end of this discussion Brad stated, "Nothing thematic ever happens, which makes it impossible to for me to understand how Sammy views himself and his world."

Katherine asked Brad to take a step back and examine several underlying dynamics. First she noted that maltreated children who have experienced minimal levels of empathic connection with a caregiver are prone to feeling disconnected from others and have reduced capacity to connect empathetically to others (Malchiodi, 2010). Conversely, she noted the ability to behave empathically is critical to moral reasoning (Lawrence *et al.*, 2006). Katherine also pointed out that maltreated children, who have poorly organized sensory systems, are prone to sensory misperceptions, resulting in inadequate body awareness and integration of self, body and environment (Gaskill & Perry, 2012). During the Embodiment (E) stage of the E-P-R developmental paradigm the child's embodied experiences are essential for establishing a 'body-self' and are critical in forming security and trust (Jennings 1990, 2005, 2014). Accordingly, Katherine emphasized the importance of engaging Sammy in play interactions at the sensorimotor or *embodiment* stage, which involves messy play as well as nurturing, comforting and empathic play. During this discussion Katherine pointed out how Sammy was drawn to sensory and embodiment experiences in his play activities, whereas Brad seemed to expect more advanced play, characteristic of the *projection* or *role* stages. Relating this back to treatment goals, Katherine reminded Brad that embodiment activities help build the neural network necessary for continued growth in empathy.

Table 30.1 Degree of immersion: therapist use of self scale

Immersion relates to ways and degree to which you use yourself during a play therapy session with a child. Using the scales that follow, mark on the line the degree to which you evaluate your "immersion" on the following factors.

1. Verbal Use of Self

1.1 Here and Now Discussion (Quadrant II, Open Discussion and Exploration)

During the session what was the degree to which you were involved in verbal discussion 'about' the child's life or with the child outside of the play activity?

1 **LOW**	**2**	**3** **MODERATE**	**4**	**5** **HIGH**
Almost not at all. Stayed in play activity and/or followed child's lead.		Some discussion observed, usually led by the child. Discussion included spontaneous information sharing about school, and activity, a family member, etc.		Spent significant part of session outside of metaphorical play in direct discussion.

Child's Response: Please rate the effectiveness of **your use of** *Here and Now Discussion*

☐ **Low** (Play was shut down. Child pivoted away as a defensive response.)
☐ **Medium** (Child elaborates play – i.e., adds verbalizations and or play themes/ actions)
☐ **High** (Child incorporates/assimilates language or actions facilitated by therapist, indicating a greater awareness of self or circumstances.)

Provide an example that supports your rating of the child's response.

1.2 Reflecting and Tracking Statements (Non-intrusive Responding/Co-facilitation)

During the session to what degree were you using reflecting and tracking statements? Tracking refers to what the client is doing or what the play objects are doing. Reflecting statements refer to the guesses or statements about what the therapist thinks the client is experiencing such as *"you seem really happy right now"* or what the character is experiencing, "that one is very angry."

1 **LOW**	**2**	**3** **MODERATE**	**4**	**5** **HIGH**
Used few or no reflecting or tracking statements.		Some use of reflective or tracking statements. Reflections included feelings or non-verbal actions in play.		Primarily used reflecting and tracking statements related to child's emotions, non-verbal behavior, play actions, verbal content play sequences or metaphors.

Child's Response: Please rate the effectiveness of **your use of** *Reflecting and Tracking Statements*

☐ **Low** (Play was shut down. Child pivoted away as a defensive response/or child annoyed by your use of tracking and reflecting statements.)
☐ **Medium** (Child elaborates play – i.e., adds verbalizations and or play themes/ actions.)
☐ **High** (Child incorporates/assimilates language or actions facilitated by therapist, indicating a greater awareness of self or circumstances.)

Provide an example that supports your rating of the child's response.

1.3 Restating Content (Non-intrusive responding/Co-facilitation)

During the session to what degree did you paraphrase what the child said during the play without adding meaning or interpretation? Children talk in therapy directly about the play media and through the play media.

1	2	3	4	5
LOW		**MODERATE**		**HIGH**
Used few or no restatements of verbal content.		Some/moderate use of restatements or paraphrases made to child after child's comments.		Frequent use of restatements/paraphrases of child's verbalizations during the play.

Child's Response: Please rate the effectiveness of **your use of** *Restating Content*

☐ **Low** (Play was shut down. Child pivoted away as a defensive response/or child annoyed by your use of tracking and reflecting statements.)
☐ **Medium** (Child elaborates play – i.e., adds verbalizations and or play themes/actions.)
☐ **High** (Child incorporates/assimilates language or actions facilitated by therapist, indicating a greater awareness of self or circumstances.)

Provide an example that supports your rating of the child's response.

1.4 Interpretations (Active Utilization)

During the session to what degree did you utilize interpretive statements? Interpretations are verbal comments made by the therapist (after observing repetitive play themes and play scenarios). It is a function of utilizing play material to assist a child to develop new understanding/meaning by bringing a link between play and the child's lived life to conscious awareness. Types of interpretations may include reflective interpretations, linking interpretations and/or bridging interpretations (Yasenik & Gardner, 2004, 2012).

1	2	3	4	5
LOW		**MODERATE**		**HIGH**
No use of interpretations too soon to utilize an interpretation. Tracking and restatements used. Still formulating hypotheses/child not ready or able to work at high levels of consciousness.		Some/moderate use of interpretations. First level: reflective interpretations used. Some soft hypotheses are formed and use of characters to test hypotheses observed. Interpretations made primarily through the play metaphor.		Used one or more types of interpretative statements (reflective, linking or bridging). A child's current or past experience embedded in play is raised to a higher degree of consciousness through the interpretation made by therapist to child.

(*Continued*)

Table 30.1 (Continued)

Child's Response: Please rate the effectiveness of **your use of** *Interpretations*

☐ **Low** (Play was shut down. Child pivoted away as a defensive response/or child annoyed by your use of tracking and reflecting statements.)
☐ **Medium** (Child elaborates play – i.e., adds verbalizations and or play themes/actions)
☐ **High** (Child incorporates/assimilates language or actions facilitated by therapist, indicating a greater awareness of self or circumstances.)

Provide an example that supports your rating of the child's response.

2. Emotional Use of Self

2.1 Emotionality (Non-intrusive responding/Co-facilitation)

During the session what was the degree of the emotional intensity that you assigned to either reflective statements or by inserting an emotion or emotional meaning to a character or characters in the play metaphor? If you were in role play, also consider your intensity of use of emotions re: the character that was assigned the emotion(s) including tone, prosody, duration, volume, facial expressions (therapist).

1 LOW	2	3 MODERATE	4	5 HIGH
Primarily observed the child/ rarely reflected emotions of the child or the characters in the play.		Some/moderate emotional intensity utilized and focused on one or more characters or emotionality mirrored the child's emotional expression.		High use of emotionality. Many reflective statements or inserted emotional comments noted (both verbal comments and vocalizations). The overall use of self was intense on a number of levels and was either therapist led or directed by the child.

Child's Response: Please rate the effectiveness of **your use of** *Emotionality*

☐ **Low** (Play was shut down. Child pivoted away as a defensive response/ or child annoyed by your use of tracking and reflecting statements.)
☐ **Medium** (Child elaborates play – i.e., adds verbalizations/emotions and or play themes/ actions.)
☐ **High** (Child incorporates/assimilates language or actions facilitated by therapist, indicating a greater awareness of self or circumstances.)

Provide an example that supports your rating of the child's response.

2.2 Emotional Self (All Quadrants)

During the session what was the degree to which you were personally emotionally involved during the session? This may include your awareness of feeling a particular intense emotion or you notice yourself shut down or become numb or that you may temporarily lose track of following the client. You may become aware of a personal experience or be triggered to a personal memory.

1 **LOW**	2	3 **MODERATE**	4	5 **HIGH**
I did not feel particularly emotionally involved but remained empathic and had a clear sense of neutral but present feeling.		I felt moderately emotionally involved. I noticed I had some of my own feelings related to the client material.		If felt highly emotionally involved and affected by the client's presentation or play scenario and/or child's disclosure. I felt flooded or shut down during some part of the session.

I became aware of one or more of the following emotions:

☐ Anger ☐ Sadness ☐ Fear ☐ Confusion ☐ Joy ☐ Worry ☐ Frustration
☐ Other____________________

Child's Response: Please rate the effectiveness of **your use of** *Emotional Self*

☐ **Low** (Play was shut down. Child pivoted away as a defensive response/or child annoyed by your use of tracking and reflecting statements.)
☐ **Medium** (Child elaborates play – i.e., adds verbalizations/emotions and or play themes/actions.)
☐ **High** (Child incorporates/assimilates language or actions facilitated by therapist, indicating a greater awareness of self or circumstances.)

Provide an example that supports your rating of the child's response.

3. Physical Use of Self

3.1 Physical Self (All quadrants)

During the session what was the degree to which you were physically involved during the session? Physical self includes physical movement in play activities, physical proximity or touch, level of physical energy.

1 **LOW**	2	3 **MODERATE**	4	5 **HIGH**
Very little physical involvement. Primarily observed the child and followed child in the space. Did not engage in activities even when approached by the child..		Engaged in physical play only when directly invited to do so. Some physical play and movement with moderate contact.		Fully engaged in the play. Physical contact part of the play. High energy output. You may consider touch therapies as high, such as Theraplay.

Child's Response: Please rate the effectiveness of **your use of** *Physical Self*

☐ **Low** (Play was shut down. Child pivoted away as a defensive response/or child annoyed by your use of tracking and reflecting statements.)
☐ **Medium** (Child elaborates play – i.e., adds verbalizations/emotions and or play themes/actions.)

(*Continued*)

Table 30.1 (Continued)

☐ **High** (Child incorporates/assimilates language or actions facilitated by therapist, indicating a greater awareness of self or circumstances.)

Provide an example that supports your rating of the child's response.

4. Self-System

4.1 Embodiment (All quadrants)

During the session what was the degree to which you were aware of your self-system in the presence of the child? Self-system includes your body/energy awareness in relation to 'other' (the child). It is the ability to be consciously aware of your internal state(s) of being.

1 LOW	2	3 MODERATE	4	5 HIGH
Disembodied: I was mostly operating from a cognitive space. I was not specifically aware of my internal state(s) of being during much or most of the session. **Lost in embodiment:** I became unaware of my body and internal states of self and self-responses during the session.		Moderate and intermittent awareness of self-system. There were points during the process where the I-self/ Me-self were operating in a reflexive and conscious way. I could identify at least one point in the session where I could identify my internal experience in relation to the child.		Highly aware of my self-system. I was mostly aware of a reflexive process of the back and forth of the "me-self" informing the "I-self" and vise versa. Body-mind awareness "on-line."

Child's Response: Please rate the effectiveness of **your awareness of** your *Self-system*

☐ **Low** (Play was shut down. Child pivoted away as a defensive response/or child disengaged from you or the play.)

☐ **Medium** (Child elaborates play – i.e., adds verbalizations/ emotions and or play themes/ actions and appears to connect non-verbally.)

☐ **High** (Child appeared highly connected in a non-verbal or verbal way.)

Provide an example that supports your rating of the child's response:

TOTAL IMMERSION SCORE ☐ ______________

8–16 Low Immersion 17–24 Moderate Immersion 25–40 High Immersion

IMMERSION SUMMARY

1. Verbal Use of Self

Here and Now Discussion	☐ Decrease	☐ Increase	☐ On track
Reflecting and Tracking Statements	☐ Decrease	☐ Increase	☐ On track

Restating Content ☐ Decrease ☐ Increase ☐ On track
Interpretations ☐ Decrease ☐ Increase ☐ On track

2. Emotional Use of Self

Emotionality ☐ Decrease ☐ Increase ☐ On track
Emotional Self ☐ Decrease ☐ Increase ☐ On track

3. Physical Use of Self

Physical Self ☐ Decrease ☐ Increase ☐ On track

4. Self-System

Embodiment ☐ Decrease ☐ Increase ☐ On track

Provide a rationale that supports your rating.

Adapted from Yasenik, L., & Gardner, K. (2012). *Play Therapy Dimensions Model: A Decision-Making Guide for Integrative Play Therapists*. London: Jessica Kingsley Publishers.

Next, Katherine and Brad examined *The Degree of Immersion: Therapist Use of Self Scale*. The first category, Verbal Use of Self, (Table 30.1) contains four sub-categories. Subcategory 1.1 – Here and Now Discussion (Quadrant II – Open Discussion and Exploration) asks the therapist to reflect on the degree to which they were involved in verbal discussion 'about' the child's life or with the child outside of the play activity. Brad noted he sometimes attempted to engage Sammy in brief discussions about his day at the beginning of each session, rating this sub-category as 1.5. Brad described Sammy's response as 'low,' underscoring that Sammy often ignored his questions. After some discussion, Brad and Katherine agreed that lower levels of immersion were likely 'on track' as Sammy seemed excited and 'driven' to enter play. However, by making brief statements or inquiries, based in the here-and-now, Brad also demonstrated interest in Sammy's world.

Sub-category 1.2 (Table 30.1) (within the *Verbal Use of Self* category) is the use of reflecting and tracking statements, which generally occur in Quadrant III, Non-Intrusive Responding, or when the therapist enters Quadrant IV, Co-facilitation. Brad rated his use of reflecting and tracking statements low-to-moderate (2 rating). Brad commented that the play seemed 'bound' to Quadrant III and therefore he limited his tracking statements to behavioral actions such as, "You moved that one over there." In contrast, Brad seldom made reflections regarding what Sammy, or characters/objects in the play, might be experiencing such as, "You seem upset right now." Upon reflection it became clear that Brad struggled to enter or elaborate the play, and as there were no clear play themes, he felt 'stuck' in Quadrant III. Brad also recalled that on one occasion he made a reflection such as, "That one was really sneaky – attacking without any warning." Brad felt this statement intensified aggressive swordplay, to the point where he then needed to set a therapeutic limit. No surprisingly, Brad rated Sammy's response to tracking and reflecting statements as Low to Moderate – 2.

Brad tried to pace how often he re-stated content (sub-category 1.3) (Table 30.1) and rated his level of immersion in the moderate range – 3. Brad further commented that he generally paraphrased Sammy when re-stating, often worrying he might unintentionally add meaning or content beyond what Sammy intended. Brad judged Sammy's response as low, emphasizing Sammy's play actions and verbalization seemed unaffected – Sammy neither deepened the play nor added new content. Finally, as play interactions never entered Quadrant 1 (Table 30.1) – Active Utilization, Brad rated his use of interpretive reflections as 1 (sub-category 1.4).

Following this discussion, Katherine invited Brad to reflect on decisions he made regarding his verbal use of self. It seemed to Katherine that Sammy's play communicated a range of underlying needs, such as a need for power, control or protection, as well as a need to experience his own body and boundaries. Brad had become stuck on getting the content of his restatements "right" and had not considered the larger context of what Sammy was communicating, verbally and non-verbally through play actions. Feeling "stuck" in Quadrant III also seemed to diminish Brad's use of self – particularly his playful self. Katherine further reflected that as Sammy progressed to the *role* stage it might be important to bring certain needs to higher levels of consciousness to assist in self-development.

The next category, *Emotional Use of Self,* was explored dyadically and experientially. The first sub-category in this area is emotionality (2.1), which relates to the degree of the emotional intensity the therapist assigned to their reflective statements, or by inserting an emotion or assigning emotional meaning to a character in the play metaphor. If involved in role-play, emotionality relates to the intensity of emotions ascribed to the character, including voice tone, prosody, duration, volume, facial expressions. While working in Quadrant III, Brad purposefully kept his level of emotionality in the low-to-moderate range, rating this sub-category (2.1) as 2. Brad's rationale was that increasing the emotionality of the play only increased aggressiveness in Sammy's play behaviors. Brad remarked that when he (infrequently) made statements such as, "He's too strong and powerful, what am I going to do?" or "He just wants to hurt others," Sammy would state, "I am going to kill you . . . cut you to pieces." Sammy would accompany these statements with harder sword strikes.

Sub-category 2.2, *Emotional Self,* asks therapists to examine the degree to which they were personally involved during the session. This could include a wide range of intrapersonal experiences, such as an increase or shutting down of an intense emotion, a recollection of a personal experience or triggering of a difficult memory. In exploring sub-category 2.2, Brad said he sometimes needed to "step out" of any feelings that came up for him in the play as he felt a need to remain present in case limit setting was needed. Thus, Brad rated this sub-category a 2.5, and categorized Sammy's response as *low*. Katherine wondered whether the play communicated something about past 'hurts' for Sammy. When Katherine asked Brad to specify emotions he experienced in the play (as listed in section 2.2), Brad identified the following feelings: worry (that it would get out of control), confusion (not understanding where the play was going or what Sammy needed) and frustration (at himself, for not 'knowing' what to do).

Katherine invited Brad to engage in swordplay. Katherine assumed the role of Sammy while Brad directed the intensity of her play. After a few moments of swordplay Katherine invited Brad to switch roles, and demonstrated how to regulate the intensity of play by moving slower/faster and accompanying this movement with verbal markers (e.g. "going slower and slower, down and down"). At one point, Katherine also demonstrated how to insert a brief play action, such as using a shaker or noise-making object as a barrier for protection. In debriefing this role-play Brad remarked on the effectiveness of down regulating actions, without feeling the play stop. The brief insertion of an object also did not seem disruptive. In the role of Sammy, Brad noticed Katherine's energy, facial expressions and playfulness! After further discussion, Brad wondered whether Sammy's sword fighting intensified at times when he sensed Brad pulling out of the play by constraining or limiting play actions. Brad's insight was that Sammy was looking to make contact and explore boundaries of himself and others, and what he needed Brad to bring forward related to his playful self.

Brad's new understanding of Sammy's play, and the underlying dynamics, shifted his ratings for the two remaining categories. While discussing category 3, Physical Use of Self, Brad asked to change his rating as he now thought his physical play was at a moderate level (rating – 3), and should increase or at least be varied in service of Sammy's need for making contact and self-expression.

Brad provided low ratings for the last category, Self-System. Within this category, the therapist is asked to reflect on and rate embodiment experiences (sub-category 4.1), which relates to the degree the therapist was aware of their 'self-system,' including body/energy awareness, in the presence of the child. Brad rated this sub-category a 1.5 (Table 30.1). Essentially, Brad was operating mostly from a cognitive perspective, blocking awareness of the self-system. Brad not only struggled to access his playful self, but he disengaged from his feelings and body/energy levels, in relation to the child. Katherine pointed out how a caregivers' reflective functioning, their ability to reflect on their experience and the experience of their child, is critical to attunement; she then emphasized this same process is necessary in vitalizing the relationship between the therapist and child.

A total immersion score was calculated, placing it in the Low Immersion range with a score of 13. Again, a low score is neither right nor wrong! It is a question of the 'just right' amount.

Conclusions

The Play Therapy Dimensions Model is an integrative decision-making model that invites the therapist to view the therapy process on a moment-to-moment basis, particularly child-therapist interactions. Play and expressive therapists should constantly work on knowing the self and use their self-understanding in the most appropriate and meaningful ways, therapeutically (Yasenik & Gardner, 2019). When used in supervision, *The Degree of Immersion: Therapist Use of Self Scale* supports a recursive, self-reflection process that deepens therapist access to knowledge of the self. Through careful observation of the degree of therapist immersion and the child's response, the therapist becomes more attuned and responsive to the child, bringing forward new opportunities for self-expression and self-exploration.

References

Andrews, J. (1991). *The active self in psychotherapy: An integration of therapeutic styles*. Boston: Allyn and Bacon.

Badenoch, B. (2008). *Being a brain-wise therapist: A practical guide to interpersonal neurobiology*. New York: W. W. Norton.

Brown, S. (2009). *Play: How it sharpens the mind, opens the imagination, and invigorates the soul*. New York: Bantam Books.

Cashdan, S. (1988). *Object relations therapy: Using the relationship*. New York: W. W. Norton.

Chown, A. (2015). *Play therapy in the outdoors: Taking play therapy out of the playroom and into natural environments*. London: Jessica Kingsley Publishers.

Crouch, A. (1997). *Inside counseling: Becoming and being a professional counsellor*. London: Sage.

Edwards, J. K., & Bess, J. M. (1998). Developing effectiveness in the therapeutic use of self. *Clinical Social Work Journal*, 26, 89–105.

Erwin, E. (1997). *Philosophy and psychotherapy*. London: Sage.

Fonagy, P., & Target, M. (1997). Attachment and reflective function: Their role in self-organization. *Development and Psychopathology*, 9, 679–700.

Gaskill, R., & Perry, B. D. (2012). Child sexual abuse, traumatic experiences, and their impact on the developing brain. In P. Goodyear-Brown (Ed.), *Child sexual abuse: Identification, assessment, and treatment* (pp. 29–47) Hoboken, NJ: Wiley.

Gergen, K. J. (1991). *The saturated self*. New York: Basic Books.

Gergen, K. J. (1996). The healthy, happy human being wears many masks. In W. T. Anderson (Ed.), *The Fontana postmodernism reader*. London: Fontana Press.

Gergen, K. J. (2009). *Relational being: Beyond self and community*. New York: Oxford University Press.

Goodyear-Brown, P. (2010). *Play therapy with traumatized children: A prescriptive approach*. Hoboken, NJ: Wiley.

Goodyear-Brown, P. (2019). *Trauma and play therapy: Helping children heal*. New York: Routledge.

Hewson, J. (2001). Integrative supervision: Art and science. In M. Carroll & M. Tholstrup (Eds.), *Integrative approaches to supervision* (pp. 65–75). London: Jessica Kingsley Publishers.

Jennings, S. (1990). *Dramatherapy with families, groups and individuals*. London: Jessica Kingsley Publishers.

Jennings, S. (2005). Embodiment-projection-role: A developmental model for the play therapy method. In C. Schaefer, J. McCormick, & A. J. Ohnji (Eds.), *The international handbook of play therapy: Advances in assessment, theory, research and practice* (pp. 65–76). Lanham, MD: Jason Aronson.

Jennings, S. (2014). Applying an embodiment-projection-role framework in groupwork with children. In E. Prendiville & J. Howard (Eds.), *Play therapy today: Contemporary practice with individuals, groups and careers* (pp. 81–96). London: Routledge.

Kaduson, H., & Schaefer, C. (Eds.). (2003). *101 favorite play therapy techniques* (vol. III). Northvale, NJ: Jason Aronson.

Kihlstrom, J. F. (1993). What does the self look like? In T. K. Srull & R. S. Wyer (Eds.), *The mental representations of trait and autobiographical knowledge about the self book: Advances in social cognition* (vol. 17, pp. 2–40). Hillsdale, NJ: Erlbaum.

Kottman, T. (2011). *Play therapy: Basics and beyond* (2nd ed.). Alexandria, VA: American Counseling Association.

Kottman, T., & Meany-Walen, K. K. (2018). *Doing play therapy: From building the relationship to facilitating change*. New York: Guilford Press.

Lawrence, E. J., Shaw, P., Giampietro, V. P., Surguladze, S., Brammer, M. J., & David, A. S. (2006). The role of shared representations in social perception and empathy. *NeuroImage*, 29, 1173–1184.

Leiberman, J. N. (1977). *Playfulness: It's relationship to imagination and creativity*. New York: Academic Press.

Lowenstein, L. (2006). *Creative interventions for bereaved children*. Toronto, ON: Champion Press.

Malchiodi, C. (Ed.). (2008). *Creative interventions with traumatized children*. New York: Guildford Press.

Malchiodi, C. (2010). *What a rubber duck and empathy have in common*. Retrieved from http//tlcinstitute.wordpress.com/2010/08/02/what-a-duck-and-empathy-have-com.

Markus, H. R., & Cross, S. (1990). The interpersonal self. In L. A. Pervin (Ed.), *Handbook of personality: Theory and research* (pp. 576–608). New York: Guilford Press.

McInnes, K. (2019). Being a playful therapist. In P. Ayling, H. Armstrong, & L. Gordon Clark (Eds.), *Becoming and being a play therapist. Play therapy in practice* (pp. 98–109). London: Routledge.

Moyles, J. R. (1989). *Just playing?* Buckingham: Open University Press.

Nalavany, B., Ryan, S., Gormory, T., & Lacasse, J. (2004). Mapping the characteristics of a "good" play therapist. *International Journal of Play Therapy*, 14(1), 27–50.

O'Connor, K. (2000). *The play therapy primer* (2nd ed.). New York: Wiley.

Piaget, J. (1977). *The language and thought of the child*. London: Routledge and Kegan Paul.

Ray, D. (2004). Supervision of basic and advanced skills in play therapy. *Journal of Professional Counselling: Practice Theory and Research*, 32, 38–41.

Rosenberg, S. (1988). Self and others: Studies in social personality and autobiography. In L. Berkowitz (Ed.), *Advances in experimental social psychology* (vol. 21, pp. 56–96). New York: Academic Press.

Rubin, L. (Ed.). (2008). *Popular culture in counseling, psychotherapy, and play-based interventions*. New York: Springer.

Schaefer, C., & Cangelosi, D. (2002). *Play therapy techniques* (2nd ed.). Northvale, NJ: Jason Aronson.

Siegel, D. J. (2007). *The mindful brain: Reflection and attunement in the cultivation of well-being*. New York: W. W. Norton.

Wilson, K., Kendrick, P., & Ryan, V. (1992). *Play therapy: A non-directive approach for children and adolescents*. London: Ballerie Tindall.

Wosket, V. (1999) *The therapeutic use of self: Counselling practice, research and supervision*. London: Routledge.

Yasenik, L., & Gardner, K. (2004). *Play therapy dimensions model: A decision-making model for play therapists.* Alberta: Rocky Mountain Play Therapy Institute.

Yasenik, L., & Gardner, K. (2012). *Play therapy dimensions model: A decision-making guide for integrative play therapists.* London: Jessica Kingsley Publishers.

Yasenik, L., & Gardner, K. (2019). Turning points and the understanding of the development of self through play therapy. In L. Yasenik & K. Gardner (Eds.), *Turning points in play therapy and the emergence of self: Applications of the play therapy dimensions model* (pp. 15–42). London: Jessica Kingsley Publishers.

31

WHAT AM I DOING OUT HERE?!

Exploring the challenge for play therapy in the outdoors

Alison Chown

Over the past decade or more there has been a growing concern about the disconnect our children have from nature and the outdoors and the impact this can have on all aspects of their development. MacFarlane (2015:323) notes that:

> We live in an era of diminishing childhood contact with nature, and landscapes outside the urban . . . the so-called 'roaming radius' – tightened from generation to generation. . . (has) cinched right down to house, garden and pavement.

Louv (2005) described this as a 'nature deficit disorder' and is credited by many as starting the movement to reconnect children and nature. Here in the UK there has been a rapid rise in the number of organisations offering play activities, wild play and therapeutic play in nature as a way to build resilience in children and young people and to encourage them back outside. With the publication of my books *Play Therapy in the Outdoors* (2014) and *A Practical Guide to Play Therapy in the Outdoors* (2017), the play therapy movement in England is catching up with these developments and, more importantly, there is a growing number of play therapists who are either working in outdoor spaces or interested in taking that step through the playroom door. The Collaboration of Outdoor Play Therapists network (COOPT), set up in late 2017 as a space to consider best practice and exchange ideas, already has around 50 members with more joining at each new event. One of the topics of conversation has been around how we differentiate what we as play therapists do when we are outside from what organisations offering therapeutic or nature play activities do.

As a qualified and experienced play therapist working in schools since 2004, and in outdoor spaces since 2010, I have often heard both teachers and support staff say something like "Oh yes, we do play therapy when we have golden time" or "But we already do play therapy in our nurture group," and I have to explain that whilst play can be therapeutic, play therapy is very different, not least because I had to train part time for two years at post-graduate level, write lots of assignments and have two years of personal therapy! But these conversations do raise the question of what the perception is of play therapy, therapeutic play or playwork and what is seen to be the difference. And for those of us who have taken our practice outside, such comments might cause us to ask "**Are we still 'doing' play therapy if we're not in the playroom**

behind a closed door?" We do need to have a shared understanding of the terminology we use in order to be clear about what is being offered to children and young people. Whilst it is beyond the scope of this chapter to delve in depth into definitions of therapeutic play and indeed play itself, I will consider some working definitions and attempt to make some basic distinction between play, therapeutic play, therapeutic playwork and play therapy.

In defining playwork, the Playwork Foundation webpage suggests: "Playwork is an approach to working with children that aims to support and facilitate the play process" but also acknowledges that "Different play or playwork organisations define or describe playwork in slightly different ways" (playworkfoundation.org n.d.). For example, Play Scotland, Play for Health define it as "the work of creating and maintaining spaces for children to play" (playscotland.org n.d.) whilst Play England suggests it is "a highly skilled profession that enriches and enhances children's play" (playengland.org n.d.). However, looking at the differing perspectives, some common ground emerges:

- Children's play needs are at the centre of the work.
- Playwork enhances the quality and the range of the play experience.
- Play is led by the children.
- Playworkers plan for play, observe and reflect.
- Playwork is a highly skilled profession.
- Playwork is not driven by predetermined outcomes.

Interestingly, in 1998, Gordon Sturrock and Perry Else delivered a paper to the IPA/USA Triennial Conference entitled 'The Playground as Therapeutic Space: Playwork as Healing' and suggested that "playspace should now be seen as therapeutic space and playwork advanced as having an unexplored, healing potential" (Sturrock and Else 1998), which added the therapeutic dimension to playwork that is found in the literature today.

Fraser Brown, the world's first Professor of Playwork defines the role of the playworker as "creating flexible environments which are substantially adaptable and controllable by the children" (Brown 2018). When seen in the context of Sturrock and Else's paper, this seems to parallel the role of the play therapist, and, yet, in the 2018 Longfellow Lecture, he also states that in his mind, playworkers are the **only** people who work non-directively and to the child's agenda. Reflecting on this, I felt it rather ignored in the world of the non-directive play therapist. In discussion with Professor Sue Jennings (2019), play and dramatherapist, supervisor and author, we reflected that this may well be true if one sees the play therapist as someone who encourages children to play in order to resolve difficult life experiences through play, but with commentary and reflection from the therapist so that the intervention in this way is part of the process. If we put this alongside Principle 8 of the Playwork Principles (Playwork Principles Scrutiny Group 2004) we can see there may indeed be a nuanced difference in our two approaches.

> Playworkers choose an intervention style that enables children and young people to extend their play. All playworker intervention must balance risk with the developmental benefit and well-being of children.
>
> *(Principle 8)*

If we also take Brown's further suggestion, that therapeutic playwork makes use of the therapeutic aspects of play to help children comes to terms with traumatic events, deal with psychosocial difficulties and realise optimal growth and development, then we can begin to see a simple

Play → Therapeutic Play → Therapeutic Playwork → Play Therapy

Figure 31.1 Play to play therapy continuum

continuum where at one end there is just play and at the other play therapy with therapeutic play and therapeutic playwork either side of the mid-point (Figure 31.1).

Play Therapy UK (PTUK) suggests that therapeutic play is useful when a child has slight to moderate problems, preventing these from becoming worse and as a means to assessing whether or not a child may need to be referred on to a play therapist (playtherapy.org). This would sit comfortably within the preceding continuum, particularly since Fraser Brown in the Longfellow Lecture referred to earlier, discusses the very positive impact of therapeutic playwork with children from a Romanian orphanage, where, one might suppose, the complexity of the presenting issues was fairly significant, and certainly not mild to moderate.

In considering the PTUK definition of therapeutic we also need to be clear that it isn't the play itself which becomes more complex and so needs a play therapist as opposed to a therapeutic play worker but the nature and complexity of the child's experiences (Jennings 2019).

Another group of practitioners who work with children and young people to develop resilience and promote healthy development is the network of Forest School practitioners. The Forest School Association notes that:

> Forest School is an inspirational process, that offers ALL learners regular opportunities to achieve and develop confidence and self-esteem through hands-on learning experiences in a woodland or natural environment with trees.
>
> Forest School is a specialised learning approach that sits within and complements the wider context of outdoor and woodland education.
>
> *(Forest School Association, n.d.)*

There is no obvious reference here that what is on offer is therapeutic play and yet those who know the work of Forest School are likely to acknowledge that it does offer therapeutic opportunities, since both play and outdoor environments have a therapeutic potential, and the principles of the Forest School allow for child-led and child-directed activities, environments that are as safe as is practicable to allow for risk taking and acknowledges that trust is central to the process.

Indeed, if we look at the Forest School Elements, the equivalent to Axline's Principles for Play Therapy (1969), we can see there are a number of distinct parallels in approach, which are also shared by Brown's description of the Therapeutic Playwork Practice Fundamentals, most notably:

- The creation of a safe space
- Being non-directive and working to the child's agenda
- Creating a 'bond' between the adult and the child
- Allowing for the child to freely express thoughts and feelings in their own individual way

If we consider the idea of the creation of a bond based on trust and acceptance, we can add something about the type and quality of the relationship to the simple continuum (Figure 31.1) to begin to create a clearer picture of the differences between the various types of play.

Play

When children are at play in home environments or unstructured time in school, the adults around are likely to step in and give direction to the play, what playworkers refer to as '**adult**erated' play. This may be in the form of an instruction, a comment or observation or an intervention as a result of an assumption they have made about the quality or type of play. In other words, it is a direction. In school settings, play-based activities may be both directed, including having to meet pre-determined outcomes, or they may be freer with some therapeutic qualities, such as in a nurture group. Our own Early Years Foundation Stage (EYFS) statutory guidance (DfE 2014), whilst suggesting children will be taught using a games and play has assessment points between age two and three and again at five. The statutory framework talks of 'standards' that must be met and this kind of language does not make for an easy bedfellow with the idea of play that has therapeutic qualities. It is more likely to create anxiety.

Therapeutic play

Whilst there is an argument to say all play can be therapeutic in that children may be resolving tensions, contradictions and past experiences through their play, for play to have a truly therapeutic effect, we might say that there needs to be an adult present to offer some reflection such as in play within a nurture group. The nature of the relationship between the adult and child, whilst it may be less directive and more reflective, may still not allow the child to fully lead without some adult intervention and control over the environment for the play. Whilst there may be some reflective commentary or conversation with the child, there is still likely to be a certain amount of guidance, too, in the resources available, the play space and the play itself.

Therapeutic playwork

The National Playing Fields Association publication 'Best Play: What Play Provision Should Do for Children' (2000) gives the core function of playworkers as being "to create an environment which will stimulate children's play and maximise their opportunities for a wide range of play experiences." So we might say that the therapeutic aspect of playwork is almost a by-product of the process of facilitating high quality play experiences and environments as the playworker responds to the play cues issued by the child. That is not to diminish its importance or efficacy but to distinguish it from play therapy where the therapist's role is explicitly to facilitate the healing process through play, commentary and reflection. For playwork to be therapeutic, the quality and nature of the responses to play cues will need to be non-directive, and to allow for the children to see the playworker as an ally in their adventures as opposed to a director of them.

Play therapy

The British Association of Play Therapists (BAPT) defines play therapy as

> an effective therapy that helps children modify their behaviours, clarify their self-concept and build healthy relationships. In Play Therapy, children enter into a dynamic relationship with the therapist that enables them to express, explore and make sense of their difficult and painful experiences.
>
> *(BAPT 2014a)*

Therefore, the core purpose of the play therapists is to work with children and young people who are experiencing social, emotional or mental health difficulties. The play is the vehicle for the process and the purpose is to heal rather than facilitate play.

Gettins (2014) notes that while playworkers focus on children's right to play and to direct their own experiences, most of this play is recreational. Play therapy on the other hand is remedial, supporting children to "resolve issues through the therapeutic relationship" (p. 68).

Looking at the simple continuum in Figure 31.1, we can add the adult role to begin to further illustrate the difference between each type of play (Figure 31.2).

Play →	Therapeutic Play →	Therapeutic Playwork →	Play Therapy
Adult as director	Adult as guide	Adult as ally	Adult as co-constructor

Figure 31.2 Different forms of play

In looking at the purpose of our differing roles, we need also to look at the nature of the relationships we form in those roles. Play therapists work at a psychodynamic level to gain insight into client's thoughts, feelings and actions and how these might relate to their early experiences – we bring what is in the unconscious into conscious awareness so that our clients may find their own resolutions. Those engaged in therapeutic play or playwork with children are likely to be working with conscious states and whilst they may be empathic, and gain some insight into children's feelings and motivations, their level of affect attunement is unlikely to be the same as the play therapist. To sit with client's most overwhelming feelings and not be overwhelmed oneself is the essence of our work and we are supported to be able to do this by personal therapy and supervision.

Does all of this really matter? Well yes, I rather think it does, because as playworkers note, they are a profession and have skills and qualities that are different to other people who use play in their work, just as Forest School practitioners have a different set of skills to play therapists. We need to maintain these differences, not so that we can construct a hierachy of practitioners, but because those with whom each of us works need to have access to the right kind of play experience to meet their differing needs. Those nuanced differences referred to earlier in this chapter may be small but they are significant. They may mean the difference between a child finding a space to be encouraged to play, develop and learn and where adult intervention is minimal and a space where their play is controlled and directed to such an extent that their creativity is crushed and their development stultified.

And it matters to our play therapy clients who deserve to have the appropriate intervention to meet their needs. With the rise in the number of young people with mental health issues from one in ten a decade or so ago to one in eight now and the raising of the threshold for treatment by CAHMS services as a result of cuts to public spending, young people are in even greater need of access to independent therapeutic treatment. The mental health charity Young Minds notes that three in four children with a diagnosable mental health condition do not get access to the support they need (Young Minds Impact Report 2018:2). To counter this, many more schools are now seeking to commission the work of play and creative arts therapists. With the nature and complexity of the issues many of our young people are experiencing, they need the skills and expertise of therapists who have an understanding of complex mental health issues and are supported by relevant training and supervision of their work. This is underlined by the Ethical Basis for Good Practice in Play Therapy Principle B: Beneficence;

> Play Therapists strive to benefit those with whom they work, acting in their best interests and always working within their limits of competence, training, experience and supervision. . . . An obligation of the Play Therapist is to act in the best interests of clients and this is the paramount consideration for (them).
>
> *(BAPT 2014b)*

And it matters if we are moving our practice into the outdoors so we can be clear with ourselves and others that we are 'doing' play therapy, wherever we practise.

> I am clear that what I do when working outside in nature is play therapy – it does not change just because the location changes. This is so whether I am working with and individual client or a small group. We may be accompanied by other adults, some supervisory and others working therapeutically in the broad sense, but I am the play therapist and how I interact with the children is different.
>
> *(Chown 2017:3)*

Having noted these differences, it is also important for me to state that in the early development of my own work, the Forest School approach was very influential, as was the early years education Reggio Emilia approach. I first came across Forest School when I was working in primary school in Somerset, which was in one of the most socio-economically deprived areas of the county. I had been working with a group of boys whose behaviour was very challenging to staff and who had had a range of what we now call Adverse Childhood Experiences, or ACEs, placing them at risk of exclusion. They were offered places in one of the local Forest School groups and being outdoors, in a non-school environment where they were trusted with a knife and with fire was very affirming for them. Most of the group responded very quickly to the two male leaders and developed a much more positive self-concept which began to have a pay off back in school. The opportunity to experience more democratic, mutually satisfying relationships with adults, was a new experience for them but showed them the potential they had for a different kind of relationship with their peers and teachers back at school.

It was not always plain sailing but the experience of Forest School had a significant impact on their social and emotional development. It highlighted for me the advantage of working in a democratic space which belonged to neither and yet to both and being in an outdoor environment where the intensity of everything can be mediated by nature. I also think that being in a bigger, more open space without restrictive walls focused the relationships as these were what 'held' them and allowed them to feel safe – they were symbolic walls, the container for the work (Chown 2014).

The Reggio Emilia approach to learning in the early years grew out of a town of the same name in Northern Italy where the raising and educating of the children in a progressive and co-operative way was seen as a community responsibility. It recognises the child's uniqueness and individuality and the 'hundred languages' they have for expressing themselves. It also places great significance on the environment around the child.

> The physical environment of the schools is therefore much more than a simple container for learning and teaching. Rather, the environment can be seen to be a central

component of the learning and teaching relationship. Indeed, it is easy to see why Reggio educators call the environment "the third teacher."

(Learning and Teaching Scotland 2006:19)

This concept of the environment as being more than just the space we might work in backed up my experience with the boys at Forest School where the change of environment had a significant impact and not just because it wasn't school or because it was where the Forest School took place. Being outside, in nature, seemed in itself to enable them to contain some of their more overwhelming feelings despite the 'bigness' of the space. It felt counter-intuitive. Curiously, however, I love being by the sea and find both the sight and sound of the water can still me inside, and the bigger the waves, the greater the stillness. It is as if seeing energy and action without, takes away the same that is within. Perhaps for the boys, being in a big space where, if their feelings exploded and tumbled out, they would be dissipated by the openess, reduced the intensity of what was within them. And of course, the boundaried, trusting and affirmative relationships they had developed with the Forest School Leaders who were not fazed by the overwhelming nature of their feelings also enabled them to feel the containment they had not experienced during their early developmental stage of childhood. It seemed to me, that just as there was the potential play space, the intermediate playground referred to by Winnicott (1971:47), between the boys (the 'infants') and the Forest School staff (the metaphorical mother figure), so there was a potential second intermediate playground between the therapeutic alliance and nature, "one which reflects the fluid and dynamic nature of the process of play therapy through its own transformations as weather and seasons change" (Chown 2014:136).

Beyond the playroom walls

I started this chapter with reference to Louv's idea of 'nature deficit disorder.' There is little doubt that the increasing use of 'screens' on games consoles, computers, laptops, phones and tablets and the increasing commercialisation of play spaces have had a significant impact on the amount of time children spend outdoors (see ch. 3, Chown 2017). Lack of access to nature impacts negatively on children's development and has been credited with affecting issue as diverse as from obesity and short sightedness to poor mental well-being and sleep deprivation. Many children want to be outdoors more often and a 2016 survey by the National Trust showed that 84% of parents believe that playing outdoors makes their children more imaginative and creative, while 96% felt it was important for children to have a connection with nature. However, paradoxically, some of the reasons why children do not access outdoor play are parental fears, primarily around traffic and strangers, and their own perception that time is short and spaces to play out in are limited. One of the more worrying aspects of reduced time in the outdoors is that children are growing up unable to calculate risk since most of what happens indoors has been made physically, if not psychologically safe. Much of the 'adventurous play' which young children access, unless they go to something like Forest School is generally commercialised and therefore made so safe as to be devoid of any real risk. The opportunities to experience the exhilaration and psychological growth when a fear is faced and overcome by using your own resources is a rare feeling for most children today.

But the need to reconnect children with nature was not the primary driver for my deciding to work in outdoor spaces. As a play therapist, I accept that through play, children can 'speak' about difficult and unspeakable experiences and with reflection and commentary from me, they can begin to process these and heal themselves. They do not need words. I am also a non-directive, client-centred therapist, in the humanist tradition of Carl Rogers and Virginia Axline.

To lead the play and to direct the activities, except where this is a chosen intervention based on the client's needs, would be deeply contradictory.

What I have found working as a play therapist is that there are those clients who gravitate towards and need access to an outdoor space. So, if I am to be child-centred and allow the child to lead the play, I feel I need to allow for that play to be in places other than the playroom. Provided we are physically safe, I can go where the child needs to go because the therapeutic alliance we have created becomes the container for the work; it allows for symbolic walls.

Case study

I worked with a 16-year-old client in a school for those with profound and multiple learning needs but who was only placed there because of her behaviour and developmental needs, because the local authority had no provision for girls at risk of exclusion. She would sit on the window sill of the playroom threatening to jump out – a drop of some six feet. This usually happened after she had completely trashed the room, turning over the tables and chairs and hurling the resources around. She needed to escape from the toxic chaos that she had created but which was a metaphor for her life as she saw it. At that point early in my play therapy career I had not considered working outside the playroom and I was very concerned that she would jump out and hurt herself in the process. In discussion in supervision I was torn between her desire to go out and my battle to keep her in to keep her safe. I felt that she needed to step out in order to move on from the inner and external chaos but also felt that this would break ethical boundaries. I continued to work inside with her, and she continued to want to leave via the window! In the end, I made it a game, and was eventually able to reduce the intensity of her need to go out through being playful but empathic. However, I am now convinced that I should have followed her lead, remained child centred, holding her needs as central to my approach and let her out – through the door! I think it would have been a real demonstration of my trust in her and it would have strengthened the therapeutic alliance and her sense of trustable self. Later, still working in the same establishment, I did venture out with two other clients with profound and multiple learning needs, Chloe whose case is presented in *Play Therapy in the Outdoors* (Chown 2014), and Bobby, who returning after a school holiday, refused point blank to enter the building and received all of his education in the driveway beyond the reception area doors for several weeks, in January, until he was able to allow himself back inside. For his sessions with me, he sat on my lap, in his hat, coat and scarf, and wrapped in a large blanket and I sang songs to him. He had no verbal communication but responded physically to being rocked like a baby as I sang. If I stopped rocking, he would nudge me hard and rock himself, relaxing only when I started again.

Networking

When I wrote *Play Therapy in the Outdoors*, it was the culmination of a number of years of what might be very appropriately termed, 'wandering in the wilderness!' What has been quite profoundly encouraging on my 'pioneering' journey is that there are a considerable number of experienced play therapists who have arrived at this point, almost at the same time, as well as students who are interested in working outdoors. Establishing the COOPT network has allowed us to come together and share our enthusiasm, concerns and experiences and ensure our practice is the best it can be, and adheres to BAPT's Ethical Basis for Good Practice in Play

Therapy. Although we have not all arrived at this destination for exactly the same reasons, we share thoughts and feelings about working in outdoor spaces.

We feel that outside, in nature, we enter a more democratic space which belongs to neither of us but to both. This enhances the child's feelings of ownership of the space and the resources in it, particularly in a setting where the landscape is familiar to the child and this makes for a much more equal relationship. In the playroom, however objective we may try to be in selecting the resources we use, we bring something of ourselves to that selection and therefore guide the process to some extent.

We live in a decade where children spend much of their free time indoors or in organised activities in commercial play spaces. There are fewer opportunities for stillness and just 'being' in nature and all that it has to offer. We can be part of the movement to reconnect to nature but within the context of still being a play therapist and doing play therapy. We see nature as a co-therapist; if we expand the 'borders' of the playroom, we expand the possibilities for transformation. Nature provides us with additional resources which our clients may use in their interaction with us or we may reflect on their relationship with nature and what this brings to the therapy.

We know that sometimes children's feelings can be so intense and overwhelming that they need to 'escape' them in order to begin to feel they can contain them. Working in a bigger, outdoor space allows for a mediation of the intensity of these feelings, in the psychological safety provided by the symbolic walls of the therapeutic alliance.

We also acknowledge that attention needs to be paid to the spiritual aspect of development. By this, I do not mean the development of a faith or religious aspect but how we see the world and understand our experiences of all it has to offer. I think it is about the true connectedness with what we might describe as the non-human world and how we come to understand the things that pass. Nature provides concrete examples of transformation – the potential for change; day into night, summer to winter, water into ice, earth into mud, bare trees into leafy canopies and a calm sea to an angry one and back again.

We agree that working in outdoor spaces is not appropriate for every child and that we need to take each client to supervision so we can be sure the need is theirs and not our own when we feel it is right to step through the playroom door.

Settling, transition and transformation – an early framework

My framework of settling, transition and transformation (Chown 2014) charted my experience of how my clients moved from the playroom to the outdoors and the changes that occurred as a result. Although I work in outdoor spaces, I still feel that the playroom offers the first safe space for my clients and I to get to know each other and '**settle**' together to build a strong therapeutic alliance which will contain our work together if and when we move outside. When I have my first session with a child, I make reference to any outdoor space just as I would make reference to the other resources in the room, and tell them once we get to know and trust each other, we can step out there. Many choose to do this, some don't, but the choice is theirs.

The **transition** point, when a child chooses to step through the playroom door, can be an indication of a readiness to test new found aspects of self, of developing confidence and of trust and allows for the establishment of a core physicality that helps to process embodied, pre-verbal trauma. It provides for a time of reflection by the therapist, a time to guage the strength of the therapeutic alliance and her own level of 'confident therapist self.' This latter is crucial to effective work in the outdoors, for any anxiety about such things as whether the client will remain in the prescribed area or return to the playroom when the session is over will impact on her ability to be fully present for the client and interfere with the child's own processes.

Case study

Jamie was seven and a bit when we started sessions together. He did not know his birth father who left when he was tiny and he carried a 'big sad' around with him. His role in life seemed to be to ensure everyone else in the family was 'OK,' whilst desperately needing them to 'see' that he was not OK. He was often in trouble for not listening to the teacher and found it very difficult to concentrate in class. He had friends, but there were frequent arguments and fallings out. He needed to be noticed. He spent a lot of time playing hide and seek in the room, finding the smallest space in the cupboard to squeeze into. After he had done this, he would want to go outside to play football, when he would teach me skills and tricks (or try to!). He needed this in order to counter the feelings of being small and unnoticed that hide and seek brought up. Outside, he was in charge and felt skilful and important and the feelings of being insignificant he could not tolerate were dissipated and homeostasis restored. His venturing into the outside world was the catalyst for transformation although this would take time. He still combined both activities, hide and seek indoors and football outside but needed the hide and seek less and less. He began to be able to talk about his 'feelings,' something he had blocked. When his grandad's dog died, he cried openly and bitterly, grieving for his own loss – his father. Without his football, which we could not have played inside, he would have struggled to find another activity which gave him such a feeling of being skilful and having some power.

I have found that **transformations** occurring outside the playroom are more visceral, more deeply embodied than any experienced within it. I think they are more real for the child, enhanced by the sensory feedback from the elements which heightens awareness, and as the therapist, I feel more deeply attuned to both the client and their process.

Case study

David is ten, a serious boy who lives with his grandmother. His mother became pregnant after she had left home and she returned for a time to give birth and begin to look after him. However, she missed her other life and returned to the South, visiting somewhat irregularly, even when she moved nearer. Most sessions, we went on adventures in the school grounds and built imaginary homes just for us two. My role was mostly passive, with David being the one who went out hunting or scouring the woods for whatever we needed and whose job was to keep us both safe. When our sessions were due to end, he told me he was very sad and for several sessions he stayed inside and did things on the computer, his default activity. We spoke about being sad, and how we might feel very sad for one thing happening now because it reminded us, deep inside, of other great 'sads' that had happened. In our final session, he asked to go on another adventure. We built our last home together at the back of the school, admiring the view we had and constructing a sophisticated security system which kept out the ghosts and zombies. It was all rather jolly and playful.

As we returned to the playroom he stopped me and said "Wait, I've got some TNT, I'm going to blow it up!" This was something we had never done before. As we 'watched' it explode into the air, he said "There's a bit left, here, you blow that up" and I did. We marvelled at the vast cloud of smoke and all the tiny pieces that were descending, then he turned his back to it, and said "There, it's gone!" It was a moment of great poignancy for us both.

Although simple, I still feel this framework of settling, transition and transformation helps to focus my reflections on my client's processes and key points along the way. There be more than one transition point for those who choose to move in and out of the playroom and any number of transformations can occur. It helps to mark the pathway each child takes, both literally, in where and how they choose to play and metaphorically as they move through the therapeutic process. I hope others working in the outdoors may adapt it or find a new frame of reference for themselves and that through our networking we can find opportunities to exchange thoughts and ideas and that the movement to work outdoors grows ever stronger.

References

Axline V. (1969) *Play Therapy*. New York: Ballantine Books.

BAPT (2014a) *Information for Professionals and Employers: What Is Play Therapy?* Available at www.bapt.info/play-therapy/info-professionals-employers/ [Accessed 23 June 2019].

BAPT (2014b) *Ethical Basis for Good Practice in Play Therapy*. Available at www.bapt.info/play-therapy/ethical-basis-good-practice-play-therapy/ [Accessed 12 April 2019].

Brown F. (2018) *Longfellow Lecture – Giving Children Hope: The Value of Therapeutic Play*, online video, YouTube. Available at www.bing.com/videos/search?q=fraser+brown+2018+longfellow+lecture&view=detail&mid=53456FDDC5393DBCB43A53456FDDC5393DBCB43A&FORM=VIRE [Accessed 24 June 2019].

Chown A. (2014) *Play Therapy in the Outdoors: Taking Play Therapy Out of the Playroom and into Natural Environments*. London: Jessica Kingsley Publishers.

Chown A. (2017) *A Practical Guide to Play Therapy in the Outdoors: Working in Nature*. London: Routledge.

Department for Education (2018) *Early Years Foundation Stage Statutory Framework*. Available at www.gov.uk/government/publications/early-years-foundation-stage-framework-2 [Accessed 29 June2019].

Forest School Association (n.d.) *What Is Forest School?* Available at www.forestschoolassociation.org/what-is-forest-school/ [Accessed 29 June 2019].

Gettins T. (2014) Therapeutic play as an intervention for children exposed to domestic violence. In Prenderville E. and Howard J. (eds) *Play Therapy Today: Contemporary Practice with Individuals, Groups and Careers*, pp. 64–78. Abingdon: Routledge.

Learning and Teaching Scotland (2006) *The Reggio Emilia Approach to Early Years Education*. Available at https://education.gov.scot/improvement/Documents/ELC/ELC35_ReggioEmilia/ELC35_ReggioAug06.pdf [Accessed 14 May 2019].

Louv R. (2005) *Last Child in the Woods: Saving Our Children from Nature Deficit Disorder*. London: Atlantic Books.

MacFarlane R. (2015) *Landmarks*. London: Hamish Hamilton.

National Playing Fields Association (2000) *Best Play: What Play Provision Should Do for Children*. Available at www.freeplaynetwork.org.uk/pubs/bestplay.pdf [Accessed 2 May 2019].

Play England (n.d.) *Playwork*. Available at www.playengland.org.uk/playwork-2/ [Accessed 25 June 2019].

Play Scotland (n.d.) *Play for Health*. Available at www.playscotland.org/what-is-play-playwork/ [Accessed 25 June 2019].

Play Therapy UK (2017) *Therapeutic Play and Play Therapy – What's the Difference?* Available at www.playtherapy.org.uk/ [Accessed 15 June 2019].

The Playwork Foundation (n.d.) *What Is Playwork?* Available at https://playworkfoundation.org/about// [Accessed 25 June 2019].

Playwork Principles Scrutiny Group (2004) *Statement by the Playwork Principles Scrutiny Group: Playwork Principles*. Available at https://www.playengland.org.uk/playwork-2/playwork-principles/ [Accessed 21 June 2019].

Jennings S. (2019) Supervision conversation by Professor Jennings, 26 June.

Sturrock G. and Else P. (1998) *The Playground as Therapeutic Space: Playwork as Healing: The Colorado Paper*. Available at https://ipaewni.files.wordpress.com/2016/05/colorado-paper.pdf [Accessed 27 June 2019].

Winnicott D. (1971) *Playing and Reality*. Hove: Bruner-Routledge.

Young Minds Impact Report (2018). Available at https://youngminds.org.uk/media/3396/impact-report-2018-19-low-res.pdf [Accessed 27 June 2019].

32
SCHOOL-BASED PLAY THERAPY

Athena A. Drewes

Children spend more time at school than in any other setting. It is the most natural setting to receive not only an education but also mental health services to help address increasing numbers of trauma and learning difficulties. Although not all children experience trauma directly, most are exposed to violence through the media, and in the past several years rising numbers of children are bringing handguns and rifles to school, shooting and killing peers, teachers, and ultimately themselves. One out of five children experience impairing emotional problems, approximately six to eight million (about 12%) of all children and youth in the United States, although only one-third of these children get the help needed (Mental Health America, 2009), with serious mental health problems and behavioral difficulties untreated and even undiagnosed. The most common student referral is for disruptive classroom behaviors (Abidin & Robinson, 2002), and children's aggressive behavior is the most common presenting problem (Cochran et al., 2010). When a child becomes disruptive, the critical relationships between the teacher and student, and with other students, often become damaged and strained (Hamre et al., 2007; Myers & Pianta, 2008).

When children have to deal with traumas, it becomes difficult to expend the energy needed to learn and excel academically. A downward spiral ensues whereby academic ability slips further behind, and risk for secondary emotional and behavioral problems, as well as academic and peer/social difficulties increases. Without intervention these disruptive behaviors often lead to serious problems over the child's lifespan, which include psychiatric diagnoses, antisocial behavior, violence, drug abuse, and juvenile delinquency (Barkley, 2007; Mental Health America, 2009), as well as school suspensions and school dropout (Mayer, 1995). In turn, many children become chronic and costly burdens to society as adults (addicts, poorly skilled workers, welfare cases, etc.). It is often difficult for families to seek mental health services given cost, distance, time, and social stigma. Thus, having school-based play therapy services available makes the most sense.

Advantages of using play therapy

There are many advantages to offering mental health services within the school: (1) the child's comfortableness with the setting and school staff, (2) lack of stigma, (3) cost effectiveness, (4) likelihood of treatment compliance, (5) easy accessibility to the child and flexibility of scheduling. There is also access to an interdisciplinary team of the school nurse, school social worker, physical or occupational therapist, speech/language therapist, behavioral coach, special education teachers, music and art teachers, as well as administrative and custodial staff.

Schools are an ideal setting to implement interventions and proactive/preventive measures to help children and families. It can also become a bridge to link services to children coming from special education preschool or therapeutic nurseries (Drewes, 2001a). The school setting allows more children and families to have access to services and professionals than through private agencies. Plus, children are in school for longer periods of the day, attend daily and for a full year, which lends itself to flexibility of scheduling due to holidays, emergency meetings and absenteeism, opportunities to access services without the parent needing to transport the child to another setting, as well as not having the financial burden and loss of work time (Drewes, 2001a).

Play therapy is a natural way for children to express themselves; it is developmentally appropriate, helps prepare the student to learn; it is fun and it works! (Drewes, 2001a). School counselors can utilize play therapy and play-based techniques to help children deal with trauma and reactions to natural disasters, neglect, sexual abuse, homelessness, poverty, divorce, bullying, community and domestic violence, and peer difficulties, among other issues.

The use of toys and play materials makes counseling an inviting and comfortable place that conveys that talking is not always required or expected, establishes that the setting and play therapist are different from the academic environment, and allows the child to be less performance-oriented and more relaxed (Drewes, 2001a). The child quickly understands that the school counselor is there to truly focus on them and understand their communications through play in a nonjudgmental way (Drewes, 2001a).

> Since play is the language of the child, play provides a medium for building the essential relationship between counselor and child. The counselor is able to enter into the child's emotional world as it is freely revealed and acted upon by the child.
>
> *(Landreth, 1983, p. 202)*

Finally, school play therapists can use their knowledge of play as a useful tool with school personnel and parents, in evaluating and providing clinical interventions for children. They can share their knowledge of developmentally appropriate play-based curriculums along with implementing play-based programs or training for teachers and counselors. Play observations and play-based assessments can be incorporated into the child's individual evaluation giving a broader and more comprehensive view of the child's development on all levels. And, the use of play-based interventions and techniques can also be used to help to communicate with students, develop a wide range of skills, improve peer relationships and social skills, prevent bullying and school violence (especially through group classroom lessons), address the needs of at-risk students, and remove emotional and behavioral obstacles to learning.

Challenges of using play therapy/techniques in the school

Specific barriers to facilitating play therapy in schools include lack of training in play therapy; time available with students; facilities; supplies and administrative support; time consumed with non-counseling duties; and belief that play therapy is not effective (Drewes & Schaefer, 2011; Ray, 2010). Sixty-seven percent of counselors surveyed had no university-level courses in play therapy, highlighting a lack of training (Perryman & Doran, 2010). Now, through the efforts of the Association for Play Therapy (www.a4pt.org) there are state and national conferences, online and distance training, play therapy centers and universities that offer graduate courses, an abundance of approved providers offering training, and numerous Registered Play Therapist-Supervisors available for supervision.

School administrators may discourage using play therapy because it crosses the boundary into being therapy and no longer counseling. Often the term 'therapy' is an emotionally laden one that many parents associate negatively to deep emotional disturbances or stigma of mental illness along with judgments of their parenting skills. School administrators might consider play therapy outside the scope and expertise of the school counselor (Drewes, 2001a). Ways to get around the term 'therapy' is by using developmental play, play counseling (or play counselor), play development, counseling with toys, emotional growth through play (Drewes, 2001b; Landreth, 1983).

Limited time

School counselors are consistent in identifying lack of time spent with children as the primary barrier to facilitating play therapy. School psychologists are limited by assessment needs and report writing, with school counselors mired in hall and bus duty, test administration, substituting in the classroom, administrative duties and clerical tasks, referral and classification meetings, classroom observations, traveling to other buildings to offer services, along with frequent interruptions during sessions by loud speaker announcements, students or staff entering the office, phone calls or emergencies in dealing with behavioral outbursts by children (Drewes, 2001a; Ray, 2010). Assuring privacy and guarding the importance of the counseling session is vital for the child to gain trust and that their time and play space is respected and honored. Interruptions can be handled by having a sign or door hanger stating "In Session. Please Do Not Disturb," turning off phone ringers, asking office staff to intercept phone calls, and politely but firmly limiting intrusions (Drewes, 2001a; Ray, 2010).

In addition, teachers may have the expectation that the child be 'fixed' or 'straightened out,' seeking a 'quick fix.' They may feel resentment toward play in sessions, seeing it as a reward, just 'playing around,' or being too permissive, and not productive in eliminating behaviors. The school counselor will need to explain how play therapy helps accomplish the school's educational objectives; that it is a process that takes time; and that working through play helps the child develop self-control, assume responsibility for their actions, and express their feelings (Drewes, 2001a). Offering in-service presentations to school staff on what play therapy is, demonstrating some play-based activities, and inviting school staff to see your office and materials can lessen negative views about what you are doing (Drewes, 2001a; Perryman & Doran, 2010; Ray, 2010).

During open school evenings when parents meet their child's teachers, put down footprint shaped colorful stickers on the floor heading toward your office, so parents can see your room, materials, and understand what you do. Having a parent newsletter, brochure, or handout sharing what play therapy is; a monthly tip about the benefits of play; or a game the parent can use with their child can help dispel any stigma or negative views about play therapy being used with their child (Drewes, 2001a).

Teachers may also resent having a child just seen in counseling disrupt the classroom once back. The school counselor needs to take time at the end of the session for re-integration back into the classroom through use of mindfulness activities, deep breathing, and affect regulation techniques so the child returns calm. If clearly agitated, then an alternate "cool down" location such as the principal's office, library, or crisis room can be utilized, along with notifying the teacher of the delay in returning back to class.

Difficulty getting a child released from academics for counseling sessions, depending on time of year, is a major problem. Research reports that scheduling counseling was more difficult during standardized testing semesters than during alternate semesters (Ray et al., 2008). Scheduling issues can be reduced by seeing children before the start of the school day, during lunchtime or reducing sessions from 30 minutes to two 15-minute sessions weekly especially for children with lots of energy, poor impulse control, or short attention span. Teachers can also be helped to utilize techniques in the classroom to address anger management and release of feelings, as well as having the school counselor conduct group lessons on bullying or feelings, thereby giving the teacher a 'break' during the day. This collaborative effort and shifting of services to the classroom can help lessen referrals made by teachers as well as lessen reluctance by teachers to allow a child to be pulled out for counseling.

Counseling time requires structure, predictability, and safety in the therapy room. I find it most effective having sessions broken down into components: (1) a short beginning (checking in, sharing updates from parents, teachers, reviewing homework tasks from the last session, following up on anything unfinished or noteworthy from the last session), (2) structured task (play-based activity working on treatment goal) of about five to ten minutes, (3) some 'free' play utilizing play stations in the room (around ten minutes), (4) structured ending of deep breathing or affect regulation activity while cleaning up (five minutes), and then (5) transition back to class.

Space limitations

The ideal space should be 150–200 square feet (Ginott, 1961), but often this is a luxury with no adequate space at all to have a separate play room and there may be no space for materials or resistance toward having materials left in the space shared by other school personnel. Lack of locking file cabinets, direct phone line, comfortable chairs, and privacy of work can be daunting. Flexibility and creativity can overcome these hurdles. Use of a vacant office or classroom, part of the cafeteria or library, a corner of a classroom, a hallway area, a large storage closet, school stage, cafeteria, or auditorium can be transformed into a play space with a portable play therapy kit containing essential materials, put into a suitcase with wheels. One school counselor converted a school bus into a play therapy office used to travel to various schools where services were given. Grants or special funding may provide money for a central location where children can be bused for counseling. Confidentiality and limits on the amount and volume of noise by children being seen needs to be considered, especially if you are using a public area like a hallway or cafeteria where people walk through. The child should know that what they say and do may be heard or seen by others (Drewes, 2001b).

An office can be divided, using a bookcase or other divider to separate work area from play area. Even putting a sheet over the desk area to limit distractions and touching personal items can help to delineate which area of the room is for play. Use of a blanket on the floor can also create a boundary for play, especially for more active children and preschoolers. Use of bins, baskets, or Tupperware containers can hold items and keep the room clutter-free. Use of a bookcase to store play materials can be covered with a sheet attached with Velcro so that it is out of sight during meetings with staff or parents. Overall, the room should be designed to be warm, inviting, and comfortable.

A shoe bag, with separate slots to hold shoes can be hung from the inside of a door, to visibly store puppets and be inviting. Bins with wheels can store miniatures for a sand tray, and use of a blue plastic container (sweater bins or a litter box) with a lid and two or three inches of sand in it, on the floor with a plastic mat underneath, can double as a sand tray. Children can design their own placemats for use when pounding clay or messy artwork. I have the child draw pictures or designs on 11 × 17-inch paper and add an affirmation ("I am lovable," "My smile is magical," etc.) they can make up or take from affirmation cards or books. The paper is laminated on both sides. The placemats are available as needed, and at termination the child gets their placemats to keep. The personal affirmations are reviewed with the child and reinforced when they use the placemats.

Specific areas can be designated, such as as a family corner, art corner, block corner, and feelings corner, with storage bins similarly labeled. The family corner would have dolls, bottles, doll clothes, toy dishes, and play foods; the block corner could have blocks, Legos, a mat that has roads and towns on it, small cars, trucks, emergency vehicles (police, fire, ambulances, military), airplanes; the feelings corner would have a variety of puppets (at least two each of people and animals – domestic and wild), dress up clothes, different hats, small family people, bubble wrap to stomp on and pop, clay to pound, affirmation mats, a non-breakable mirror to see how their faces look when angry, sad, etc., sand tray and additional miniatures, dinosaurs and aggressive and domestic animals; the art corner would have a variety of expressive arts materials and paper (Drewes, 2001b). The child then can pick at least one area to play in during the nondirective time, and then items would be returned to their designated area and bins at the end of the session.

Expense of materials

Lack of monies for materials can hamper conducting play therapy. However, there are many creative ways to obtain materials: garage/yard sales, hand-me-down toys from siblings or relatives, thrift stores, Dollar Stores, going to a local hardware store or chain (like Home Depot or Lowe's) and asking for donations of dust-free, pebble free sand, the school's Parent-Teacher Association, which might be able to give a grant for a particularly expensive item (like a Sandtray), and sending notes to parents asking for particular toys. However, be selective in what is used. Think of the therapeutic value and whether it can be easily replaced (Drewes, 2001b).

Types of play materials

The literature abounds with theoretical reasons regarding which toys should be selected, why, as well as how to utilize them (Drewes, 2001b). The general consensus is that whatever material or toy is placed in the room, it is there for a specific purpose. The objects should be selected so that the child is free to focus on internal processes and toys allow for maximum expression. No object should be included just because it is a popular fad. The toys are placed in a consistent location within the room to create (1) a therapeutic atmosphere of trust (toys will be there each week), (2) predictability (always in the same location), and (3) a child's personal capability (no help is needed) (Drewes, 2001b). By removing of misplaced, broken, or complicated toys, the room becomes stress-free and ready for the child to resume where they left off last session and getting to issues. The toys should be selected rather than accumulated (Landreth, 1991). Landreth states that:

Toys and materials should be selected to facilitate the seven essentials in play therapy:

> 1) the establishment of a positive relationship with the child, 2) the expression of a wide range of feelings, 3) the exploration of real-life experiences, 4) reality testing of limits, 5) the development of a positive self-image, 6) the development of self-understanding, and 7) an opportunity to develop self-control.
>
> *(1991, p. 117)*

Kottman (2003) categorizes materials into five general areas that include family/nurturing, scary toys, expressive toys, and pretend/fantasy toys. Most play therapists utilize puppets, dolls with baby bottles, family figures, folding doll house with furniture, toy dishes and utensils, play food, culturally sensitive expressive arts materials (crayons, markers, paper, play dough, clay/plasticine or Model Magic or Crayola's Dough (a soft, marshmallow texture, but will air dry hard and can be painted)), scissors, glue sticks, tape, toy phones, small cars and trucks, emergency vehicles, small airplanes, small people figures (family group, police, army, action figures), zoo and farm animals, medical kit (with stethoscope), play money, and blocks. If you are also utilizing directive approaches (Prescriptive, Adlerian, Integrative Play Therapy, CBT) board games (chess, checkers, Sorry, Uno, Talking, Feeling, Doing Game, deck of cards, etc.) should be included for older children.

Although child-centered play therapists may include toy guns, rubber knives, play swords, handcuffs, and a punching bag/blow up vinyl Bobo doll, schools often have a zero-tolerance toward aggression by children. A child using their thumb and forefinger to mimic the shape of a gun in class to shoot the teacher can result in being expelled from school. The school counselor needs to consider the possible conflict within the child in being allowed to aggressively play out issues using forbidden school items, and possibly announcing this in the classroom or to a parent. The school administration and teachers should be made aware that these materials are used in play therapy and why, to avoid any possible negative consequences to either the child or school counselor.

There needs to be a level of comfort with messy materials and chaos, especially when running groups or working with impulsive children. Toys to avoid for safety reasons include sharp, pointed items, glass, highly personal items that could be destroyed, and complicated or mechanical toys.

Cultural competence and sensitivity

With the rapid increase of immigrant children in schools, school counselors must acquire knowledge and skills, as well as materials to work with them. Diversity in schools includes not only cultural but also ethnic, sexual, and socioeconomic diversity, along with physical special needs. Gil (2006) suggests in order to be cross-culturally competent and responsible, the play therapist needs to build sensitivity, obtain knowledge responsibly, and develop active competence. She recommends "that our collection contain as many representations as possible of the world and its diverse and plentiful cultures" (Gil, 2006, p. 21). Cultural differences can also show in play, language, social interactions, and methods of child rearing, all of which can easily be misunderstood and cause problems (Drewes, 2005). Consequently, we need to be aware of what we say that could convey a value judgment, and also gain an understanding of the student's ethnic and cultural background in order to understand the specific cultural meaning of their play.

Culturally appropriate and diverse items should also be included with the play materials. The Crayola company has a separate line of crayons, markers, paper, and paints that offer a range of skin tones to match that of children being seen. Objects representative of various cultures the children are from should include food items, religious representations, types of housing, multilingual posters, flyers, and information leaflets and brochures about play therapy and your work, ethnic photographs, prints and paintings, music, rugs, fabrics, pottery, etc. Include a map of the world or world globe for reference when children talk about countries of origin, where relatives live, and to see distances between where they may have lived and currently live.

The therapeutic powers of play

The school counselor should have an understanding of what the therapeutic factors of play are, in order to explain to school administrators and parents the need for play and play therapy in the school setting.

Therapeutic factors are the actual mechanisms that effect change in clients (Yalom, 2005). They represent (1) levels of abstraction between general theories and concrete techniques; (2) offer a framework for understanding the origin and treatment of problematic behaviors; (3) refer to specific clinical strategies for obtaining desired change of a dysfunctional behavior; (4) techniques to implement the therapeutic factors (Schaefer & Drewes, 2014; O'Connor, 2010).

There are core agents of change identified by Schaefer and Drewes (2014):

Self-expression and access to the unconscious

Among the major therapeutic powers are its communication power, whereby children express conscious and unconscious thoughts and feelings through play activities than words. Next are teaching powers whereby children attend and learn better when play and games instruct, and the power of abreaction, allowing reliving past stressful events and release of associated negative emotions in a safe environment.

Through specially chosen toys, games, and materials for therapeutic and neutral stimulus qualities, the child can reveal unconscious conflicts via defense mechanisms of projection, displacement, and symbolization (Klein, 1955). With support, the child can transform and integrate unconscious wishes and impulses into conscious play and actions.

Metaphorical insight and narrative creation

Storytelling and use of narratives through puppets, writing, drawing, and mutual storytelling, the child joins in interactive fantasy play. This in turn results in learning a lesson or solution to a problem (Gardner, 1971) in a gradually paced, indirect method, with room for repetition with less emotional arousal than direct confrontation (Frey et al., 2000). Play enables the child to organize fragmented memories and experiences into cohesive, meaningful narratives.

Abreaction, stress inoculation, gradual exposure, and counterconditioning

Traumatic experiences are reenacted and relieved through gradual exposure and desensitization. The child mentally digests and gains mastery over horrific thoughts and feelings.

Attachment and relationship enhancement

Positive emotional bonds and behavior management between parent and child are facilitated. Studies in filial therapy (Ray et al., 2005; VanFleet & Guerney, 2003) and parent-child interaction therapy (Brinkmeyer & Eyberg, 2003) have shown success in relationship enhancement and child behavior management. Through step-by-step, live-coached sessions, the caregiver and child create positive affective experiences which result in a secure, nurturing relationship, and helps with effective limit setting and consistency in discipline.

Power/control

Children feel powerful and in control during their play, especially over disasters and feelings of helplessness. They can make the play world conform to their wishes and needs as they tower over the play materials and determine what and how to play.

Competence and self-control

Play provides children with unlimited opportunities to create through stories, through sand tray or drawings, gaining mastery and skill development resulting in enhancement of self-esteem. In addition, children learn self-control through thought and behavior stopping, helping stop, think, and plan ahead, anticipating consequences of various potential behaviors and actions. Skills are mastered through practice opportunities and positive reinforcement that will generalize into other settings.

Sense of self and accelerated development

A child can begin to experience complete acceptance and permission to be oneself without fear of judgment, evaluation, or pressure to change (Axline, 1947). Through reflection on the play, they can understand inner thoughts, feelings, develop an inner self-awareness, and achieve emotional development.

Creative problem solving, divergent thinking, and coping ability

Fantasy play is a cognitive skill associated with the ability to control impulses, delay gratification (Singer & Singer, 1990), increase creativity and divergent thinking (Schaefer, 1999), and cope with stress. Without fear of consequences, children freely come up with novel combinations and discoveries which can aid them in problem-solving of social and emotional issues.

Improvement in cognition and literacy

Play-based techniques teach children more adaptive thoughts and behaviors, social and emotional skills, affect regulation, and impulse control. Since play is enjoyable, interest and attention are enhanced by inclusion in learning tasks.

Reality testing and behavioral rehearsal

Play experiences allow children to practice reading social cues to help differentiate fantasy from reality situations. In social pretend play, children switch back and forth between roles they are playing and their real selves (Schaefer, 1999). Pretend play allows for better discrimination between reality and fantasy (Singer & Singer, 1990).

Rapport building and the role of the therapeutic relationship

Perhaps one of the most potent factors is the relational component of rapport-building that occurs in response to the playful and fun-loving therapist. And, since "play is the language of the child, play provides a medium for building the essential relationship between the counselor and child" (Landreth, 1983, p. 202).

Practical applications

There is an arsenal of play-based programs utilizing paraprofessionals, teachers, groups within the classroom, and parents. Following is a brief summary of some of the options:

Primary Project: An evidence-based approach. The play therapist teaches and supervises paraprofessionals, child associates, as direct service providers, to use child-centered play therapy as a preventive measure and to maximize school adjustment and other related competencies, as well as reduce social, emotional, school difficulties from preschool through third grade (Peabody et al., 2010). It is most effective for children just beginning to show difficulties or in anticipation of difficulties that will impact learning, rather than children with numerous and deeply entrenched difficulties. A play therapy room is used, and children are seen individually for 12–15 weeks then into play pair groupings.

Child-Teacher Relationship Training (CTRT): Adapted from Landreth and Bratton's (2005) ten-week evidence-based Child-Parent Relationship Therapy (CPRT), using 30-minute sessions. It is for children experiencing mental health difficulties in early childhood, utilizing teachers and paraprofessionals trained and supervised in child-centered play therapy to work within the classroom as well as individually in a playroom. CTRT also focuses on enhancing the child-teacher relationship (Morrison & Pretz Helker, 2010).

Trauma-Focused Group Play Therapy: A child-centered play therapy group paradigm to specifically address traumas from natural and human-made disasters and abuse. Groups are no larger than five children, with toys and materials selected based on their ability to facilitate creative, emotional, exploratory, and expressive play. The school counselor is group leader facilitating play, but does not lead or structure the group, on a weekly basis, over the school year (Shen, 2010).

Group Sandplay in Elementary Schools: Designed specifically utilizing both parallel and joint play in a therapeutic environment, encouraging safety, a sense of belonging, and exploration of self in relationship to others. Called *Sand Tray Friendship Groups*, they meet weekly for 10 to 20 sessions, with the school counselor as witness and facilitator. Each child has an individual sand tray, chooses from a collection of miniature figures and creates world views. Children may tell stories about the worlds they create or participate without using any words (Kestly, 2010).

Group play interventions for children with attention deficit/hyperactivity disorder

Using enjoyable Developmentally Appropriate Games (DAGs), skill development, motivation, social skills, impulse control, and anger/stress management are targeted, resulting in increased self-confidence through group acceptance. Groups meet for 90 minutes (even 30 or 60 minutes), for ten weeks (in or out of classroom), with eight to ten children per group. For highly impulsive children, one adult for every two children is recommended (Reddy, 2010).

Individual play therapy

Multicomponent interventions are needed for reducing challenging disruptive behavior implemented over time and across environments to produce generalizable increases in prosocial behavior. Thus, no single treatment approach can address the complex, multifaceted aspects of children's behavioral problems. Thus, a prescriptive and integrative treatment approach is needed, including family members, teachers, classroom and school environments, teaching strategies, and peer interactions, along with individual and group treatment approaches (Drewes, 2019).

These approaches are typically based on cognitive-behavioral techniques and contingent management strategies (Boxer & Frick, 2008) addressing information-processing deficits and distortions. Child-centered play therapy (CCPT) and Adlerian play therapy also facilitate a relationship and enhance coping skills. By allowing expression of aggressive feelings or behaviors in the presence of an empathic and caring school counselor who reflects, accepts, and limits strong emotions and negative behaviors, the child learns socially acceptable behaviors and their expression (Ray et al., 2015).

Play-based techniques

Collaborative treatment planning (Drewes, 2019)

Treatment is more successful when the child (and family) participate. Creating measurable treatment goals in cooperative and collaborative ways helps ensure the child will work on lessening problem areas. Rip a piece of paper into ten strips, select three issues each about home, school, and family. Leave one piece of paper blank, allowing the child to spontaneously address something not covered, or a 'free' session to skip issues. The child writes on each strip selected goals, and decorates an envelope to hold them. Each session the child picks a piece of paper to focus on, but can decide not to work on it and select another, that then must be addressed. After using a play-based technique, rip a piece of the paper off, then put it back in the envelope. In this way the child sees progress is made toward the goal, although not finished with it.

Feelings balloon (Drewes & Cavett, 2012)

Helps explain what therapy is, as well as quantify feelings. Using a balloon, the child is encouraged to blow it up (or therapist if child is unable or allergic to latex), putting in anxious, angry, and upset feelings. Once it is blown up, using a marker, the child or counselor writes on the balloon various feelings "inside," along with names of people whom the child has negative feelings about. The counselor explains the balloon is like a person's head and heart filled with many feelings and explores what would happen if the balloon kept getting bigger and bigger. It would pop, but in real life all the angry feelings building up might make the child "pop" by becoming aggressive or physical toward others. These feelings get in the way of learning and feeling happy. The child then lets out some air, a little at a time and sees how much smaller the balloon is getting. The analogy is that in each session the child lets out feelings a little at a time, not all at once, in a slow and safe way that the child can handle, thereby leaving room for positive feelings to enter. The child is encouraged to let out the rest of the air, but there's a little residual air left in the balloon. It is explained that we cannot get all of it out, just as we cannot erase the past. The experience is still with us, but the feelings and memories do not have to take up all our energy and emotional space. Later when the child comes into a session agitated or upset, the counselor can refer back to the balloon analogy, stating "I see your feelings balloon is getting filled up again. We can work on letting some of that out in here, safely."

Identification/quantification of feelings – gingerbread person feelings map (Drewes, 2001c; Drewes & Cavett, 2012)

A variation of Color-Your-Life technique (O'Connor, 1983), using a drawing of a gingerbread person with arms outstretched, eyes, nose, smile. For use with teens, younger children and caregiver – child dyads. Helps assess (1) overall range of feelings, (2) where emotions are perceived, and (3) how well integrated the emotions are. It takes a few minutes and can be used in the first session or any time during treatment.

Next to the shape of the gingerbread person, list "happy," "sad," "afraid," "angry," "love," and "worried." Ask the child to choose a few other feelings to include, expanding the feelings vocabulary, and giving the child control over the task as an active participant. (Sometimes surprising words come up: "petrified," "stupid," "anxious," "wishing I was somebody else"). Add the child's word(s) underneath the others. Next, the child chooses a color for each feeling and puts a line of the color next to the feeling word. It does not matter what colors are chosen. The child colors in the part of the gingerbread person where they physically experience each feeling listed. Once complete, the drawing is processed with the child with particular attention paid to whether anger is noted or not, and how anger might manifest in situations. Note how many feelings are integrated, how much color is used and where, any discrepancies (a spot of color representing anger outside the figure; or draw a heart of love but "walled off" by layers of scared, hurt or angry feelings).

Guided relaxation exercise – safe place (Drewes, 2006, 2015)

An effective way to teach deep breathing – as relaxation used any time. The child sits with eyes closed or stares off at something in the distance and imagines being a movie director, using their imagination to make a movie. They are directed to breathe slowly in and out, and think of a time and place when they felt safe. It could be lying in the sun at the beach, hiding under their covers, snuggling with a favorite pet, etc. The camera zooms in and films the location, taking in all that is there. Then the camera shot 'freezes,' with continued breathing slowly in and out, and looking around noticing what they see, smell, hear, and feel, and whether or not people or animals are there, and feel how safe they are, how relaxed their bodies feel in their safe place. Next the safe place is named, preferably one word. It is the key to get back to the safe place any time they wish to go, just by saying the name. The child continues to breathe in and out slowly, focusing on how relaxed and safe they feel. After a few minutes, they are to move the camera back slowly, and in another minute, they will be back in the room with their eyes open. The goal is to create an operant-conditioned link between the deep breathing, feeling safe, and the word chosen. When the child is upset, anxious, angry, they can remember the word, and their bodies will automatically begin to relax (Drewes, 2015). Small pinwheels and blowing bubbles can also be used at the end of sessions to practice deep breathing.

Schools are the ideal setting to utilize play therapy, although there are limitations and difficulties in implementation, the benefits far outweigh them.

References

Abidin, R., & Robinson, L. L. (2002). Stresses, biases, or professionalism: What drives teachers' referral judgments of students with challenging behaviors? *Journal of Emotional and Behavioral Disorders, 10*, 2014–2112.

Axline, V. (1947). *Play therapy: The inner dynamics of childhood*. Boston: Houghton Mifflin.

Barkley, R. A. (2007). Attention deficit/hyperactivity disorder. In E. J. Mash & R. A. Barkley (Eds.), *Treatment of child disorders* (pp. 55–110). New York: Guilford Press.

Boxer, P., & Frick, P. (2008). Treating conduct problems, aggression, and antisocial behavior in children and adolescents: An integrated view. In R. Steele, T. Elkin, & M. Roberts (Eds.), *Handbook of evidence-based therapies for children and adolescents: Bridging science and practice* (pp. 241–259). New York: Springer.

Brinkmeyer, M. Y., & Eyberg, S. M. (2003). Parent-child interaction therapy for oppositional children. In E. Kazdin & J. R. Weisz (Eds.), *Evidence-based psychotherapies for children and adolescents* (pp. 204–223). New York: Guilford Press.

Cochran, J. L., Cochran, N. H., Nordling, W. J., McAdam, A., & Miller, D. T. (2010). Two case studies of child-centered play therapy for children referred with highly disruptive behavior. *International Journal of Play Therapy, 19*(3), 130–143.

Drewes, A. A. (2001a). The possibilities and challenges in using play therapy in schools. In A. A. Drewes, L. M. Carey, & C. E. Schaefer (Eds.), *School-based play therapy* (pp. 41–61). New York: Wiley & Sons.

Drewes, A. A. (2001b). Play objects and play spaces. In. A. A. Drewes, L. M. Carey & C. E. Schaefer (Eds.), *School-based play therapy* (pp. 62–80). New York: Wiley & Sons.

Drewes, A. A. (2001c). Gingerbread person feelings map. In H. G. Kaduson & C. E. Schaefer (Eds.), *101 more favorite play therapy techniques* (pp. 92–97). Northvale, NJ: Aronson.

Drewes, A. A. (2005). Play in selected cultures. In E. Gil & A. A. Drewes (Eds.), *Cultural issues in play therapy* (pp. 26–71). New York: Guilford Press.

Drewes, A. A. (2019). Play therapy for children with school behavior problems. In H. G. Kaduson, D. Cangelosi, & C. E. Schaefer (Eds.), *Prescriptive play therapy: Tailoring interventions for specific childhood problems* (pp. 251–268). New York: Guilford Press.

Drewes, A. A., & Cavett, A. (2012). Play applications and skills components. In J. A. Cohen, A. P. Mannarino, & E. Deblinger (Eds.), *Trauma focused CBT for children and adolescents: Treatment applications* (pp. 106–123). New York: Guilford Press.

Drewes, A. A. & Schaefer, C. E. (2011). *School based play therapy* (2nd ed.). Hoboken, NJ: Wiley.

Frey, K. S., Hirschstein, M. K., & Guzo, B. A. (2000). Second step: Preventing aggression by promoting social competence. *Journal of Emotional and Behavioral Disorders, 8*(2), 102–112.

Gardner, R. (1971). *Therapeutic communication with children: The mutual storytelling techniques.* New York: Science House.

Gil, E. (2006). *Helping abused and traumatized children: Integrating directive and nondirective approaches.* New York: Guilford Press.

Ginott, H. G. (1961). *Group psychotherapy with children.* New York: McGraw-Hill.

Hamre, B. K., Pianta, R. C., Downer, J. T., & Mashburn, A. J. (2007). Teachers' perceptions of conflict with young students: Looking beyond problem behaviors. *Social Development, 17*, 115–136.

Kestly, T. (2010). Group Sandplay in elementary schools. In A. A. Drewes & C. E. Schaefer (Eds.), *School-based play therapy* (2nd ed., pp. 257–283). New York: John Wiley & Sons.

Klein, M. (1955). The psychoanalytic play technique. *American Journal of Childhood Education, 49*, 73–74.

Kottman, T. (2003). *Partners in play: An Adlerian approach to play therapy.* Alexandria, VA: American Counseling Association.

Landreth, G. L. (1983). Play therapy in elementary school settings. In C. E. Schaefer & K. J. O'Connor (Eds.), *Handbook of play therapy* (pp. 200–212). New York: Wiley& Sons.

Landreth, G. L. (1991). *Play therapy: The art of the relationship.* Muncie, IN: Accelerated Development.

Landreth, G. L., & Bratton, S. (2005). *Child parent relationship therapy.* New York: Routledge.

Mayer, G. R. (1995). Preventing antisocial behavior in the schools. *Journal of Applied Behavior Analysis, 28*, 467–478.

Mental Health America. (2009). *Factsheet: Recognizing mental health problems in children.* Retrieved from www.mentalhealthamerica.net/farcry/go/information/get-info/children-s-mental-health/recognizing-mental-health-problems-in-children.

Morrison, M. O., & Pretz Helker, W. (2010). Child-teacher relationship training. In A. A. Drewes & C. E. Schaefer (Eds.), *School-based play therapy* (2nd ed., pp. 1818–195). New York: John Wiley & Sons.

Myers, S. S. & Pianta, R. C. (2008). Development commentary: Individual and contextual influences on student-teacher relationships and children's early problem behaviors. *Journal of Clinical Child & Adolescent Psychology, 17*, 600–608.

O'Connor, K. J. (1983). Color-Your-Life technique. In C. E. Schaefer & K. J. O'Connor (Eds.), *Handbook of play therapy* (pp. 251–258). New York: Wiley.

O'Connor, K. J. (2010). *Beyond the power of play: Using therapeutic change processes in play therapy.* Presented at the Association for Play Therapy Conference, Louisville, KY.

Peabody, M. A., Johnson, D., & Hightower, A. D. (2010). Primary project: An evidence-based approach. In A. A. Drewes & C. E. Schaefer (Eds.), *School-based play therapy* (2nd ed., pp. 163–180). New York: John Wiley & Sons.

Perryman, K., & Doran, J. (2010). Guidelines for incorporating play therapy in the schools. In A. A. Drewes & C. E. Schaefer (Eds.), *School-based play therapy* (2nd ed., pp. 61–86). New York: John Wiley & Sons.

Ray, D. C. (2010). Challenges and barriers to implementing play therapy in schools. In A. A. Drewes & C. E. Schaefer (Eds.), *School-based play therapy* (2nd ed., pp. 87–106). New York: John Wiley & Sons.

Ray, D. C., Armstrong, S. A., Balkin, R. S., & Jayne, K. M. (2015). Child-centered play therapy in the schools: Review and meta-analysis. *Psychology in the Schools, 52*(2), 107–123.

Ray, D. C., Bratton, S., Rhine, T., & Jones, L. (2005). The efficacy of play therapy with children: A meta-analytic review of treatment outcomes. *Professional Psychology: Research and Practice, 36*, 376–390.

Ray, D. C., Henson, R., Schottelkorb, A., Brown, A., & Muro, J. (2008). Effect of short-and long-term play therapy services on teacher-child relationship stress. *Psychology in the Schools, 45*(10), 994–1009.

Reddy, L. A. (2010). Group play interventions for children with attention deficit/hyperactivity disorder. In A. A. Drewes & C. E. Schaefer (Eds.), *School-based play therapy* (2nd ed., pp. 307–332). New York: John Wiley & Sons.

Schaefer, C. E. (1999). Curative factors in play therapy. *Journal for the Professional Counselor, 14*(1), 7–16.

Schaefer, C. E., & Drewes, A. A. (Eds.). (2014). *The therapeutic powers of play: 20 core agents of change* (2nd ed.). Hoboken, NJ: Wiley.

Shen, Y. (2010). Trauma-focused group play therapy in the schools. In A. A. Drewes & C. E. Schaefer (Eds.), *School-based play therapy* (2nd ed., pp. 237–256). New York: John Wiley & Sons.

Singer, D. G. & Singer, J. L. (1990). Imaginative play in childhood: Precursor of subjunctive thoughts, daydreaming, and adult pretending games. In A. Pellegrini (Ed.), *The future of play theory* (pp. 187–219). Albany: State University of New York Press.

VanFleet, R., & Guerney, L. (Eds.). (2003). *Casebook of filial therapy*. Boiling Springs, PA: Play Therapy Press.

Yalom, I. D. (2005). *The theory & practice of group psychotherapy* (5th ed.). New York: Basic Books.

33

THE HEALING POWER OF PLAYFUL ANIMALS

Animal Assisted Play Therapy® as an intervention for childhood bullying in South Africa

Monique Van Loggerenberg and Risë VanFleet

Children throughout the world are drawn to animals. Research has shown that they talk about them, draw them, think about them, and have dreams about them (Jalongo, 2004; Melson, 2001; Melson & Fine, 2015). Children seem to enjoy having animals as companions and often describe them as "friends." Guerney's (1991) study of latchkey children (those at home alone, such as after school and before parents came home from work) showed that companion animals helped them cope with being alone. Other studies have shown that children with companion animals have more interactions with peers, greater empathy, and demonstrate more prosocial behaviors and higher self-esteem (Jalongo, 2004, 2014). The involvement of animals in therapy work is increasing rather dramatically in a number of countries around the world, partly because of this natural and seemingly universal interest of children.

Play is sometimes referred to as the "language of children." While children can express themselves more effectively in words as they grow older, they tend to express far more through their play. Play therapy represents one of the most developmentally appropriate forms of intervention for children who are distressed or experiencing psychosocial difficulties of many kinds. Play therapy, too, has been growing rather dramatically throughout the world. Play is universal among children the world over, as long as they are permitted to play.

Combining these two areas of vital interest to children, animals and play, has great potential for helping children with a range of difficulties. The field of Animal Assisted Play Therapy® (AAPT) has been developed to integrate these approaches in a systematic manner. This chapter provides an overview of the practice of AAPT, and then explores the problem of bullying in middle childhood and how AAPT and, more specifically, Equine Assisted Play Therapy® (EAPT) can be applied to address it. The program and research described here have taken place in South Africa and are run by the first author who is certified in AAPT.

Overview of Animal Assisted Play Therapy®

Animal Assisted Play Therapy® (AAPT) is defined as

> the integrated involvement of animals in the context of play therapy, in which appropriately trained therapists and animals engage with clients primarily through systematic playful interventions, with the goal of improving clients' developmental and psychosocial health, while simultaneously ensuring the animal's well-being and voluntary engagement. Play and playfulness are essential ingredients of the interactions and the relationship.
>
> *(VanFleet & Faa-Thompson, 2017, p. 17)*

The approach fully integrates animal-assisted therapy with various forms of play therapy, as well as several other fields such as animal behavior and ethology. It requires considerable training for both humans and animals to conduct properly, but it is very useful in breaking down barriers, increasing motivation, and providing a template of healthy relationships within which clients can explore, express, and evolve to meet their therapeutic goals. AAPT can be used with clients of all ages, and it can be conducted with individuals, families, and groups. A number of different species can be involved, including dogs, horses, cats, goats, rabbits, and more. Therapists must become well-versed in the species with which they work, including the ability to read their body language fluently.

Guiding principles of AAPT

The practice of AAPT is based heavily on relationship. Therapists create healthy, reciprocal, mutually beneficial, kind, and fun relationships with their own animals, as their interactions with their animals are on display to clients during sessions. These interactions serve as a model and a metaphor for clients as they work to develop healthy attachments and relationships of their own (VanFleet & Faa-Thompson, 2017). Co-founders of AAPT VanFleet and Faa-Thompson (2017) have articulated the nature of these relationships in their book. The guiding principles of AAPT reflect the values and principles of the approach. These are briefly described here.

Respect. AAPT seeks to show equal respect of both clients and animals. The needs of all living beings involved in the sessions are considered of equal importance.

Safety. AAPT interactions and interventions are kept physically and emotionally safe. Unsafe activities are stopped and interactions are redirected to ensure safety.

Enjoyment and Choices. It is a requirement that AAPT sessions are enjoyable and pleasant for the animals as well as the clients. While difficult material certainly comes up in the course of therapy, the overall atmosphere of lightness is maintained. Furthermore, both clients and animals have a choice as to whether or not to participate. The animals also sometimes select the activities, just as clients do. This unique approach allows the development of the client-animal relationship that can be so helpful.

Acceptance. The AAPT therapist accepts the clients and the animals for who they are. They attend to the feelings of both and respond accordingly. A "goodness of fit" model is used to match animal personalities and preferences with the type of work they are invited to do.

Training. In AAPT, only animal-friendly, non-aversive, and humane positive training and handling methods are used. This is important not only for animal welfare but also for the message given to clients about relationships.

Relationship. The AAPT process focuses in mutual relationships rather than control. That applies to the therapist-animal relationship, the client-animal relationship, and the therapist-client relationship.

Empowerment. This approach encourages growth and independence, and the strengthening of confidence and competence. The same processes to engage clients fully in their own development apply to the animals as well. If the animals choose not to participate, they are not pushed to do so.

Process. AAPT is a process-oriented type of therapy. Sessions do focus on specific activities or goals, but the overall process is of greatest concern. Any time that animals are involved there can be unexpected behaviors or situations, and AAPT therapists learn how to be prepared for them and to handle them.

Foundations. AAPT is grounded in well-established theories as well as thoughtful consideration of clinical best practices, a wide range of play therapy interventions, lifespan development, ethical professional practice, and humane treatment of animals. These and other foundational knowledge and skills need to undergird all that happens within AAPT. Training, supervision, and animal welfare are valued and required so that the foundations are strong on which AAPT is then built.

Professional competencies

AAPT can be used to enhance therapeutic work toward a wide range of goals, including very challenging ones. Appropriate animals can be enlisted to assist with problems such as poor self-regulation, trauma and attachment difficulties, relationship problems, anxiety, depression, selective mutism, autism spectrum challenges, school issues, conduct problems, personality disorders, and even with clients who have harmed animals. There are many more. In AAPT one does not rely on some nebulous "magic" of the animals, but still engages in therapeutic interventions just as if no animal were available. Then, the therapist considers ways in which the animals can assist or enhance efforts. Creative interventions can involve calming activities with the animals, but also very active play such as going through obstacle courses, teaching the animals new behaviors, or active free play.

To do all this well, therapists need competencies that take some time and effort to develop. VanFleet and Faa-Thompson (2017) have described these competencies in their book. These are listed here and defined more fully in the book.

- *Therapy skills* – setting goals, conducting sessions, processing client material
- *Play therapy competence* – solid knowledge of play therapy methods, ability to conduct play therapy with a range of presenting problems
- *Animal behavior* – clear knowledge of animal behavior, including ethology, classical and operant conditioning, animal body language
- *Animal handling* – knowledge and skills for caring for and moving with animals, including how to walk with the animal on a leash/rope, avoidance of intrusive human behaviors, selection of animal-friendly equipment, ability to anticipate stressful situations from the animal's point of view
- *Split attention* – the ability to simultaneously watch client and animal behavior to determine what they each might be feeling and experiencing, thereby enabling one to take appropriate action and prevent miscommunication and interaction problems
- *Facilitation of client-animal interactions* – knowledge and skills for selecting appropriate interventions that involve clients and animals, helping clients process what is happening in the

session to reach their goals, and commenting and behaving in ways that help the clients develop healthy and enjoyable relationships with the animals, all in the service of client goals

- *Ethics and welfare* – thorough knowledge and decision-making skills to practise ethically vis-à-vis the clients and with full attentiveness to animal welfare
- *Relationship* – an understanding of what makes secure attachments and healthy relationships, and the ability to place relationship at the core of all that occurs in AAPT

Training workshops and a full certification program are offered through the International Institute for Animal Assisted Play Therapy® (IIAAPT) at www.iiaapt.org. This involves online courses, in-person hands-on workshops, supervised practice, and demonstrated competencies. This is an active international program with professionals from four continents certified to date.

AAPT in action: the use of EAPT in South Africa as a method to address being bullied during middle childhood

The rest of this chapter focuses upon the work of the first author (Monique Van Loggerenberg) in using Equine Assisted Play Therapy® (EAPT) in South Africa to address bullying problems. This includes a background of the problem, how AAPT in general can be helpful, the research project that explored the application of EAPT with this population, and a case study.

Bullying is an increasing global phenomenon (Rezapour, Khanjani & Soori, 2019; Sikhakhane, Muthukrishna & Martin, 2018; Kumar, Stern, Subrahmanian, Sherr, Burton, Guerra, Muggah, Samms-Vaughan, Watts & Mehta, 2017) with an estimated figure of 246 million children worldwide being exposed to bullying annually (Greene, Robles, Stout & Suvilaakso, 2013). Specifically, statistics from a South African study (Trends in International Mathematics and Science study (TIMSS) from 2015), revealed that amongst the 49 countries that took part in this study, South African Grade 5 children were bullied the most (Makou & Bourdin, 2018). Although this study focused on trends in mathematics, a segment dealt with aspects that influence students, such as bullying.

Bullying is defined as a form of intentional, repetitive, aggressive behaviour, with the aim to hurt someone else physically or emotionally (Lohmann & Taylor, 2013; Coloroso, 2011) with the perceived occurrence of an imbalance of power (Lohmann & Taylor, 2013). A bully is considered a person who habitually seeks to harm or intimidate those whom they perceive as vulnerable (Oxford Dictionary, 2018). A victim on the other hand is perceived as someone who has been bullied by another person but who does not elicit bully behaviour on more than one occasion over a two-month period (Olweus, 2010), also known as passive victims (Fleischer & Schwartz, 2003). Another category exists known as bully-victims who are considered victimised children who become bullies themselves displaying disruptive and impulsive emotional reactions that may act as reinforcement to their perpetrators (Panayotou, Fanti & Lazarou, 2015) thus perceived as fluctuating between being the bully and the victim depending on the situation (Nepgen, 2016). The bully-victim is also called the aggressive victim (Boswell, 2016).

For the purpose of this contribution, victims include both passive and aggressive victims as victims of bullying do not intentionally invite bullying (Van Loggerenberg, 2019).

Being a victim of bullying can result in several psychosomatic symptoms such as stomach aches, headaches, or disturbances in sleep patterns. Bullying not only affects the victim but the bully as well (Powell & Ladd, 2010). Victims reported disturbances in social functioning (van Dijk, Poorthuis & Malti, 2017), resulting in parents or teachers often referring the victim for therapy.

When referred for therapy, it has been noted that during conventional forms of therapy children often display resistance to being "treated" due to being bullied; perhaps due to them getting teased and labelled by peers even more when it becomes known that they are receiving treatment. This observation is supported by Barish (2018) who notes that resistance to therapy is a problem for child therapists of all theoretical orientations. To overcome this hurdle, different therapeutic methods believed to be "child friendly" were considered for this study. One clear choice was play therapy, as Landreth (2012) emphasises that for children, toys are their words and play is their conversation. Thompson (2009) posits that children often display emotionally significant experiences during play, and work through them with their play themes (VanFleet, Sywulak & Sniscak, 2010). Even with play therapy, resistance to therapy has been noted, as traumatised children can perceive play therapy as threatening and avoid engaging with the therapist (Ayling, Armstrong & Clark, 2019). The inclusion of a live animal during play therapy shifts the focus from the child and therapist to a mutual companion – the animal (VanFleet & Faa-Thompson, 2017).

The benefits of play during therapeutic sessions is well known, but the ethical inclusion of a live animal during play therapy sessions is a relatively new phenomenon in South Africa. In South Africa animals are included in therapeutic sessions for various reasons, e.g. occupational therapists involve horses during hippotherapy, social workers include dogs during prejudicial events to calm children down and prepare them for court procedures, therapists and counsellors also include animals during therapeutic sessions. Even so, little formal training is available such as that provided by the International Institute of Animal Assisted Play Therapy® (IIAAPT). This conceptualization of the needs of bullied children led to a formal PhD study to highlight the need for an international body such as the IIAAPT that can provide ethical guidelines on how animals should be incorporated during therapy sessions, and rather than being seen as "tools", viewed as sentient beings whose needs should be considered during therapy just as much as human needs are taken into account.

The PhD study by the first author of this contribution was conducted according to the ethical clearance provided by the Nelson Mandela University in Port Elizabeth, South Africa. This research project posed to determine if Equine Assisted Play Therapy® can be perceived as an effective therapeutic method in addressing bullying. Equine Assisted Play Therapy® (EAPT) is considered a subdivision of Animal Assisted Play Therapy® (AAPT) whereby at least two horses are included during play therapy sessions. AAPT is a relationship-focused therapeutic intervention (VanFleet & Faa-Thompson, 2017) and bullying points to an asymmetrical relational problem (Limber, Olweus, Wang, Matthew & Breivik, 2018) whereby the victim experiences the dire consequences of being bullied. Specifically, horses are chosen for EAPT sessions due to their big and powerful appearance. During EAPT sessions no forceful behaviour is allowed by anyone, and horses choose if they want to partake in activities or not. The children are given the same consideration in allowing them to choose whether or not they want to involve the horse in activities. Therefore, AAPT was considered a good therapeutic fit whereby the therapist could assist the victim in rectifying false beliefs regarding relationships with self and others. "Goodness of fit" within a therapeutic context suggests that the child should be screened and evaluated to determine if the suggested therapeutic intervention would best serve the child and the therapeutic goals (McClowry, Rodriguez & Koslowitz, 2008).

One of the consequences of being bullied is a lack of self-esteem due to the bullied children being at a loss about how to handle the situation, as posed by Smith (2012). Recent literature is beset with confusing terminology trying to explain the term self and self-esteem. For clarity Figure 33.1 displays how self-esteem was regarded for this study.

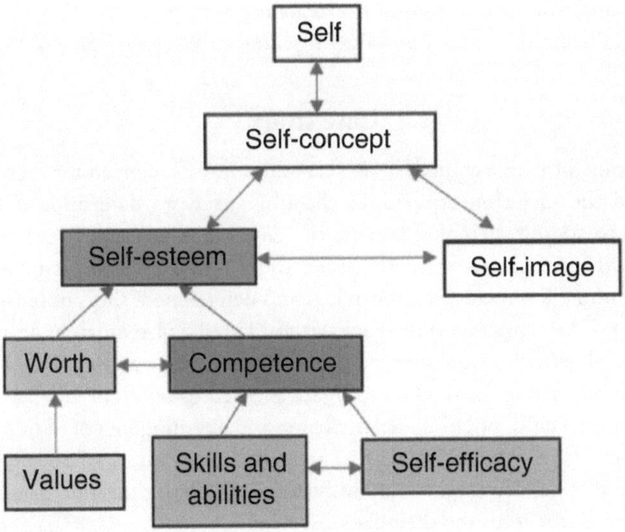

Figure 33.1 Schematic description of self-esteem

The ***self-concept*** (Figure 33.1) consists of a global overarching view of self, relative to the rest of the world (Geldard, Geldard & Yin Foo, 2017; Akande, Akande & Odewale, 2015; Tracy, 2013). ***Self-esteem*** relates to how a person feels about him/herself (Jordaan, 2013). It is at the self-esteem level that therapists can influence children's wrongly formulated perceptions. Foster, Kernis and Goldman (2007) suggest that the best time for intervention programmes are periods during which rank-order stability is relatively low, as self-esteem is particularly malleable during such periods of relative upheaval in self-concept. Such a period is during middle childhood, which is approximately between the ages of 6 and 12 years. Although the play of a 6-year-old child usually differs from the play displayed by a 12-year-old, the common ground during this study was that all participants were victims of bullying and suffered from low self-esteem.

Therapeutically, the goal was to improve the child-participants' self-esteem and equip them with skills to curb being bullied. Since EAPT can be adapted to various therapeutic orientations (VanFleet & Faa-Thompson, 2017) and in this study was conducted in individual sessions, each session could be adapted to the individual child's age and needs.

The parents of children included in this study were not directly involved in the therapy sessions, but they formed an important part of this research project. Since children cannot attend therapy without the consent of their parents, it is important that the perceptions of the parents regarding the therapeutic intervention could be obtained. Parents experience their children in a different context than the therapist and can provide valuable information regarding the behavioural changes as displayed by the child.

During the exit interview at the conclusion of the tenth therapy session, the parents revealed that their children did not perceive therapy sessions as "therapy" and referred to them as "visiting the horses." Statements like these indicate the shift in focus from the child's perspective. The children perceived the sessions positively while reaching their therapeutic goals, pointing to one of the IIAAPT values – that therapy should be enjoyable for both therapy client and animal whilst not deducting from the seriousness of reaching therapy goals. To illustrate how EAPT was implemented during this study the following illustration of a therapy session is provided.

Case study

Joshua (name changed for anonymity), a 10-year-old boy, was referred for therapy due to being bullied. He fitted the inclusion criteria for the PhD study to determine if EAPT could be deemed effective to address the consequences of being bullied. At the onset of the therapeutic sessions Joshua's self-esteem was measured using the Culture Free Self-esteem Inventory, 3rd edition. This inventory was included to curb bias and determine if shifts in self-esteem occurred after the ten-week EAPT intervention. Each session lasted approximately 45 minutes. During these sessions several activities were performed, always involving at least one of the horses.

An example of one activity was when Joshua was asked to build his safe place. He was given the opportunity to make use of poles, tables, chairs, and anything he could find to build a place which would make him feel safe. He was also given the opportunity to decide how big his safe place should be and if the therapist should assist him in building this safe place. He decided to build the safe place alone, with the therapist observing.

Upon completion he was given a bucket with carrots. The carrots were a que to the horses that they may form part of the activity and play along. To the child the carrots signified aspects of himself he considered valuable. He named each carrot, e.g. I like my eyes; I like that I can play good rugby. Joshua needed to protect his self-esteem, in the form of the carrots, while the horse was described as a "bully trying to breach Joshua's safe place." The therapist reflected the interactional process between Joshua and the horses. For example, the therapist would awaken awareness that Joshua was "teasing" the horses by holding the bucket of carrots close and then quickly taking it away. Parallels were drawn afterwards between Joshua's interaction with the bully at school. The child's awareness grew about how he might be eliciting bullying behaviour from the bully at times by the way he behaved when the bully was nearby.

The emotional reaction the child experienced when thinking about the bully was addressed using several therapeutic methods such as hitting papers to release some of the anger. During this activity the horses were not involved due to their sensitivity to the environment, and they could have been startled by the sounds of the paper being boxed and the child voicing his anger. Once the child expressed that he had no anger left, the child and therapist returned to the horses.

Different ways of behaving around the bully were then explored. The reactions of the horses upon experiencing the new way the child behaved was explored by both the therapist and child, focusing on the body language the horses displayed. (In AAPT and EAPT, children's awareness of animal body language is enhanced so they are more aware of animals' feelings.) The therapist and child then evaluated if the horses reacted differently when the child displayed the new behaviour. They both considered the possibilities of incorporating this new behaviour around the bully at school. The session ended with the child building relationship and rewarding the horses for good behaviour by giving them carrots, which also provided the opportunity for the therapist to enquire how the child can reward himself if he got to display the new behaviour at school.

This example of one of the sessions demonstrates how effective EAPT can be in providing immediate feedback to children in therapy, something that is often not possible in an office without another sentient (animal) being expressing and giving feedback via body language to the child's behaviour. Although children often don't recognise the displayed emotional disposition of the animal, the therapist acts as interpreter, voicing the disposition of the horse, pointing out body language, and awakening awareness in the child regarding his/her behaviour and the influence it has on the animal. The therapist should therefore be able to split her attention between the child's needs and the animal's needs (VanFleet & Faa-Thompson, 2017).

The study concluded that there was an increase in all the child-participants' global self-esteem. Battle (2002) posits that an increase in global self-esteem indicates that a child perceives him/herself as acceptable and likeable and embraces a sense of belonging. Unfortunately, the results of this study cannot be generalised to the greater population due to it incorporating only seven case studies.

The goal of this research was to gain deeper insights regarding the therapeutic process and how it was perceived by parents, teachers, and the bully victims. All three groups reported that the children perceived EAPT as positive. If parents do not deem this therapeutic method to be effective, they will not allow their children to attend this type of therapy. This study, therefore, focused on the in-depth perceptions and recommendations of parents to make EAPT, as a relatively new therapeutic intervention, even more effective. The parents and teachers were given a voice during this study whereby they could highlight which aspects they deemed valuable and effective and if the overall process was considered worthy of their financial and emotional efforts.

It was recorded that all the parents deemed EAPT as "value for money" and "effective." Some of the parents commented that although they did not form part of the sessions, they valued that they were given feedback after each session. The parents were asked to encourage changes in behaviour and document these behaviours as they occurred. Some of the remarks of parents also included that they didn't know why the therapy was so successful, but that they considered it to be the combination of the horses and therapist. As one of the children commented: "I like the therapist even more when the horses are there." This comment points to the triangular relationship that was present during the therapy sessions, whereby the therapist was considered more approachable when the child saw the positive relationship between horses and therapist.

References

Akande, A., Akande, B., & Odewale, F. (2015). Putting the self back in the child – an African Perspective. *Early Child Development and Care, 103*, 103–115.

Ayling, P., Armstrong, H., & Clark, L. G. (2019). *Becoming and being a play therapist: Play therapy in practice*. London: Routledge.

Barish, K. (2018). How to be a better child therapist: An integrative model for therapeutic change. New York: W. W. Norton.

Battle, J. (2002). *Culture-free self-esteem inventories. Examiner's manual* (3rd ed.). Texas: Pro-ed Inc.

Boswell, M. A. (2016). *School level predictors of bullying among high school students*. Lexington: University of Kentucky Press.

Coloroso, B. (2011). Bully, bullied, bystander . . . and beyond: Help students choose a new role. *Education Digest: Essential Readings Condensed for Quick Review*, 77(4), 36–39.

Fleischer, W. P., & Schwartz, L. (2003). Mental health sequelae of bullying: A review and case report. *The Canadian Child and Adolescent Psychiatry Review, 12*(1), 13–17.

Foster, J. D., Kernis, M. H., & Goldman, B. M. (2007). Linking adult attachment to self-esteem stability. *Self and Identity, 6*(1), 64–73.

Geldard, K., Geldard, G., & Yin Foo, R. (2017). *Counselling children: A practical introduction*. (5th ed., p. 10). London: Sage.

Greene, M. E., Robles, O. J., Stout, K., & Suvilaakso, T. (2013). *A girl's right to learn without fear: Working to end gender-based violence at school*. Retrieved June 27, 2019, from https://plan-uk.org/file/plan-report-learn-withoutfearpdf/download?token=HMORNNVk.

Guerney, L. F. (1991). A survey of self-supports and social supports of self-care children. *Elementary School Guidance and Counseling, 25* (4), 243–254.

Jalongo, M. R. (Ed.). (2004). *The world's children and their companion animals: Developmental and educational significance of the child/pet bond.* Olney, MD: Association for Childhood Education International.
Jalongo, M. R. (Ed.). (2014). *Teaching compassion: Humane education in early childhood.* New York: Springer.
Jordaan, H. J. (2013). *Self-esteem, coping and the psychology of infertility.* Port Elizabeth: Nelson Mandela University.
Kumar, A. K. S., Stern, V., Subrahmanian, R., Sherr, L., Burton, P., Guerra, N., . . . Mehta, S. K. (2017). Ending violence in childhood: A global imperative. *Psychology, Health & Medicine, 22*(1), 1–16.
Landreth, G. (2012). *Play therapy: The art of the relationship* (3rd ed.). New York: Routledge.
Limber, S. P., Olweus, D., Wang, W., Matthew, M., & Breivik, K. (2018). Evaluation of the Olweus bullying prevention program: A large scale study of U.S. students in grades 3–11. *Journal of School Psychology, 69*, 56–72.
Lohmann, R., & Taylor, V. (2013). *The bullying workbook for teens: Activities to help you deal with social aggression and cyberbullying* (1st ed.). New York: Raincoast Books.
Makou, G., & Bourdin, J. (2018). *Reality check: Are SA pupils the 'most bullied' in the world?* Retrieved May 27, 2019, from https://africacheck.org/reports/reality-check-sa-pupils-bullied-world/.
McClowry, S. G., Rodriguez, E. T., & Koslowitz, R. (2008). Temperament-based intervention: Re-examining goodness of fit. *NIH Public Access, 2*(1–2), 120–135, June.
Melson, G. F. (2001). *Why the wild things are: Animals in the lives of children.* Cambridge, MA: Harvard University Press.
Melson, G. F., & Fine, A. H. (2015). Animals in the lives of children. In A. H. Fine (Ed.), *Handbook on animal-assisted therapy: Foundations and guidelines for animal-assisted intervention* (4th ed., pp. 179–194). San Diego, CA: Academic Press.
Nepgen, A. (2016). *Groepwerk met slagoffers van boeliegedrag: 'n maatskaplike werk intervensie.* Potchefstroom: North-West University.
Olweus, D. (2010). *The handbook of bullying in schools: An international perspective.* New York: Routledge.
Oxford Dictionary. (2018). Retrieved March 25, 2019, from https://en.oxforddictionaries.com/definition/bully.
Panayotou, G., Fanti, K. A., & Lazarou, C. (2015). Fearful victims and fearless bullies? Subjective reactions to emotional imagery scenes of children involved in school aggression. *Personality and Individual Differences, 78*, 29–33.
Powell, M. D., & Ladd, L. D. (2010). Bullying: A review of the literature and implications for family therapists. *The American Journal of Family Therapy, 38*, 189–206.
Rezapour, M., Khanjani, N., & Soori, H. (2019). The types of bullying behaviors and its association with general life satisfaction and self-rated health among Iranian pupils. *Journal of Research in Health Sciences, 19*(1), 1–5.
Sikhakhane, H., Muthukrishna, N., & Martin, M. (2018). The geographies of bullying in a secondary school context. *South African Journal of Education, 38*(1), S1–S11.
Smith, H. (2012). A strength-based approach to bullying. *Educational & Child Psychology, 30*(4), 44–58.
Thompson, M. (2009). Animal-assisted play therapy: Canines as co-therapists. In G. Walz, J. Bleuer, & R. Yep (Eds.), *Compelling counselling interventions: Vistas* (pp. 199–209). Alexandria, VA: American Counselling Association.
Tracy, B. (2013). *Motivation.* New York: AMACOM Division of American Management Association.
van Dijk, A., Poorthuis, A. M. G., & Malti, T. (2017). Psychological processes in young bullies versus bully-victims. *Aggressive Behaviour, 43*, 430–439.
Van Loggerenberg, M. (2019). *Perceptions on the efficacy of equine assisted play therapy™ in addressing low self-esteem of young bullied children.* Port Elizabeth: Nelson Mandela University.
VanFleet, R., & Faa-Thompson, T. (2017). *Animal Assisted Play Therapy.* Sarasota, FL: Professional Resource Press.
VanFleet, R., Sywulak, A. E., & Sniscak, C. C. (2010). *Child-centered play therapy.* New York: Guilford Press.

34

CREATING A SAFE SPACE IN A SAFE PLACE

Working creatively with young people in a secure setting

Larry Galloway

Introduction

Over many years of working therapeutically with looked-after young people in the residential care sector, I began visiting children in secure units who had been made the subject of a section 25 of the Secure Accommodation Order (Children's Act 1989).

All the secure units that I visited around the country were different in size and appearance and initially seemed very formidable places with high fences, locked doors and stringent checks on everyone who entered the unit. They were often viewed negatively, but over time I came to respect and admire much of the work that I observed.

Nearly all the young people that I have worked with who have spent time in a secure unit told me that although they hated losing their freedom, for the first times in their lives they were able to feel safe – "it does what it says on the tin, it keeps you safe". Some of those young people at difficult times in their lives even conspired to get themselves back to the safety of the secure unit, not because they were institutionalised but because it bought them time to process what had happened to them in the past, removed them from distractions and kept them safe from harm.

To illustrate the unique experience of working in this setting, I will review the work that I undertook with Ava over a four-year period, covering three separate admissions to secure units during her creative journey.

Changes to the law on deprivation of liberty of children

Over the years several changes have taken place, but it has always been recognised that to deprive a child of their liberty is a serious matter with far-reaching ramifications. Young people have rights under Article 5 of the European Convention on Human Rights (ECHR) not to be deprived of their liberty without legal authorisation, for example, under:

- Section 25 of the Children Act 1989, which provides for the placement of looked-after children in secure accommodation;
- The Mental Health Act 1983;

- The Youth Remand provisions of the Legal Aid, Sentencing and Punishment of Offenders Act 2012.

A child who is being "looked after" by a local authority under a care order or is being provided with accommodation under section 20 of the Children's Act (1989) cannot be placed or kept in accommodation which has the purpose of restricting the child's liberty unless the requirements of section 25 are met. They are:

- That the child has a history of running away (absconding) and is likely to run away from accommodation which isn't secure, and
- If she/he runs away, she/he is likely to suffer significant harm, or
- If she/he isn't in secure accommodation she/he is likely to injure themselves or someone else.

A brief background history of Ava

Ava's parents separated early in her childhood; her dad was a drug addict and her mum, with whom she lived, had poor parenting skills. Ava was on the Child Protection Register for neglect. Mum's parents were closely involved with Ava. They felt that they couldn't offer long-term care, but they were able to provide a degree of consistency and stability as well as affection. The family lived in various parts of the Mediterranean before settling in London. The family continued to present as chaotic and unable to address Ava's challenging behaviour and repeated absconding when she would put herself at serious risk, including being gang raped.

Ava became a witness against a woman who was grooming young girls for sexual exploitation and serious threats to Ava's life were made by the suspect and her associates. A residential care placement was sought for Ava that would put her at a distance from London and the suspect, but Ava continued to abscond back to London from wherever in the country she found herself. Eventually the decision was made to seek a section 25 Secure Accommodation Order so as to maintain her safety, especially whilst the suspect was awaiting trial.

Therapeutic journey

I met Ava at aged 13 when she was placed in residential care in the West of England, and whilst she was not at this point ready to engage with therapy we did form a positive relationship so that when she was placed in a secure unit she said that she did want to start therapy but would only work with someone she knew already and not a stranger. Despite the secure unit being in the South East, which entailed a four-hour journey there and a four-hour journey back, her local authority decided that they should support Ava's request and asked me to begin working with her. The unit was relatively small, only taking eight young people, and trusting relationships could be formed as a 1:1 approach was possible.

Special arrangements had to be made and agreed upon to allow me to bring in my play therapy "tool kit." Normally stringent conditions were in place concerning what would be allowed to be brought into the unit, including leaving my car keys in a locker and removing staples out of the middle of magazines. I wanted to build trust with Ava and provide her with materials that would encourage her creativity and imagination, including play items such as cap guns, handcuffs and scissors.

During her first session we discussed three simple ground rules:

1 I would never hurt her,
2 She would not hurt me, and
3 She wouldn't deliberately damage any of the creative materials.

These rules she felt made sense to her and seemed fair and she said that she would "never hurt me" and felt that she could "trust me." We also covered confidentiality and my having to pass on any information that she gave me that indicated that either she or another child might be at risk of significant harm.

Ava initially showed very little curiosity in looking through the materials that I had brought, preferring me to take things out and show her; this lack of curiosity can indicate poor attachment in early childhood when it wasn't safe to explore one's environment? The need to develop a "safe place" from which to explore the world is critical in a child's inner world and emotional development. Ava's world had lacked consistency and predictability so that she couldn't anticipate what would happen to her next, a very scary place to be. Even after just a short time in the secure unit Ava seemed more confident and relaxed and she was even engaging with education, which she hadn't done for some time – she was beginning to develop her "safe space."

During that first session Ava chose to make a "sorry" card for her Grandad as he had been very upset by her absconding and ending up in the secure unit. Ava now had time to reflect on past behaviours and how she could start to make things right with her family. At the end of the session she asked me to bring a cash till and shopping items next time as it reminded her of happy times playing with her Nan.

In our second session Ava focused mostly on creating a shop with me as a customer and she spent a long time setting out the shop and the money, displaying her need for order and control in her life. Ava wanted me to "steal" an item and when I was caught and crying she turned away saying "go whilst I'm not looking and before the police arrive." This may indicate that Ava does have a moral code and is aware of right and wrong; some young people in this play situation can become punitive towards the adult and remain unforgiving, but Ava "turned a blind eye" to my misdeeds and let me go. This may indicate a growing empathy and awareness of how difficult it can be for adults to do the right thing and her willingness to forgive or at least overlook adult frailties?

Given Ava's age (now 14) and the various behaviours that she has been involved with in the past, being in her safe space allowed her to enjoy and experience play from a younger age where no shame is felt at this pleasure because she is away from distractions and threats.

Our sessions continued at the secure unit until Ava was due back in court to assess whether she needed to remain on the secure order. Another positive for Ava was that the same judge always tried to hear her case, which offered great continuity for her, and over time she had developed trust in him and he showed that he respected and liked her and always listened to what she had to say. This relationship proved to be beneficial for her as he recognised the importance to Ava of continuing with her therapy as she had begun to engage with the therapeutic process with me. On one occasion he directed that a residential placement that was a long distance away should not be pursued because of the difficulties this would create for her Nan to visit as well as Ava being able to continue with her therapy.

Ava returned to her previous residential care home and I continued to work with her on her life story and creating a memory book and initially she settled well, re-engaged with her education and resisted the temptation to abscond. Another role I was able to provide for her was as a secondary attachment figure and to provide continuity from one placement to another each time she found herself with adults she didn't know or had not developed trust in; a link that was important to her as it meant she didn't have to keep telling her story all over again.

As the criminal court case came closer Ava became more fearful, especially when an attempt was made by the suspect's gang to snatch her from the residential unit, which was foiled by her carers. Once more Ava resorted to her survival strategy of running away and eventually her local authority sought a further secure accommodation order and she was placed in a different and larger secure unit and I was asked to continue the therapeutic work with her.

This time in our first session Ava displayed a lot of anger about her loss of liberty, but due to the trust that we had built up between us she was prepared to begin working on some critical issues which she hadn't been prepared to do outside of the secure setting. I showed Ava a paper exercise called "heartfelt feelings," which had a blank heart shape in the middle and a list of feelings underneath so that she was able to select a colour to represent each emotion. The value of this exercise for Ava was that it didn't depend on her having the words to explain her feelings and it provided her with a powerful visual record that she could, if she wanted, share with others.

Ava had three issues that she wanted to explore using the heartfelt exercise:

1. Being in secure unit – for this the emotions she chose were *scared, worried, sad and angry, and she left the centre of the heart empty to represent her "emptiness."*
2. Going back to my home (i.e. residential home) – the emotions she selected were *happy, love, sad and glad.*
3. Giving evidence – for this situation Ava chose only three feelings: *worried, which filled half of the heart shape; scared, which filled a quarter of the shape; with the other quarter representing sadness.*

Ava found the work unsettling at first and then visually striking and she wanted me to give copies to her social worker, barrister and the judge hearing the criminal case, which I agreed to do.

During her first period in this secure unit Ava felt safe enough for us to explore what giving evidence in the case might involve and how best she could be supported in order to lessen her real feelings of fear.

Ava made great progress and her views supported by her "heartfelt feelings" work were acknowledged and acted upon so that she could give evidence via video link and did not have to go to the court in London. Later Ava was surprised when told by her barrister that the suspect had now changed her plea to guilty, and so Ava did not now have to give evidence even by video link, which was a huge relief for her; what impacted most on her was that the woman had been found guilty and received a long custodial sentence and so was no longer a threat to her. With this danger removed the court decided that Ava could leave the secure unit earlier than planned and she happily returned to the residential home.

Unfortunately Ava was moved to a home that was in the middle of the town and shared the house with an 11-year-old girl. I was then on leave for two weeks and when I returned I found that Ava was back in the secure unit following incidents where the child had been put at risk by absconding with Ava. The decision to disrupt Ava's placement at the single occupancy house and have her share with a much younger child in an urban setting failed to recognise the level of trauma that Ava had experienced and how she was emotionally at a very young stage and still learning to self-regulate by first needing to co-regulate like a toddler. Concerns were also expressed around whether Ava was emulating the female who had groomed her, but I felt it was more likely that she did not recognise that there was any real difference between her age and that of the 11-year-old?

I again continued to work with Ava at the secure unit where I remained as her continuity thread, linking the various places in her life. Being in her safe space once more allowed Ava to reflect on recent events and how she could have done things differently and a look towards what the future might hold for her – she was depressed at the thought that she could drift into the prison system. The work we were able to do in this setting helped Ava to acknowledge that she is in secure because of the choices that she had made, and from the work we did it was clear that being *powerful* was important to her and being *powerless and lonely* was the most worrying. Ava showed courage by trying to be honest about her feelings and was open to using a variety of creative ideas, including asking for poetry books. She especially liked *Drowning Not Waving* by Stevie Smith as it showed how easy it is for people to misread our signals and that life can be a series of opportunities to be misunderstood?

This time when Ava left the secure unit she moved to a new placement some distance from where she had been living and her request to continue with her therapy was again agreed and I remained her continuity thread for the next year until, against Ava's wishes, her local authority decided to end her therapy sessions because of her chronological age as she was deemed to no longer need therapeutic support. Up until this point her local authority had always tried to listen to her and had recognised the benefit to her of ongoing therapy, and this enabled her therapeutic journey to take place in different settings and areas. With limited resources it is likely that funding was diverted to younger children coming into the care system?

Ava remained in contact with me for some time afterwards by telephone, text and email and in one email thanked me for "always being there" for her, and when she was ready to get a job she asked if I would give her a reference, which I did. She would even keep in touch when she went missing from the home to let me know she was safe and "off hard drugs" and gave me permission to pass this information to her social worker.

Some reflections on whether being in secure helped or hindered the therapeutic journey

Today, for the majority of young people being placed in some form of secure accommodation often a long way from their family and friends will be a negative experience, so any decision to seek a section 25 Secure Accommodation Order is taken very seriously. There are now alternative systems, such as (if age appropriate and in a care facility) being placed on a DOLs order (deprivation of liberty safeguard order) whereby the young person can remain within the community, including attending school, but must be under adult supervision at all times; not an easy option but preferable for many.

Given that Ava, before going into a secure unit, had always resisted engaging with therapy then it is likely that the therapeutic work would have taken far longer to begin. By limiting where and what Ava could do in secure, our relationship was of a more intense nature, as she had physical evidence that she was seen as worthwhile for someone to continue a relationship with her even though it meant covering long distances. This helped her to develop resilience as a major factor is the belief that "someone has your back" and is looking out for you no matter what. My consistent, predictable visits became something she looked forward to and could rely upon as I kept to the same day and the same time for each session.

Living with real fear of harm meant that Ava had to be hypervigilant and ready for "flight or fight" at all times. Within the secure unit, Ava was able to reconnect with the everyday tasks of life such as her education, which gave her a great sense of achievement. It also allowed her natural curiosity and creativity to flourish and she was open to ideas and to taking creative risks with some impressive results through her therapy sessions.

As each new period of time in secure began, Ava would tell me that she was very "angry" about being deprived of her liberty, but her body language told a different story – she would be smiling, relaxed and playful as she said this? The issues that Ava was prepared to look at in her safe space were more challenging and at times very uncomfortable for her, but she knew that afterwards she would still be protected and safe and this was different from the therapeutic work that took place in the residential units.

Ava found her voice in the secure system and she was able through her therapeutic work to tell her story in her own way and found that people did listen to her, believed her and in her.

Ava and I were given a unique opportunity due to a set of circumstances coming together, including her local authority being willing to support her therapeutic needs by funding me to travel long distances to provide her with continuity and predictability as well Ava's very real need for protection against a determined criminal group.

For this young person, at a difficult time in her life, being in secure did what it promised to do and kept her safe.

References

Children Act (1989) London: UK Government.

Deprivation of Liberty Safeguards (2015) *Amendment to the Mental Capacity Act*. London: UK Government.

European Convention on Human Rights (1948) *Article 5*. Brussels: General Assembly of the European Union.

Legal Aid, Sentencing and Punishing of Offenders Act (2012) London: UK Government.

Mental Health Act (1983) London: UK Government.

35
HEALING THROUGH THERAPEUTIC PLAY IN MALAYSIA

Priscilla Ho

Introduction

It has been an enriching journey since 2006 when I started to train as a play therapist. I have worked with children and young people age three years to 18 in the community. Concerns around the mental health well-being of young people in the community here in Malaysia are on the increase. However the level of education in acquiring therapeutic play skills among the allied health professionals is still low and lacking in training and support.

I operate from my creative arts centre Creativity at Heart. We advise on the mental health well-being of young people. In the centre we offer creative arts and play therapy. We also conduct creative parenting training. I work closely with the community, schools and the local psychiatry unit.

The children that I worked with are referred from the following agencies and individuals:

1 Child and adolescent psychiatrists
2 Clinical psychologists
3 School counsellors
4 Speech therapists
5 Occupational therapists
6 Teachers

Problems surface for the young people who are not able to cope at home and in schools. In some situations children refuse to attend school. In the local environment, if the young people fail to attend school over a period of time, they will be dismissed from school. Parents or carers are sometimes advised to transfer the young person to another school, hoping that they may cope better.

The main issue seen over the years seems to be mainly caused by insecure or ambivalent attachments. Looking at the core area of needs, trust is most crucial for a child in the first stages of psychosocial development. Many children are sent to day care centres from the age of two to six years old before they enter into formal primary education from age 7 to 12 years old. They then move on to high school from age 13 to 19 years old.

For some families who are able to afford a maid in the house, the children are being nurtured by the maid instead of the parents. Thus the maid becomes the primary parent. Soon the child speaks the language of the maid. As most of the maids are on two-year renewable contracts, children experience repeated loss when the maids leave the family.

When a child encounters difficulties in school or at home, the concerned parent would normally take the child to see the local hospital psychiatrist for a diagnosis. Or it can be that the child is not able to cope well in school and the recommendation is to see a children's psychiatrist or clinical psychologist. Most referrers were from the local psychiatrist to Creativity at Heart, a non-profit centre.

My main focus facilitating the therapeutic sessions are using art, sand play, and creating and writing stories. To date Creativity at Heart has designed a therapeutic activity book titled *I Am Special* and published 13 story books written by children in the community.

Preparing the child for play therapy

Individual children

An appointment is made with the play therapist. The first meeting will be for the parents/carer and the play therapist. The therapist takes a detailed history of the child's development from birth to the current age. The developmental milestones are very critical as they allow us to see the possible difficulties faced by the child and his/her family. Parents or carers are given a copy of the SDQ (strengths and difficulties questionnaire) for assessment. The child is asked to do the SDQ too, if they are 12 or older.

The child will then be scheduled to have the first session with the therapist.

On that day when child arrives, the therapist will ask the child whether he or she would like to have their mother in the play space. We have noticed over the years that three- to ten-year-olds prefer to have their parent in the session with them. For those 11 years old and older, they can choose whether to invite their carer or parent into the session. Children over 12 will complete their own SDQs. Most children age 12 and older choose not to have their parents in a session except in unique cases.

Sessions for children from three to eight years old are witnessed by their parents, who will often play with the children at the appropriate time, encouraged by the play therapist. Parents learn therapeutic play skills during these sessions with their child. This further improves the bonding between the child and parent. We have noticed that while the parents often rated their child high in hyperactivity, the child's report will show high in emotion. The parents will learn to get to know the child better during the sessions. The parents are also encouraged to plan special time with their child at home, once a week, for an hour during six sessions of therapy.

In a family where there is a sibling, the child will invite the sibling for a therapeutic play session on the fifth or last session. We use a simple exercise allowing them to draw a mandala together. It is also helpful to see the family dynamics with the parents and children. After the end of the sixth session, the parents and the therapist hold a review: we consider how the child is coping in school and at home, with feedback from the family and the teachers. If there is a need to continue sessions, we take a three-week break to allow the child space before coming back for further sessions. At the same time the parents will have therapeutic play sessions with the child at home, on the same day at the same time they would have had therapy.

I Am Special – therapeutic group sessions in children's homes

Requested by the care home manager

Most of the children in the care homes grow up feeling angry, sad and insecure due to early separation from their families. Many worry whether they will ever have an opportunity to meet their families again. Most of the children come from broken families; some may meet up with their families. The purpose of group play is to allow the children to come together and to let their voices be heard. They support one another due to their shared experience of being separated from their parents at a young age.

SDQs will be taken from the carers as well as the children if they are over ten years old. The therapeutic group play will be planned over eight sessions, each session being two hours long. There is a weekly theme for each session as this encourages the children to help and nurture one another. Some children may need additional therapeutic play or play therapy and the sessions will be scheduled accordingly.

Each group is composed of six children with one child being the lead. The sessions are a combination of both non-directive and directive activities. There will also be a group discussion, where, most importantly, everyone's views will be heard. This helps the group dynamics and the children coming together. The weekly two-hour sessions are planned over eight weeks. The book *I Am Special* is used to allow children to express their thoughts and feelings in words and drawings (see Figure 35.1).

Children who are chronically ill

Doctors in the paediatric wards will refer chronically ill children for therapeutic play sessions. Sessions are one to one and for a duration of 40 minutes. It also depends on the state of health of the child on the day. Sessions are held once or twice a week depending on the child's emotional needs. Most children are affected with cancer, renal failure or other chronic neurological illnesses.

Figure 35.1 Activities focus on sensory, rhythmic, storytelling and role play

Figure 35.2 Chronically ill children

Children who are chronically ill often experience fear of death and loneliness, not being able to go back to school and an indefinite stay in the hospital (Figure 35.2). They long to go home and be a normal child again. For some long-term cases, the children are even able to learn the medical procedures they have to go through and share with other children. Children are given a copy of *I Am Special* for them to express their thoughts and feelings. Some of the comments written by children were 'when will I receive a new kidney', 'dear god, help me to get well', 'I want to be a doctor to help others'.

Working with chronically ill children requires the therapist to be able to put herself in the child's shoes to empathise with the pain and fear. It is important that we reflect on what was presented. The medium of play helps a child to learn and overcome their fears. From my personal experience working with the different groups of children over the years, these children are indeed my teachers. Their resilience levels are higher and helping them to cope through play is a blessing.

Examples of practice

On the following pages are some examples of the work we carry out at the 'Creativity at Heart' centre.

'Play is the child's language and toys are their words'

Play therapy with children from different environments in the community use a range of mediums and creative arts play: art making, storytelling, sand tray, clay, sound and movement, puppets and masks and small world play using objects. (See case study on page 404 and Figure 35.4).

Working with abused and traumatised children

Figure 35.3 shows drawings created by abused and traumatized children. Besides drawing the images the children will be asked to tell the story of the characters they have created.

Figure 35.3 Examples of images created by abused and traumatised children

Case study

A five-year-old child whose parents were divorced when he was four years old and the initial custody was given to the father as the mother was not able to take care of him. After one year the mother applied to the court to have custody of the child. He had been brought up by his paternal grandmother from a baby till he was five years old.

Since he was to appear in court, he would have six play sessions in order to prepare him for this as the judge would ask him for his opinions. This would help to achieve a fair custody. During his play sessions he expressed anxiety, anger and fear.

The Sessions:

His favourite activity was always telling the story, which he had drawn and created in the sand area. The child's story goes as follows: 'Once there was a fish named Little Fei who wished to be able to see his papa fish. It was locked in a cage and Papa fish was in the big ocean' (see Figure 35.4). During his sessions in the sand tray he presented the good animals and the bad animals. He felt that the animals must not fight and should be able to share out their food with one another. At the final results of the court hearing, custody was awarded to the father. The grandfather of the child told me he was able to answer all the questions asked by the judge without fear or anxiety.

Figure 35.4 Little Fei locked in his cage with Papa fish in the big ocean

Healing, memory or resolution

Healing and grief work play a very important role for children who have lost someone they loved dearly. The creative activity is to design a healing or resolution box, which is a very helpful technique. While creating the box it is very important to stay connected. Use a recycled box and coloured papers (Figure 35.5). Once complete the child writes a message and puts it inside the box.

Every day the child is encouraged to write a message and put it into the box. For younger children the child and the parent can make the healing box together. Then they can write a message for one another.

Figure 35.5 Example of a healing box

Final reflections

Therapeutic play helps children to express their fears through a creative medium. It provides the child with a safe place and helps the child to cope better as the play sessions continue. What the child cannot do in reality, they can do in their creative and therapeutic sessions. If the parent is in the session it helps them to understand and ultimately connect with their children.

AFTERWORD

Mooli Lahad

Writing an Afterword after such a fulfilling book is almost impossible. What can one add to the many ideas born from passion, compassion and much love to creativity, play and playfulness?

Playfulness is a unique phenomenon as it is both a psychological term referring to style or attitude, and also an action or behavior. A person with playful attributes is governed by internal motivation, internal orientation and is unbound to external rules and is actively involved in their environment. And my feeling is that this description was shining from all the chapters.

There are various ways to end a story:

- **Sealing the circle:** The ending may remind us of the commencing by returning to an important part, place or character. Or it may **tie-back:** The ending connects to some odd or unusual element earlier in the book.
- **The construing:** not necessarily a 'happy ending,' but a satisfying one, a reward for a journey concluded, a mystery revealed, an ambiguity solved.
- **Problem and resolution:** This common structure suggests its own ending. Framing the problem at the top and then offer readers possible solutions and resolutions.
- **Behold the forthcoming:** Most stories are about things that have already happened. But what do people say will happen next? What is the likely consequence of this decision or those events?
- **The apposite quote**: Often clichéd, this technique remains a sturdy tool for ending stories. Some characters just speak in endings, summarizing in their own words an immaculate summary or refinement of what has come before. In most cases, you can write it better than the source can say it. But not always. . .

Now the question is what you, the reader, will do with all this wealth?

Perhaps my way to end this book is by offering you a small gift, an old Zen story what I call the opposite quote.

An old monk was travelling in the hills with his small bag in which he kept the piece of bread a farmer down the hill gave him a few days ago. Every so often he sat down and opened his small bag and took some bread and drank spring water. At one of his stops he saw a beautiful gem right by where he sat for his humble dinner. He picked the precious stone up looked at it and put it in his bag. Not long after when he sat again a young man came and asked the old man if he has some food to share. The old man opened his bag and as he pulled the last piece of his bread to share, the gem fell out too. The fellow looked at the precious stone and immediately realized how much food, drink and even a whole farm he could buy with this stone, and he asked the old man if he will give him this stone. The old monk said 'yes indeed but would you like some of my bread', 'No', said the young man, 'no it is stale and I'd rather not take the last piece from you'. The young man left the monk on the side of the trail and walked away happily with the gem in his pocket, so content and delighted, dreaming of all the good things that can happen to him now.

Some days later the monk was sitting by the lake when suddenly he saw the young man approaching. In his hand he held the precious gem. 'What happened?' asked the old man? The young man looked at the monk and said: 'please, please take this gem back and please share with me your generous heart'.

Now after you have read this 'gem' of a book, full of creative, novel, interesting and well-crafted methods and models, you have a 'gem' in your own hand, and this will hopefully remind you, when you feel content and delighted of the humble old monk . . . share these gems with everyone, you will receive more than gems by sharing.

CLOSURE

Clive Holmwood

As Mooli Lahad states in his Afterword, it is almost impossible to know what to say to close the journey we have been on in this book. This volume, in particular, is unique, in that it has embraced all aspects of play from truly rich and diverse social and cultural backgrounds from around the world, without judging any particular philosophy or approach to play.

This book has no doubt extolled that play is central to our very lifeblood; and central to the development of a child's (and adults), physical, psychological, mental and spiritual health and development. It also reminds us of the latent, natural, creative power that exists in our capacity to make meaning, comprehend, develop and learn using play, no matter who and where we are on the planet. For children we are also reminded of the vital importance of the role of the parental figure in their lives to support them in the play process. Play is global and essential and its importance should be championed and promoted.

Our hope is that readers have enjoyed and embraced this 'gem' of a book, as Mooli Lahad has called it, as much as Sue and I have enjoyed the challenge of putting it together (A huge thanks to my fellow editor, Sue Jennings, for her knowledge, wisdom and insight). We hope that along with our previous volume, The *Routledge International Handbook of Dramatherapy*, and our third volume (to follow), The *Routledge International Handbook of Stories and Therapeutic Storytelling*, this will provide the reader with an inspiring series of chapters that cover play, drama and storytelling. We hope this will be of benefit to all play, creative and arts based therapists, practitioners, teachers, students, academics, parents and policy makers and anyone from around the world, who shares an interest; and that it reminds us that all creativity, the arts, storytelling, play and drama is central to human development thinking and sustenance, is it what makes us ultimately – human.

INDEX

Note: page numbers in *italics* indicate a figure and page numbers in **bold** indicate a table on the corresponding page.